An Introduction to

Our Dynamic Planet

This is the first undergraduate textbook to integrate fully results from geophysics, geochemistry and petrology to describe the structure, composition and dynamic processes that operate throughout the solid Earth. It presents an Earth system science approach to studies of the Earth's interior and develops a global view of solid Earth cycles to explain geodynamic and plate tectonic processes.

The opening chapters explore the formation and evolution of the early Earth and the development of its layered structure. The book then considers the driving forces for plate tectonic movements at the surface of the Earth, with separate chapters on processes operating at constructive and destructive plate boundaries and continental collision zones. The focus then moves to global cycles within the deep Earth and their effect on the surface environment. Particular consideration is given to interactions between the geosphere, hydrosphere and atmosphere and how these interactions influence and moderate conditions at and beneath the Earth's surface.

An Introduction to Our Dynamic Planet provides concise but comprehensive coverage of the solid Earth for courses with an Earth system science perspective. It is designed for use on intermediate undergraduate courses and incorporates a wealth of features to support student learning.

NICK ROGERS is a Senior Lecturer in Earth Sciences at The Open University, Milton Keynes, UK. His research interests are in the field of trace element and isotope geochemistry applied to igneous petrology, the composition of the mantle, and the interaction of mantle plumes with the continental lithosphere. Dr Rogers is a Council Member of The Geological Society of London and is also their Publications Secretary.

Cover image: View into an active lava tube with a skylight in Kilauea Volcano, Hawaii. Lava tubes form when the surface of a lava flow cools and solidifies, while the interior remains molten. The molten lava then flows through a hollow passage of cooled rock. Here, part of the roof of the tube has collapsed, allowing a view into the tube.

An Introduction to Our Dynamic Planet

Edited by Nick Rogers

Authors:

Stephen Blake

Kevin Burton

Nigel Harris

Ian Parkinson

Nick Rogers

Mike Widdowson

CAMBRIDGE UNIVERSITY PRESS

Cambridge, New York, Melbourne, Madrid, Cape Town, Singapore, São Paulo, Delhi

Cambridge University Press
The Edinburgh Building, Cambridge CB2 8RU, UK

IN ASSOCIATION WITH

The Open University, Walton Hall, Milton Keynes MK7 6AA, UK

www.cambridge.org
Information on this title: www.cambridge.org/dynamicplanet

First published 2007

This co-published edition first published 2008

Edited and designed by The Open University.

Typeset by SR Nova Pvt. Ltd, Bangalore, India.

Printed and bound in the United Kingdom by The University Press, Cambridge.

This book forms part of an Open University course S279 *Our dynamic planet: Earth and life*. Details of this and other Open University courses can be obtained from the Student Registration and Enquiry Service, The Open University, PO Box 197, Milton Keynes MK7 6BJ, United Kingdom: tel. +44 (0)845 300 60 90, email general-enquiries@open.ac.uk

http://www.open.ac.uk

British Library Cataloguing in Publication Data available on request

Library of Congress Cataloguing in Publication Data available on request

ISBN 978 0 521 494243 hardback; ISBN 978 0 521 729543 paperback

1.1

Contents

An introduction to the structure and composition of the Earth

Chapter 1

The Earth is a special place, not least because it is our home planet. It is the only place in the Universe where we are sure we can survive and thrive and, as far as we know, the only place where life has evolved. Despite the efforts of numerous space probes and ground-based searches for both extraterrestrial life and intelligence, at the time of writing (2006), this is all there is.

Perhaps the failure of projects designed to discover intelligent life elsewhere, such as the Search for Extraterrestrial Intelligence (SETI), is not so surprising if you consider the scale of the Universe and the time it takes for signals to travel between stars. Our closest stellar neighbour, for example, is four light-years away and even the briefest text message would have to wait eight years for an answer, assuming that it was decipherable by the recipient. But distance is not the only factor in our apparent isolation. Life needs the right conditions to start and an environment that is not so extreme that it is annihilated before it has had time to evolve. Planet Earth provides such an environment; an environment that all too often is taken for granted and exploited at great risk to our future viability, both as a civilisation and as a species.

But how did that environment evolve? Where did the atmosphere, the oceans, and the rocks come from? Why are there oceans and continents? Has the Earth always looked like it does today or has it changed through time? These and many other similar questions are fundamental to our understanding of the evolution of Earth and life and to the foundations of modern Earth system sciences – the study of Planet Earth as a whole, functioning system.

Our everyday experience of the Earth is probably one of underlying constancy and unchanging stability. The mountains and hills appear unchanged, rivers continue to flow to the sea and the oceans are always in the same place. Only plants and animals respond to the seasons, but even those follow an apparently unchanging annual cycle controlled by the constant orbit of the Earth about the Sun. Geology, it seems, remains a solid foundation on which biological activity relies. But the simple scene of geological inactivity is not the same everywhere and at all times. Consider, for example, cliff erosion and the landscape where the land meets the sea. There are many examples where buildings and land that were once thought to be secure are now precariously perched on cliff tops ready to slip into the sea during the next severe storm.

At other times, a tranquil landscape is disturbed not by weather and climate but by internal forces that change the Earth's surface for ever. Earthquakes lay waste to large areas, reducing buildings, towns and sometimes cities to piles of rubble at terrible cost to individuals and society at large. Dormant volcanoes come to life in paroxysmal explosions, inundating the surrounding area with thick volcanic ash and lava. One such event happened in May 1980 when Mount St Helens in Washington State, USA, came to life after a period of over 150 years of quiescence and within a very short period of time destroyed itself and large swathes of the surrounding forests. Spirit Lake, so beloved of postcard photographers and tourists alike, was emptied of its contents and filled with volcanic rock debris in a matter of minutes as the side of the volcano slipped away (Figure 1.1).

(a)

(b)

(c)

Figure 1.1 Mt St Helens in the USA: (a) before, (b) during and (c) after the eruption of May 1980. By volcanological standards, this was a modest event, erupting ~1 km^3 of rock in an eruption column that reached about 10 km altitude. The geological record reveals numerous volcanic eruptions in the recent past (up to 1 Ma ago) that were 100 or even 1000 times larger than this one. (USGS)

1 Ma = 10^6 years

Volcanic eruptions and earthquakes are probably the most dramatic of natural events that illustrate the powerful forces that shape the fabric of the landscape and the topography of the planet. Individually, they may prove catastrophic to the local environment, but they represent the surface expression of the processes that have produced the continents, atmosphere and oceans over the history of the Earth. If we are to understand these processes we need to expand our mental horizons beyond the everyday and explore the depths of the Earth and the duration of geological time. On these different levels and timescales we see a very different planet: one in a state of dynamic activity where the continents move across the surface, ocean basins are created and destroyed and mountain belts rise and fall in concert to the pulse of the Earth's internal energy.

Such, then, is the aim of this book – to stretch your mental horizons by:

- describing and analysing the way the solid Earth works
- investigating the composition of the Earth and its constituent layers
- revealing how the Earth has evolved from a cloud of gas and dust via an ocean of magma to a planet with a solid crust, oceans and a breathable atmosphere.

To do that you will need to call on physics and chemistry, mathematics and geology – these disciplines provide knowledge and techniques that should help you to understand our dynamic planet.

In this first chapter, you will be introduced to your home planet. The first part explores very briefly what is unique about the Earth when compared with the Moon and other terrestrial planets: Mercury, Venus and Mars. The second part of the chapter investigates the structure of the Earth: how it is divided into core, mantle and crust, and the evidence on which that division is based. Finally, the lines of evidence concerning the composition of the Earth and its constituent layers are drawn together from various sources, such as meteorites and samples of the mantle. This should give you the necessary background information to embark on your journey into the depths of the Earth and back through geological time.

(a)

(b)

(c)

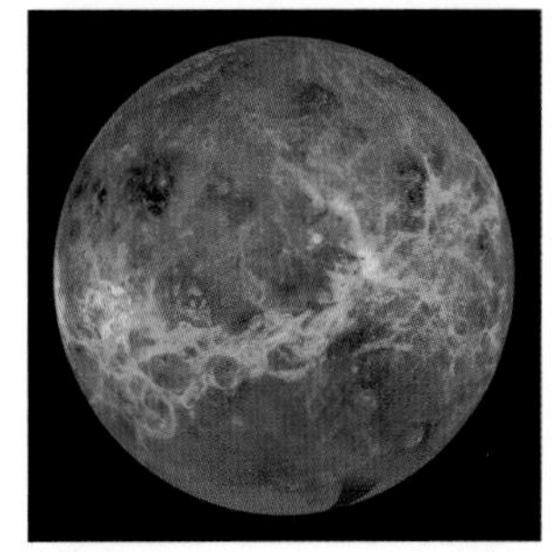
(d)

(e)

▶ **Figure 1.2** Images of the surface/planetary discs of (a) the Earth; (b) the Moon; (c) Mars; (d) Venus; (e) Mercury. (NASA)

1.1 Planet Earth is unique

Many everyday aspects of our environment that we take for granted are most unusual in the Solar System – yet they contribute to making the Earth as amenable to life as we know it. Our knowledge of the surface features and conditions of all four terrestrial planets, i.e. Mercury, Venus, Mars and the Earth, together with the Moon, is becoming ever more detailed thanks to the spectacular results from numerous interplanetary probes (Figure 1.2).

Study the images in Figure 1.2, and then consider the following question.

■ What are the characteristics of the Earth's surface that make it so favourable for life?

■ The presence of liquid water and the availability of oxygen in the atmosphere.

None of the other terrestrial planets currently has liquid water on its surface and, although Mars has water locked away in its ice caps and reveals geological evidence for the presence of water in the past, its atmosphere is thin and dominated by carbon dioxide (CO_2) and nitrogen (N_2). Only Venus has a thick atmosphere, but it is dominated by CO_2, sulfur dioxide (SO_2) and surface temperatures and pressures that would be intolerable for life. The other two bodies, Mercury and the Moon, are small, dry and have insignificant atmospheres.

These differences, although critical to life, appear superficial to the solid planets that lie beneath the atmospheres (see Box 1.1). Consequently, the question that arises is whether or not the solid planets share other common features and structures.

Box 1.1 Earth's interacting systems

In this book, the different components of the Earth are divided according to their physical state into the atmosphere, hydrosphere and geosphere. The term **atmosphere** should be familiar to you: it describes the envelope of gases that surrounds the Earth. The **hydrosphere** may be less familiar, but it refers to all of the water on the planet – in the oceans, seas, lakes and rivers. Part of the hydrosphere is locked up in the form of ice and is sometimes referred to separately as the **cryosphere**. The solid Earth, which makes up the bulk of the planet by mass, is known as the **geosphere**. The geosphere is the main subject of this book, but the atmosphere and hydrosphere also interact with the solid Earth through a variety of processes that will be explored later. Finally, all forms of life are collectively described as the **biosphere**.

It is useful to consider the relative sizes of the different Earth systems and their influence on our lives. The Earth has a radius of 6371 km, yet the atmosphere becomes too thin to breathe just above the top of the highest mountains, which is about 10 km above sea level. Even though atmospheric physicists extend their range of study to the boundaries of space some 300 km above the surface, atmospheric pressures for most of this distance are less than 1% of that at sea level.

The oceans appear vast and the dark ocean floor is more alien to us than the face of the Moon, yet on average it is only 3 km below the surface.

■ The Atlantic Ocean is about 5000 km across at its widest. If that width was reduced to a puddle 5 m across, how deep would the puddle be to maintain the true scale of the ocean?

■ As 1 km = 1000 m and 1 m = 1000 mm, the scale reduction is 1:1 000 000, so our puddle should be

$$\frac{3\text{ km}}{1\,000\,000} = 0.000\,003\text{ km} = 3\text{ mm deep.}$$

While the oceans appear very deep, on the scale of the whole Earth they represent a very thin film on the Earth's surface – which is about 0.05% of the Earth's radius deep.

And what of the biosphere? Life extends to about 10 km above the surface. It also appears that life can exist at the greatest depths yet explored in the oceans, again about 10 km below sea level, making the biosphere about 20 km thick. But, as with the oceans, on the scale of the whole Earth this represents just 0.3% of the Earth's radius. No wonder then that the biosphere has been described as nothing more than 'a thin green smear' on the surface of the Earth.

1.1.1 Surface features

The Earth's surface is characterised by continents and oceans, but is there evidence for similar features on the other planets?

■ Study Figure 1.2 again. Are the surfaces of the different planets uniform or diverse?

■ Diverse: with the possible exception of Venus, each image has dark and light regions and areas with many craters and other areas with relatively few.

While only the Earth has a permanent hydrosphere, there is clearly a contrast on the other planets between the highland regions, which may be equivalent to the continents on Earth, and the lowlands or basins, which may be analogous to terrestrial oceans. Some planetary surfaces are also pockmarked with craters, while on others such features are rare.

The Moon

The Moon orbits the Earth; some of the other planets have moons.

Our nearest neighbour, the Moon, is the Earth's constant companion. Results from the Apollo missions in the 1960s and 70s revealed just how old the Moon is – it is in excess of 4.4 Ga and is as old as the Earth itself. However, the Moon is exceptional, not because of its age – all planetary bodies orbiting the Sun are of

1 Ga = 10^9 years.

roughly the same age – but because of its size. In relation to the mass of the Earth, the Moon is the largest of any of the satellites of the other planets. Ganymede and Callisto, which are moons of Jupiter, and Titan – the largest moon of Saturn – may be slightly larger than the Moon, but they are tiny in relation to their parent bodies. Of the terrestrial planets, the only other planet with satellites is Mars. These small objects, named Phobos and Deimos, appear to have been captured from the **asteroid belt** lying between Mars and Jupiter, and are a relatively late addition to the Martian system.

The Moon has two different surfaces: the **lunar highlands**, which are pale-coloured regions with a dense covering of impact craters, and **lunar maria**, which are darker, flatter regions with fewer craters. This suggests a division into elevated regions that are analogous to continents and depressions that were once thought to be filled with water – hence *maria* (Latin for sea). The number of craters on a planet's surface is related to its age – more craters means an older surface – so, even before the Apollo missions in the 1960s and 70s, scientists realised that the highlands are older than the maria. What was not known was the absolute age of the rocks until samples were returned and subjected to radiometric dating – the determination of the age of a rock using naturally occurring radioactive isotopes.

Direct dating of samples from different areas of the Moon revealed that the highlands date from 4.4 Ga, whereas the maria are younger at 4.2–3.8 Ga, confirming and quantifying the conclusions from crater counting. The study of Moon samples also indicates that these two surfaces are made of different rock types:

- the maria are dominated by **basalts** that are similar to those found on Earth
- the highlands are made of a rock known as **anorthosite**, which is an uncommon rock type on Earth that is rich in **plagioclase feldspar**.

The craters and the surface ages show that the Moon experienced a period of intense meteorite bombardment early in its history, but that this had declined by the time the maria formed.

Mercury, Venus, Earth and Mars

Similar cratering studies of the other terrestrial planets have revealed an even greater diversity in surface ages. Mercury, for which there is only limited information, appears very similar to the Moon with a highly cratered ancient surface. The different surfaces of Mars also appear similar to those of the Moon: cratered highland regions surround basins with fewer impact craters. Although this division is less easy to see in Figure 1.2, and the absolute ages of these different surfaces are as yet unknown, ages of 3.8 Ga have been suggested for the highlands – assuming a similar impact history to the Moon.

Radar images of the surface of Venus reveal a somewhat different story, with a lower cratering density suggesting a much younger surface age. Impact craters on Earth are much less easily recognised. Certainly impact features similar to those on the surface of the Moon are rare.

■ What do these observations suggest about the mean age of the surface of the Earth compared to that of the Moon?

■ The Earth's surface must be considerably younger than that of the Moon.

This conclusion is confirmed by many observations on the ages of rocks at the Earth's surface today. Even though the oldest terrestrial rocks are almost 4 Ga old, most rocks at the surface of the Earth are much younger, with ages of tens to hundreds of millions of years. Most notably, no rocks recovered from the oceans are older than 200 Ma.

The cratering record on the Moon, Mercury and parts of Mars suggests an extraordinary violent birth and early life for these planets, involving innumerable impacts of asteroid-sized objects – a violent early history that must have been shared with the Earth and Venus. This period of time on Earth, between its formation at 4.55 Ga and the ages of the earliest rocks at ~4.0 Ga, is known as the **Hadean**. As there are no rocks on Earth older than ~4.0 Ga, much of our knowledge of the processes that affected Earth during this period is derived from comparison with other planetary bodies.

■ Why do you think there is no direct record of the earliest surfaces of the Earth?

■ They have been removed by the effects of surface processes.

In part, the lack of a terrestrial record of this time (i.e. the Hadean) is a consequence of surface processes. Unlike the Moon, which has no atmosphere, the Earth's surface, and to a lesser extent the surface of Mars, is being continually reworked by chemical and physical interactions with the atmosphere and hydrosphere by the processes of **weathering** and **erosion**. On Mars, the thinness of the atmosphere and the ephemeral nature of the hydrosphere allow older features such as impact craters to be preserved for longer than on the other planets.

■ Do the same arguments concerning weathering and erosion apply to Venus?

■ In part yes, but Venus does not have a hydrosphere, and as many terrestrial surface processes require a flowing liquid, the relative youth of the surface of Venus may have a different explanation.

There are other processes that relate to the deeper working of the planets that also renew planetary surfaces on geological timescales, for example volcanic activity. These are clearly apparent on Earth and are the subject of scientific investigation on Venus, but have not been dominant in the evolution of Mercury or the Moon and are only occasionally active on Mars. Such internal processes, as you will discover later in this book, are driven by the internal heat of a planet and so are related to the planet's size. Given that Earth and Venus are of a similar size and are the largest of the terrestrial planets, the relatively young ages of their surfaces also relate to how much heat each holds deep in its interior.

■ Assuming the Earth dissipates its heat through the processes of plate tectonics (Chapter 3), which of the terrestrial planets might also be expected to show plate-like features?

■ Venus, because of its similar size.

Figure 1.3 and Figure 1.4 (overleaf) show the surfaces of the Earth, Venus and Mars in false colours to denote the topography of each planet. Shades of blue represent depressions or basins with elevations below the mean altitude of the surface while shades of red denote highlands. If the water from the oceans could be removed to reveal the topography of the ocean floors, you would see something like the image in Figure 1.3.

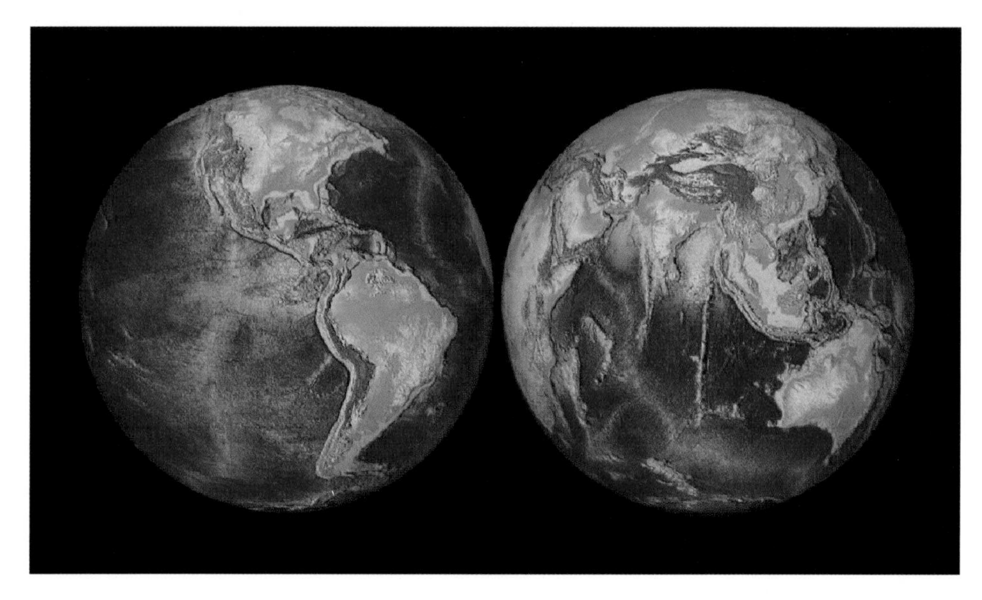

Figure 1.3 Topography of the Earth with the oceans removed. The blue areas corresponding to the Pacific and Indian Oceans indicate low elevations – the deeper the colour, the lower the elevation. Pale-blue areas close to shorelines are equivalent to the continental shelf; land is represented by shades of colour – green: 0–500 m (e.g. Amazon Basin); yellow: 500–1000 m (e.g. Western Australia); brown: 1000–2000 m (e.g. Iran and Afghanistan); red: 2000–5000 m (e.g. Andes); and grey: >5000 m (e.g. Himalaya). (NOAA/NASA)

The ocean basins are far from flat or uniform. Beneath the sea are hidden some of the most dramatic topographic features on Earth – huge volcanoes, deep trenches and the longest mountain chain on Earth, the ocean-ridge system, which links all the ocean basins together. The ocean ridges mark regions on the Earth's surface where new crust is being created. If new crust is being created then older crust must be being absorbed back into the Earth's interior (assuming that the Earth is not expanding), and the deep ocean trenches that surround the Pacific Ocean and lie south of Indonesia in the Indian Ocean mark the places where this is happening. So if similar recycling processes are happening on other planets, then similar surface features should also be visible.

■ Compare Figure 1.3 with Figure 1.4a. Do you see any structures or features on Venus that resemble ocean ridges or trenches?

■ No. There are no such features on Venus.

Superficially, none of the major features on the surfaces of either Venus or Mars suggest plate tectonics is an active process on either planet. Both planets have volcanic features – Olympus Mons on Mars is the largest volcanic structure in the Solar System and such features are the external expressions of internal processes, either active now or in the past. But major features that might be related to plate tectonics, such as ocean ridges, are not apparent. Taken with the ages of the surfaces from the cratering record, the evidence indicates that plate tectonics is not an active process on either planet.

(a)

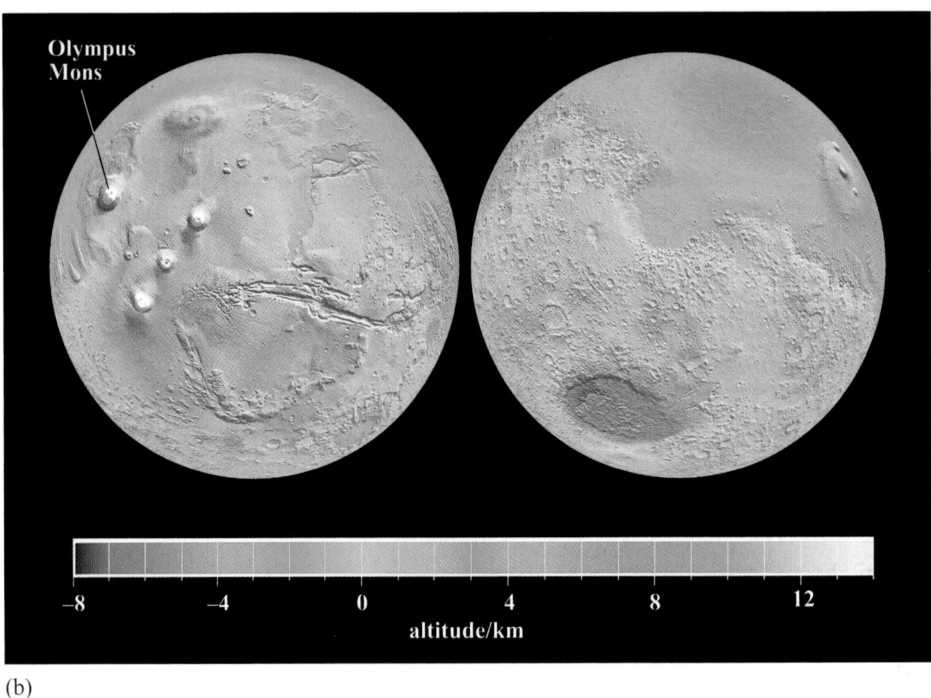

(b)

Figure 1.4 False-colour images of (a) Venus and (b) Mars, colour-coded to provide elevation information across the total topographic range of each planet. The blue (e.g. circular region in centre of left Venus image) denotes areas 1–2 km below the mean surface elevation (mse); green denotes 0–1 km below mse; yellow denotes 0–1 km above mse and red denotes areas 1–2 km above mse. Higher elevations are in shades of pink to white (small area below centre of left Venus image). The total topographic range on the image of Venus is much less than that of Mars. (NASA)

While the cratering record can be used to provide information about the relative ages of surfaces and the intensity of processes that modify the surface, it does not tell us anything about the internal structure, such as whether other planets have continents and oceans with Earth-like compositions and structures. Do these different surfaces reflect the differences between continents and oceans, as on Earth?

We can investigate the similarities and differences quantitatively, using a device known as the **hypsometric plot**. The hypsometric plot is a histogram of topographic height over a whole planet, with heights determined above or below a reference altitude. The plot for Earth is shown in Figure 1.5a.

- The hypsometric plot for Earth is obviously bimodal. By comparison with Figure 1.3, what do you think each of the two peaks in Figure 1.5a represents?

- The peak at relatively low height (i.e. the one on the left) represents the floor of the oceans (which are typically at 3–6 km below sea level) and the peak at greater relative height represents the surface of the continental crust (which is mostly less than 1 km above sea level).

The hypsometric plot clearly reveals the difference in elevations between the oceans and the continents on Earth, and we shall return to the causes of this difference later in the chapter.

- How does the hypsometric plot for Venus (Figure 1.5d) compare with that for the Earth?

- It is different. Although both have a peak between 0 and 1 km, this is the only peak in the Venusian data (Earth has two peaks) and the range of elevation is less.

Venus is often described as the Earth's sister planet because the two planets have similar sizes. Yet the contrasting hypsometric plots strongly suggest that their crustal structures are fundamentally different. This comparison may be flawed because of the different surface processes that operate on Venus compared with on Earth, largely because of the lack of a biosphere and a hydrosphere on Venus. Nevertheless, it is difficult to see how a unimodal distribution of heights could be transformed into a bimodal distribution, or vice versa, simply as a consequence of surface processes.

The hypsometric plot of Venus, therefore, suggests that the surface features are very different from those on Earth. The division into continents and oceans that is apparent on Earth is lacking on Venus and also on Mars. Continents and oceans (with or without water) appear to be another unique feature of Earth, and later in this book you will discover how they came into being and the processes responsible for their formation.

In summary, in addition to an oxygen-rich atmosphere and a permanent and active hydrosphere, the differences between the Earth and other terrestrial planets are more than just skin deep. Continental and oceanic crust, youthful surfaces, the presence of plate tectonics and the coupling of the Earth with the Moon are all features that are unique to our planet. In addition, the Earth has a strong magnetic field and a relatively rapid rate of rotation. In the following chapters you will explore how some of these unique features have come about, the processes that produced them, and the processes they control.

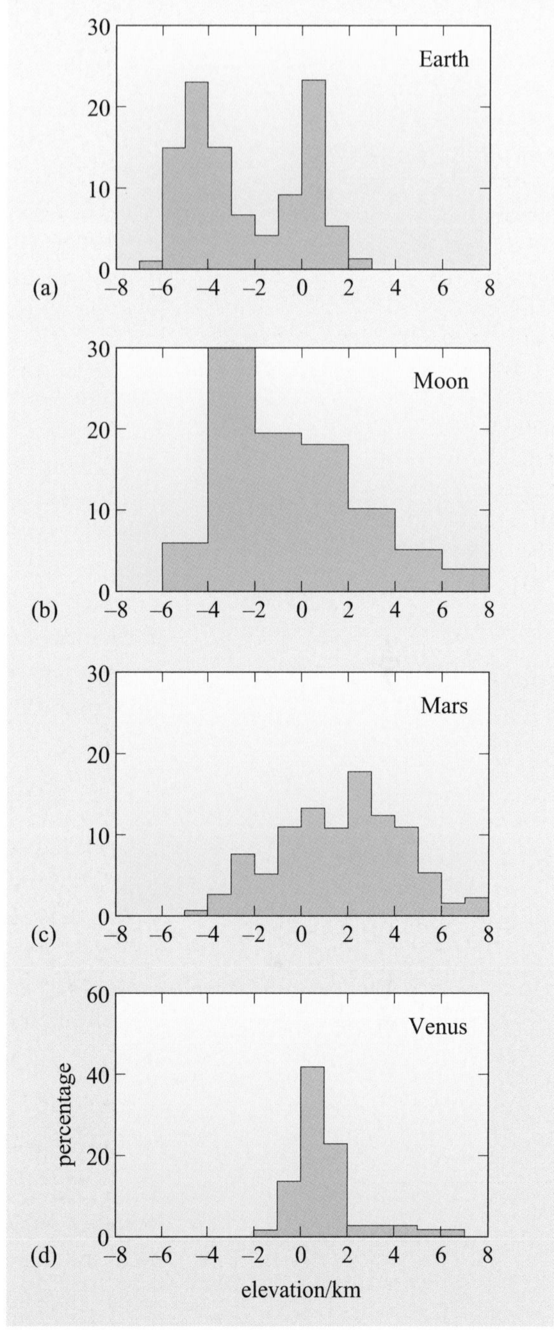

Figure 1.5 Hypsometric plots for (a) the Earth, (b) the Moon, (c) Mars and (d) Venus showing the percentage of the solid surface above and below sea level for the Earth, and mean surface elevation for the Moon, Mars and Venus. (Watts, 2001)

1.2 The structure of the Earth

Our knowledge of the structure of the Earth comes from a variety of sources, but the most fruitful line of discovery is through the study of earthquakes and seismic waves – the science of **seismology**. Earthquakes are familiar to most people, if not through first-hand experience, then via reports in the media when they lay waste to areas of population and result in the deaths of sometimes many thousands of people. It is perhaps ironic that these most violent of natural phenomena, that have killed so many people throughout history, are eagerly awaited by seismologists, providing yet more opportunities to probe the interior of our planet.

Earthquakes result from the release of strain energy that builds up in parts of the Earth where large bodies of rock are moving relative to each other. One such example is the San Andreas Fault in California, where a sliver of western California is moving northwards relative to the rest of North America. While the relative movement continues gradually, no earthquakes occur, but should the fault become locked for some reason, then the rocks become increasingly deformed or strained. When the fault unlocks, the energy stored as strain is released almost instantaneously and much of it dissipates in the form of seismic waves. It is these seismic waves that both deliver the destruction and provide information on the structure of the Earth.

1.2.1 Seismic waves – Earth probes

Seismic waves are of four main types: P, S, Love and Rayleigh. The latter two types are confined largely to the surface of the Earth and, while they are highly destructive and can be used to measure the magnitude of an earthquake, they do not provide information about the interior of the Earth, so they will not be considered further here. By contrast, P-waves and S-waves are both **body waves** and travel through the interior of the Earth.

P-waves involve the transmission of pulses of compression and dilation (expansion), in which the particles of the medium vibrate backwards and forwards in the direction the wave is travelling (propagating). P-waves are like sound waves and become audible when they reach the surface, producing rumbling sounds.

S-waves involve shear displacements, meaning that the motion of particles in the medium is at right angles (perpendicular) to the direction of wave propagation. The motion can be vertical, horizontal, or a combination of the two, depending on the nature of the earthquake. S-waves are similar to the motion induced when a taut rope is shaken.

The speed of a seismic wave is related to the physical properties of the medium through which the wave travels. In general, the wave speed is a function of two properties – the **elastic modulus** and the density – and is of the form:

$$\text{speed} = \left(\frac{\text{elastic modulus}}{\text{density}}\right)^{\frac{1}{2}} \tag{1.1}$$

The elastic modulus is a measure of how much deformation or strain occurs in a material when it is subjected to a given amount of stress, and is defined as the ratio of stress to strain. Because materials can be strained in a number of ways, they can have more than one elastic modulus. For example, the *bulk modulus* relates to the compressibility of a medium whereas the *shear modulus* describes how rigid that medium is. For P-waves, the speed depends on both the bulk modulus and the shear modulus, whereas for S-waves it depends only on the shear modulus.

The equation for the speed of P-waves, v_P, is:

$$v_P = \left(\frac{K + \frac{4\mu}{3}}{\rho} \right)^{\frac{1}{2}} \tag{1.2}$$

where K is the bulk modulus, μ is the shear modulus and ρ is the density. The equation for the speed of S-waves, v_S, is:

$$v_S = \left(\frac{\mu}{\rho} \right)^{\frac{1}{2}} \tag{1.3}$$

where μ is the shear modulus and ρ is the density.

- Can you see why v_P is greater than v_S?
- Given that both K and μ are positive numbers, $K + \frac{4\mu}{3}$ is always greater than μ, so v_P will be greater than v_S.

- What do Equations 1.2 and 1.3 infer about P- and S-wave speeds in liquids of similar densities to solids?
- Given that a liquid has no rigidity (it occupies the shape of any containing vessel), it cannot be deformed. Hence μ must be zero. Therefore, liquids cannot transmit S-waves. Also, for a given density, the P-wave velocity is lower in a liquid than in a solid.

Seismic waves are detected using instruments known as seismometers. A simple example of the output from such an instrument, called a seismogram, is shown in Figure 1.6. Because P-waves always travel faster than S-waves, P-waves always arrive at a seismic station first. The speed of the seismic waves through the Earth can be measured in the laboratory, so the difference in their arrival times can be used to calculate the distance a seismometer is from the centre of an earthquake using the simple relationship between speed, distance and time.

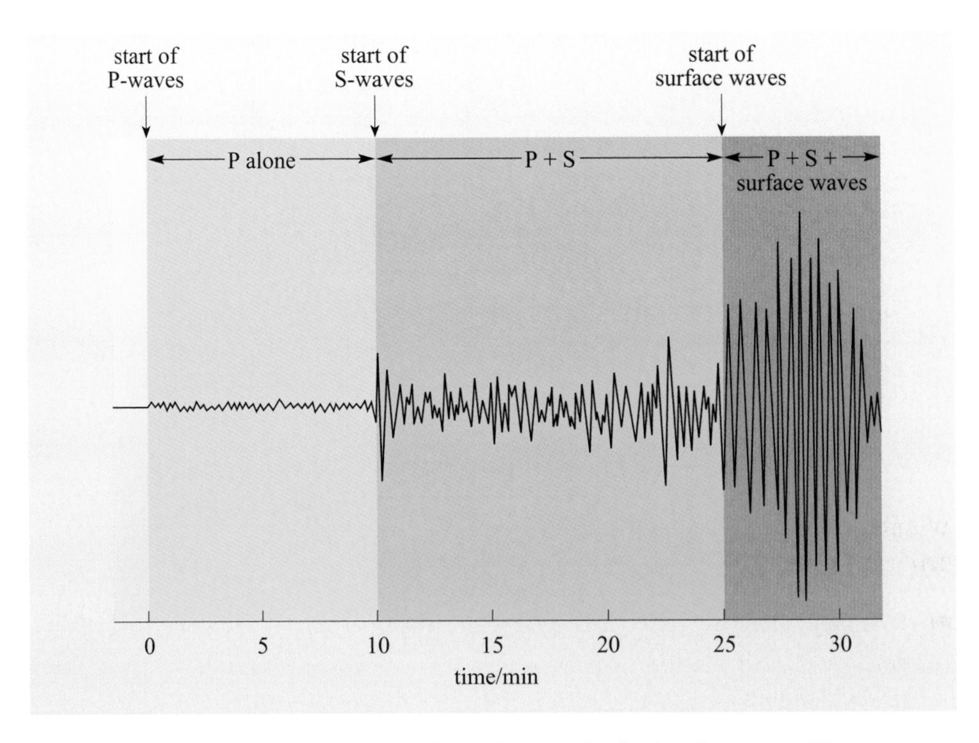

Figure 1.6 A simple seismogram, in this case produced by an earthquake in Turkey and recorded in the USA (Cambridge, Massachusetts). Notice the difference in the displacements caused by the different types of waves – the first arrival of P-waves followed by S-waves and then the surface waves (Love and Rayleigh).

So far, all references have been made to the speed of seismic waves. However, objects and waves have two attributes to their motion – a magnitude of motion and a direction. These are the two attributes of velocity. In everyday speech the two terms are frequently used interchangeably, but in science the difference between speed and velocity is quite definite. Speed refers to the rate at which an object is moving while velocity includes a reference to the direction in which the object is moving. Subsequently, all references to seismic waves will use the term 'velocity', as the propagation of seismic waves through the Earth involves changes in both magnitude and direction.

1.2.2 Seismic refraction

Seismic velocity is a function of both density and the physical state of the deep Earth, and determining the variation of seismic velocity with depth in the Earth has been a major factor in the determination of the structure of the Earth's interior.

The rules that govern the propagation of seismic waves through the Earth are similar to those that govern the passage of light through transparent media, such as glass and water. You have probably noticed the way in which a stick appears to be bent when partially immersed in water. The light from the stick below the water is bent or **refracted** at the water's surface because the velocity of light in water is different from that in air. So it is with seismic waves as they pass through the different layers of the Earth. This is illustrated in Figure 1.7. The angle of refraction (r) is directly related to the angle of incidence (i) and the ratio of the seismic velocities of the two layers, such that:

$$\frac{\sin i}{\sin r} = \frac{v_1}{v_2} \tag{1.4}$$

This equation summarises what is known as **Snell's law** and applies to all forms of wave energy, including light and sound as well as seismic waves. Boundaries also reflect some of the incident energy; a possible reflected wave path is shown in Figure 1.7.

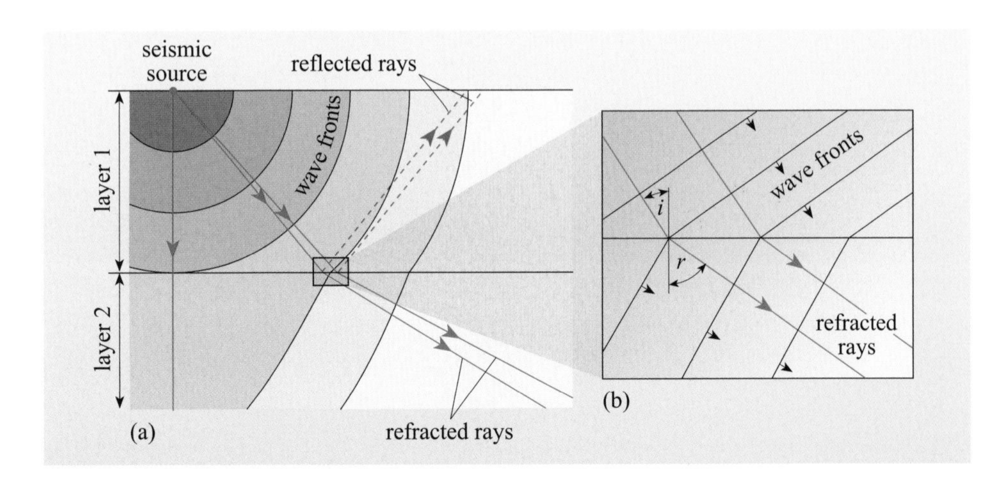

Figure 1.7 Refraction of seismic waves. In (a), evenly spaced wavefronts spread out from a seismic source. When they enter the lower layer with a higher velocity, they change orientation and some of the energy is also reflected. In (b), two wave paths (in red) are also shown tracing two different points on the wave front (in black).

Within the Earth, most changes in seismic velocity take place continuously with depth, such that refraction occurs as a continuous process.

- Imagine the Earth consists of a series of thin concentric shells each with a seismic velocity slightly greater than the one above. What will be the shape of the ray path?
- The ray path will be refracted to a shallower angle at each boundary and curve back towards the surface.

The trace of a ray path will be similar to that in Figure 1.8. At each boundary between successive layers, the ray path is refracted to increasingly shallow angles. Eventually, as seismic velocities increase and the angle of incidence increases, a particular angle is reached at which

$$\frac{v_1}{v_2} = \sin i, \text{ i.e. } \sin r = 1, \text{ meaning that } r = 90°.$$

At this angle, the refracted wave propagates along the boundary between the two layers. This angle is known as the **critical angle** and the wave that propagates along the boundary between the two layers is known as the critically refracted ray or the head ray.

- What do you think happens to the wave path when it strikes the boundary at an angle greater than the critical angle?
- It is reflected rather than refracted.

Ray paths with an incident angle greater than the critical angle trace a path back to the surface some distance from the original source (Figure 1.8).

The distance of a seismic receiver from a seismic source can be measured in kilometres but is usually expressed in terms of the angle subtended at the centre of the Earth (as shown in Figure 1.8) and is denoted by Δ (delta). A plot of arrival times for different seismic waves against Δ (the **epicentral angle**), known as a **travel-time curve**, is shown in Figure 1.9.

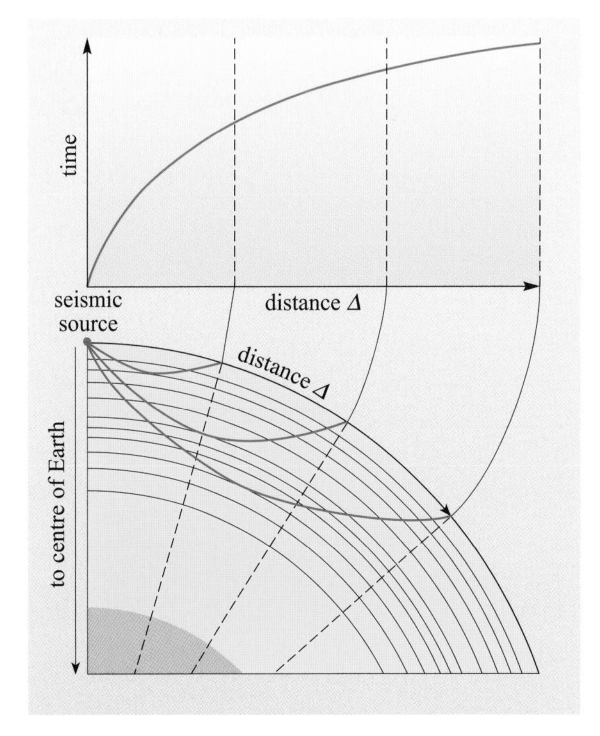

Figure 1.8 Schematic paths of seismic waves in a multiple-layered Earth. Rays are refracted and reflected at each boundary and the overall ray path curves towards the surface. The distance from the earthquake is measured as the angle subtended at the centre of the Earth between the receiving station and the earthquake source.

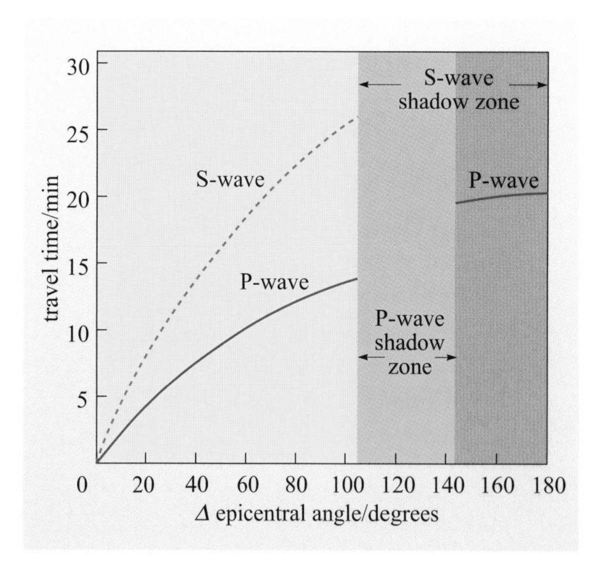

Figure 1.9 P-wave and S-wave travel times for the whole Earth. There is a P-wave shadow between epicentral angles 105° and 142° and an S-wave shadow at all angles greater than 105°.

- Study Figure 1.9. Do the arrival times of S- and P-waves vary continuously for all epicentral angles?

- No. Arrival times of P-waves increase up to an epicentral angle of ~105°, after which there is a gap before they appear again at about 142°, but at much later times. Arrival times of S-waves also increase up to ~105° but there are no S-waves at epicentral angles greater than 105°.

The areas of the Earth's surface at these epicentral angles where there are no seismic waves are known as **shadow zones** and they are more or less the same for any earthquake, no matter where on the Earth it originates. This simple observation shows that the internal structure of the Earth must be radially symmetric.

An everyday analogy for the structure of the Earth is that of a peach, with a core (the peach stone) surrounded by what is known as the mantle (the peach flesh). (One might pursue this analogy further and suggest that the Earth's crust is represented by the peach skin.)

You have seen from Figure 1.8 that waves which emerge at the greatest epicentral angles have penetrated to the greatest depths. In Figure 1.9 you can also see that at epicentral angles of less than 105° the gradient of the travel-time curve decreases as Δ increases.

- What does this tell you about seismic velocity changes with depth in the Earth?

- The gradient of the curve is equal to time divided by distance, which is the inverse of speed. As the gradient decreases with epicentral angle this must mean that, up to a certain depth represented by an epicentral angle of 105°, seismic P- and S-wave speeds increase.

- What does the S-wave shadow suggest about a part of the Earth's interior?

- You will recall from earlier in this chapter that S-waves do not travel through liquids. The S-wave shadow, therefore, shows that a significant part of the Earth's deep interior is liquid.

This important break in the seismic signal at 105° is known as a **discontinuity** and this one, by definition, marks the boundary between the Earth's core and the overlying mantle. Figure 1.10 shows how this discontinuity comes about by tracing the paths of a number of seismic waves through the Earth. The abrupt velocity decrease at the core–mantle boundary causes rays entering the core to be refracted towards the vertical and so producing the P-wave shadow starting at an epicentral angle of ~105°. The refracted wave passes through the outer core and emerges, much later than would be expected if the core were not there, at a much greater value of Δ. Other wave paths that strike the core–mantle boundary at lower incident angles pass through the outer and inner core and the minimum epicentral angle at which they can emerge is 142°.

Figure 1.10 also shows other wave paths through the Earth and labels them according to a particular code that is related to the different Earth layers through which they have passed, and whether or not they were reflected or refracted. For example, PKP represents a P-wave that passed through the mantle, followed by the outer core and then back to the surface through the mantle as a P-wave. SKP denotes a shear wave that converted to a P-wave at the core–mantle boundary, was refracted through the core and was then refracted back through the mantle to the surface.

- What do you think PKIKP denotes?
- A P-wave that passed through the mantle, through the outer core and the inner core, and then back through the outer core and mantle to the surface.

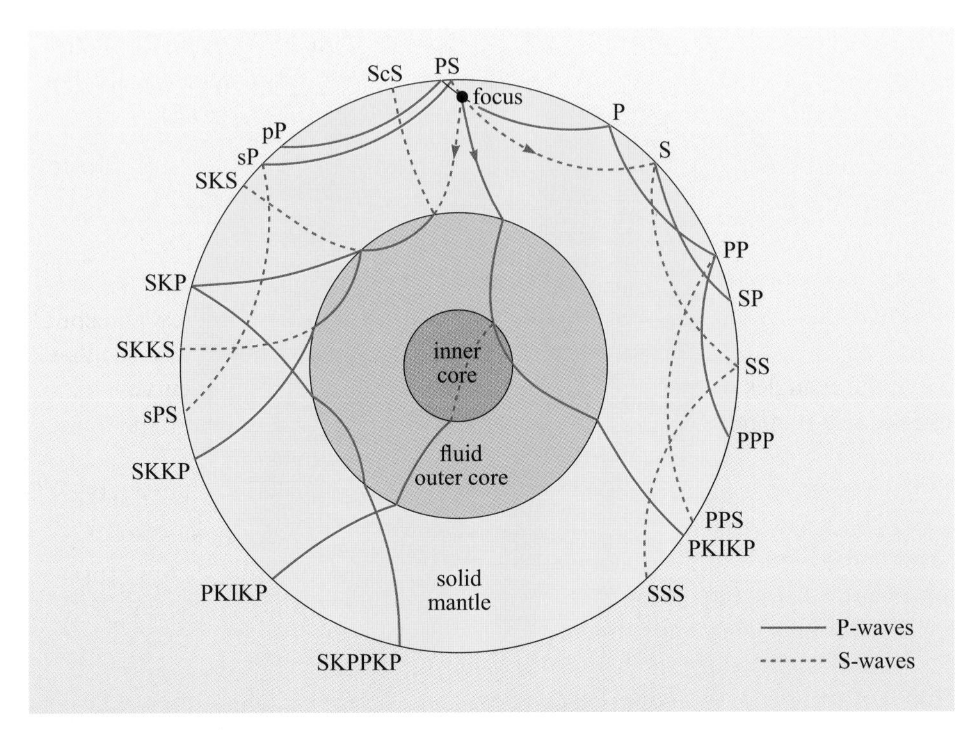

Figure 1.10 A selection of P- and S-wave paths through the Earth. The waves are emitted from the earthquake focus near the top of the diagram.

You should now realise that Figure 1.9 is a gross oversimplification of the real data from seismology and, to give you a flavour of the complexity that seismologists deal with, a compilation of over 60 000 arrival time records relating to 2995 earthquakes is shown in Figure 1.11. In addition to simple P- and S-wave arrivals, this diagram shows arrival times for waves variously refracted and reflected within the Earth.

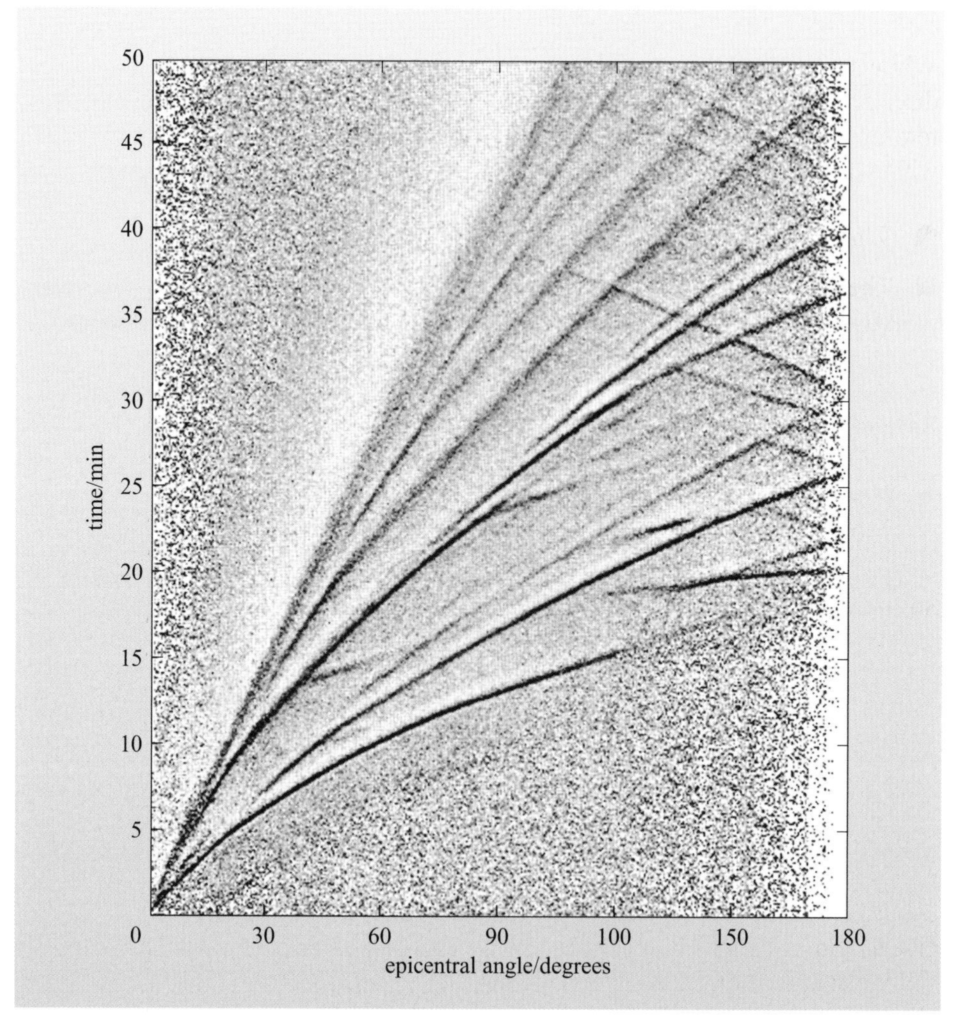

Figure 1.11 Travel-time curves from over 60 000 arrival records relating to 2995 seismic events recorded digitally between 1980 and 1987. (Earle and Shearer, 1994)

The above paragraphs describe how seismic waves pass through the Earth and how arrival times are determined by the Earth's structure. In real life, however, all the seismologist has to go on are the arrival times of P- and S- waves and a reasonable knowledge of the seismic velocities of various rock types as they occur at the Earth's surface. The process of converting these data into information on the internal structure of the Earth is complex and was originally achieved using a laborious mathematical procedure known as inversion. (The details of this procedure are beyond the scope of this book, but see Fowler (2005), Appendix 3, for a detailed explanation.) The application of inversion techniques to numerous earthquakes gave results that were so good that they formed the basis of a more modern approach, based on the use of fast computers to **forward-model** earthquake events. In forward modelling, an initial estimate of the depth–velocity profile from earlier inversion results is used to predict the arrival times of a given seismic event. These predicted arrival times are compared with the actual arrival data and adjustments are then made to the velocity profile until the model matches the observed arrival times. This approach allows seismologists to identify regions within the Earth with anomalous seismic velocities with a good degree of accuracy.

1.2.3 The density of the Earth and Earth layers

While seismic refraction data can inform us of both the depth of seismic discontinuities in the Earth and the variation in seismic velocities with depth to quite remarkable accuracy, it is difficult to extract the density of the different layers within the Earth with the same degree of certainty from seismic velocities alone.

- What do you think is the reason for this?
- The two equations that relate P- and S-wave velocities to density (Equations 1.2 and 1.3) also contain other variables: the two elastic moduli, K and μ. Hence there is no unique solution.

It is possible to eliminate μ from Equations 1.2 and 1.3 by combining them:

$$v_P{}^2 = \frac{K + \frac{4\mu}{3}}{\rho} \text{ and } v_S{}^2 = \frac{\mu}{\rho}$$

So:

$$\mu = \frac{3}{4}(\rho v_P{}^2 - K) = v_S{}^2\rho$$

Rearranging gives:

$$\frac{K}{\rho} = v_P{}^2 - \frac{4v_S{}^2}{3} \qquad (1.5)$$

But Equation 1.5 still requires a knowledge of K, and how it varies with depth, to give any information about density, ρ. In 1923, two geophysicists, Adams and Williamson, formulated another equation that eliminated K. They considered how gravity and density would change through the Earth if density changed only as a result of the mass of the overlying layers, by what is known as **self-compression**. The so-called Adams–Williamson equation takes the form:

$$\frac{\Delta\rho}{\Delta r} = \frac{\text{change in density}}{\text{change in depth}} = \frac{\rho G m}{r^2\left(v_P{}^2 - \frac{4v_S{}^2}{3}\right)} \qquad (1.6)$$

Equation 1.6 is a differential equation that describes the change of density ($\Delta\rho$) with the change of depth (Δr) as a function of distance from the Earth's centre (r), the seismic velocities (v_P and v_S), density (ρ) and total mass (m) at that distance. G is the universal gravitational constant. It is applied by evaluating the right-hand side of the equation at the surface, followed by successive layers at increasing depths. In practice, the integration begins at the top of the mantle, based on a density of 3200 kg m^{-3} – taken from occasional mantle samples found at the Earth's surface – and the process is repeated to the base of the mantle, where there is a change from solid to liquid, and a new density is chosen. Choosing a density for the core is more hit-and-miss than it is for the mantle but, with the added constraint that the total mass assumed in the calculation cannot exceed the known mass of the Earth, density variations through the core can be determined with a degree of certainty. The results of these calculations are summarised in Figure 1.12, which shows how density, gravity and pressure vary with depth in the Earth.

1.2.4 Earth reference models

The information summarised in Figure 1.12 is known as an **Earth reference model**. It describes the important physical properties within the Earth from the surface to the core. Such reference models, and there are more than one, provide the background structure against which anomalies can be compared and recognised. The agreement between different Earth reference models is now so good that seismic velocities throughout most of the mantle and core are known to ±0.01 km s^{-1}, while density, pressure and gravity are defined equally well.

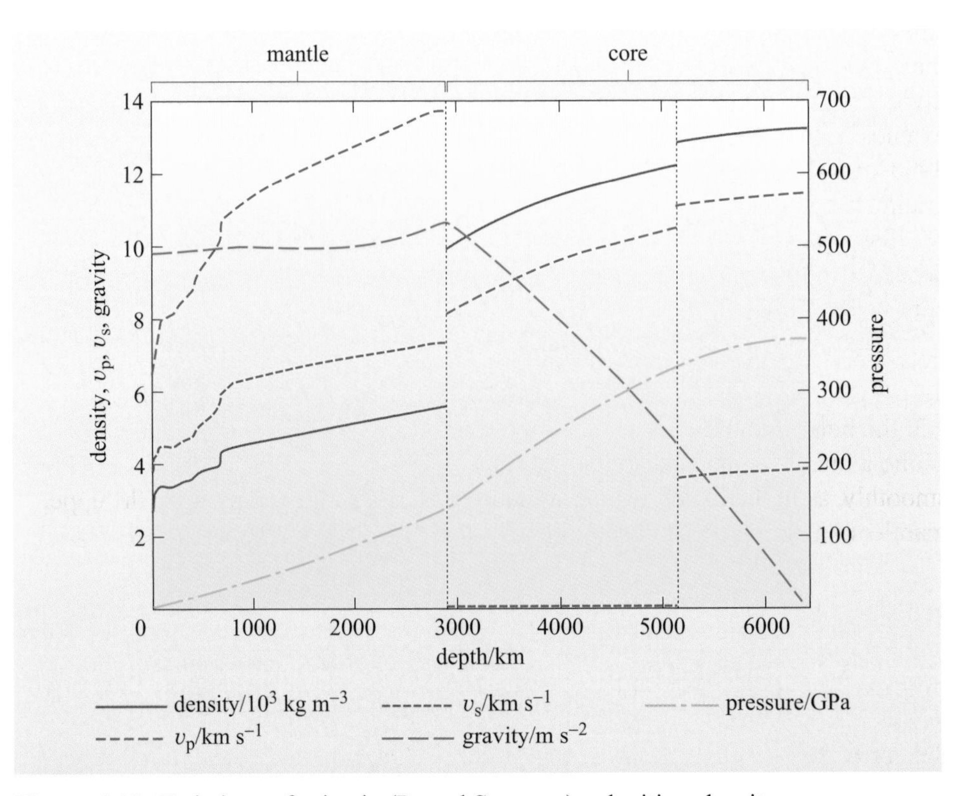

Figure 1.12 Variations of seismic (P- and S-waves) velocities, density, pressure and gravity in the Earth. (Based on the PREM reference model; Dziewonski and Anderson, 1981)

A rapid tour from the surface through the mantle reveals just how thin the Earth's crust is. The crust beneath the continents varies, with an average thickness of about 35 km, but ranging between 15 km and 75 km. Beneath the oceans the crust is much more uniform and thinner (7–8 km in thickness). Overall, the density of the continents is lower than that of the oceanic crust, with typical continental crustal densities of 2600–2800 kg m^{-3} compared with 2800–3000 kg m^{-3} for oceanic crust.

The crust is separated from the mantle by a major seismic discontinuity, the **Mohorovičić discontinuity**, which is almost invariably abbreviated to the **Moho**. The change in seismic velocity across the Moho is greater than in any other region of the Earth, except for the boundary between the core and mantle.

Below the crust is the mantle, which extends down to its boundary with the core at 2900 km. It is far from uniform, especially the uppermost part of the mantle, which in places shows low seismic velocities. The region beneath the lithosphere and down to 220 km is sometimes referred to as the **low velocity zone**. Below 220 km, seismic velocities increase smoothly to a depth of about 400 km where there is the first of two small discontinuities, the first at 410 km followed by a second at 670 km. Both involve increases in P- and S-wave velocities of 5–7%, and this region is sometimes referred to as the **mantle transition zone**. This is a region within the mantle where the crystal structures of the constituent minerals, olivine and pyroxene, are transformed into dense mineral phases with a structure similar to the mineral **perovskite** (Chapter 7). The 670 km discontinuity marks the conventional boundary between the upper and lower mantle, which is also divided into two layers, denoted D′ and D″ (D-prime and D-double prime). D′ extends from 670 km down to 2700 km and D″ extends from 2700 km down to 2900 km. Although relatively thin, the D″ layer is marked by reduced velocity gradients compared with the monotonic increase in seismic velocity through D′. Seismic velocities in the D″ layer also vary by up to 4% and there is much debate as to whether these relate to differences in composition, temperature, or a combination of the two.

Invariably, increases in seismic velocity mirror increases in density because, despite the inverse relationship between velocity and density in Equations 1.2 and 1.3, the bulk moduli increase with depth more rapidly than density. Near the Moho a mantle density of around 3200 kg m^{-3} is typical, and this increases either smoothly, as in the lower mantle, or in a series of discrete jumps, as in the upper mantle, to a density of almost 5000 kg m^{-3} at the core–mantle boundary.

The core is distinguished from the lower mantle by the **Oldham–Gutenberg discontinuity**, named after its discoverers. This and the Moho are the two major seismic discontinuities in the Earth. It is marked by a drop in P-wave velocity from 13.7 km s^{-1} to about 8.1 km s^{-1}, while S-waves drop from about 7.3 km s^{-1} to zero. The curves in Figure 1.12 show that there is a marked density change at the core–mantle boundary, the density almost doubling from 5000 kg m^{-3} to 10 000 kg m^{-3}, in addition to a phase change from solid to liquid. Seismic variations within the outer core are smooth because it is liquid, and the outer core extends to a depth of 5150 km. Finally, the innermost layer of the Earth is the inner core. This is known to be solid, with P-wave velocities of about 11 km s^{-1} and a density of 12 000–13 000 kg m^{-3}.

1.3 The Sun, meteorites and the bulk composition of the Earth

1.3.1 Starting conditions

In order to understand how the Earth works today, it is important to know its bulk composition. To gain that knowledge you need to know how the planet formed initially: both the process of formation and the raw materials used. The dramatic changes at the core–mantle boundary, in seismic velocity, density and phase (solid to liquid), almost certainly mean that the composition changes with depth. To determine just what that change is requires an investigation of the original material that made up the Earth, and for that we need to change our direction of observation outwards from the Earth into the Solar System – to the Sun and meteorites.

The Earth is one of only eight planets (Pluto was downgraded to a 'dwarf planet' in 2006) in the Solar System, which also includes innumerable smaller bodies such as **asteroids**, **comets** and **planetary satellites**, all of which formed, as far as we know, about 4.6 Ga ago out of roughly the same primordial material. Today, there are two types of planet:

- the so-called terrestrial planets, which are small rocky bodies orbiting close to the Sun (e.g. Earth)
- the gas giants that follow more distant orbits in the colder outer reaches of the Solar System.

Planetary formation is linked to the formation of stars themselves. Stars are now widely considered to form from clouds of gas, mostly hydrogen and helium, which are found throughout the Galaxy. Such clouds also contain small amounts of ices and dust composed of elements heavier than helium. Figure 1.13 shows one such example of a stellar nursery. It is one of the best-known images from the Hubble Space Telescope and shows pillars of gas and dust in the Eagle Nebula in the constellation Serpens.

(a)

(b)

Figure 1.13 (a) Columns of cold gas and dust in the Eagle Nebula. The columns protrude from the wall of a vast cloud of molecular hydrogen and are up to four light-years long. In places the interstellar gas is dense enough to collapse under its own weight, forming young stars that continue to grow as they accumulate more and more mass from their surroundings. (b) A close-up of the so-called evaporating gaseous globules (EGGs) at the tips of the gas 'fingers'. (NASA)

To give you some idea of the scale of the gas clouds shown in Figure 1.13, the individual columns are up to four light-years long and emerge from an even larger and more diffuse cloud of gas and dust (shown towards the bottom of Figure 1.13a). Of importance to stellar and planetary formation are the small globules protruding from the ends of the gas 'fingers' (Figure 1.13b) that have been dubbed evaporating gaseous globules (EGGs) within which the density of the gas has increased to such an extent that the cloud is locally collapsing to form a star. These EGGs are roughly the size of our Solar System. Eventually, when the embryonic star in the centre of the EGG starts to shine by its own nuclear energy, the remnant gas and dust is blown away and the system stops growing.

What goes on in an EGG? How does a star and its planetary system form from a collapsing and rotating cloud of dust and gas? One possible sequence of events and processes in the formation of our Solar System is outlined in Box 1.2.

Box 1.2 Formation of the Solar System

The evolution from condensation of grains in the solar nebula through planetary embryos and into planets occurred about 4.6 Ga ago and is summarised in Figure 1.14 (overleaf).

Condensation of the solar nebula

The first phase of the formation of the Solar System involved the gravitational collapse of an interstellar dust cloud dominated by hydrogen and helium, but also containing traces of metallic and silicate dust and ices, such as water (H_2O), methane (CH_4) and ammonia (NH_3) (Figure 1.14a). Once the Sun started to shine, the heat vaporised most of the dust and ices and the vapour was transported further away from the Sun by the early intense solar wind to the cooler regions of the forming Solar System where it cooled and condensed (Figure 1.14b).

The start of accretion

Condensed particles of dust and ice collided with each other and tended to stick together (Figure 1.14c). These random collisions increased the particle size to about 10 mm over a period estimated to have been about 10 000 years. In the inner orbits, these clumps would have been dominated by silicates and metal; their composition is preserved in the most primitive meteorites.

Formation of planetesimals

Particle collisions continued increasing the size of clumps and decreasing the number of objects in orbit around the newly formed Sun, eventually producing a profusion of bodies ranging in size from 0.1–10 km in diameter, termed **planetesimals** ('tiny planets') (Figure 1.14d).

Accretion and the development of planetary embryos

Once formed, the larger planetesimals would become the focus of accretion because of their larger gravitational attraction. At this time, the larger bodies would begin to heat up because of the release of kinetic energy as smaller bodies impacted. Above 10 km this focusing of accretion on the larger planetesimals would have led to the relatively rapid growth of larger objects, known as planetary embryos, with diameters of up to a few thousand kilometres (Figure 1.14f). It is estimated that the planetary embryos would have swept up any remaining planetesimals within a few thousand years.

Planetary embryos and giant impacts

The next stage of growth would have been slower, involving fewer, chance collisions between planetary embryos. Such giant impacts probably fragmented both of the impacting bodies, with the debris subsequently recombining to form a new, larger body (Figure 1.14g). The heat released was also probably capable of melting the newly combined mass, creating a molten mantle of silicate material known as a magma ocean. Metallic material sank through the magma ocean and formed a dense metallic core, thus producing a differentiated planetary embryo.

Differentiation of planetary embryos and the assembly of a planet

Giant impacts would have continued to occur on an ever increasing scale between these partially differentiated planetary bodies and it is estimated that it would have taken about 10 Ma for the terrestrial planets to reach half their current mass and about 100 Ma to complete their growth and build an Earth-sized planet (Figure 1.14h).

Completion of terrestrial planet formation

Once the last giant impact had occurred (for the Earth this resulted in the formation of the Moon; see Section 2.3), there were just five surviving planetary bodies (Mercury, Venus, Earth, Mars and the Moon). Of course, accretion continues to the present day in the guise of meteorite impacts, but fortunately most of these are too small as to be noticeable; but even so, the estimated accretion rate to the Earth today, including dust particles, is about 10^7 kg y^{-1}.

▶ **Figure 1.14** Artist's impression of the stages of planetary growth. See text for explanation. Figure continues overleaf.

(a)
(b)

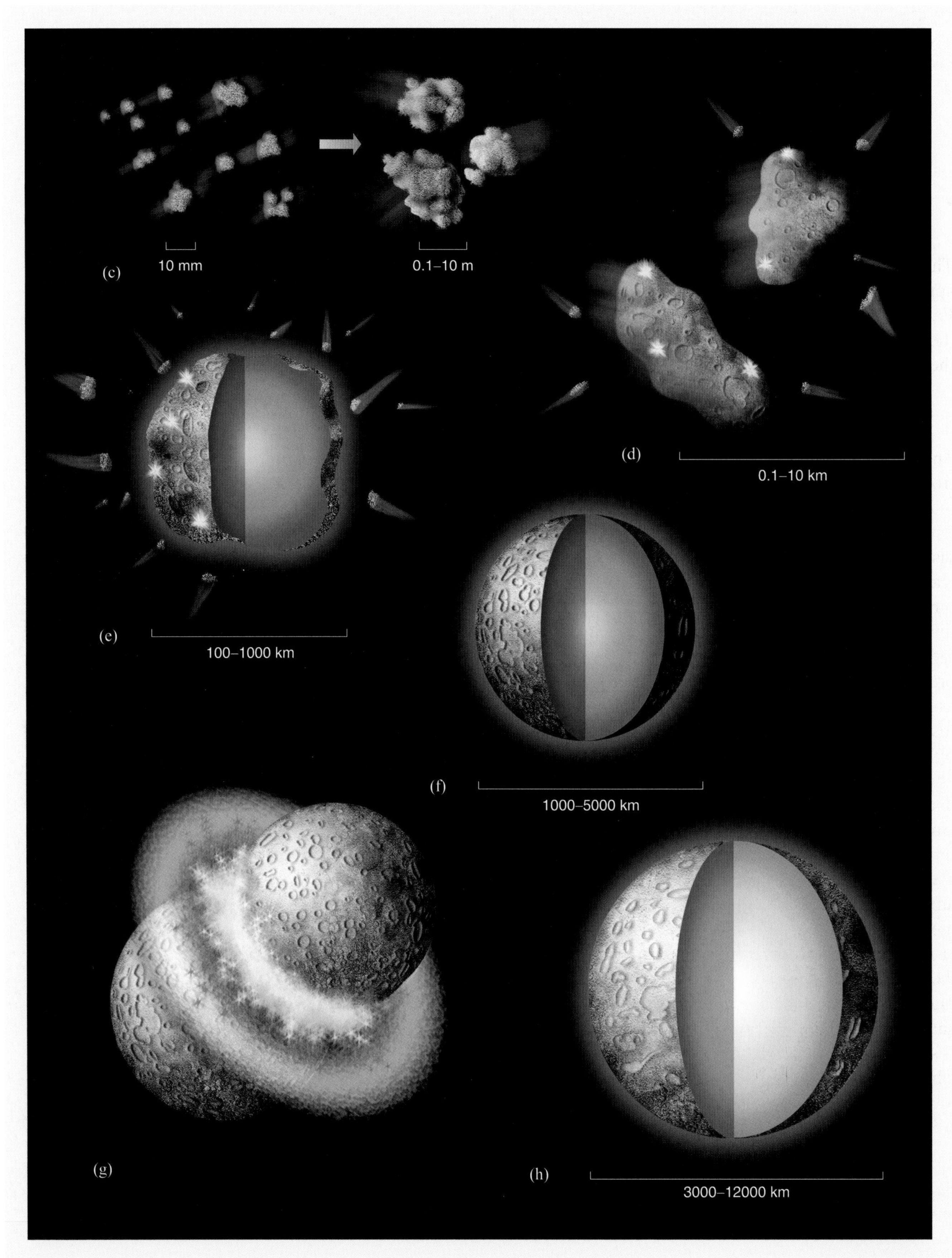

(c)
10 mm
0.1–10 m
(d)
0.1–10 km
(e)
100–1000 km
(f)
1000–5000 km
(g)
(h)
3000–12000 km

Images like those shown in Figure 1.13 offer a visual link between astronomy and Earth sciences, providing clues as to our own astronomical origins. Those origins involved:

- the condensation of gas to form dust particles
- the accretion of dust to form planetesimals, followed by decreasing numbers of increasingly larger collisions between planetesimals to form planetary embryos
- collisions between planetary embryos during a period of giant impacts to form the planets.

The evidence for the violence of planet formation is nowhere more graphically apparent than on the surface of the Moon (Section 1.1). The composition of the Earth is therefore related to the composition of the solar nebula and, as most of the material from the solar nebula condensed to form the Sun, the composition of the Earth is also related to that of the Sun.

1.3.2 The Sun

The composition of the Sun is known from studying the electromagnetic radiation it emits, which is dominantly white light – a continuous spectrum at visible, near-ultraviolet and infrared wavelengths. As this light passes through the Sun's atmosphere, the different chemical elements absorb radiation at specific wavelengths, producing dark bands in the spectrum as recorded by earthbound instruments. This is the same method by which astronomers gain knowledge of the compositions of distant stars and galaxies. The amount of light absorbed is proportional to the amount of an element present in the Sun's atmosphere and, while such measurements do not readily give the absolute abundances of elements in the Sun, they do provide valuable information on the relative abundances. Some of this information is summarised in Figure 1.15, in which element abundances are shown relative to a nominal abundance of silicon (Si), in this case 10^6 atoms.

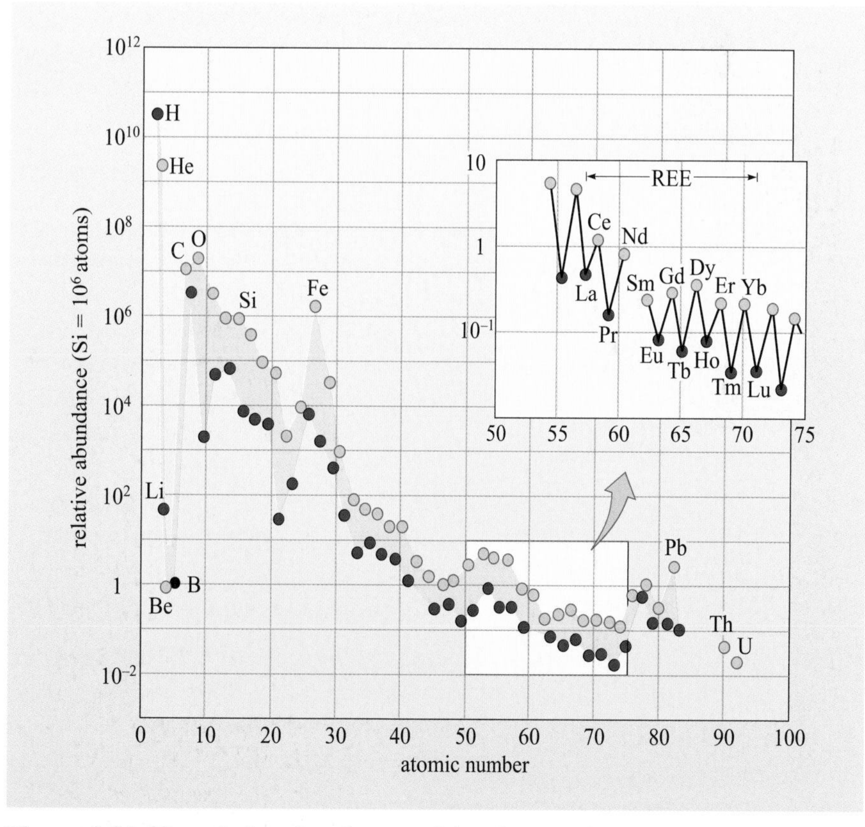

Figure 1.15 The relative abundances of the elements in the Sun. The abundances are expressed relative to silicon (Si), the abundance of which is arbitrarily set at 10^6 atoms. Thus, in the Sun, for every million atoms of Si there are ~5×10^{10} atoms of hydrogen (H) and about 1 atom of beryllium (Be) and boron (B). The elements are ordered by atomic number from left to right and selected elements are labelled with their chemical symbols. Note that the vertical scale is logarithmic, such that each increment on the scale is 10 times that of the previous one. REE = rare earth elements. (Gill, 1982)

Study Figure 1.15 and compare the relative solar abundances of two everyday elements, such as iron (Fe) and lead (Pb). Both of these elements are used in many modern-day applications, both have to be won from ore bodies by industrial processes, yet iron is almost 1 000 000 times more abundant in the Sun than is Pb. Indeed Pb is only slightly more abundant in the Sun than elements such as gold (Au; atomic number 79) and platinum (Pt; atomic number 78) which, on Earth, are regarded as being much rarer and of greater value.

- Why do you think that Pb seems to be more abundant than the precious metals?
- The Earth's crust is the source of our mineral wealth and its composition is the product of many processes that have occurred during and since planetary accretion.

Different elements have different geochemical properties and so respond in contrasting ways during geological and planetary processes. It so happens that Pb has become concentrated in the Earth's continental crust, whereas Fe has remained largely in the mantle and core, as you will see below. Our everyday view of the Earth is dominated by what we observe and find in the crust, but the crust, as you already know, is but a thin skin on the surface of our planet. If we could see right through to the core, our view of what the Earth is made of and which elements are abundant and which are rare would be very different.

1.3.3 Element condensation from the solar nebula

The solar nebula is considered to have been hot to start with and as it cooled the different chemical elements gradually condensed according to their individual volatilities. Many of the chemical elements are metals and most have boiling points much too high to be considered significant in our everyday experience. But under the low-pressure conditions of the solar nebula, differences in their boiling points become important. Because the pressure in the nebula was low relative to that experienced at the Earth's surface today, the temperatures at which different elements vaporised (or condensed on cooling) were lower than they are on the present-day surface of the Earth. For example, alumina (Al_2O_3) has a vaporisation temperature (boiling point) of about 3500 K under atmospheric conditions (10^5 N m^{-2}), but in the lower pressures of the solar nebula (10 N m^{-2} or 10 000 times lower than atmospheric pressure) this is reduced to 1700 K.

Figure 1.16 shows the temperature at which various compounds and elements condense out of a nebula with a solar composition, plotted against the fraction of the nebula condensed. As you can see, some elements condense as specific compounds, whereas others condense as the elements themselves. This sequence allows the subdivision of the different elements and compounds into those that belong to an early condensate, a metal–silicate fraction, and volatiles. At even lower temperatures, compounds such as water, methane, ammonia and nitrogen condense as solids, all termed ices.

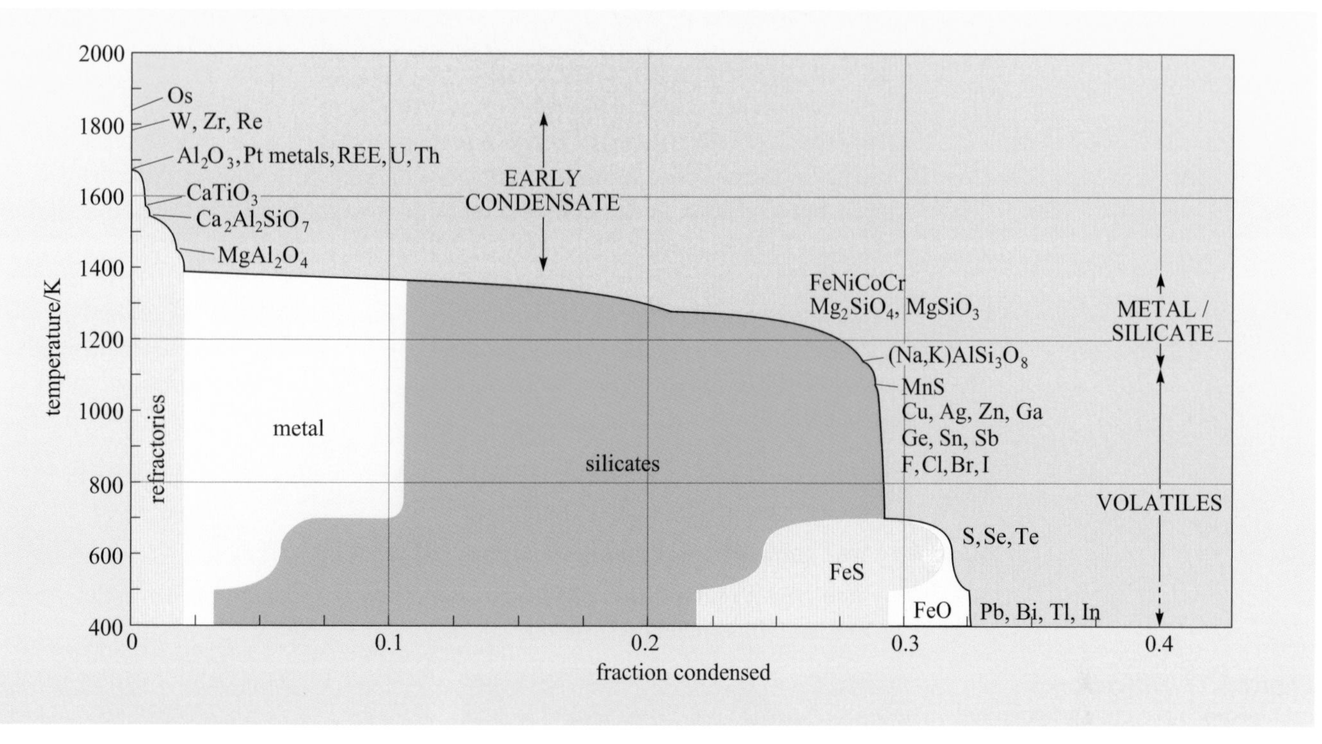

Figure 1.16 The condensation sequence of the solar nebula at a pressure of about 10 N m^{-2}. The *x*-axis represents the fraction of the nebula that has condensed at the temperature given on the *y*-axis. The curve is annotated to show which elements and compounds condense at which temperatures. At this low pressure, all materials condense from gases to solids directly without an intervening liquid phase. (K = kelvin = T °C + 273) (Morgan and Anders, 1980, modified using data from Lodders, 2003)

■ What are the proportions of the dominant components of the condensate?

■ The early condensates, or refractories, comprise 2% of the total. This is followed by metals, which comprise 12% of the total, and silicates, which comprise 19% of the total. The remainder includes volatiles and ices, such as H_2O, CH_4 and NH_3.

■ If the temperature in the solar nebula decreased outwards from the proto-Sun, what mechanism might have produced the terrestrial planets (Mercury, Venus, Earth and Mars) and the gas giant planets (Jupiter, Saturn, Uranus and Neptune)?

■ Given the sequence of condensation in Figure 1.16, volatile elements and ices would only condense when the temperature of the solar nebula dropped below 300 K. Perhaps such cold conditions only occurred beyond the orbit of Mars.

Element (and compound) volatility is one of the important properties that controls the behaviour of the elements in planetary formation and, consequently, the composition of all planetary bodies.

1.3.4 Meteorites

A second source of information about the composition of the Solar System and the primary material of the planets comes from meteorites. These are astronomical objects of varying sizes, made of metal or rock, that fall to Earth. They have been studied for many years, not only because they are of intrinsic interest but also because they contain material thought to be representative of the early solar nebula. Meteorites are derived from the asteroid belt, which comprises many small planetary bodies (possibly planetary embryos) that did not accrete into a single larger body. This probably resulted from gravitational disturbances caused by Jupiter, perturbing the orbits of individual asteroids and causing repeated violent collisions that resulted in further fragmentation rather than accretion.

Meteorites come in a variety of compositions, but can be broadly classified into:

- **stony meteorites**, consisting dominantly of silicate minerals
- **iron meteorites**, primarily composed of metallic iron
- **stony-iron meteorites**, a hybrid of the other two, containing variable amounts of silicates and metal.

Figure 1.17 Different types of meteorites: (a) a cut and polished surface of an iron meteorite that has been etched with acid to reveal details of its internal texture; (b) a carbonaceous chondrite; (c) a stony meteorite with a basaltic composition; (d) a stony-iron meteorite, or pallasite, with olivine crystals set in a metal matrix; (e) a relatively large mm-sized chondrule from a carbonaceous chondrite (Bokkeveld meteorite), showing individual crystals within it; (f) a collection of individual chondrules, each less than 1 mm in diameter, from an ordinary chondrite (Sharps meteorite). (Natural History Museum)

The iron and stony-iron meteorites have undergone an amount of chemical processing and are described as **differentiated**; their compositions have been affected by removal of some or all of the silicate minerals to leave a bulk composition enriched in iron (Fe) and nickel (Ni), the major metallic elements.

Stony meteorites account for up to 95% of all known meteorite falls and are subdivided into **chondrites** and **achondrites**, depending on whether they contain **chondrules** or not. Chondrules are small, roughly spherical globules of silicate minerals that are between 0.1 mm and 2 mm in diameter (Figure 1.17e and f). Their shape and crystallinity suggest that they were once molten in a low gravitational field, indicating that they formed away from major planetary bodies, either on the surface of planetesimals or even within the solar nebula.

Achondrites constitute only about 10% of all stony meteorite falls. Their textures are reminiscent of terrestrial igneous rocks and they are classified as differentiated meteorites, along with the irons and stony-irons.

Chondrites are subject to further subdivision and classification based on their mineralogy. The three main classes are:

- **ordinary chondrites**, so-called because they are the most abundant type
- **enstatite (or E-) chondrites**, which are rich in the magnesium silicate mineral enstatite ($MgSiO_3$)
- **carbonaceous (or C-) chondrites**, which contain non-biogenic carbon-rich organic compounds in addition to silicate minerals.

The ordinary chondrites are further subdivided according to their iron contents and their oxidation state, reflected in the amount of iron in silicate minerals relative to that in chemically reduced phases such as metallic iron and sulfides. These groups are referred to as **H-**, **L-** and **LL-chondrites**. This rather complex classification is summarised in Figure 1.18.

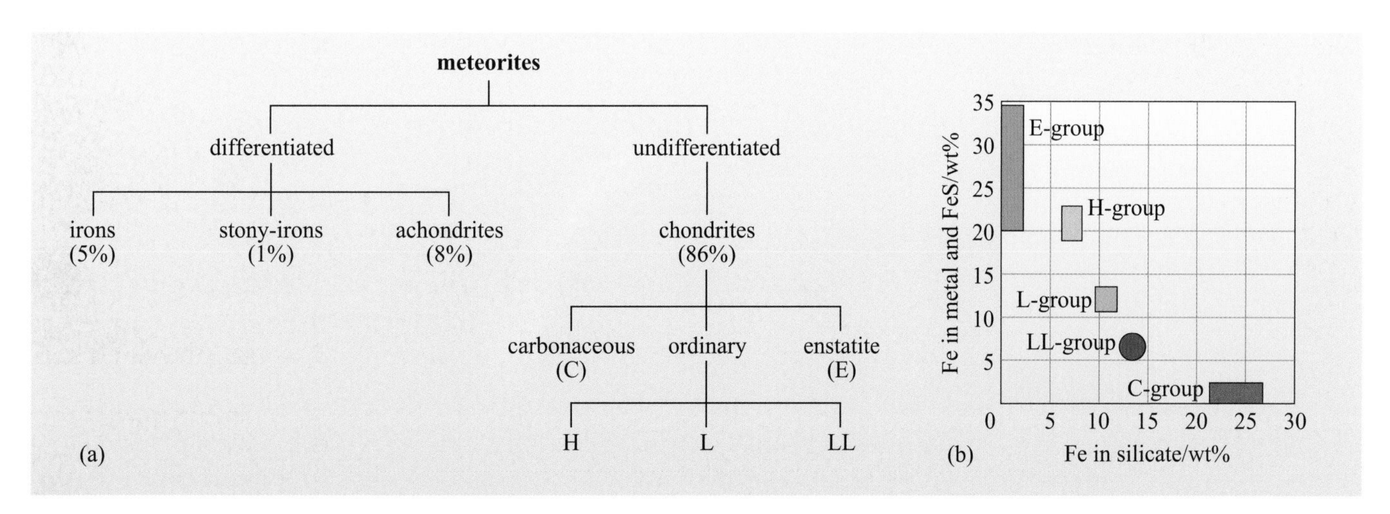

Figure 1.18 (a) Meteorite classification based on observable mineralogy and textures. (b) Classification of chondritic meteorites based on the oxidation state of iron.

- Considering this overall classification, what do you consider to be the dominant process in meteorite differentiation?
- The separation of a metallic phase from the silicate minerals.

The compositions of iron meteorites reflect processes that have taken place within a planet-sized body (or planetary embryo) where temperatures were at one time high enough to melt the silicates and the metallic phases and allow them to separate as a result of gravity. Thus, in our search for the most primitive composition, the differentiated meteorites can be rejected, leaving the chondrites. But which of these represents the most primitive material?

- If you return to a consideration of the processes that occur within the solar nebula, which are the elements that would be most easily lost from a meteorite?
- The volatile elements and ices, because they have the lowest condensation temperatures.

Assuming that condensation from the solar nebula continued without disturbance to low temperatures, then the most primitive meteorites should contain abundances of the volatile elements in the same proportion as in the Sun. A comparison of the compositions of the different meteorite groups reveals an overall similarity in the abundances of elements found in high-temperature early condensates and increasingly diverse concentrations of the more volatile elements. This feature is illustrated in Figure 1.19, where the abundances of selected elements in enstatite chondrites and ordinary chondrites are compared with those in average carbonaceous chondrites.

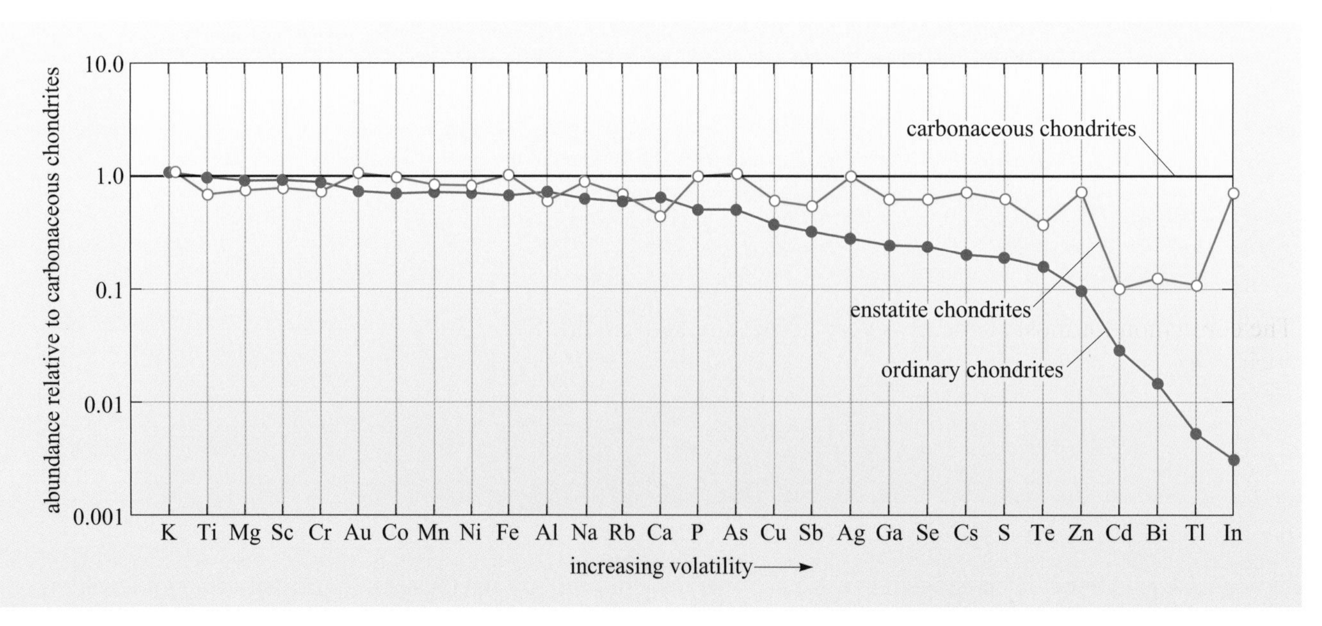

Figure 1.19 Comparison of the abundances of selected volatile elements in enstatite, ordinary and carbonaceous chondrites. Element abundances are normalised to carbonaceous chondrites. (McSween, 1987)

- Which types of chondrite do you consider to be the most differentiated and which type the least differentiated?
- The ordinary chondrites have lost a greater proportion of their volatile elements and so are more differentiated than the E-chondrites. The C-chondrites are therefore the least differentiated because they have not lost their volatile elements.

The C-chondrites thus appear to have been least affected by processes after their formation. This conclusion is borne out when their elemental compositions are compared with those of the **solar photosphere**, which is the outer layer of the Sun. This is shown in Figure 1.20 for a wide range of elements with different volatilities and different chemical properties.

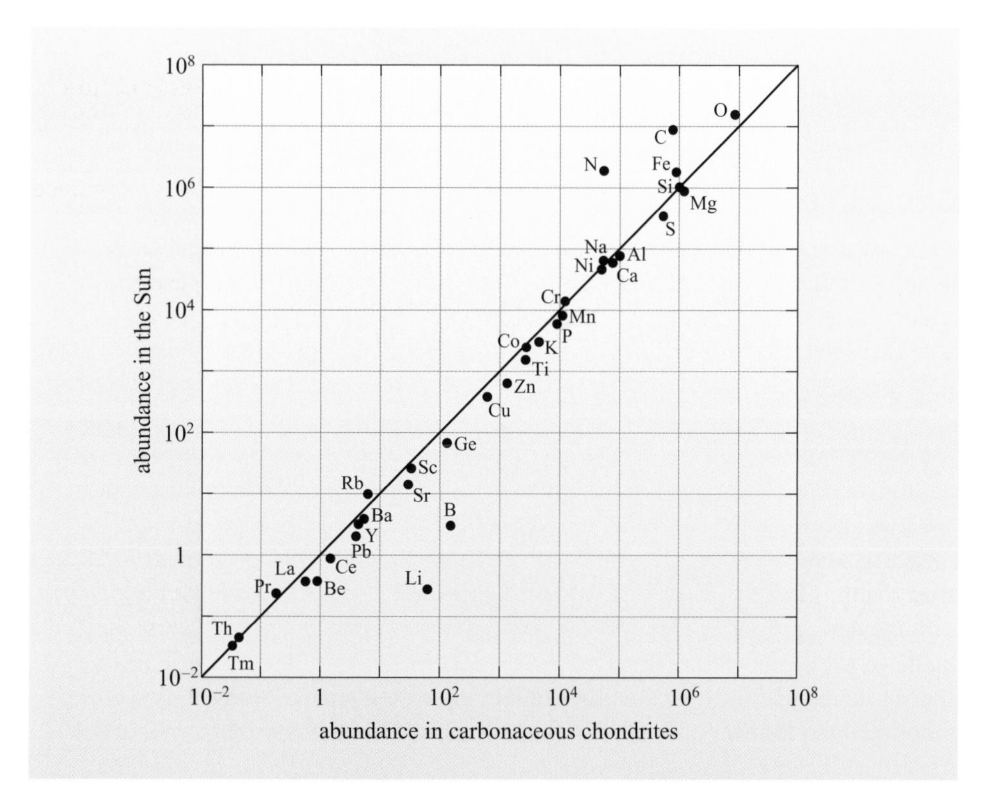

Figure 1.20 The relative abundances of the elements in the solar photosphere plotted against those in C-chondrites, based on Si = 10^6 atoms. (McSween, 1987)

The correlation for most elements is very good, ranging from the most abundant, such as oxygen, magnesium and iron, to the least abundant such as thorium (Th) and thulium (Tm). There are some exceptions – for example, lithium (Li) and boron (B) are depleted in the solar composition relative to C-chondrites because they are used up in nuclear fusion reactions in the Sun. Conversely, carbon (C) and nitrogen (N) are relatively depleted in carbonaceous chondrites, possibly because they are among the most volatile elements. Despite these differences, the match between carbonaceous chondrites and the Sun is remarkable and gives us some confidence that this really is the composition of the primordial material of the Solar System.

1.4 The composition of Earth layers

The results of seismic investigations into the Earth reveal a twofold division into a dense core and an overlying mantle, with the mantle comprising about 68% by mass of the whole Earth. By contrast, the crust comprises about 0.4% of the Earth's mass and is insignificant in this context, even though it is the part of the Earth with which we are most familiar.

■ Recalling the variety of meteorite compositions, which types of meteorite do you think will give the best indications of the compositions of planetary layers?

■ The Earth is a differentiated planetary body, so the differentiated meteorites will give the best indications of the compositions of planetary layers.

The likely compositions of the core and mantle should now be fairly obvious – the core is made of a dense alloy of iron and nickel while the mantle is composed of silicates which, by analogy with achondrites, will be rich in magnesium (Mg). How can we test these ideas? Fortunately, there are places where mantle rocks are brought to the surface by volcanic and tectonic processes and so at least these samples can be compared with meteorites. However, there are no such samples of the core so comparison of the core and meteorites is less direct, relying instead on properties such as seismic velocity and density.

1.4.1 The Earth's mantle

The Earth's mantle is almost everywhere obscured by the crust, but occasionally geological processes carry mantle rocks to the surface. Mantle samples occur in two main geological situations. The first is as accidental fragments of rock known as **xenoliths** brought to the surface in volcanic rocks that erupt from great depth. They are distinctive because they have a green colour that contrasts with the dark grey or black of the host volcanic rock and, for many years, their significance was unclear. However, as ideas of the composition of the Earth's interior developed they became the subject of more intensive study. These rocks are termed **peridotites** and are rich in the minerals olivine and pyroxene, which are both silicates rich in magnesium and iron.

Similar rock types are also found in the deeper layers of rock sequences called **ophiolites**. Within an ophiolite, a thin layer of sedimentary rocks overlies volcanic lavas, dykes and gabbros, all of basaltic composition, which in turn overlie peridotite. Ophiolites are found in young mountain belts and are interpreted as being segments of old oceanic crust that have been trapped between colliding continents.

■ How could you confirm that a sample of peridotite from a xenolith or ophiolite is from the upper mantle?

■ The density or seismic velocities could be measured and compared with the Earth reference model (Section 1.2).

Laboratory measurements of the density of peridotites (about 3200 kg m^{-3}), and their P-wave speeds (approximately 8.0 km s^{-1}), are a close match for the measured properties of the upper mantle.

Direct comparisons between the composition of peridotites and carbonaceous chondrites are less easy because the mantle represents the silicate residue after core formation. However, some compositional parameters, particularly those involving element ratios, can be exploited to make a useful comparison. An example is shown in Figure 1.21, which plots the Mg/Si ratio against Al/Si in both meteorites and mantle samples.

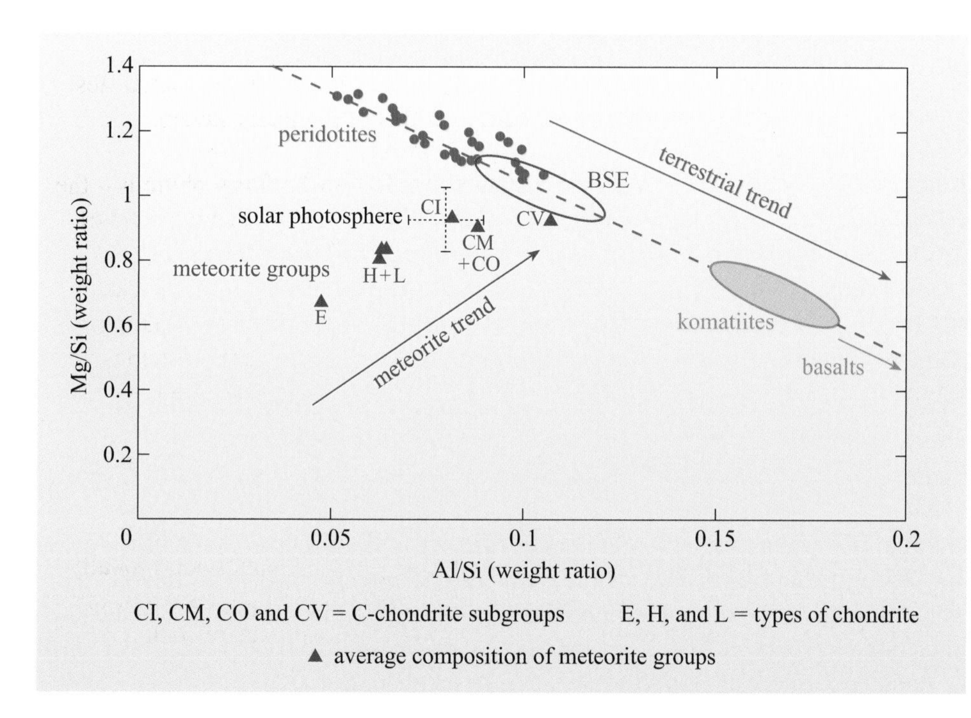

Figure 1.21 The variation of the Mg/Si ratio with the Al/Si ratio in terrestrial mantle samples and different classes of meteorites. The composition of the solar photosphere is shown for reference. The negative trend defined by the terrestrial mantle (peridotite) samples is related to the generation and extraction of basaltic **magma**. Komatiites are Mg-rich, ultramafic magmas that were produced by extensive melting of the mantle during the Archaean. BSE represents the range of estimates of the composition of the bulk silicate Earth (sometimes referred to as the primitive upper mantle). (Adapted from Drake and Righter, 2002)

In Figure 1.21, peridotites define a negative trend along which the Mg/Si ratio decreases as the Al/Si ratio increases. Firstly, this shows that the composition of the mantle as revealed by peridotites is not constant – the mantle is heterogeneous. The causes of this variation are indicated by the position of basalts and **komatiites** on this diagram, which lie on an extension of the peridotite trend at higher values of Al/Si and lower values of Mg/Si. Basaltic magma is derived from the mantle by a process known as partial melting, and this will be explored in greater detail in Chapter 4. For now, it is sufficient to appreciate that the variation in peridotite composition is a direct consequence of the extraction of basaltic magma.

■ What does this relationship between basalts and the mantle indicate about the most probable location of primitive mantle within the array of peridotite data?

■ Mantle samples with the most primitive compositions will be those that have not experienced the extraction of basaltic magma and so they will plot at the low Mg/Si end of the peridotite array, closest to the composition of basalt.

Average meteorite compositions, in contrast to peridotites, define a positive trend on Figure 1.21: both Mg/Si and Al/Si increase together from E-chondrites that have the lowest values of both ratios, increasing through H- and L-chondrites and then into the C-chondrites, which have the highest values of both ratios. Notice that the C-chondrites are further subdivided into a number of different groups, designated CI, CM, CO and CV. Although, the origin of this variation remains unclear, the intersection of the positive meteorite trend with the terrestrial array has been used to define the composition of the primitive mantle.

- Do the variations illustrated in Figure 1.21 suggest that the Earth has a composition directly comparable with C-chondrites?

- No. C-chondrites have generally lower values of both Mg/Si and Al/Si than the most primitive mantle composition.

Although the chondritic Earth model has been used as a standard for many decades, it is now well established that for certain elements, of which Si is one, the **bulk silicate Earth** does not have a composition that is represented by any meteorite group and so it is not directly comparable with the solar photosphere and hence the Sun. The only class of meteorites that do fall on the terrestrial array are an unusual group of meteorites designated CV.

Other elements can be investigated in a similar way, and the abundances of selected elements in the mantle relative to CI meteorites are illustrated in Figure 1.22. The abundances of the different elements are all normalised to that of Mg, and it shows that elements such as aluminium (Al), calcium (Ca) and uranium (U) retain the same relative abundances in the Earth's mantle as they have in CI meteorites and hence the Sun. Other elements, such as Si, the alkali elements (sodium (Na), potassium (K), rubidium (Rb) and caesium (Cs)) and transition elements such as copper (Cu), silver (Ag) and zinc (Zn) are all depleted by up to two orders of magnitude relative to CI meteorites.

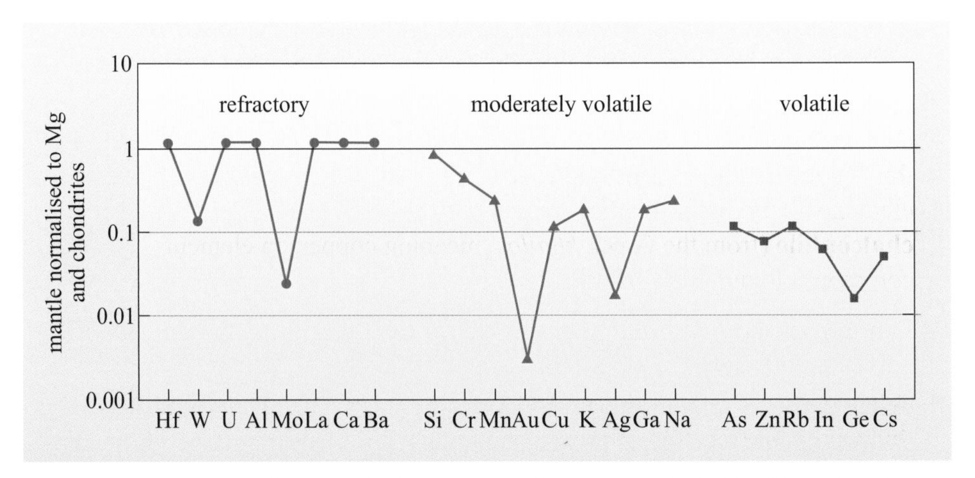

Figure 1.22 The abundances of selected trace elements in the bulk silicate Earth normalised to the composition of CI meteorites, in order of increasing volatility from left to right. (Abundances taken from McDonough and Sun, 1995)

- Referring back to the dominant processes in the solar nebula, suggest a property that might cause these depletions.
- All of the elements listed as depleted are more volatile than Al, Mg, Ca and U so their depletions may relate to conditions in the solar nebula or to a subsequent loss of a volatile component after planetary formation.

Relative to CI meteorites, the Earth's mantle is depleted in volatile elements and this may in part explain the difference between the estimated bulk composition of the Earth and the different meteorite classes – the Earth has lost a greater proportion of its volatile components compared with meteorites. But that is not all.

- Consider the element tungsten (W). Using the information in Figures 1.16 and 1.22, does its relative abundance relate to its volatility?
- No. W is a refractory element with a condensation temperature similar to Al, Ca and U, yet it is depleted in the mantle with an abundance of only ~13% of that expected, assuming a chondritic starting composition.

Clearly, the abundances of elements in the bulk silicate Earth are not controlled solely by their volatility, and other properties govern their distribution within the Earth. Box 1.3 describes these properties in relation to the position an element occupies in the Periodic Table.

Box 1.3 Element properties and distribution coefficients

The distribution of the elements between the different phases (metal, silicate and sulfide) in meteorites and other natural materials has led to a broad classification of their geochemical (and cosmochemical) properties. Elements that are found in the metallic **phase** of a natural system are referred to as **siderophile** (from the Greek, *sideros*, meaning iron and *philos*, meaning like or love – literally iron-loving). They contrast with those elements that preferentially bond with oxygen, especially in silicate or oxide structures, which are known as **lithophile** (from the Greek *lithos*, meaning stone). There is also a subgroup of lithophile elements that tend to be gaseous at the Earth's surface, notably H, C, N, O and the noble gases, that are referred to as **atmophile**. A third major grouping refers to elements that frequently occur bound with sulfur and these are known as **chalcophile** (from the Greek *khailos*, meaning copper, an element commonly found as a sulfide).

Figure 1.23 shows how these properties are distributed across the Periodic Table and that some elements can behave in different ways. For example, in the absence of a separate metal phase, cobalt (Co) will behave as a lithophile element, but it partitions preferentially into the metal liquid as soon as a molten metal is introduced into the system. Similarly, iron (Fe) can be siderophile, lithophile or chalcophile, depending on the prevailing conditions. Despite this overlap of properties, the broad distinction between siderophile, lithophile and chalcophile has proven useful in the development of geochemistry.

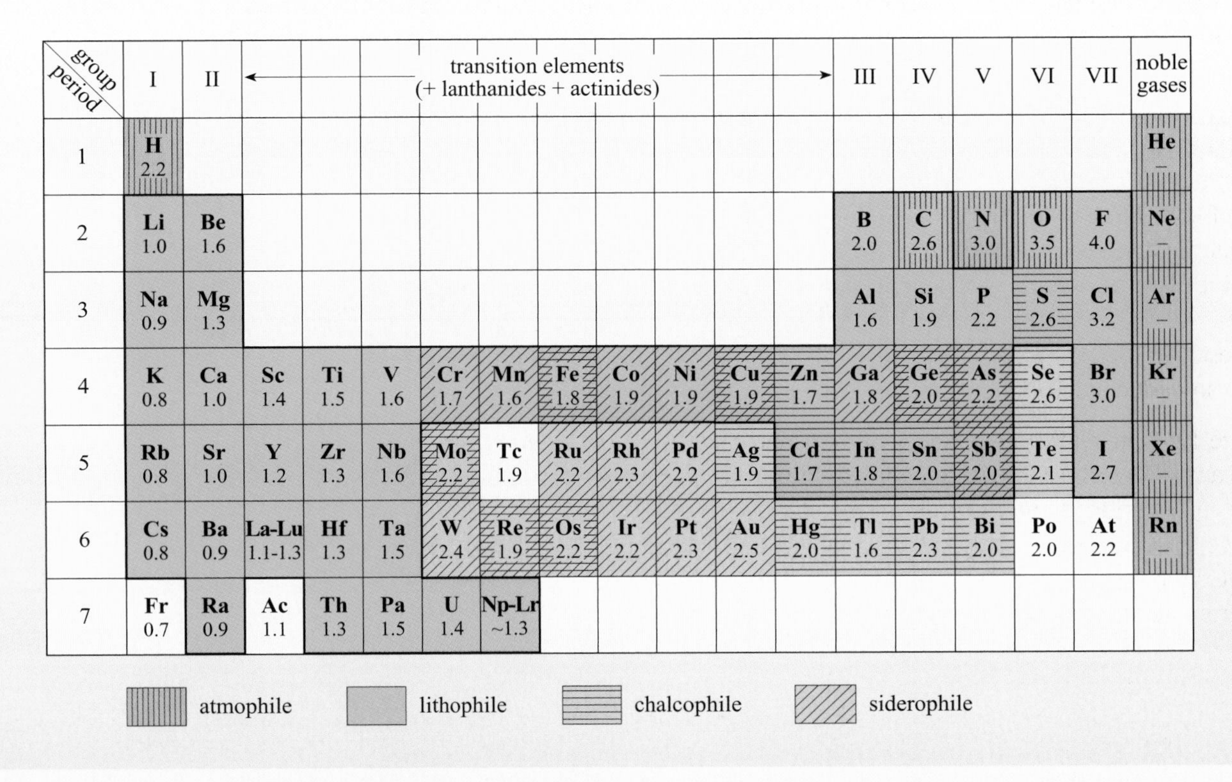

Figure 1.23 The Periodic Table subdivided according to the geochemical properties of the elements. Colour shows the dominant property and the shading shows subsidiary properties. The number beneath each element symbol refers to electronegativity (Pauling scale).

The differences in behaviour are related to a property known as **electronegativity (*E*)**, which is based on a relative scale from 0 to 4.0 and related to the ability of an atom to attract electrons and so form a negatively charged ion. Broadly speaking, those elements with low electronegativities ($E < 1.7$) are lithophile, forming positive ions that bond with negative oxygen ions. Those elements with $1.7 < E < 2.1$ tend to be chalcophile, because they more readily form covalent bonds with elements such as S. Finally, siderophile elements have $2.1 < E < 2.5$ and most readily form metallic bonds. Elements with $E > 2.5$, such as O, N and the halogen elements, F, Cl, Br and I, readily attract electrons to form negative ions and form ionic bonds with the lithophile elements with low electronegativity.

The way in which elements distribute themselves between solid and molten states is known as **element partitioning** and is quantitatively expressed in the form of a number known as a **partition** or **distribution coefficient**, variously defined as K, K_d, K_D or D. It is defined by the following simple equation:

$$D = \frac{\text{concentration of element } i \text{ in phase } a}{\text{concentration of element } i \text{ in phase } b}$$

$$D = \frac{C_a^i}{C_b^i} \tag{1.7}$$

In the case we are considering in this chapter, the distribution coefficient is between molten metallic iron and solid silicate minerals. The convention is for the numerator (C_a^i) to be allocated to the liquid and the denominator (C_b^i) to be allocated to the solid. Thus, a lithophile element will have a D value of <1 and a siderophile element will have a D value of >1.

Most commonly, the distribution coefficient is used to describe the way in which an element partitions between crystals and liquids in silicate systems. In contrast with this chapter, in these systems elements that partition into a solid phase ($D \geq 1$) are said to be **compatible,** whereas those that are excluded and concentrate in the silicate liquid ($D < 1$) are said to be **incompatible**.

- Having studied Box 1.3, suggest why elements such as W, Mo, Au and Ge have lower abundances in the bulk silicate Earth than expected for their volatilities.

- These elements can behave as siderophile elements in the presence of a molten metal. Therefore they have probably been concentrated into the Earth's core.

Core formation involved the gravitational separation of molten metallic iron from silicate melt or solid crystals. By analogy with iron meteorites, which are considered to be representative of the cores of small planetary bodies, this iron selectively removes a number of other elements that, because of their chemical properties, bond more easily with iron than they do with silicates; W is one of these so-called siderophile elements (Box 1.3).

Figure 1.24 shows the abundances of the siderophile elements in the mantle, once again normalised to CI meteorites. In this case, the most volatile elements have been excluded and the order of the elements is according to their increasing siderophile properties as determined by laboratory experiment, so that elements on the right are increasingly siderophile. Clearly, the more siderophile elements are more depleted in the Earth's mantle.

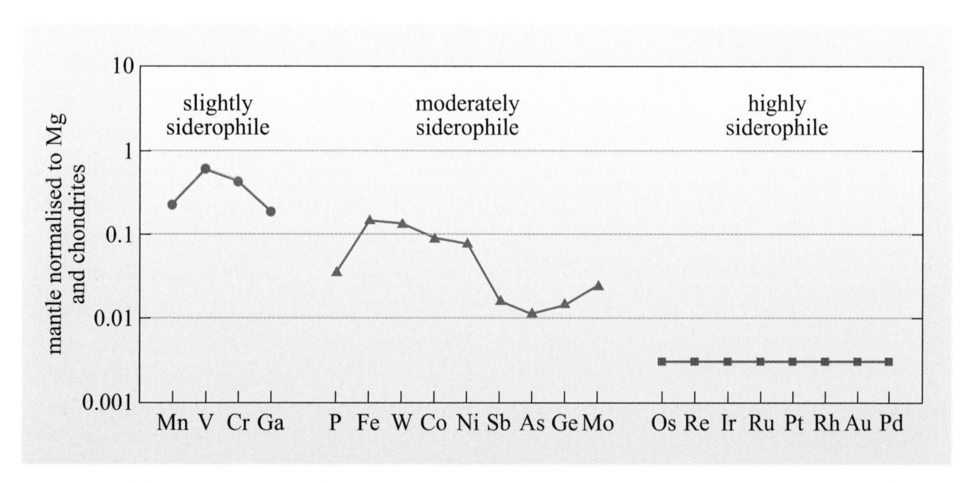

Figure 1.24 Abundances of the siderophile elements in the Earth's mantle relative to CI meteorites. (Data from McDonough and Sun, 1995)

To summarise all of this information, the present-day composition of the Earth's mantle may be similar to that of some C-chondrites. The ratios of refractory lithophile elements (e.g. Ca, Al, Mg, U, Th and rare earth elements) in the mantle are similar to those in chondritic meteorites generally, indicating a close link with primordial (solar) abundances. But the mantle is depleted in both volatile and siderophile elements. The latter are now presumably concentrated in the core, whereas the volatile elements were probably lost during accretion or even earlier during condensation from the solar nebula. The important result, however, is that the abundances of elements within the mantle are controlled systematically by their geochemical and cosmochemical properties.

1.4.2 The Earth's core

From the various lines of evidence, it is clear that the Earth's core is primarily made of iron, according to both the model developed above and the requirements of the seismic Earth reference model (Section 1.2). Samples of the core are not available at the Earth's surface, so any scientific investigation of the composition of the core has to be done either remotely (e.g. with seismology) or by using laboratory analogues and simulations. The latter involves subjecting materials to the temperatures and pressures of the core in the laboratory, measuring their properties and comparing them with Earth reference models.

One of the most useful properties is density; Figure 1.25 shows the variation of density with depth, and hence pressure, for the Earth's core and for pure iron.

■ Do the curves match?

■ No. The density of the core is less than that experimentally determined for pure iron.

The mismatch between the curves for the core and pure iron requires the core to include other elements that reduce its overall density.

■ Which other elements might the core contain?

■ Siderophile elements.

The core must contain significant amounts of siderophile elements, but can these elements account for the lower density of the core? In short, the answer is no. Most of the siderophile elements have atomic masses, and hence densities, either similar to or even greater than that of iron – if anything, siderophile elements make the problem of the core's low density even worse! The unavoidable conclusion (assuming experiments have given the correct results) is that the core includes a significant fraction of an element (or elements) with an atomic number and density less than liquid Fe–Ni alloy.

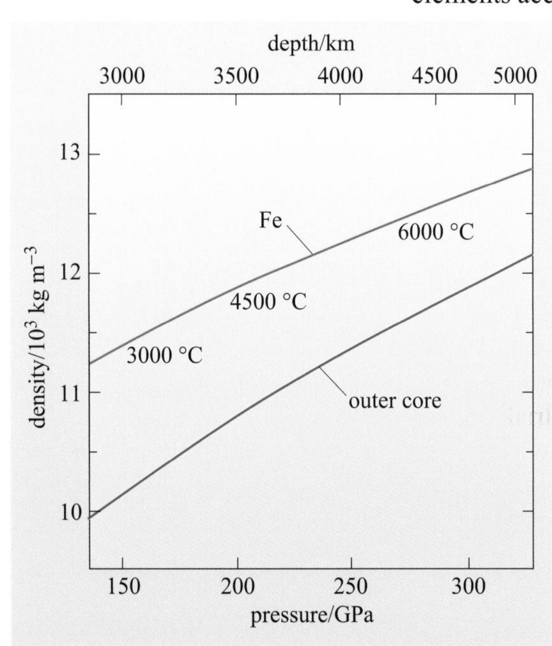

Figure 1.25 Variation of density with pressure in the outer core and the density of iron as determined experimentally for conditions in the core. Temperatures are indicated along the curve for pure iron.

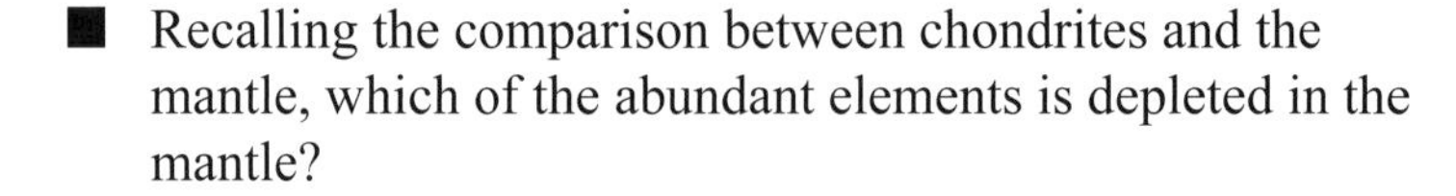

■ Recalling the comparison between chondrites and the mantle, which of the abundant elements is depleted in the mantle?

■ Silicon.

Figures 1.21 and 1.22 show that Si is depleted in the mantle by ~10% and although that depletion may be related to its volatility, it is also possible that Si could have partitioned into the core. Certainly some Si has been identified in the metal phase of iron and stony-iron meteorites, but the concentrations are far too low – and are measured in ppm rather than the 10–15% required to reduce the density sufficiently. However, some experiments reveal that Si has increasing siderophile tendencies at very high pressures. Other elements that have been suggested include O and S, in the forms of FeO and FeS respectively. Both FeO and

FeS can dissolve in molten Fe in sufficient quantities to reduce its density enough to match that of the core.

- From what you have understood so far about the behaviour of the different elements, can you see why either of these suggestions might *not* be reasonable?

- Sulfur is a volatile element and so may have been lost from the Earth before the core formed, whereas the very high electronegativity of oxygen (Box 1.3) means that it is dominantly lithophile.

A major problem that must be overcome in determining whether or not any particular element could be incorporated in the Earth's core relates once again to its volatility during accretion. Sulfur is a particularly volatile element, condensing out of the solar nebula in the form of FeS at a temperature of ~600 K (Figure 1.16). Its relative abundance in the Earth's mantle is broadly similar to that of other similarly volatile elements, such as the halogens, and more volatile elements such as Pb, Bi and thallium (Tl). These observations indicate that S is behaving as a typical volatile element, that it is neither more nor less depleted in the mantle than it should be and that, by analogy with these similarly volatile elements, it is unlikely to have been present in the early Earth in a high enough concentration to reduce the density of the core sufficiently.

This leaves oxygen. Experimental studies have shown that, at the conditions appropriate for the core–mantle boundary, oxygen can dissolve in molten Fe as FeO. The mantle contains abundant oxygen bound in silicates that dissociate into dense oxide phases in the deep mantle, and a variety of reactions between these oxides and oxygen-free molten iron have been suggested, all of which lead to the production of FeO (wüstite). (You will encounter these high-pressure transformations in Chapter 6, but for the moment it is sufficient to know that they exist.) Thus, mechanisms for transferring oxygen from the mantle to the core can be demonstrated on the basis of high-pressure experiments, and oxygen remains the main candidate for the light element in the Earth's core.

The Earth's core and the magnetic field

Although the core is a very remote component of the Earth, it has a profound effect on the surface environment as it is the source of the Earth's magnetic field. The Earth has one of the strongest magnetic fields in the Solar System and certainly the strongest magnetic field associated with any of the terrestrial planets. Continuing on the theme of what makes the Earth unique in the Solar System, it is an important aspect of maintaining and preserving a benign environment on the Earth's surface by protecting the surface from cosmic rays and high-energy particles.

The Earth's magnetic field cannot result from the presence of a solid magnet in the interior because permanent magnetism cannot remain above the **Curie point** of any magnetic material. For example if a solid iron magnet is heated to about 600 °C it loses its magnetism. Given that the Earth's core is considerably hotter than this (<4000 °C) the field must be generated by some other mechanism.

Some clues to where and how the field is generated can be found in the record of observations of the magnetic field over time. For example, the position of magnetic North moves at a perceptible rate over periods of decades to centuries, and the geological record reveals rapid flips in the Earth's magnetic polarity many times in the recent and more distant geological past (Chapter 3). These observations suggest that the magnetic field must originate in a region of the Earth that can move and respond rapidly – in other words, it must originate in a very fluid part of the Earth, the outer core. However, the mechanism whereby the field is generated is still not fully understood. It is thought to be related to an interaction between the Earth's rotation and convection within the outer core that, in turn, is driven by a combination of solidification of the inner core and secular cooling.

Traces of ancient magnetism, known as **remnant palaeomagnetism**, have been frozen into some of the Earth's oldest volcanic rocks and demonstrate that the Earth's magnetic field has been in existence for at least ~3.8 Ga. If **compositional convection** is the principal means of generating the Earth's magnetic field, then the inner core must be at least as old as the magnetic field. Compositional convection caused by crystallisation of the inner core also involves cooling of the core. However, the estimated cooling rate of the core is too great for the core to maintain heat for 3.8 Ga, and suggests that the inner core cannot be older than about 1 Ga. In order to account for the age of the magnetic field and this cooling rate, it has been suggested that there may be significant radioactive heating of the core (for example, from the decay of the light element potassium (K), which may be incorporated in the core in trace amounts). Alternatively, it has been suggested that cooling rates, and compositional convection, may have been slower earlier in Earth's history.

1.5 The Earth's crust: continents and oceans

To bring this introductory chapter to an end, we return to the differences between the continents and the oceans, not just that one is covered with water and the other not, but the differences in the composition and physical structure of oceanic and continental crust and how they relate to the bimodal hypsometric curve described earlier in this chapter.

The geology of the Earth's surface is known so well in places that it is difficult to treat it as a single compositional entity as is routinely done for the mantle and core. Geological observations show that the continental crust is highly complex and variable on a variety of length scales – from a single outcrop a few metres across to the width of a continent. The variation in rock types has kept field geologists busy for over 200 years, and still large tracts remain unmapped. However, where the deeper layers of the continents have been exposed by tectonic processes and erosion, the predominant rock type is a variety of granite. Similarly, seismic velocities in the uppermost layer of the crust are highly variable, reflecting its geological complexity, but below this surface layer velocities and densities become more uniform, implying that a single composition might be justified.

But that is not the whole story because within many crustal sections a marked increase in seismic velocities occurs at ~15 km depth. Known as the **Conrad discontinuity**, it separates the upper crust from the lower crust and, while not apparent in all sections, it is best developed in regions of older rocks of Archaean or Proterozoic age. Examples of seismic crustal profiles are shown in Figure 1.26.

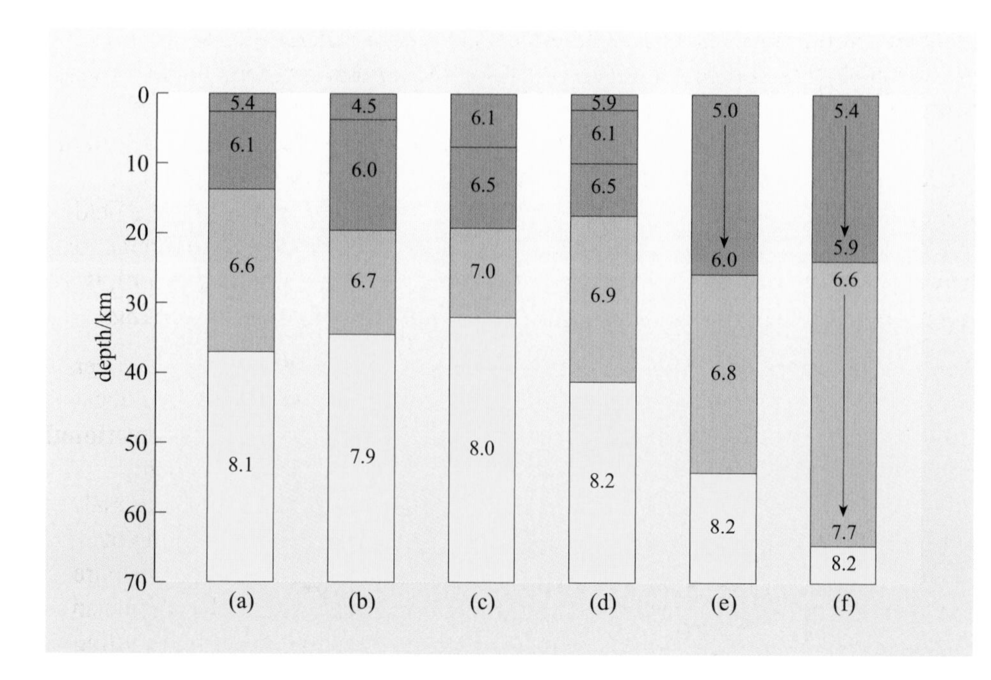

Figure 1.26 Sections through six representative segments of the continents: (a) a stable continental shield, Wisconsin, USA; (b) a modern continental rift, Basin and Range Province, USA; (c) a 400 Ma continental collision zone in northern Scotland; (d) a 100 Ma old ocean–continent plate boundary in southern California; (e) a modern continent–ocean plate boundary, Central Andes, South America; (f) a modern continent collision zone in the central Alps. The arrows in (e) and (f) indicate smooth velocity increases between the values indicated at either end. Seismic velocities of 6.0–6.5 km s^{-1} are typical of the upper crust and are comparable with those of granodiorite, a variety of granite. Higher velocities of 6.6–7.0 km s^{-1} are typical of the lower crust, while a sharp increase to velocities of about ≥8 km s^{-1} marks the Moho at the base of the crust.

The general consensus is that the upper crust has a composition approximating **granodiorite**, which is a type of granite containing plagioclase and alkali feldspar in roughly equal proportions, 20% quartz and small amounts of biotite and hornblende. The lower crust, by contrast, is thought to be made up of granulites with a slightly more mafic composition, i.e. more Fe and Mg and less Si than in granodiorite. **Granulites** are metamorphic rocks that have been subjected to high pressures and temperatures such that they have lost most of their volatile components (largely water) and their mineralogy is dominated by plagioclase feldspar, pyroxenes and garnets, which give the lower crust its higher density. Overall, the bulk composition of the continental crust is thought to be close to that of the intermediate rock type, andesite, or to its plutonic equivalent, diorite.

The oceanic crust contrasts with the continental crust in almost all physical and compositional characteristics. Whereas the continental crust is of variable thickness, the oceanic crust is much thinner and of a more uniform thickness. Where the continents show a diversity of seismic structure, the oceans reveal a structure that can be consistently applied across most of the ocean basins.

The most important difference between continental and oceanic crust is its thickness. Beneath the oceans, the Moho is located on average at < 10 km depth, compared with an average of 35 km beneath the continents. Secondly, the oceanic crust has a pronounced layered structure, as illustrated in Figure 1.27. You should recall from Section 1.4 that one of the sources of mantle samples was ophiolites, which are slices of oceanic crust and mantle emplaced by tectonic processes in mountain belts.

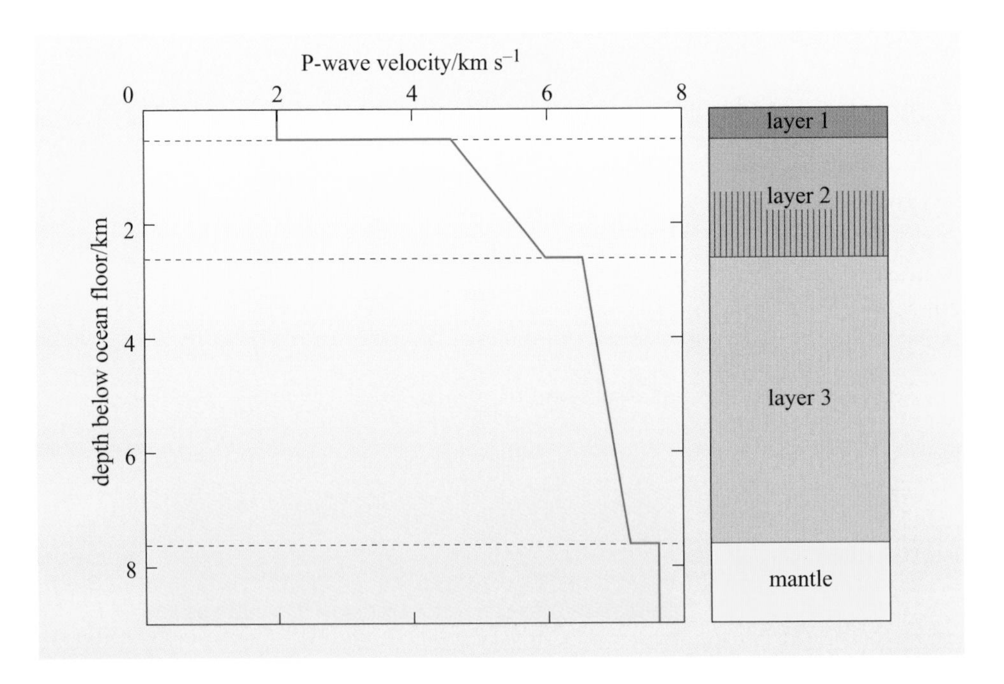

Figure 1.27 Variation of seismic velocities with depth in a typical section through the oceanic crust.

■ Given the brief description of an ophiolite in Section 1.4, what are the likely compositions of the different seismic layers in Figure 1.27?

■ Layer 1 is made up of sediments, layer 2 of basaltic lavas and small intrusions (dykes) and layer 3 of gabbros. These three overlie mantle peridotite.

■ What is the average composition of the oceanic crust?

■ Disregarding the thin layer of sediments, both layers 2 and 3 have a basaltic composition.

Overall, oceanic crust is basaltic compared to the andesitic composition of the continental crust. Importantly, the continental crust contains much less Fe, Mg and Ca and more Na, K and Si than the oceanic crust.

The combination of low density and greater thickness has important implications for the average elevation of regions underlain by continental and oceanic crust. Superficially, the Earth is composed of solid rocks that behave in a rigid way, and that is certainly correct on length scales of metres and kilometres, but that style of behaviour breaks down over longer length scales and in the mantle where temperatures are much higher. At depth, rocks begin to behave more as fluids than as rigid solids, and when loaded with ice sheets, volcanoes or mountains they begin to flow. This leads to a new mechanical, as opposed to compositional, division of the outer layers of the Earth, the importance of which will be developed in later chapters. The rigid, cold surface layer is known as the **lithosphere** and it includes the crust and the uppermost mantle. Below this is the more mobile mantle known as the **asthenosphere** (from the Greek *asthenos*, meaning weak), which can deform like a viscous fluid when under stress. When a load is applied to the Earth's surface, the lithosphere subsides into the underlying asthenosphere; by contrast, when the load is removed the lithosphere rises again. In essence this is simply Archimedes principle applied on a grand scale, but when considering the Earth it is known as **isostasy** and it controls the surface topography of the Earth.

1.5.1 Isostasy

The principles behind isostasy can be illustrated by considering blocks floating in a liquid, such as wood floating in water or, as in Figure 1.28, metal blocks floating in a container of mercury. Quite clearly, the situation illustrated in Figure 1.28 is probably one that few people will ever witness for real, but it is a representation of an experiment carried out by an eminent Earth scientist, W. Bowie (1872–1940), to illustrate the effects of isostasy for other Earth scientists who found the concept problematic!

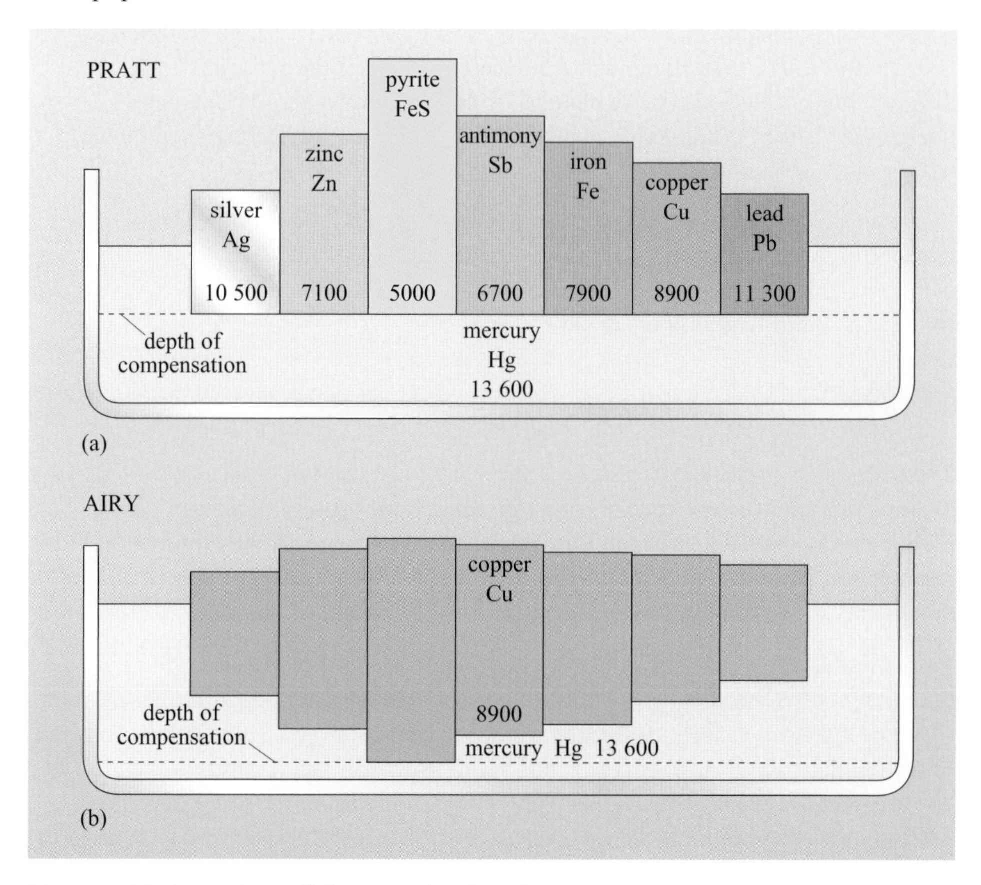

Figure 1.28 The principles of isostasy: (a) a series of metal blocks with different densities (numbers indicate densities in kg m^{-3}) floating in a container of mercury; (b) copper blocks of varying lengths floating in mercury. Note the differences in the depth of compensation. (Watts, 2000)

Figure 1.28 shows how differences in elevation, as reflected by the height of the upper surfaces of the blocks above the surface of the mercury, can be achieved in two ways. In Figure 1.28a, a series of metal blocks with the same cross-sectional area have been cast from identical masses of different metals with different densities; as a result, the blocks have different lengths (heights). When placed in the mercury, according to Archimedes principle each block displaces the same mass and hence the same volume of mercury. Consequently, the blocks exert the same pressure (mass per unit area) on the mercury on which they rest and the base of each block is at the same level in the mercury. This depth is known as the **depth of compensation**. In Figure 1.28b, only one metal is used, in this case copper, but the blocks have different heights. As a result, the longer blocks not only stand higher than the shorter ones, they also project deeper into the mercury. Defining the depth of compensation in this case is more difficult, but one way is to relate it to the depth of the lower surface of the longest block. The main difference between this example and that in Figure 1.28a is that the pressure at the compensation level beneath all of the blocks except the longest is made up of the mass of the blocks *and* any overlying mercury.

Figure 1.28 illustrates the principles behind the two models of isostasy as applied to the Earth, proposed in the 19th century by J.H. Pratt (1809–1871) and the then Astronomer Royal, G.B. Airy (1801–1892). At the time, the level of compensation was thought to be the base of the Earth's crust and the Airy model was invoked to explain the presence of a thick, low-density crustal root beneath the Himalaya and Tibet. Seismic profiles beneath this region and other mountainous areas have shown that this is the case and that high mountains are indeed underlain by thicker crust, as shown in Figure 1.26. However, it is now known that the level of compensation in the Earth is probably considerably deeper than the Moho and corresponds at least to the base of the lithosphere or even deeper in the underlying asthenosphere. Thus, within the continents, different elevations could be explained by greater than average lithosphere thicknesses beneath mountains, with thinner lithosphere underlying sedimentary basins, such as shelf seas.

In the oceans, by contrast, the structure is different and the Pratt model of isostatic compensation becomes significant.

■ Can you see why this is the case?

■ The oceanic crust has a higher density than the continental crust so the comparison is more like Figure 1.28a.

The differences in density between oceanic and continental crust and the average thickness of the oceanic crust produce a lithospheric column with a higher average density than the lithospheric column beneath the continents. Assuming a similar depth of compensation beneath both continents and oceans this results in the overall lower elevation for the surface of the oceanic crust.

Thus oceanic crust reaches isostatic equilibrium at a lower elevation than continental crust because it is both thinner and denser. And this is the underlying cause of the terrestrial hypsometric plot discussed earlier in Section 1.1. The lack of a similar bimodal plot for Mars and Venus, suggests that continental and oceanic crust as we understand them in the geological sense are absent from both of these planets. This almost certainly reflects a difference in the behaviour of their interiors from that of the Earth. It may be, for example, that the mantles of both Mars and Venus are much more viscous than the Earth's mantle and do not allow isostatic adjustments in surface elevation in response to different rock densities. Alternatively, as you will discover later in this book, since both continental and oceanic crust are the products of plate tectonic processes, the lack of crustal differences suggests that the internal processes of those two planets are different. Either way, this primary observation suggests that the internal processes of Planet Earth are unique in the Solar System and have contributed significantly to the way the Earth and its surface environments have evolved over geological time.

Summary of Chapter 1

Having reached the end of this introduction to the Earth, you should now appreciate that the differences between the Earth and the other terrestrial planets are more than just skin deep. The uniqueness of the Earth goes beyond its atmosphere and hydrosphere, being expressed in the contrasts between continental and oceanic crust, the dominance of plate tectonics and the presence of a strong magnetic field. Other characteristics not covered in detail here include the presence of an unusually large satellite, the Moon, and the Earth's relatively rapid rotation rate. These all influence the surface environment of the Earth and the evolution of the biosphere.

After this introduction, you are now in a position to explore aspects of the Earth's structure and evolution in more detail. The following chapters develop some of the themes established here in addition to introducing new ideas concerning the movement of mass and energy within the Earth and how that leads to the dominance of plate tectonics and the cycling of material between the interior of the Earth and its surface and back again.

Learning outcomes for Chapter 1

You should now be able to demonstrate a knowledge and understanding of:

1.1 The physical and chemical features of the Earth that make it unique compared with other terrestrial bodies in the Solar System.

1.2 The internal structure of the Earth and how it is divided into the core, mantle and crust.

1.3 How seismic refraction can be used to identify and define different layers within the Earth.

1.4 The principle of isostasy and why the Earth's crust is divided into continental highlands and oceanic basins.

1.5 The classification scheme used for the different types of meteorites.

1.6 How the chemical composition of the bulk Earth and each of its constituent layers have been obtained from various sources including meteorites and samples obtained directly from the mantle.

The early Earth

The Solar System is thought to have begun after one or more local supernova explosions about 4.6 Ga ago. In one widely accepted scenario of the later stages of formation of the Solar System, it is thought that there were hundreds of planetesimals in the region occupied by the inner planets. Once the planetesimals began to attain the proportions approaching those of planetary embryos, it is likely that the heat generated by collisions would have been sufficient to allow both melting and, as denser materials began to sink inwards, separation of the original constituents. However, the development of a more evolved layering, such as that seen on Earth, would require this separation to have been more or less complete once giant impacts had ceased. This is because from that point onwards further evolution of the Earth would be mainly driven by processes from within the planet.

The first 660 million years of the Earth's existence, known as the Hadean, was the stage during which the metallic core separated from the silicate mantle, the atmosphere and the hydrosphere were formed, and melting of the silicate mantle produced the earliest crust. The early Earth was violent and hot, and giant impacts would have been devastating. The heat released would have been capable of melting the outer part of the Earth to form a global **magma ocean**. Rapid motion of the magma ocean and degassing would have also destroyed any solidified surface areas. Consequently, there is virtually no direct rock record of the Hadean. This means that we have to turn to theoretical modelling and geochemistry to reveal the mechanisms of formation of the different layers in the Earth and the timescales involved.

In this chapter, you will be introduced to the evidence for the development of the early Earth. The first part reviews the various heat sources necessary to drive planetary differentiation. The second part investigates the mechanisms of core formation, the evidence for the presence of a magma ocean, and the timing of accretion and core formation. The third part explores the evidence for the origin and age of the Moon. In the fourth part, you will learn about the formation and evolution of the atmosphere and the hydrosphere, and finally, the evidence for the nature and formation of the earliest continental crust will be reviewed. You will discover that the processes that formed the layering in the early Earth, atmosphere and hydrosphere effectively shaped the planet as we know it today, and ultimately provided conditions suitable for life to develop. You will also see that water played a key role, not only in the formation of the atmosphere and the hydrosphere, but also in the early development of the solid Earth.

2.1 Heating and differentiation of the Earth

Differentiation is the process by which planets develop concentric layering, with zones that differ in their chemical and mineralogical compositions. The generation of such zones results from a differential mobility of elements due to differences in their physical and chemical properties.

One obvious way of mobilising constituents in a planetary body is by allowing them to melt. When a rock is heated, different minerals within the rock will melt at different temperatures. This phenomenon is known as **partial melting** and is a key process in the formation of liquid rock or magma, as you will see in Chapter 4. Once elements have been mobilised in this manner, they will begin to migrate under the influence of pressure or gravity.

■ Imagine a body the size of a planetary embryo that had accreted from nickel–iron and silicate minerals. Nickel–iron has a density of about 7.9×10^3 kg m^{-3} (compared with ~3.0×10^3 kg m^{-3} for silicate minerals) and a melting point some hundreds of degrees higher than silicates. What would happen if temperatures within this planetary embryo were increased to a point at which silicates began to melt?

■ Since nickel–iron has a higher melting point, it would remain solid after the silicates had begun to melt and, because it is much denser than any silicate minerals, it would begin to sink towards the centre of the body.

In addition to the separation of a metallic core from a silicate mantle, partial melting of planetary bodies also produces residual solid silicate minerals in contact with a silicate melt that have different compositions, depending on how extensive the melting is. In such a system, incompatible elements are partitioned into the melt more readily than compatible elements (Box 1.3). This process, known as element partitioning, is the principal mechanism by which incompatible elements first become concentrated into the melt. Then, if the magma is buoyant, they migrate upwards to form the overlying crust. If element partitioning continues over time, this concentration and migration will gradually create a crust and mantle with different compositions.

2.1.1 Heat sources

If partial melting is the principal cause of differentiation, then the Earth needs to be heated before layering can begin to develop. There are several sources of heat that can arise during Earth's evolution. The most important are:

- the so-called primordial heat sources, which develop in the early stages of planetary evolution (i.e. those associated with accretion, collision and core formation)
- tidal and radiogenic heating processes, which can operate long after the planet has been formed.

Accretional heating

During accretion, any planetesimal falling towards the Earth will acquire a velocity because of gravitational attraction from the Earth. The body will thus have a kinetic energy due to its motion. If the velocity of the body immediately before impact is υ, the kinetic energy (E) at that point will be:

$$E = \frac{1}{2}m\upsilon^2 \qquad (2.1)$$

where m is the mass of the body. Upon hitting the Earth, if all the kinetic energy of motion is converted into heat, then the increase in temperature, ΔT, can be calculated:

$$\Delta T = \frac{mv^2}{2(m + M)C} \tag{2.2}$$

where the body (of mass m) impacts the Earth (of mass M) and C is the specific heat capacity of Earth material (i.e. the amount of heat required to raise the temperature of 1 kg of material through 1 K).

Question 2.1

(a) A planetesimal of mass 10^{15} kg impacts the Earth with a velocity of 10 000 m s^{-1}. Calculate the rise in temperature in the Earth assuming that the heat generated by the impact spreads rapidly and uniformly throughout the whole Earth. Because m is much smaller than M, the effect of m is negligible and can be ignored, so Equation 2.2 can be simplified to:

$$\Delta T = \frac{mv^2}{2MC}$$

Take the total mass of the Earth to be 6×10^{24} kg, and the average specific heat capacity of the Earth to be 750 J kg^{-1} K^{-1}. (*Note*: 1 J = 1 kg m^2 s^{-2})

(b) Suppose that the Earth was constructed entirely of 10^{15} kg planetesimals, each of which generated the temperature rise obtained in part (a). What would be the total temperature rise?

In practice, the various uncertainties involved make it difficult to determine how much heating occurred during accretion. For instance, as you will see later in this chapter, not all of the impacting material arrived at the same time: accretion took place over ~10^7 years. More importantly, not all of the kinetic energy would be converted to heat. For example, some of the impact energy would be used in the excavation of large craters. Finally, much of the heat would have been radiated into space. Nevertheless, most estimates predict temperatures to have risen above the melting point of silicate minerals and iron–nickel, which means that the Earth is likely to have gone through an early molten stage.

Core formation

If the Earth went through an early molten phase, allowing the metals and silicates to separate, then the 'falling inwards' of the nickel–iron-rich fraction to form the core would have released **potential energy**.

■ What do you think would have happened to this gravitational potential energy?

■ The gravitational energy lost by the inward movement of nickel–iron would have been converted first to kinetic energy and then into thermal energy. It is estimated that the core-forming process would have contributed significantly to the Earth's primordial heating (although it would still have been an order of magnitude less than that generated by collision and accretion).

As you will see later in this chapter, geochemical evidence suggests that accretion and core formation were completed very early in the Earth's history but, rather than being a single 'catastrophic event', it is likely to have been a gradual process, with progressive segregation of the core as more material was accreted.

Since both accretion and core formation relate to events that occurred early in the history of planetary evolution, they are primordial processes and the heat generated by these processes was primordial heat. However, if these primordial heat sources had remained the only way of heating the Earth, their intensity would have waned through time due to continual radiative heat loss to space. Since heat drives fundamental processes such as volcanism, the fact that the Earth has remained volcanically active to the present day requires additional processes of internal heat generation.

Tidal heating

One heat source known to be generated within planetary bodies is tidal heating, which is created by the distortion of shape resulting from mutual gravitational attraction. These effects are readily observed upon the Earth's oceans where the attraction created by the Sun and the Moon produces 'bulges' in the ocean water masses that are then dragged around the planet as the Earth rotates. This process produces the ebb and flow of tides seen around the coast. In much the same way, the solid Earth is also distorted by these forces and produces tides that reach a maximum amplitude of about 1 m on the rocky surface. This deformation causes heating within the planet, although precisely where this heating is concentrated depends on the planet's internal properties. In Earth's case, it is thought to occur largely within the crust and mantle.

Question 2.2

The current rate of heating generated within the Earth by tidal distortion is estimated at 3×10^{19} J y^{-1}. Given the mass of the Earth is approximately 6×10^{24} kg, determine the rate of tidal heating. Express your answer in W kg^{-1} (1 W = 1 J s^{-1}) and to an appropriate number of significant figures.

Radiogenic heating

During the latter half of the 19th century, the eminent physicist Lord Kelvin (1824–1907) attempted to determine the age of the Earth. He believed the Earth had cooled slowly after its formation from a molten body and assumed the main sources of energy were from primordial heat and tidal friction. Taking many factors into account, including the mass of the Earth, the current rate of surface heat loss, and the melting points of various rock types, he concluded that the planet could not be much older than about 20–40 Ma.

This conclusion was contrary to the ideas of eminent geologists such as James Hutton (1726–1797) and Charles Lyell (1797–1875), who had already argued that immense spans of time were required to complete the changes produced by the action of tectonic, erosional and depositional processes. It was also greatly at odds with the then emerging theories of the evolution of life because scholars such as Charles Darwin (1809–1882) also argued for much longer periods based upon the evidence of biological speciation. As a consequence, a 'heated' controversy continued for many years. Even though Kelvin's calculations did not gain wide acceptance with geologists, he was a powerful scientific influence of the time, and it was not until much later in the 1950s that accurate radiometric dating experiments eventually proved him wrong. These experiments were conducted on primordial lead in meteorites and demonstrated that the formation of the Earth occurred about 4.6 Ga ago (see Section 2.2).

So why did Kelvin get the age of the Earth so wrong? The answer lies partly in the decay of unstable isotopes of certain radioactive elements, the discovery of which was not made until some decades after Kelvin's initial calculations. John Joly (1857–1933), an Irish physicist, was one of the first to suggest that radioactive decay, leading to radiogenic heating, was an important heat source within the Earth. It is now known that radioactive decay creates an important independent source of heat within the Earth that supplements that remaining from primordial sources (Box 2.1). This radiogenic heating is something that Kelvin could not possibly have known about when making his calculations.

Box 2.1 Radioactive decay and heating

Most elements have different isotopes (i.e. atoms having the same number of protons but different numbers of neutrons). Some of these isotopes are unstable and decay to stable forms. For example, isotopes deficient in protons decay by the transformation of a neutron into a proton and an electron, which is expelled from the nucleus. During this process, known as beta-decay, the mass of the nuclide does not change significantly. In contrast, during alpha-decay, heavy atoms decay through the emission of an α-particle, which consists of two protons and two neutrons (He^{2+}). This process reduces the overall mass of the nuclide. α- and β-particle collision with adjacent nuclei during decay causes heating through the loss of kinetic energy.

The rate of decay of a radioactive parent nuclide to form a stable daughter product is proportional to the number of atoms, *n*, present at any time, *t*:

$$\frac{dn}{dt} = -\lambda n \qquad (2.3)$$

where λ is the **decay constant** characteristic of the radionuclide in question (expressed in terms of reciprocal time). The decay constant states the probability that a given atom of the radionuclide will decay within a stated time.

The term $\frac{dn}{dt}$ is the rate of change of the number of parent atoms, and is negative because this rate decreases with time. Equation 2.3 can be integrated from $t = 0$ to t, where the number of atoms present at time $t = 0$ is n_0:

$$n = n_0 e^{-\lambda t} \qquad (2.4)$$

A useful way of referring to the rate of decay of a radionuclide is by its **half-life** ($t_{1/2}$), which is the time required for half of the parent atoms to decay.

On substituting $n = \frac{n_0}{2}$ and $t = t_{1/2}$ into Equation 2.4:

$$t_{1/2} = \frac{\ln 2}{\lambda} = \frac{0.693}{\lambda}, \text{ where ln 2 is the natural log of 2.} \qquad (2.5)$$

The number of radiogenic daughter atoms formed (D^*) is equal to the number of parent atoms consumed (Figure 2.1). So:

$$D^* = n_0 - n \qquad (2.6)$$

The total number of daughter atoms (D) consists of those produced by radioactive decay after time t (i.e. D^*) plus those already present at time $t = 0$ (i.e. D_0), that is:

$$D = D_0 + D^*. \qquad (2.7)$$

As $D^* = n_0 - n$ (Equation 2.6) we can substitute D^* in Equation 2.7:

$$D = D_0 + n_0 - n$$

Since $n_0 = ne^{\lambda t}$ (from Equation 2.4):

$$D = D_0 + ne^{\lambda t} - n$$

$$D = D_0 + n(e^{\lambda t} - 1) \qquad (2.8)$$

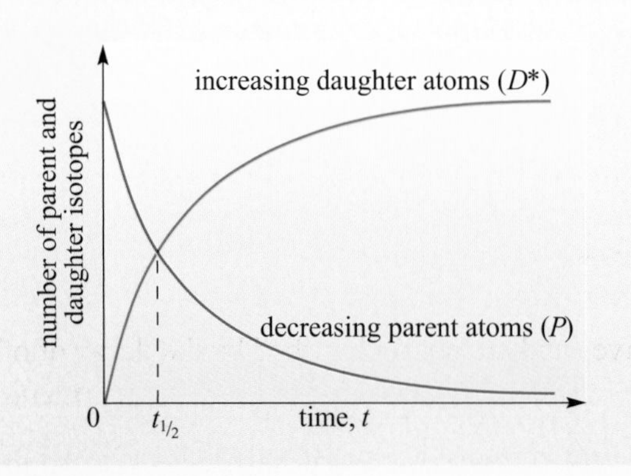

Figure 2.1 The changing number of parent and daughter atoms during radioactive decay. This illustrates that the growth of radiogenic daughter atoms (D^*) is the mirror image of the decay curve of the number of parent atoms (n_0). The half-life ($t_{1/2}$) is the time taken for half of the parent atoms to decay (at which point $n_0 = D^*$).

Table 2.1 lists a number of isotopes that either occur in the Earth or for which there is evidence of their having been active at some time during Earth history.

Table 2.1 Naturally occurring radioactive decay systems of geochemical and cosmochemical interest.

Parent	Decay mode*	Decay constant (λ)/ y^{-1}	Half-life/y	Present heat production‡	Daughter isotopes	Measured ratio
^{40}K	β+, e.c., β−	5.54×10^{-10}	1.28×10^{9}	2.8	^{40}Ar, ^{40}Ca	$^{40}Ar/^{36}Ar$
^{87}Rb	β−	1.42×10^{-11}	4.88×10^{10}		^{87}Sr	$^{87}Sr/^{86}Sr$
^{147}Sm	α†	6.54×10^{-12}	1.06×10^{11}		^{143}Nd	$^{143}Nd/^{144}Nd$
^{187}Re	β	1.59×10^{-11}	4.35×10^{10}		^{187}Os	$^{187}Os/^{188}Os$
^{232}Th	α	4.95×10^{-11}	1.39×10^{10}	1.04	^{208}Pb, ^{4}He	$^{208}Pb/^{204}Pb$, $^{3}He/^{4}He$
^{235}U	α	9.85×10^{-10}	7.07×10^{8}	0.04	^{207}Pb, ^{4}He	$^{207}Pb/^{204}Pb$, $^{3}He/^{4}He$
^{238}U	α	1.55×10^{-10}	4.47×10^{9}	0.96	^{206}Pb, ^{4}He	$^{206}Pb/^{204}Pb$, $^{3}He/^{4}He$
^{26}Al	β+	9.5×10^{-7}	0.73×10^{6}		^{26}Mg	$^{26}Mg/^{24}Mg$
^{129}I	β−	4.41×10^{-8}	1.57×10^{7}		^{129}Xe	$^{129}Xe/^{130}Xe$
^{146}Sm	α	6.73×10^{-9}	1.03×10^{8}		^{142}Nd	$^{142}Nd/^{144}Nd$
^{182}Hf	β−	7.78×10^{-8}	8.9×10^{6}		^{182}W	$^{182}W/^{184}W$
^{244}Pu	α, SF	8.45×10^{-9}	82×10^{6}		^{n}Xe	$^{n}Xe/^{130}Xe$**

*α = alpha decay (^{4}He); β^{-} = beta decay (electron or positron); e.c. is electron capture; SF is spontaneous fission.

†The production of ^{4}He from ^{147}Sm decay is insignificant compared with that produced by decay of U and Th.

‡Heat production averaged for the whole Earth in units of 10^{-12} W kg^{-1} of Earth material (not of the isotope).

***n* can be 124, 126, 128 or 129, all of which are produced by ^{244}Pu fission. Element symbols are listed in the Appendix.

■ Which of the isotopes listed in Table 2.1 remain active today and which are extinct?

■ All those with half-lives significantly less than the age of the Earth, i.e. 4.6 Ga, are extinct, namely: ^{26}Al, ^{129}I, ^{146}Sm, ^{182}Hf and ^{244}Pu. The others, principally isotopes of ^{40}K, ^{87}Rb, ^{147}Sm, ^{232}Th, ^{235}U and ^{238}U, are still active today.

If radionuclides have short half-lives and are not replenished by the decay of other isotopes, then they may be lost altogether. One such **short-lived extinct nuclide** is ^{26}Al, which has a half-life of 0.73 Ma.

■ What evidence would you look for to support the presence of ^{26}Al in the early Solar System?

■ ^{26}Al decays to ^{26}Mg (Table 2.1), so you would expect to see anomalously high abundances of ^{26}Mg relative to other isotopes of Mg in materials from the early Solar System.

Studies of primitive meteorites, notably carbonaceous chondrites, do indeed show slightly high $^{26}Mg/^{24}Mg$ ratios, which suggests that a significant proportion of the aluminium present at the time of condensation of the solar nebula was the unstable isotope ^{26}Al. This was originally created during a supernova explosion pre-dating the birth of our Solar System, and cannot be replenished by the spontaneous decay of other radiogenic elements.

- What does the observation that ^{26}Al was present in chondritic meteorites tell you about the timescale of the formation of the Solar System?

- The half-life of ^{26}Al is only 0.73 Ma, so the time between the supernova explosion that generated the ^{26}Al and the accretion of the meteorite parent body must have happened on a similar timescale of a few million years.

Given that after 10 half-lives only 2^{-10} (or $\frac{1}{1024}$th) of the original number of ^{26}Al atoms remain, then for any measurable amount of radiogenic ^{26}Mg to be found, chondritic meteorites must have formed within, at most, 7.3 Ma of the supernova. Similar arguments are used later in this chapter for the timescales of formation of other planetary bodies and layers within the Earth. The decay of ^{26}Al may have contributed significantly to the heating of planetary embryos but, because of its short half-life, any remaining ^{26}Al has long since vanished within the terrestrial planets.

While such short-lived isotopes may have been important heat sources during the early stages of terrestrial planet evolution, study of Earth material indicates that it is the isotopes of the elements U, Th and K that are responsible for most of the radiogenic heating that has occurred throughout the history of the planet. These isotopes, which all have particularly long half-lives (see Table 2.1), are termed **long-lived radiogenic nuclides** and were present in sufficient quantities after condensation and accretion to ensure that they have remained abundant within present-day Earth.

Most common minerals contain small amounts of the elements with these unstable isotopes, the most important of which are summarised in the upper half of Table 2.1. Their decay to form more stable isotopes releases tiny increments of heat, as described in Box 2.1. This decay has produced a continuous source of heat within our planet since the Earth's formation. Of course, the total amount of heat produced will depend on the concentration and types of radiogenic isotope present in the parts of the Earth, and the mass of suitable material present in its different layers.

- U, Th and K are incompatible elements (i.e. they are preferentially partitioned into silicate liquids), so where will they be concentrated in the Earth?

- The elements U, Th and K (and their radiogenic isotopes) are particularly concentrated in the silicate-dominated outer layers of the Earth, and in particular within the continental crust. They are thought to be virtually absent from the core.

As a result of the incompatibility of the heat-producing elements, the radiogenic heat produced per unit mass of the continental crust is, on average, over one hundred times greater than that of the underlying mantle. But because the mantle is so much more massive than the crust, in effect this means that the overall radiogenic heating budget is roughly split equally between the mantle and the crust despite the much greater mass of mantle material. It is the decay of these long-lived isotopes that provides sufficient heat energy to keep the Earth geologically active. Therefore, the surface heat flux is not simply the slow cooling of a once molten body, as originally envisaged by Kelvin.

Finally, while the rate of radiogenic decay is constant for each isotope system (Table 2.1), the total amount of radioactive decay, and hence heat generation, will decline over time as the reserves of the original radioactive materials are gradually used up. This gradual depletion of radioactive materials is expressed in half-lives, and each isotopic decay system has its own unique half-life. For example, ^{235}U decays through a series of α-particle emissions to the daughter isotope ^{207}Pb. The data in Table 2.1 indicate that, after 7.07×10^8 years, half of the ^{235}U originally present will have decayed to ^{207}Pb and the remainder will continue to halve every 7.07×10^8 years. Over the age of the Earth (4.6 Ga), approximately 6.5 half lives of ^{235}U have elapsed, so the heat production from ^{235}U is now $(\frac{1}{2})^{6.5} \approx 0.01$ (i.e. 1%) of what it was originally.

Question 2.3

(a) What proportions of the original ^{40}K and ^{232}Th currently remain since the formation of the Earth?

(b) Based upon the data in Table 2.1, what was the amount of radiogenic heating in the Earth at 4.6 Ga, and how does it compare with that of today?

(c) What proportion of Earth's surface heat flux loss is due to radioactive decay, compared with the 1.5×10^{-13} W kg^{-1} (the value determined in Question 2.2) created by tidal heating effects?

2.1.2 Heat transfer within the Earth

In the previous section you have seen how accretion, core formation and radioactive decay heated the Earth. But how is this internal heat transferred to the surface? Three main mechanisms of heat transfer operate within the Earth; these are **conduction**, **convection** and **advection**.

Conduction is perhaps the most familiar mechanism, since it is the process of heat transfer experienced when the handle of a pan becomes hot. Heat is conducted from the stove to the pan and then to its handle. Different materials, such as rocks of various compositions, conduct heat at different rates, and the efficiency of heat transfer in this manner is known as conductivity. This method of heat transfer is the most important in the outermost layer of the Earth (i.e. the lithosphere).

Convection involves the movement of hot material from regions that are hotter to those that are cooler and the return of cool material to warmer regions. During this transfer the material gives up its heat. It is a particularly efficient method of heat transfer, but the medium through which transfer takes place must be fluid. It should be noted, however, that the term 'fluid' describes any substance capable of flowing and is not just restricted to liquids and gases. Under the right conditions, even solid rocks can flow, albeit at a very slow rate. Over long periods of geological time, the effects of such solid-state flow become a highly significant way of transferring heat towards a planet's surface.

Advection is the final process of transferring heat when molten material (magma) moves up through fractures in the lithosphere and remains there. Advection operates when magma spreads out at the surface as a lava flow or, if it is injected, cools and crystallises within the lithosphere itself. The effect is the same in both cases, since heat is transferred by the molten rock from deeper levels where melting is taking place to shallower levels where it solidifies, losing its heat by conduction into the overlying crust. Any planetary body that exhibits, or has exhibited, volcanic activity must have lost some of its internal heat in this manner.

In Chapter 1 you learned that under the conditions prevailing deep within the Earth the solid rocks of the mantle can flow when subject to surface loads, leading to isostatic readjustment of surface elevations. The mantle can also flow when subject to temperature differences in a process known as **solid-state convection** and, while rates may be no more than a few centimetres per year, it is the most efficient form of heat transfer within all but the outermost part of the mantle.

Near the Earth's surface the rocks are too cold and rigid to permit convection, so conduction is the most significant process. You should recall from Chapter 1 that this zone of the uppermost mantle and all of the overlying crust is called the lithosphere. In the underlying asthenosphere the principal process of heat transfer is convection and this process of heat transfer applies to much of the mantle thickness down to the core (Figure 2.2a). The marked difference in strength between the lithosphere and the asthenosphere is caused by increasing temperature with depth below the Earth's surface (the **geotherm**) (Figure 2.2b). So the thickness of the lithosphere depends on the rate at which temperature increases with depth.

Finally in this section, it is of interest to return to Kelvin and the debate concerning the age of the Earth. Kelvin assumed that the Earth cools by conduction alone. You can see from Figure 2.2b that the geotherm for the convecting mantle is much shallower than that in the lithosphere. In this way, convection in the asthenosphere maintains a *higher* geothermal gradient within the lithosphere than would occur by conduction alone, thus mimicking the geotherm of a planet that is much younger

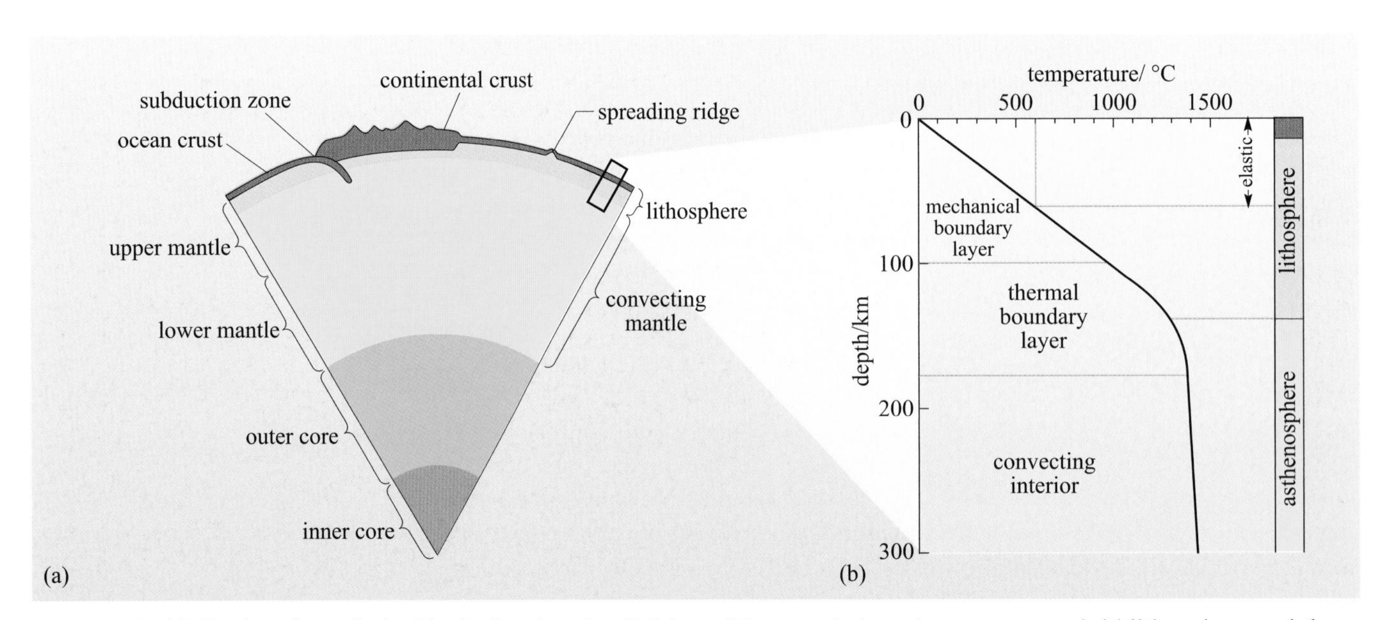

Figure 2.2 (a) Section through the Earth showing the division of the mantle into the uppermost rigid lithosphere and the mobile, convecting asthenosphere. (b) Geotherm through the lithosphere and uppermost asthenosphere.

than the Earth. By assuming that the surface, conductive geotherm applied to the whole Earth, Kelvin arrived at an erroneously young age for the Earth – a result he would have obtained even if he had included the correct estimates for radiogenic heat production!

2.2 The age of the Earth and its layers

Throughout this and the preceding chapter the age of the Earth has been given as being around 4.6 Ga. But where is the evidence for this? To find out just how old the Earth is we once again have to return to meteorites and radioactivity, for, in addition to being sources of heat in planetary systems, radioactivity also allows absolute ages to be determined from measurements of long-lived radioactive isotopes and their daughters.

Several isotope systems are used to date events and processes from throughout Earth history, but the three most commonly used are the K–Ar, U–Th–Pb and Rb–Sr systems. The principles of radiometric dating are most clearly illustrated using the Rb–Sr system, as outlined in Box 2.2 (overleaf), and isotope data from this and the U–Th–Pb system are most frequently illustrated on an **isochron plot** (or **isochron diagram** or **isotope evolution diagram**), examples of which are shown in Figure 2.3.

Figure 2.3a shows Rb–Sr isotope data from a series of ordinary chondrites that define an isochron age of ~4.5 Ga. This age relates to the last time the Rb and Sr were fractionated from each other by a particular process. In the case of Rb and Sr, both elements are lithophile (Box 1.3), so it is unlikely that they were fractionated by the separation of a metallic phase from a silicate fraction. However, Rb, being a Group 1 alkali metal, is significantly more volatile than Sr, a Group 2 element similar to Ca, which is one of the early condensing elements (Figure 1.16). Hence the Rb/Sr fractionation may relate to the loss of a volatile phase; the age indicates when the Rb/Sr ratio in ordinary chondrites was last disturbed.

Figure 2.3b shows a slightly more complex plot of data relating to the U–Pb system. You should notice from Table 2.1 that the U–Pb system has two parent

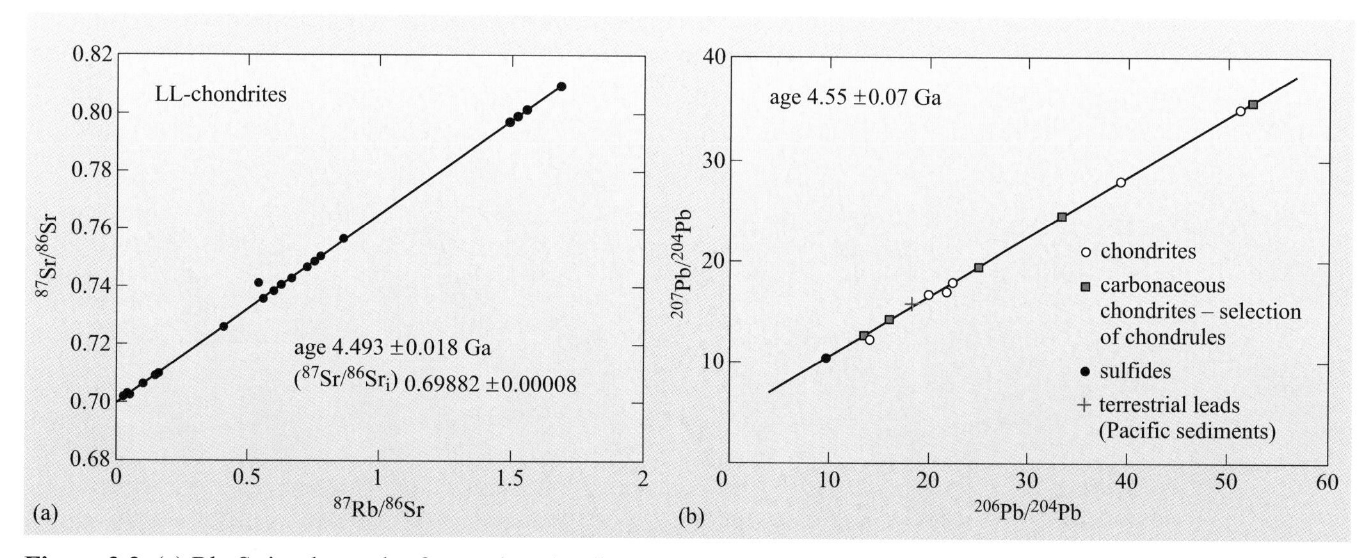

Figure 2.3 (a) Rb–Sr isochron plot for a suite of ordinary chondrites, giving an age of 4.49 ±0.02 Ga. (b) Isochron plot of $^{206}Pb/^{204}Pb$ against $^{207}Pb/^{204}Pb$ for chondritic meteorites giving an age of 4.55 ±0.07 Ga. (Minster and Allègre, 1980; Murty and Patterson, 1962)

isotopes, ^{235}U and ^{238}U, decaying to ^{207}Pb and ^{206}Pb respectively. By combining these two it is possible to eliminate the U/Pb ratio and determine an age from the plot of $^{207}Pb/^{204}Pb$ against $^{206}Pb/^{204}Pb$. In this case, the ages represent the time at which U was fractionated from Pb and, as Pb is a moderately volatile element and can be lithophile, siderophile or chalcophile in different environments, it is less easy to define the process that led to U/Pb fractionation. However, iron meteorites are rich in Pb and poor in lithophile U, so the age probably represents the timing of the separation of a metallic phase. Given that the chondrite isochron passes through the Pb isotope ratio of most iron meteorites, it adds further support to this idea.

The data illustrated in Figure 2.3 are for whole meteorite samples but, as you saw in Chapter 1, meteorites are far from homogeneous, comprising a number of different components. Primitive carbonaceous chondrites are thought to be among the least differentiated material in the Solar System. Among other things, they contain chondrules and calcium- and aluminium-rich inclusions (CAIs). Chondrules are millimetre-sized spherical droplets believed to have been produced when mineral grain assemblages were flash heated and cooled quickly. CAIs are typically centimetre-sized and consist of the first minerals to condense at equilibrium from a gas of solar composition. A detailed study of CAIs and chondrules yielded a $^{206}Pb/^{207}Pb$ isotope age for CAIs of 4567.2 ±0.6 Ma, whereas that of chondrules is 4564.0 ±1.2 Ma.

- What is the difference between the ages of CAIs and chondrules, and how old then are carbonaceous chondrites?
- The data give an interval of 3.2 ±1.8 Ma between formation of the CAIs and chondrules – carbonaceous chondrites must have formed at or after the time of formation of the chondrules, i.e. 4564 Ma.

Even though the difference between these two ages is small, it is greater than the combined uncertainty associated with the two ages – they are significantly different. The difference represents a real difference in the timing of the formation of the CAIs and chondrules.

These data show that the oldest components of meteorites, and hence the Solar System, must be close to 4.57 Ga old, but how do we know that this age also applies to the Earth?

Part of the answer to this question lies in Figure 2.3b, where the average Pb isotope ratios of Pacific sediments are compared with the data from chondrules. The sediment data fall on or close to the meteorite isochron, implying ultimate derivation from a similar source or common parent. Other evidence is found in lunar samples returned by the Apollo missions, which have ages that extend back through the Hadean. The oldest igneous rocks from the Moon are samples of anorthosite. One clast has been dated at 4.56 ±0.07 Ga using the ^{147}Sm–^{143}Nd system, placing the formation of the Moon to within 70 Ma of the start of the Solar System. But this in itself raises the question of what we are trying to date – what do we mean by the age of the Earth? In particular, given that radiometric dating systems date the time of element fractionation and therefore reflect the effects of different processes during Earth accretion and differentiation, it is probably more precise to consider the timing of major processes in the formation of the Earth.

Hence the following sections focus on determining the timing of important events in the history of the early Earth in relation to the age of the Solar System, which is given as 4.57 Ga.

Box 2.2 Radioactivity applied to dating

The rubidium–strontium isotope system provides a good illustration of the principles of isotope dating, and will be used here to demonstrate those principles. The number of ^{87}Sr daughter atoms produced by the decay of ^{87}Rb in a rock or mineral since its formation *t* years ago is given by substitution into the radioactive decay equation (Equation 2.8):

$$^{87}Sr = {}^{87}Sr_i + {}^{87}Rb(e^{\lambda t} - 1) \qquad (2.9)$$

where $^{87}Sr_i$ is the number of ^{87}Sr atoms initially present. It is, however, difficult to measure precisely the absolute abundance of a given nuclide. Mass spectrometers can, however, measure isotope ratios to very high accuracy and precision, and so it is more convenient to work with isotope ratios by dividing by the number of atoms of ^{86}Sr (which is a stable isotope and therefore remains constant with time).

$$\underbrace{\left(\frac{^{87}Sr}{^{86}Sr}\right)_P}_{y} = \underbrace{\left(\frac{^{87}Sr}{^{86}Sr}\right)_i}_{c} + \underbrace{(e^{\lambda t} - 1)}_{m}\underbrace{\left(\frac{^{87}Rb}{^{86}Sr}\right)}_{x} \qquad (2.10)$$

The present-day Sr isotope ratio $(^{87}Sr/^{86}Sr)_P$ is measured by mass spectrometry, and the $^{87}Rb/^{86}Sr$ ratio can be calculated from the measured concentrations of Rb and Sr. If the initial ratio $(^{87}Sr/^{86}Sr)_i$ is known or can be estimated then *t* can be determined, if it is assumed that the system has been closed to Rb and Sr mobility from the time *t* to the present.

Most rocks are many millions of years old, in which case it is difficult to estimate the initial Sr isotope ratio. However, examination of Equation 2.10 shows that it is equivalent to the equation for a straight line:

$$y = c + mx \qquad (2.11)$$

By plotting $(^{87}Sr/^{86}Sr)_P$ on the *y*-axis against $^{87}Rb/^{86}Sr$ on the *x*-axis, the intercept *c* is then the **initial ratio** of the system (Figure 2.4 overleaf). On such a diagram, a suite of co-genetic rocks or minerals having the same age define a line termed an **isochron**, and the diagram is called an isochron plot. The slope of the isochron, $m = e^{\lambda t} - 1$, yields the age of the rocks or minerals. If one of the rocks or minerals is Rb-poor then this may yield the initial $^{87}Sr/^{86}Sr$ ratio directly. Otherwise, the initial ratio is determined by extrapolating back to the *y*-axis using a best-fit line through the available data points.

The isotope evolution of a suite of hypothetical minerals is shown in Figure 2.4. At the time of crystallisation of the minerals all four have the same $^{87}Sr/^{86}Sr$ ratio, and plot as points on a horizontal line (AB). After each of these minerals becomes closed to the exchange of Rb and Sr isotopes, evolution begins. The sloping line (AC) represents the $^{87}Sr/^{86}Sr$ and $^{87}Rb/^{86}Sr$ ratios measured today.

- Study Figure 2.4. What has happened after time t?
- ^{87}Rb has decayed to ^{87}Sr over the period of time t to the present day. This has had two effects: (i) it has reduced the amount of ^{87}Rb present, so the $^{87}Rb/^{86}Sr$ ratios have *decreased* slightly; (ii) it has increased the amount of ^{87}Sr present and so the $^{87}Sr/^{86}Sr$ ratios have *increased*.

- But why have the $^{87}Sr/^{86}Sr$ ratios increased by differing amounts along the line AC?
- The amount of ^{87}Sr produced by radioactive decay depends on the amount of ^{87}Rb present. However, Figure 2.4 uses isotope ratios; hence the *relative* increase in ^{87}Sr (the increase in $^{87}Sr/^{86}Sr$) depends on the *relative* amount of ^{87}Rb (or the $^{87}Rb/^{86}Sr$ ratio). Thus, after a given period of time the samples with the highest $^{87}Rb/^{86}Sr$ ratio will also be those with the highest $^{87}Sr/^{86}Sr$ ratio.

The four samples shown in Figure 2.4 represent an idealised case. For an isochron to yield a slope that reflects a true age then the following assumptions must be valid.

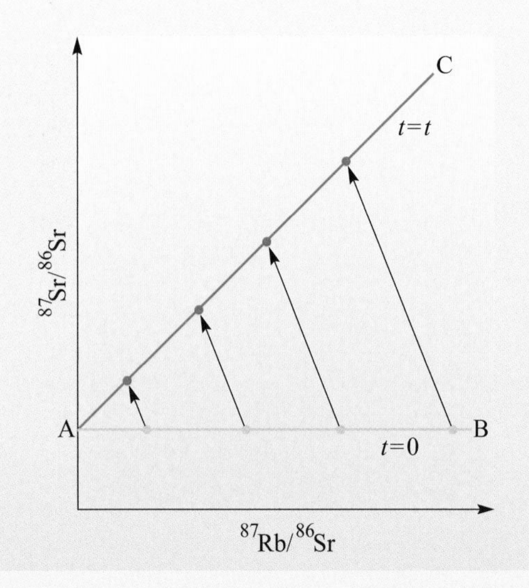

Figure 2.4 Isochron plot of $^{87}Sr/^{86}Sr$ against $^{87}Rb/^{86}Sr$ illustrating how four samples of the same age but different Rb/Sr ratios evolve from a horizontal line (AB) at the time of their formation, to plot on a straight line (AC) with a slope $m = e^{\lambda t} - 1$ at the present time.

1 All samples must be of the same age. If the samples are from the same igneous intrusion or are minerals from the same rock then this is a reasonable assumption. However, if the samples are from a large area where the geology is poorly understood then they may be of a different origin and may not be the same age and will not plot on a straight line.

2 All samples must have the same initial $^{87}Sr/^{86}Sr$ ratio when they formed, otherwise they will not plot on the same line on the isochron diagram (AB in Figure 2.4) even if they are of the same age. This assumption is likely to be valid for a suite of igneous rocks that have crystallised from the same parental melt, but it is less likely for sediments that are composed of differing amounts of recycled material.

3 The $^{87}Sr/^{86}Sr$ and $^{87}Rb/^{86}Sr$ ratios of the samples have only changed by the process of radioactive decay – no Rb or Sr has been added to or lost from them between the time of formation and the present day. Such **open-system behaviour** might occur due to the infiltration or loss of fluids, or diffusional exchange between minerals after their crystallisation.

2.2.1 Core formation and magma oceans

The formation of the metallic core is the biggest differentiation event that has affected the Earth, resulting in a large-scale change in the distribution of composition, density and heat production. One would think that such a fundamental feature would be well understood, but the physical mechanism by which metal separates from a silicate mantle and accumulates at the centre of the Earth remains poorly understood.

One potential mechanism for Fe–Ni metal separation or **segregation** is that the metal melts and forms an interconnected network. Whether or not this happens depends on a property known as the **dihedral angle**, θ (Figure 2.5). The dihedral angle is that formed by the liquid in contact with two solid grains, which in the case of the mantle will be silicate or oxide grains. If the dihedral angle θ is $<60°$, the melt will fill channels between the solid grains and form an interconnecting network, even in small melt fractions. On the other hand, if θ is $>60°$, the melt is confined to pockets at grain corners and cannot easily move, unless the melt fraction is greater than 10%.

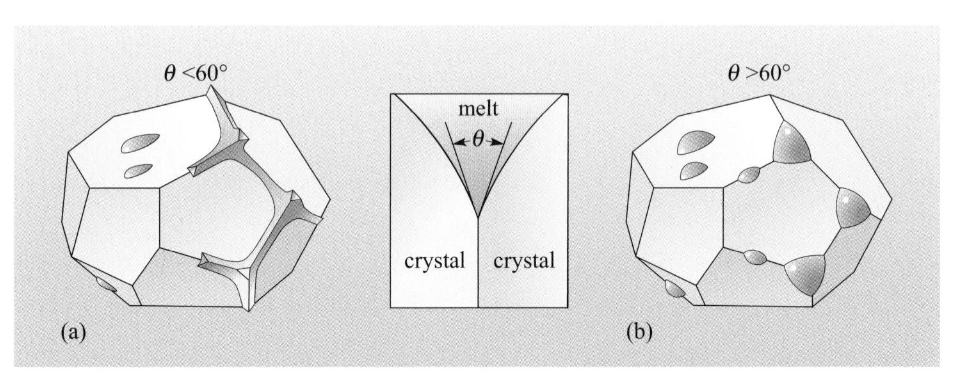

Figure 2.5 The definition of the dihedral angle θ, and the depiction of two different microstructures for static systems: (a) wetting, where melt forms an interconnected network along grain edges, and (b) non-wetting, where melt forms isolated pockets at grain corners. (Rushmer et al., 2000)

If melt is able to connect, its rate of migration is quite rapid, and can be calculated using Darcy's law:

$$v = \frac{k}{\eta}\Delta\rho g \tag{2.12}$$

where v is the velocity of the melt relative to the solid matrix, k is the permeability, η is the **viscosity** of the melt measured in Pa s, $\Delta\rho$ is the density difference between the melt and the solid, and g is the acceleration due to gravity.

Permeability can be defined as:

$$k = \frac{a^2\Phi}{24\pi} \tag{2.13}$$

where a is the mean grain radius and Φ is the melt fraction.

Question 2.4

Taking a grain radius, a, of 10^{-3} m (1 mm), Φ of 0.1 (10% volume melt), $\Delta\rho$ of 3500 kg m^{-3}, g of 9.8 m s^{-2} (the acceleration due to gravity on Earth) and a viscosity, η, of 0.005 Pa s, calculate the migration velocity of Fe–Ni metal (give your answer in kilometres per year). (*Note*: 1 Pa s = 1 kg m^{-1} s^{-1})

Such calculations show that any metallic melt that can form an interconnected network ought to sink rapidly to form a core. The key question then is the extent to which the metal connects.

- Experiments indicate that the dihedral angle θ is >60° for metallic melts. To what extent will such melts form an interconnected network?
- If θ is >60° then melts will be isolated at grain corners, creating an impermeable silicate framework through which metallic melts cannot segregate.

For this reason core formation is thought by many to occur only after the silicate framework has broken down after extensive silicate melting (>40%). At these high degrees of melting the grain boundary framework will no longer be interlocked, but rather crystals will be floating in a silicate liquid – a crystal mush. In such a mush, dense molten metal droplets would sink, but to achieve such high degrees of melting requires enormous amounts of heat.

- How might such heating have occurred on the early Earth?
- As you have previously seen, accretional heating during planetesimal–embryo Earth collisions would result in extensive melting, which is likely to have resulted in a global magma ocean.

It is important to note that there is no independent evidence that a magma ocean ever existed on Earth. Any early formed crust has long since been destroyed by impacts, erosion and plate recycling. Nevertheless, a deep magma ocean is thought to be a likely consequence if accretion included giant impacts as predicted in modern theory. The extent and depth of any magma ocean depends on many factors, including the impactor/Earth mass ratio, impact velocity and initial temperatures of the impactors.

The evidence also suggests that the Earth had a huge proto-atmosphere, formed by degassing of the Earth's interior. This would have provided a thermal blanket that retained the heat generated during accretion and sustained the magma ocean. Finally, as you will see later in Section 2.3, a giant collision is the most popular theory for the origin of the Moon, and such an impact would have delivered sufficient energy to melt the entire planet. These accretional and impact sources of heat would also have been supplemented by radioactive decay, which would have been much greater than at the present day.

If the entire silicate mantle were molten, the metal blobs would sink directly to the centre of the Earth, but if the base of the magma ocean were solid silicate then liquid metal might pool at this boundary until gravitational instability permits movement of the metal to the core in large **diapirs**. Alternatively, metal from the impactor and perhaps even the Earth's proto-core may have become highly fragmented and emulsified in the magma ocean, in which case small metallic droplets could equilibrate rapidly with silicate melt.

Cooling of a peridotite magma ocean would eventually lead to crystallisation from the bottom up. In the absence of an atmosphere, heat radiation to space would have been very efficient and the deeper part of the magma ocean would have cooled very quickly (i.e. in less than 1000 years). However, the upper part of the magma ocean could have remained hot and molten much longer, maybe for more than 10^7–10^8 years, especially if a cooled crust formed at the surface. This situation may have been complicated by the continuous supply of new material from meteorites that remelted on impact, and tidal heating.

2.2.2 Core–mantle equilibration

You should recall from Chapter 1 that one consequence of core–mantle separation is that the metal-loving siderophile elements would be strongly partitioned into the metallic core. However, trace amounts of siderophile elements are retained in the mantle and if metal segregation were an equilibrium process then these elements would provide important clues for deducing the conditions of core formation.

Figure 2.6 shows the abundances of siderophile elements subdivided into slightly, moderately and strongly siderophile, in the Earth's mantle normalised to chondritic meteorites, revealing the stepped abundance profile as discussed in Section 1.4. An assumption of many early core formation models was that metal segregation was contemporaneous with accretion, and that metal and silicate equilibrated at near-surface, low-temperature (T) and low-pressure (P) conditions. Low T–P metal silicate distribution coefficients (Box 1.3) for highly siderophile elements have been determined experimentally and found to lie between 10^{-7} and 10^{-15}. These values lead to the expected abundances in the mantle as shown in Figure 2.6.

■ How do the abundances of the highly siderophile elements observed in the silicate mantle compare with what would be expected from the experimentally determined low-pressure partition coefficients?

■ The observed data (triangles in Figure 2.6) indicate a depletion of about 2×10^{-3} relative to chondrites (i.e. present at ~0.2% of the chondritic abundance), whereas the partition coefficients predict depletions of 10^{-5} and 10^{-6} (filled circles in Figure 2.6).

As you have seen, metal segregation in a magma ocean would probably occur over a range of temperatures and pressures, and so equilibrium metal segregation at high temperature and high pressure in a deep magma ocean becomes a realistic possibility. Applying experimentally determined partition coefficients to

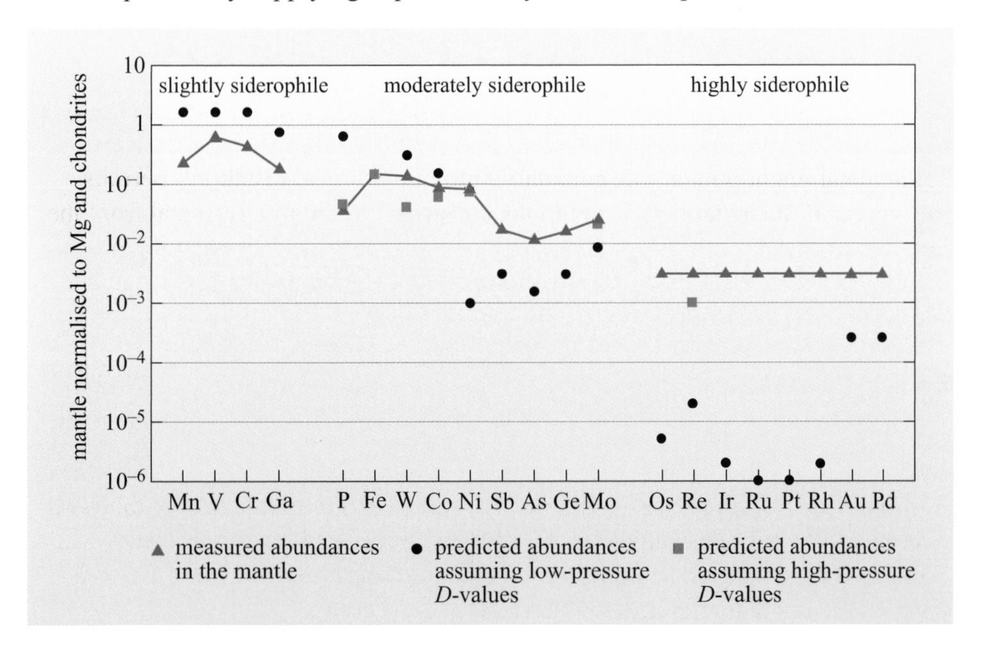

Figure 2.6 The abundance of siderophile elements in the present-day Earth's upper mantle. Also shown are the mantle abundances expected for selected elements in the mantle, assuming separation of the core as metal droplets sinking through a magma ocean and equilibrating according to low- and high-pressure partition coefficients respectively. (Modified after Drake and Righter, 2002)

metal segregation at high pressures can reproduce the abundance of Re, but as yet experimental data on other elements are lacking. However, given their dramatically different partitioning behaviour at low temperature and pressure it is unlikely that there exists any set of conditions at which all siderophile partition coefficients converged to a single value, which is required by the uniform depletion of the most highly siderophile elements.

The failure of low-temperature, low-pressure metal/silicate equilibration models to explain the siderophile excess inspired a number of alternative models, including:

- partitioning between a sulfur-rich liquid metal and silicate
- inefficient core formation, where small amounts of metal or sulfide remain behind in the mantle
- the heterogeneous accretion or so-called **'late veneer' model**. The 'late veneer' model invokes the addition of a small fraction of the Earth's mass occurring after the segregation of the core from the mantle. Core formation effectively strips all the highly siderophile elements from the mantle. Accretion of primitive material to the Earth after this process has occurred would add siderophile elements in chondritic proportions, raising the mantle concentrations to the values seen today. Currently, this is the most popular explanation for the chondritic relative abundances of the highly siderophile elements in the Earth's mantle.

In contrast to the highly siderophile elements, the abundances of moderately siderophile elements would not have been significantly altered by the later addition of meteoritic material because they would not have been so efficiently stripped from the mantle by core formation. Potentially, moderately siderophile elements can provide much more information on the conditions of core formation. Of these elements, Ni and Co provide some key constraints because their abundances in the mantle are accurately and precisely known, and their partitioning behaviour has been studied over a wide range of conditions.

■ With reference to Figure 2.6, estimate the Ni/Co ratio of Earth's mantle at the present day.

■ Figure 2.6 shows that, within uncertainty, Ni and Co are present in proportions that are close to chondritic, i.e. both at ~0.1 × chondrite.

The chondritic ratio of Ni to Co in the mantle requires the ratio of the two partition coefficients D_{Ni}/D_{Co} to be about 1.1. Experiments show that an increase in pressure and/or temperature causes both Ni and Co to become less siderophile, but at different rates (Figure 2.7).

■ Assuming a D_{Ni}/D_{Co} ratio of about 1.1 is required to explain the Ni/Co ratio of the Earth's mantle, use the experimental data in Figure 2.7 to estimate the pressure of metal–silicate equilibration.

■ From Figure 2.7, a D_{Ni}/D_{Co} ratio of 1.1 occurs at a pressure of about 28 GPa equivalent to a depth of 900–1000 km, implying high temperature and pressure metal–silicate equilibration and core segregation.

Figure 2.8 summarises the conceptual model of metal segregation and metal/silicate equilibrium in a deep magma ocean. For the model geotherm shown, the entire upper mantle would be molten (a magma ocean), whereas the lower mantle would be solid.

In the upper part of the mantle, equilibrium metal segregation from the upper mantle would occur by the 'rain-out' of small, liquid metal globules over a wide range of temperature and pressure conditions, and the metal would accumulate at the magma ocean floor. At the boundary between the lower and upper region, the metal would equilibrate a final time, giving the upper mantle its present siderophile element signature. Finally, gravitational instability would cause the formation of large metal diapirs that sink through the lower region, with or without re-equilibration. (*Note*: these divisions do not necessarily correspond to the present-day upper and lower mantle.)

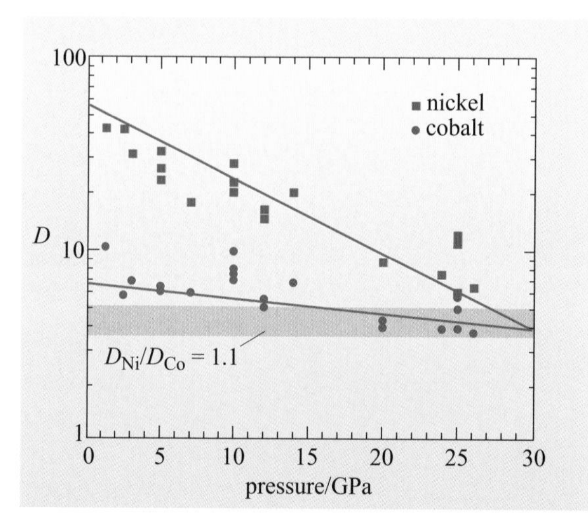

Figure 2.7 Metal/silicate partition coefficients (D) as a function of pressure for nickel (Ni) and cobalt (Co). See text for explanation. (Walter et al., 2000)

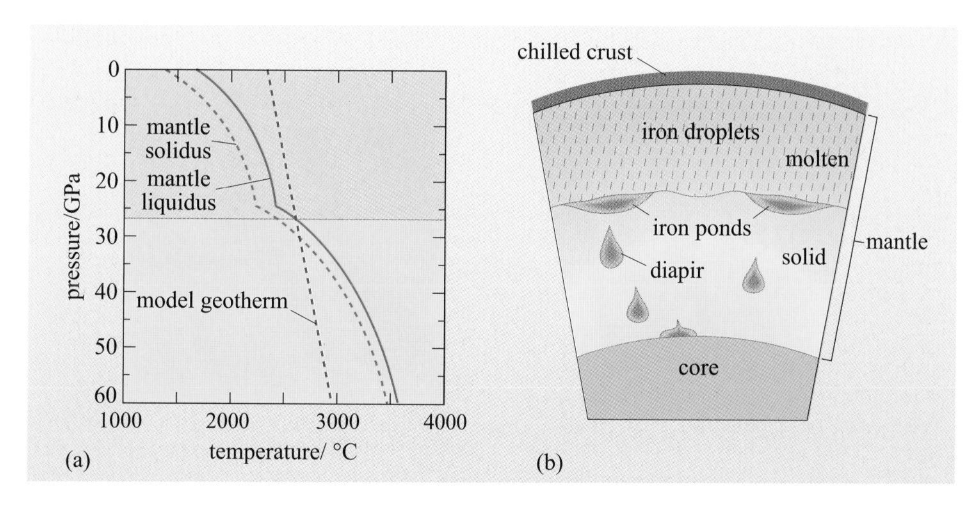

Figure 2.8 Pressure–temperature diagram showing metal/silicate equilibrium and metal segregation in a deep magma ocean. (a) Pressure–temperature plot of the mantle solidus and liquidus (Chapter 4) compared with the mantle geotherm. Where the geotherm lies above the liquidus, the mantle is completely molten (upper region); where it lies below the solidus it is completely solid (lower region). (b) Section through the Earth's mantle early in Earth history showing metal droplets falling to the base of the mantle magma. The droplets accumulate into larger bodies of dense, molten metal that eventually sink through the solid lower mantle as diapirs and accumulate in the core. (Modified from Walter et al., 2000)

2.2.3 The timing of core formation

In order for any radioactive decay system to be of use in dating a process, that process must be able to fractionate the parent element from the daughter element. Thus, in order to investigate the timing of core formation, radioactive systems are needed in which one of either the parent or the daughter elements is siderophile and the other is lithophile.

- Which of the radioactive decay schemes in Table 2.1 satisfy this criterion? (*Hint*: use Figure 1.23 to determine the dominant geochemical properties of the elements involved.)

- The two decay schemes with elements of contrasting properties are hafnium–tungsten (Hf–W) and uranium–lead (U–Pb).

Of the two systems, U–Pb involves long-lived isotopes, whereas the Hf–W scheme has a much shorter half-life – it is much easier to illustrate the principles of the approach to this problem using the Hf–W system.

The hafnium–tungsten system

In addition to the contrasting geochemical properties of Hf and W during core formation, both elements are refractory and so were accreted to the Earth in chondritic proportions, therefore the Hf/W ratio of the bulk silicate Earth is known relatively well.

■ Assuming the bulk silicate Earth has a chondritic Hf/W ratio, how will the Hf/W ratio of the core and the mantle differ from that in chondrites?

■ The core will have an Hf/W ratio lower than chondrites because W is siderophile and will be enriched in the core. Hf, being lithophile, will remain in the mantle, which will have an Hf/W ratio greater than in chondrites.

^{182}Hf decays by β-decay to ^{182}W with a half-life of 8.9 Ma (Table 2.1).

■ If the core separated from the mantle magma ocean while ^{182}Hf was sufficiently abundant, what would happen to the $^{182}W/^{184}W$ ratio of (a) the mantle and (b) the core?

■ At the time of element fractionation, the $^{182}W/^{184}W$ ratio of both core and mantle would be similar. However, with time, that of the mantle would increase rapidly because of its high Hf/W ratio, while that of the core would increase less rapidly because of its low Hf/W ratio.

The effect of Hf/W fractionation on W isotopes is well illustrated by measurements of W isotopes in iron meteorites. The data are illustrated in Figure 2.9 and reveal that the metal from iron meteorites has low $^{182}W/^{184}W$ ratios (low ε^{182}Hf) (see Box 2.3), lower than both chondritic meteorites and the silicate Earth (the mantle). The simplest explanation for this is that these metals sampled early Solar System tungsten before live ^{182}Hf had decayed.

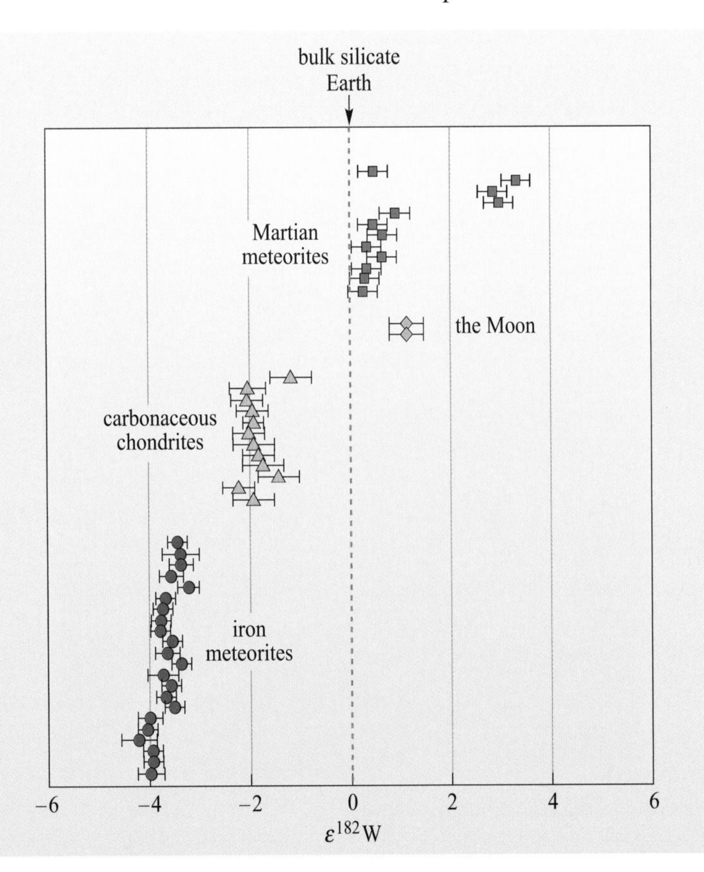

Figure 2.9 ε^{182}W for iron meteorites, carbonaceous chondrites, the bulk silicate Earth and the Moon. These show a well-defined deficiency in ^{182}W in early metals and carbonaceous chondrites relative to the bulk silicate Earth (BSE). Error bars denote analytical uncertainty on individual measurements. (Data from Foley et al., 2005; Klein et al., 2002; Yin et al., 2002; Scherstein et al., 2006; Lee et al., 2002)

Box 2.3 Tungsten isotope ratios and their notation

The beta decay of ^{182}Hf to ^{182}W ($t = 9$ Ma) has proven to be of enormous value in determining the relative timing of events during planetary accretion and core formation. Because ^{182}Hf is now extinct, evidence for its original presence is recorded in the isotope ratios of its daughter element, tungsten (W). Tungsten consists of five stable isotopes but only one of these, ^{182}W, has been partly produced by the radioactive decay of ^{182}Hf. Therefore, to use this geochronometer, geochemists need to measure the abundance ratio of ^{182}W to another of the isotopes of W, conventionally ^{184}W.

The magnitude of the variations in the ^{182}W/^{184}W ratio relate to the parent/daughter ratio, ^{182}Hf/^{184}W, in the original material. This was of the order of 10^{-4}. Consequently, variations in the ^{182}W/^{184}W ratio will be of a similar magnitude.

Modern mass spectrometers are capable of measuring isotope ratios to very high precision, in some cases to better than 10 parts in a million (0.001%). This means that isotope ratios can be and often are quoted to six decimal places. Dealing with such numbers is inconvenient, especially when significant differences occur in the fourth and fifth decimal places. To cope with this, geochemists have adopted a convention in which differences in isotope ratios are compared with a standard value, usually one that relates to the bulk silicate Earth (BSE) or meteorites. In the case of tungsten isotopes, differences in isotope ratios are measured in the number of parts in 10^4 (10 000) and are designated by the Greek letter ε (epsilon), defined according to the following equation:

$$\varepsilon^{182}\mathrm{W} = \left[\frac{\left(\frac{^{182}\mathrm{W}}{^{184}\mathrm{W}}\right)_{\mathrm{sample}}}{\left(\frac{^{182}\mathrm{W}}{^{184}\mathrm{W}}\right)_{\mathrm{BSE}}} - 1\right] \times 10^4 \qquad (2.14)$$

in which $\left(\frac{^{182}\mathrm{W}}{^{184}\mathrm{W}}\right)_{\mathrm{sample}}$ is the measured ^{182}W/^{184}W ratio of the sample and $\left(\frac{^{182}\mathrm{W}}{^{184}\mathrm{W}}\right)_{\mathrm{BSE}}$ is the measured ^{182}W/^{184}W ratio of the bulk silicate Earth.

Consider the following example. A chondritic meteorite has a measured ^{182}W/^{184}W ratio of 0.864640. In the same experiment, a standard representative of the bulk silicate Earth has a measured ^{182}W/^{184}W ratio of 0.864810.

What is the ε^{182}W value of the meteorite?

$$\begin{aligned}\varepsilon^{182}\mathrm{W} &= \left[\left(\frac{0.864640}{0.864810}\right) - 1\right] \times 10^4 \\ &= (0.999803 - 1) \times 10^4 \\ &= -0.000196 \times 10^4 \\ &= -1.96 = -2.0 \text{ (to 2 sig. figs)}\end{aligned}$$

This result shows that the chondrite measured has a slightly lower amount of radiogenic ^{182}W than the bulk silicate Earth by 0.02%, or 2 parts in 10 000. Quite clearly, saying it has a ε^{182}W value of -2 is much easier than dealing with a six figure decimal, while the sign conveys that it is less radiogenic than the BSE. A sample with a positive value has more radiogenic W than the BSE.

The ε notation also has another advantage in that it allows results from different laboratories to be compared more easily.

Question 2.5

Calculate the ε^{182}W value for the following data from another laboratory in which the same meteorite has a measured ^{182}W/^{184}W ratio of 0.864523 and a standard representative of the BSE has a measured ^{182}W/^{184}W ratio of 0.864696.

The answers to the above calculations are given to 2 significant figures because the precision with which individual measurements are made is of the order of 0.3ε units, so any result is subject to this degree of uncertainty. Both sets of original data give the same result within the limits of experimental uncertainty, despite giving different absolute values for the measured ratios. This illustrates that while different laboratories may get different absolute values of $^{182}W/^{184}W$ ratios, even though they are measuring the same samples, the calculation of $\varepsilon^{182}W$ values overcomes these differences, as long as the samples and standards are measured in the same laboratory.

But what about the Earth's core? Unfortunately we do not have access to the core but we can compare W isotope ratios in the silicate part of the Earth (the mantle and crust) with chondritic meteorites – which are representative of the bulk silicate Earth at least as far as Hf and W are concerned. The results in Figure 2.9 show that the bulk silicate Earth has a different $\varepsilon^{182}W$ from chondritic meteorites.

■ Is the difference in W isotopes between chondrites and the bulk silicate Earth consistent with core formation?

■ Yes, the bulk silicate Earth has a higher $\varepsilon^{182}W$ and hence a higher $^{182}W/^{184}W$ ratio than chondrites, signifying that it has a higher Hf/W ratio.

The tungsten isotope difference between early metals, carbonaceous chondrites and the bulk silicate Earth reflects the Hf/W ratio of the material that formed the Earth and its fractionation during the lifetime of ^{182}Hf. As a result of core formation, the bulk silicate Earth has an elevated Hf/W ratio (~15) relative to chondrite meteorites (Hf/W ratio ~1). Therefore, provided that the Earth's core formed early, i.e. during the lifetime of ^{182}Hf, an excess in the ^{182}W atomic abundance in the bulk silicate Earth relative to chondrites or iron meteorites will be generated. The conclusion from the extinct Hf–W system is that planetary differentiation occurred while ^{182}Hf was still present. As its half-life is 8.9 Ma, this implies that the core and mantle must have separated within the first few tens of millions of years of Earth history.

Of course, the detailed interpretation of W isotopes in planetary bodies is more complex than described here, but the principles remain the same. The data for chondrite meteorites, some of which are independently dated, yield an $\varepsilon^{182}W$ value of around –2, relative to the bulk silicate Earth (and the Moon), which has a much higher Hf/W ratio. These differences have been modelled with core separation between 30 Ma and 50 Ma after the start of the Solar System.

As a footnote to this section, studies of Pb isotopes in the Earth and meteorites have also been exploited to investigate the timescales of core segregation, again based on the contrasting properties of U and Pb. The U–Pb system has a long half-life and so any measurements of Pb isotopes in modern Earth materials reflect a long history of U–Pb fractionation. However, it can be argued that the major U–Pb fractionation occurred early in Earth history, that it was related to core formation, and that it occurred over a period of 70–150 Ma after the start of the Solar System.

2.3 The Moon

The Moon is a significant part of the terrestrial system and no consideration of the history of the early Earth would be complete without at least a brief consideration of its formation. One of the fundamental discoveries of the Apollo missions was that the Moon consists of a variety of igneous rock types that differ widely in their mineralogy, composition and age. The most visible evidence of these differences is the existence of two distinct terrains on the Moon:

- the light-coloured feldspathic rocks of the highlands
- the dark basalts of the maria (see Figure 1.2b).

Both rock types are igneous and indicate that at some time during its history the Moon was extensively molten. Indeed the anorthosites (Section 1.1.1), which are also very old, are thought to have formed by a process of flotation on a magma ocean, their dominant mineral, plagioclase, being lighter than the iron-rich basalts from which they crystallised.

As discussed previously, the oldest igneous rocks are generally considered to be the iron-rich anorthosites, which may be as old as 4.56 ±0.07 Ga. Since anorthosites are igneous rocks, they provide a lower limit for the age of the Moon, but again imply that the Moon must have formed within the first 70 Ma of the start of the Solar System.

It is also possible to use Hf–W model ages to date the age of the Moon. One lunar basalt yields a positive ε^{182}W, i.e. it has more radiogenic W than the bulk silicate Earth (Figure 2.9). However, the Moon is also thought to possess a higher Hf/W ratio than the Earth (see below), so the derived age is similar to that of the bulk silicate Earth, and ages for Hf/W fractionation in the Moon range between 30 Ma and 45 Ma after the start of the Solar System.

Structurally, the Moon contrasts with the Earth in that it has a very small core and is totally solid – there is no convecting outer core and hence no magnetic field. The Moon, being much smaller than the Earth and having a much larger surface to volume ratio, cooled very rapidly after formation. As a result, the lithosphere is very thick, comprising in excess of 75% of the thickness of the Moon's mantle.

The compositions of lunar rocks have been used to develop models of the bulk composition of the Moon, which can be compared with that of the bulk silicate Earth and meteorites. These are summarised for a selection of elements with contrasting properties in Table 2.2. (Element names and symbols are listed in the Appendix.)

Table 2.2 Comparison of elemental abundances in primitive meteorites, the Earth and the Moon.

	CI chondrite (primitive meteorite)	Earth (crust + mantle)	Moon (crust + mantle)	Ratio of trace element abundance Moon/Earth
Volatile elements				
K/ppm	545	180	83	0.46
Rb/ppm	2.32	0.55	0.28	0.51
Cs/ppb	279	18	12	0.67
Moderately volatile				
Mn/ppm	1500	1000	1200	1.20
Refractory elements				
Cr/ppm	3975	3000	4200	1.40
Th/ppb	30	80	112	1.40
Eu/ppb	87	131	210	1.60
La/ppb	367	551	900	1.63
Sr/ppm	7.26	17.8	30	1.69
U/ppb	12	18	33	1.83
Siderophile elements				
Ni/ppm	16500	2000	400	0.200
Mo/ppb	1380	59	1.4	0.024
Ir/ppb	710	3	0.01	0.003
Ge/ppm	48	1.2	0.0035	0.003

Note: some elements are parts per million (ppm) and some are in parts per billion (ppb).

- Relative to the Earth, what do you notice about the abundances of the volatile elements in the Moon?
- They are all lower than those of the Earth.

One of the primary observations of the Moon is that it is depleted in the most volatile elements and enriched in refractory elements. This has been interpreted as relating to a very high temperature origin for the material that makes up the Moon.

- How do the siderophile element abundances in the Moon compare with those of the bulk silicate Earth?
- They are much lower.

Explaining this difference is more complex, especially as the Moon may not possess a metallic core. The extreme depletion of the siderophile elements in the silicate portion of the Moon strongly suggests that the material of which the Moon is made was already differentiated – it had already lost a metallic fraction and hence its inventory of siderophile elements.

In addition to these characteristics, any model of lunar formation must also take into account the following:

- the **angular momentum** of the Earth–Moon system. Angular momentum is a property of rotating systems that depends on mass and its distribution, angular velocity, and radius. The angular momentum of the Earth–Moon system is contained in the Earth's rotation and the Moon's orbital motion, and is unusually high compared with the other terrestrial planets
- the Earth and the Moon have indistinguishable oxygen isotope compositions, whereas most planetary bodies have different and distinct oxygen isotope compositions.

2.3.1 The formation of the Moon

The origin of the Moon has been debated for over a century, but particularly since the Apollo mission provided samples to study. Several theories of formation have been suggested.

Co-accretion

This theory proposes that the Earth and the Moon simply accreted side by side. The difficulty with this model is that it does not explain the angular momentum of the Earth–Moon system. This model explains neither the difference in density nor the difference in the depletion of volatile elements.

Capture

This theory proposes that the Moon was originally in a heliocentric orbit and was captured following a close approach to the Earth. However, it is difficult to do this without the Moon spiralling into the Earth and colliding. It is also difficult to explain the indistinguishable oxygen isotope compositions of the Earth and the Moon with this model.

Fission

This theory proposes that the Moon split off as a blob during the rapid rotation of a molten Earth. George Howard Darwin (1845–1912), the son of Charles Darwin, originally championed this idea. At one time (before the young age of the ocean floor was known) it was thought by some that the Pacific Ocean might have been the residual space vacated by the loss of material. This model does have some attractive features: it explains why the Earth and the Moon have identical oxygen isotope compositions. It also explains, for example, why the Moon has a lower density, because the outer part of the Earth would be deficient in iron due to core formation, and it explains why so much of the angular momentum of the Earth–Moon system is in the Moon's motion. However, it is not clear why the Moon should spontaneously split away from the Earth without some large input of energy.

Impact models

Following the Apollo missions it was proposed that the Moon formed as a result of a major impact on the Earth that propelled sufficient debris into orbit to produce the Moon. Such models are now the most widely accepted (Figure 2.10).

Important information that came from sample-return was that more than 80% of the lunar crust was composed of anorthosite, indicating the presence of a magma ocean early in lunar history. The presence of a magma ocean requires significant degrees of melting, which occurred in response to an impact – provided the accreting body was large and that subsequent lunar accretion was rapid (1–100 Ma). Also, it is necessary to link the dynamics of the Moon with that of the Earth's spin. This led to the proposal of a series of single giant impact models in which the Moon was the product of a glancing blow collision with another differentiated planet (Figure 2.10a and b). A ring of debris would have been produced from the outer silicate portions of the Earth and the impactor (named Theia, the mother of Selene, the goddess of the Moon), which was roughly the size of Mars (Figure 2.10c). This model explains the angular momentum, the extensive early melting, the isotopic similarities and the density difference. The most recent model simulations indicate that the giant impact that formed the Moon probably occurred at the very end of Earth's accretion, when the Earth was 90% of its present size.

Certain features of the Moon may be a consequence of prior differentiation of Earth and Theia, and of the giant impact itself. For example, the depleted volatile elements require that both Earth and Theia had already become at least partially differentiated due to earlier heating. In addition, the low abundances of siderophile elements can be attributed to prior core separation in both the Earth and Theia.

- How is the sequence of events described above consistent with the observation that the Moon's core (if any exists) is very small relative to the size of the whole Moon?

- The Moon formed from mantle material derived from the colliding, partially differentiated planetary embryos. This mantle material was already depleted in Ni, Fe and S due to the development of cores within the embryos. Therefore, there would have been relatively little Ni, Fe and S left to differentiate inwards once the Moon had formed.

▶ **Figure 2.10** The formation of the Moon. (a)–(c) illustrate the collision and aftermath between the proto-Earth and Theia. Both bodies were large enough to have differentiated into core, mantle and primary crust as a result of accretionary heating. Following collision (a and b), the cores of the two bodies are thought to have combined and the mantles became mixed while some material was fragmented and vaporised and scattered into orbit around the Earth. (c) Some of the debris fell back to Earth while the remainder accreted under its own gravity to form the Moon. (d) The heat of accretion would again have resulted in wholesale lunar melting. The Moon then cooled and differentiated into mantle, primary crust and possibly a small core, depending on how much of the core of Theia was dispersed around the Earth and how much merged with the Earth's core. (e) The Moon and, presumably, the Earth were subject to further meteorite bombardment and the formation of large craters. Some of these impacts were intense enough to initiate melting of the lunar mantle, flooding the larger impact structures with basalt and forming the lunar maria. (f) Over time the Moon's orbit has slowly decayed, its rate of rotation becoming synchronised with its orbit as the distance between the two bodies progressively increases.

(a)
(b)
(c)
(d)
(e)
(f)

For the Moon's crust and mantle, this depletion was probably further augmented by further differentiation immediately after its formation. Moreover, the Moon's depletion in volatile elements, relative to the Earth, can be explained if the Moon accreted from the partially vaporised debris coalescing after the impact. In these circumstances, the more volatile elements would have had the opportunity to escape into space prior to accretion.

Question 2.6

Given the arguments regarding planetary accretion, volatile elements, and their behaviour following giant impacts, suggest a reason why the concentrations of Rb, K and Na differ between chondrites and the Earth's mantle.

2.3.2 The late heavy bombardment

The final major event that affected the early Moon is known as the late heavy bombardment. The chronology of impact events on the Moon has been based on telescopic observations, crater counting and disturbed radiogenic isotope systematics. The majority of impact-melt rocks returned from the Moon yield ages around 3.85 Ga (Figure 2.11). These ages have been interpreted as either representing a short and intense heavy bombardment or as the tail end of a prolonged post-accretionary bombardment. In either case the bulk of this bombardment, which produced craters many hundreds of kilometres across, preceded 3.8 Ga. Within this time span, the Earth must have been subjected to a significantly greater bombardment than the Moon, as it has a larger diameter and a much greater gravitational cross-section, making it an easier target to hit. It has been estimated that the impact rate for the Earth would have been more than 20 times greater than the Moon, with both more and larger impact events. The consequences for the hydrosphere, atmosphere and even the lithosphere must have been devastating. It has been suggested that the absence of any rocks on Earth older than ~4.0 Ga is the result of this late heavy bombardment, during which impact-induced mixing recycled early crustal fragments back into the upper mantle.

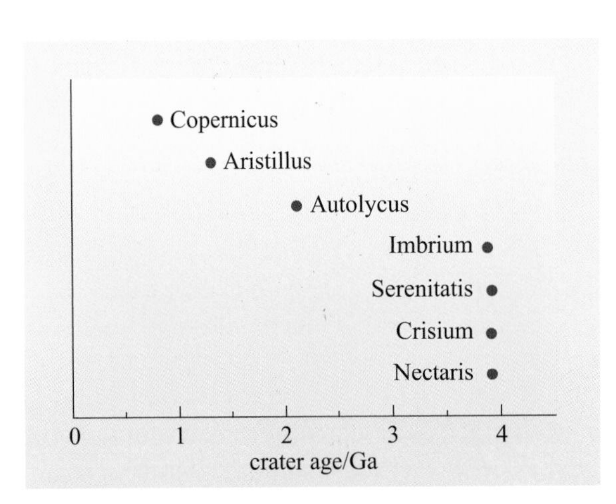

Figure 2.11 Crater ages on the Moon.

2.4 Origins of the atmosphere and the hydrosphere

The atmosphere and the hydrosphere represent the products of the outgassing of the Earth over geological time, primarily from volcanic activity. Yet the present-day composition of the atmosphere does not reflect that of volcanic volatiles. These are dominated by water vapour, carbon dioxide (CO_2) and sulfur dioxide (SO_2) with smaller amounts of nitrogen, halogens, hydrides and other more exotic volatile compounds. Most of these volatile compounds are soluble in water or, in the case of methane, are easily oxidised to water-soluble compounds. They therefore do not accumulate in the atmosphere but dissolve in the oceans and react with the oceanic crust. The remaining less-reactive and less-soluble elements and compounds, notably nitrogen and the inert gases (He, Ne, Ar, Kr and Xe), accumulate in the gaseous atmosphere.

- Which important atmospheric gas is *not* abundant in volcanic volatiles?
- Oxygen.

For about the first billion years of Earth's history, oxygen was only present in the atmosphere in trace amounts as a result of the breakdown of water vapour by UV radiation. However, this inorganic mechanism of releasing oxygen into the atmosphere produces only tiny amounts of free oxygen. Since the Earth's present atmosphere is oxygen rich and because all higher forms of life require free oxygen, there is an obvious need for some other source of oxygen. The most plausible source is oxygen-producing photosynthesis, a process that first evolved in cyanobacteria during the Archaean era (from 3.8 Ga to 2.5 Ga ago). So, in part, our currently breathable atmosphere is a by-product of life and not a primary feature of the geosphere alone.

The evolution of the hydrosphere is also intimately linked with that of the atmosphere – water is a volatile compound and is only present on the Earth's surface because the surface temperature is below 100 °C. So to understand the origins of the atmosphere and hydrosphere our attention needs to be focused on those components of the present-day atmosphere that have a limited or negligible interaction with the modern biosphere, namely nitrogen and the inert gases.

An important factor dictating whether or not an object in the Solar System can retain an atmosphere is the strength of the gravitational field at its surface – the stronger the field, the stronger the gravitational forces acting on the molecules in the atmosphere. Without the gravitational field, those molecules moving away from the planet would be lost. Even with the gravitational field, those molecules with particularly high velocities can still escape. This leads to the notion of **escape velocity**, which is the minimum velocity needed before a body has enough *kinetic energy* to escape from the surface of a planet (i.e. overcome its gravitational field). It can be shown that the escape velocity (V_{esc}) for a body of mass M and radius r is given by:

$$V_{esc} = \sqrt{\frac{2GM}{r}} \qquad (2.15)$$

where G is the universal gravitational constant. Whether atmospheric molecules have sufficient velocity depends on the temperature. As the temperature of a gas increases, its molecules move around more quickly, and the average velocity of its molecules increases. Some fraction of the molecules will always be travelling fast enough to overcome gravitational forces, allowing them to escape to space. At low temperatures, this proportion is negligible, but at higher temperatures it becomes progressively more significant, until most molecules exceed the escape velocity for the planetary body. Note that the relevant temperature is at a level in the upper atmosphere above which the atmosphere is so thin that a molecule moving outwards has little chance of colliding with another, and so *will* escape if it has sufficient velocity.

Different gases have different molecular masses, so their average velocities are different at a given temperature. In order for a planetary body to retain a particular gas in its atmosphere for a period of time of the same order as the age of the Solar System, the average velocity of the molecules in the gas should be less than about one-sixth of the escape velocity. (If the average velocity exceeds one-sixth of the escape velocity, a significant proportion of molecules will be moving faster, and will be lost.) This condition is achieved on only a few planets and satellites.

Figure 2.12 explores these relationships further. For each of the planets (and Pluto and the Moon), the temperature is plotted along the horizontal axis and one-sixth of the corresponding escape velocity on the vertical axis. Note that in order to cover the range of values needed, the scales are not linear: a particular distance along an axis corresponds to a doubling of the quantity. For the named gases, the sloping lines show at each temperature the average molecular velocities of each gas. Figure 2.12 thus defines the conditions under which a planet would lose or retain that gas over geological timescales, i.e. thousands of millions of years. The giant planets plot well above all the lines; they can therefore retain any of the gases. The Moon plots below all the lines – it cannot retain gases.

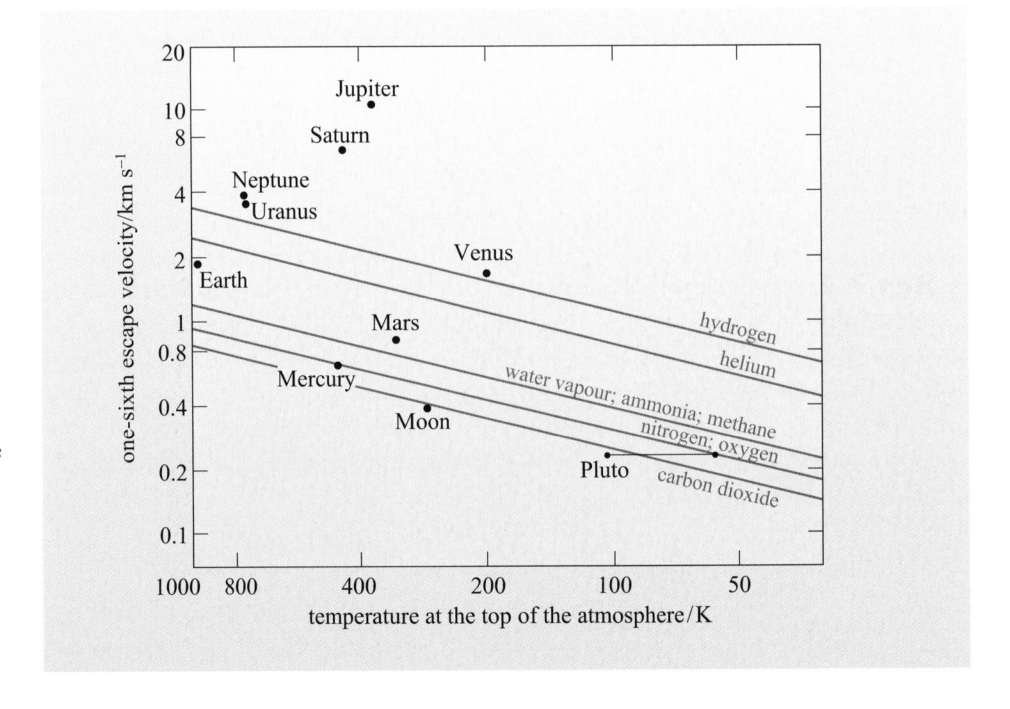

Figure 2.12 Graph summarising conditions of absolute temperature (K) and escape velocity for which planetary bodies can retain common gases in their atmospheres for long periods. For planetary bodies with substantial atmospheres, the temperature is that at the top of the atmosphere. For other bodies, it is the mean surface temperature. Pluto's temperature varies greatly according to its position around its orbit. *Note*: both axes are logarithmic scales.

- Which gases should Earth be able to retain?
- Earth plots below the hydrogen and helium lines so it cannot retain these gases. It plots above the lines for water vapour, nitrogen, ammonia, methane, oxygen and carbon dioxide, so it can retain these gases.

2.4.1 Nitrogen and the inert gases

Nitrogen is the most abundant gas (78.1%) in the atmosphere; the third most abundant is argon (0.93%), one of the inert gases. The remaining inert gases, Ne, Kr and Xe are also found in small traces in the atmosphere. All are relatively unreactive elements in inorganic systems and, although nitrogen can be removed from the atmosphere (fixed) by bacterial activity, its presence in the atmosphere in relatively large quantities is a reflection of its inorganic chemical inertness.

Much attention has been paid to the isotopic composition of the inert gases both in the atmosphere and in the mantle because they can tell us much about the sources of the Earth's volatile elements. In particular, the isotopes of Ar and Xe are significant because they are the daughter products of radioactive decay schemes (Table 2.1), which can be exploited to tell us about the timescales of planetary degassing. The reason why Ar is one of the more abundant atmospheric gases is that it is dominated by one isotope, ^{40}Ar. ^{40}Ar is the daughter of radioactive ^{40}K, which has a half-life of 1.28 Ga (Table 2.1). Xe is much less abundant than Ar, but one of its isotopes, ^{129}Xe, is the daughter of ^{129}I, but in this case the radioactive parent has a half-life of only 15.7 Ma.

- What information can these two radioactive systems provide?
- By analogy with short-lived isotopes in the solid Earth (e.g. Hf–W) and in meteorites (Mg–Al), the I–Xe couple provides information on the timescales of outgassing in the early phases of Earth evolution. By contrast, the long half-life of ^{40}K means that the isotope ratio of Ar reflects the history of planetary outgassing over the whole age of the Earth.

How does this work? The principles behind these isotopic systems are described in Box 2.4. This explains how the radiogenic isotope ratios of the atmosphere and the mantle evolve with time and why measurements of the isotope ratios of Xe and Ar in both are necessary to understand global outgassing.

Box 2.4 Ar and Xe isotopes and the evolution of the atmosphere

The effects of planetary outgassing on the K–Ar and I–Xe radioactive systems are shown graphically in Figure 2.13, in which global outgassing is regarded to be a single event that occurred early in Earth history. Prior to this event, the mantle had low I/Xe and K/Ar ratios and so the ratio of the daughter isotope to a non-radiogenic isotope increased steadily. After degassing, the two parent–daughter ratios in the mantle increased markedly, because Xe and Ar have been lost to the atmosphere and the isotope ratios of Xe and Ar in the mantle both increased at a greater rate than before.

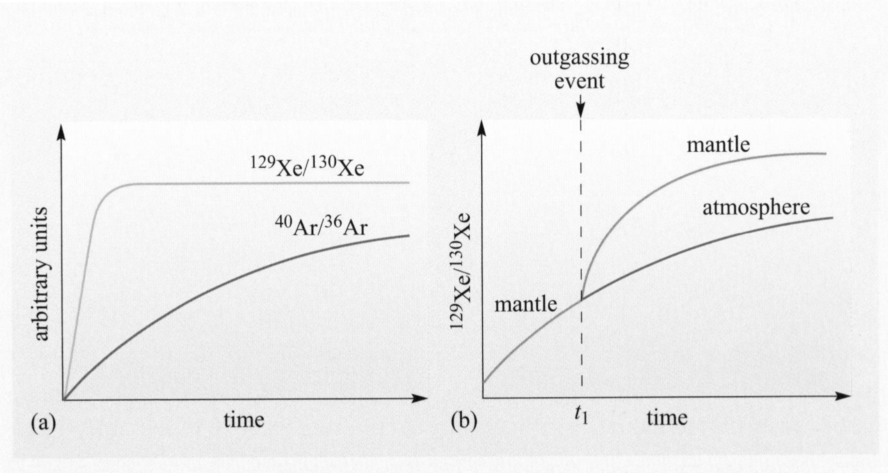

Figure 2.13 Schematic diagrams showing the effect of planetary degassing on the isotopic evolution of Xe and Ar in the Earth's atmosphere and mantle. (a) The simple situation in which the atmosphere gradually forms from the mantle by degassing. Early in Earth history there is a period when the $^{129}Xe/^{130}Xe$ ratio evolves rapidly, because of the decay of ^{129}I, until it reaches a maximum value when all the ^{129}I has decayed. Because ^{129}I has a half life of 15.7 Ma this occurs between 100 and 150 Ma after the formation of the Solar System. (b) If the atmosphere formed as the result of a major outgassing event, t_1, early in Earth history while ^{129}I was active, then the fractionation of Xe into the atmosphere and the preferential retention of I in the mantle would result in greater ^{129}Xe production in the mantle. See text for further discussion.

The critical issue concerns the timing of the degassing event relative to the half-life of ^{129}I. The example illustrated in Figure 2.13b shows how the isotope ratio of $^{129}Xe/^{130}Xe$ varies if degassing occurs before all the ^{129}I has decayed. Subsequent to degassing, the mantle and the atmosphere evolve along different paths.

■ What will be the consequences for the $^{129}Xe/^{130}Xe$ ratios of the mantle and atmosphere if degassing occurs after all the ^{129}I has decayed?

■ They will be identical.

Note that the important measurement to make is not the Ar and Xe isotope ratios of the atmosphere, but those of the mantle. The reason for this is that any outgassing event or process extracts virtually all of the Xe or Ar from the mantle and so the radioactive clock is reset. Because all the volatile elements eventually end up in the atmosphere, its isotope composition reflects the sum total of all outgassing events and processes throughout Earth history. Hence the measurement of Xe isotopes in the Earth's mantle provides a means of distinguishing between early, catastrophic planetary outgassing as opposed to gradual outgassing over the whole of Earth history.

As you may imagine, making isotope measurements of Ar and Xe in the mantle is extremely difficult and enormous care needs to be taken to avoid atmospheric contamination, both of the samples and in the measuring instruments. However, the most successful measurements show that both the $^{129}Xe/^{130}Xe$ and the $^{40}Ar/^{36}Ar$ ratios of the mantle are high relative to the atmosphere.

This observation requires early and extensive degassing of the mantle to generate high parent/daughter ratios in the mantle early enough to have an effect on the I–Xe system. Taken together, the Xe and Ar isotope ratios of the present-day mantle show that between 80% and 85% of the atmosphere was outgassed extremely early in Earth history and, given that the half-life of ^{129}I is only 15.7 Ma, this suggests that outgassing occurred within the first few tens of millions of years after accretion.

■ Which major events occurred within the first few tens of millions of years of Earth evolution?

■ Core formation and the giant impact that formed the Moon both occurred within 30–50 Ma of Earth accretion.

The short timescale of these two major events and the development of the earliest atmosphere were all part of the primary differentiation of the Earth. However, Earth history is never that simple – some outgassing continues to the present day, largely via mid-ocean ridge and within-plate volcanic activity, but the Xe and Ar isotope record of the atmosphere and the mantle show that little if any of the noble gases and, by analogy, nitrogen, are recycled via plate tectonics or any other process into the mantle.

2.4.2 Loss of the earliest atmosphere

Xenon isotope data also provide evidence that much (>99%) of the Earth's early atmosphere was lost within the first 100 Ma. The present I/Xe ratio of the Earth is an order of magnitude higher than chondritic values at the start of the Solar System. We know the abundance of ^{129}I at that time, and as all of this ^{129}I formed ^{129}Xe, it should have produced xenon that was highly enriched in ^{129}Xe, given the Earth's high I/Xe ratio. However, instead, the Earth has xenon that is only slightly more radiogenic than is found in meteorites that are rich in primordial gases. This provides evidence that the Earth had a low I/Xe ratio that kept its xenon isotope compositions close to the chondritic value. At some point xenon was lost, but by this time ^{129}I was close to being extinct so that the xenon did not become radiogenic despite a very high I/Xe ratio, i.e. the $^{129}Xe/^{130}Xe$ ratio stayed low.

You have already seen that gravitational forces are likely to prevent many gases from escaping Earth (Figure 2.12), so it is difficult to see how a large early atmosphere would have been lost. One possibility is that the atmosphere was blown off by a major impact like the Moon-forming giant impact. Alternatively, hydrogen and helium may have been present in substantial amounts in the early atmosphere but, as Figure 2.12 shows, were easily lost through thermal escape. If large amounts of these gases escaped rapidly, then other heavier gases could be lost at the same time by being carried along with them.

Thus, the noble gas data indicate that >80% of the atmosphere was formed extremely early in Earth history, and that much of this early atmosphere was lost in the first 100 million years of Earth history. This suggests that initially the atmosphere was far denser and more massive than today's atmosphere. The presence of such an atmosphere would have an insulating effect, leading to the build up of temperatures in the outer parts of the Earth, facilitating the formation of magma oceans and core formation.

2.4.3 Water

The evolution of the hydrosphere is intimately linked with that of the atmosphere because water is also a volatile compound. Until recently, there were competing views as to the origin of the Earth's water. One was that the Earth accreted as a dry body and its water was subsequently added through cometary impact. The alternative view was that the Earth inherited its water from water-bearing minerals in the undegassed interiors of planetary embryos, and that this was outgassed, along with Xe and Ar, early in Earth history. As noted earlier, the evidence suggests that the early Earth experienced intense meteoritic bombardment and must have been hot. It might be assumed, therefore, that surface conditions were too extreme for the young Earth to have a liquid hydrosphere and that much of the initial water was lost to space. However, current models suggest that with the presence of a dense early atmosphere the pressures at the surface of the Earth beneath this atmosphere would have been high enough to ensure that a significant proportion of water and other volatiles were retained in solution. This debate was partially resolved with the measurement of the deuterium/hydrogen (D/H) ratio in three comets, using both spaceprobe measurements (the Giotto probe to comet Halley) and two ground-based measurements of radio and infrared emissions. All three measurements agree within experimental uncertainty and show that deuterium (heavy hydrogen) is twice as abundant relative to hydrogen in comets as it is in the terrestrial hydrosphere. Such a major distinction effectively rules out comets as a major source of the Earth's water.

Thus the preferred model for the evolution of the hydrosphere is that it degassed from the mantle, and that this material was ultimately derived from water-bearing grains that became incorporated into planetesimals and eventually into planetary embryos.

2.5 The earliest continents

2.5.1 Isotopic evidence

As stated earlier, evidence suggests that, in its early stages, Earth may have had a magma ocean, sustained by heat from accretion and the blanketing effects of a dense early atmosphere. With the loss of the early atmosphere during planetary collisions, the Earth would have cooled quickly, the outer portions would have solidified, and it would have developed its first crust.

We have little idea of what such a crust might have looked like. This is because, unlike the Moon, the Earth appears to have no rock preserved that is more than 4.0 Ga old. There was intense bombardment of the Moon until about 3.85 Ga

(Section 2.3) and Earth's earliest crust may therefore have been destroyed by such impacts. It may also be that because the Earth was hotter this crust was highly unstable. Some have argued that the earliest crust was like the **lunar highlands** – made from a welded mush of crystals that had previously floated on a magma ocean. Others have suggested that it was made of denser rocks more like the basalts that presently form the Earth's ocean floor.

Some of the oldest rocks on Earth come from exposures in western Greenland near Isua. These include ancient sediments and volcanic lavas that have since been subjected to folding and intrusion by younger igneous rocks. Metamorphosed sediments from Isua give an age of about 3.9 Ga, which is regarded as a minimum age of sediment deposition. However, these rocks were formed by weathering and erosion of pre-existing crustal material, and may preserve clues as to the age and origin of this older crust.

Just as U–Pb and Hf–W are ideal for studying the rates of accretion and core formation, and I–Xe is useful for studying the rate of formation of the atmosphere, the ^{146}Sm–^{142}Nd (samarium–neodymium) system is useful for studying the early history of melting in the silicate Earth. ^{146}Sm decays to ^{142}Nd with a half-life of 103 million years (Table 2.1), but unlike Hf–W or U–Pb systems, both Sm and Nd are lithophile elements and remain in the mantle during core formation. However, during melting of the silicate mantle to produce the crust, Nd is more incompatible than Sm (that is, Nd is preferentially partitioned into the silicate melt) producing a crust with a low Sm/Nd ratio and leaving the mantle with a complementary high Sm/Nd ratio. If melting of the silicate Earth occurred within the lifetime of the now extinct ^{146}Sm (about 500 Ma), then this would fractionate Sm/Nd and could potentially produce differences in the abundance of ^{142}Nd in the early crust. If subsequent recycling of such crust by weathering and erosion or meteorite impact failed to eradicate the difference in ^{142}Nd, this might be detectable in early Archaean rocks.

However, such ^{142}Nd differences (sometimes termed anomalies), if they ever existed, were always likely to be small because of the low abundance of ^{146}Sm in the early Solar System. As a consequence their measurement is extremely difficult and for many years only one sample showed the hint of such an effect, and these data were questioned by many. Study of ^{142}Nd anomalies has, of course, focused on the Isua rocks and improved measurement techniques have recently confirmed the existence of differences in ^{142}Nd (from the present-day mantle) in many rock types seen at Isua, including metamorphosed sediments and volcanic rocks (Figure 2.14). These results indicate that the precursor materials of the Isua rocks were derived from material that differentiated from the silicate mantle within 50 Ma of the start of the Solar System.

Surprisingly, it was also discovered that an excess of ^{142}Nd is ubiquitous in all modern terrestrial rocks – at least all those measured so far – relative to chondritic meteorites (Figure 2.14). If it is assumed that chondritic meteorites are representative of the material from which the Earth was made, then this excess ^{142}Nd indicates that the silicate Earth must have experienced a global chemical differentiation within 50 Ma of the start of the Solar System.

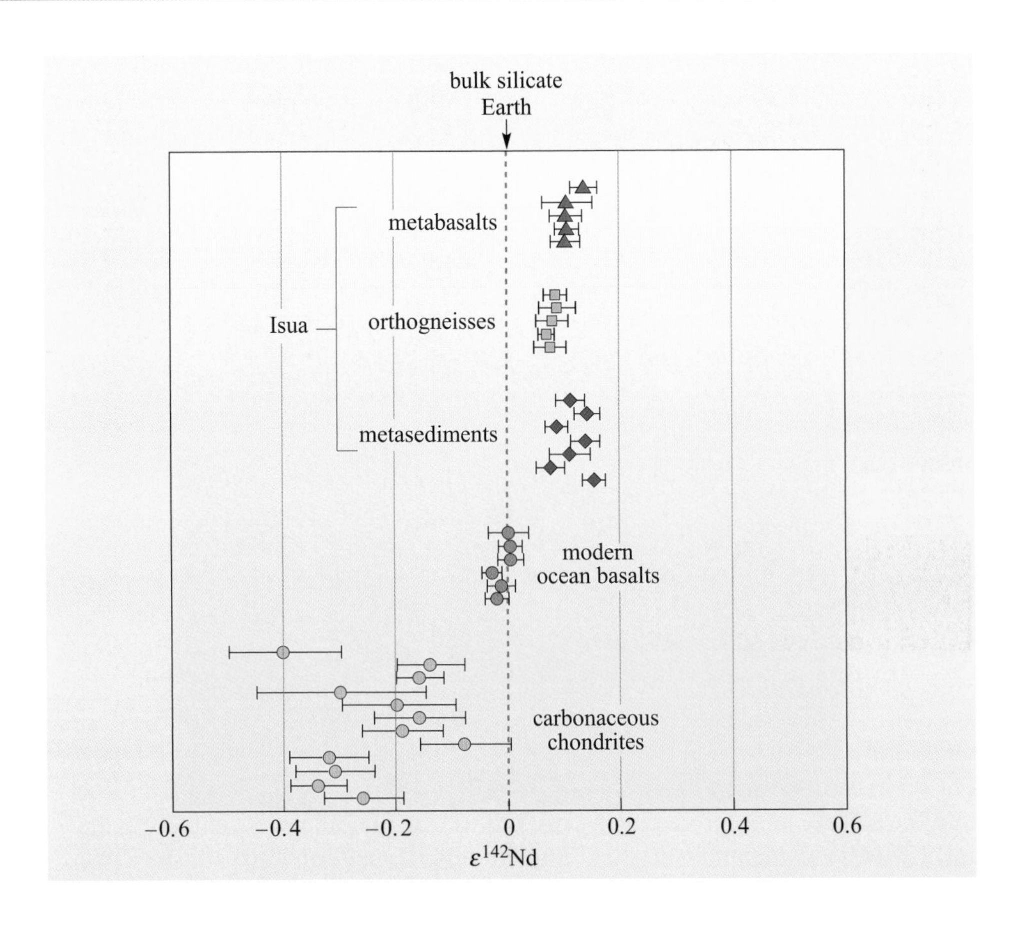

Figure 2.14 $^{142}Nd/^{144}Nd$ ratios measured for chondrites, present-day terrestrial rocks and Archaean rocks at Isua.

2.5.2 Geological evidence

The rocks themselves at Isua allow us to make some inferences about conditions on and in the Earth 3.8 Ga ago. They include lavas known as komatiites (Box 2.5) that have what is called a *pillow structure*, which indicates cooling under water, and sediments rich in quartz pebbles and volcanic debris (Figure 2.15). The observation that sediments and lavas were deposited and erupted under water indicates that there must have been bodies of liquid water at the Earth's surface, and perhaps even ocean basins by 3.9 Ga. Land areas would have to have been exposed to weathering and erosion to form such sediments. The presence of quartz pebbles implies that at least some of the land being weathered was similar to the upper parts of present-day continental areas. For example, granite is a common rock that occurs in continental areas and is rich in quartz. It seems, therefore, that the geological processes and cycles that are recognised today were operating on the Earth at least 3.9 Ga ago, albeit with some differences in detail.

■ What does the occurrence of sedimentary rocks at Isua tell us about the surface temperature of the Earth 3.9 Ga ago?

■ For weathering, erosion and deposition of sediments to occur there must have been both rain and liquid water present, implying surface temperatures of between 273 K and 373 K (i.e. 0–100 °C).

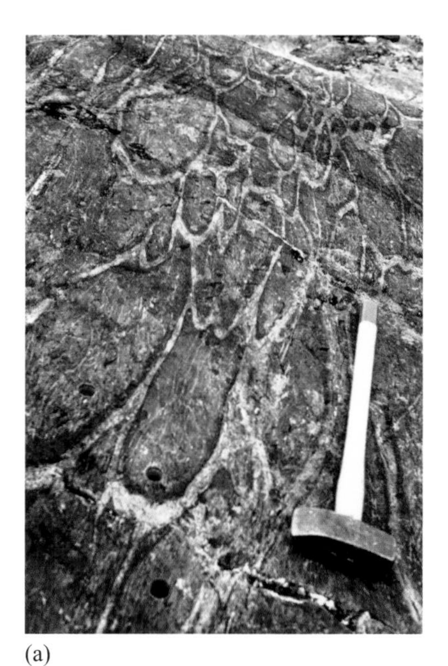
(a)

(b)

Figure 2.15 (a) Metamorphosed pillow lava at Isua showing pillows with dark rims in a matrix of chert (fine-grained silica). (b) Metamorphosed conglomerate at Isua, showing pebbles of quartz. (Steve Moorpath)

Box 2.5 Komatiites

Komatiites are rare ultramafic lavas that were produced most commonly during the Archean and are very rare in the Phanerozoic. These magmas are thought to provide a record of the thermal and chemical characteristics of the mantle through time. Komatiites are distinguished by their high MgO content (>18 wt%) and olivines showing a *spinifex* texture (elongated, skeletal, branching crystals) (Figure 2.16), which were initially thought to be indicative of the rapid quenching of magma. When komatiites were subjected to melting experiments in the laboratory they were found to possess very high liquidus temperatures – a result that was initially interpreted as indicating deep melting at high temperatures. The progressive decline of komatiites was then used as evidence for progressive cooling of the Earth's mantle.

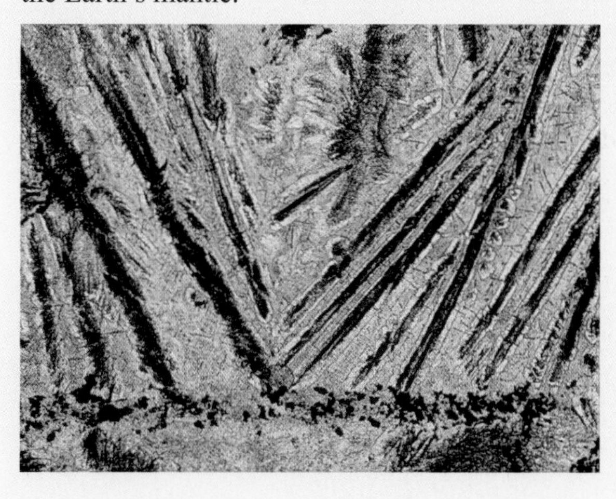

Figure 2.16 Olivine needles in a spinifex-textured komatiite. Each crystal is 2–3 cm long. (Keiko Kubo)

However, over recent years an alternative interpretation of komatiite formation has gained favour. This is because komatiites are rich in water, and experiments indicate that the liquidus temperature is dramatically reduced under hydrous conditions. Similarly, in the presence of water spinifex textures can develop even at a low cooling rate. Therefore it has been proposed that the high water contents in komatiite lavas are due either to the Archaean mantle containing a higher volatile content, left over from accretion, or that water was introduced into the mantle as hydrated crust in a subduction zone environment. This alternative interpretation of komatiites indicates that water played a key role in their formation and, as a consequence, the Archaean mantle was only slightly hotter than the present day. It also indicates that subduction may have operated in the early Archaean.

2.5.3 Continents and the hydrosphere during the Hadean

The Archaean sediments and volcanic rocks at Isua and at other locations across the world show that by about 3.9 Ga water was present on the surface of the Earth, and that a cycle of erosion and deposition of continental rocks had been established. Is there any evidence for water further back in time?

The possibly surprising answer to this question is 'yes', and the source of this evidence is from a lonely outcrop of Archaean sandstones and conglomerates in the Murchison district of western Australia. These ancient rocks, which are about 3 Ga old, are known as the Mount Narryer and Jack Hills quartzite units. They contain detrital grains of the mineral zircon. Zircons are of enormous value to Earth scientists because, once formed, they are almost indestructible and contain high concentrations of U and Th and other important trace elements. Hence they are one of the primary sources of high-precision ages from Pb isotopes of crustal rocks and their trace element contents provide further information on their formation conditions and provenance. Zircons form as accessory minerals in igneous rocks, usually granites, but, because of their indestructibility, they can survive the rock cycle of erosion, transport and deposition and are frequently preserved in coarse-grained sedimentary rocks such as sandstones, grits and conglomerates, and their metamorphic equivalents, e.g. quartzite. Thus, the inescapable conclusion is that the Jack Hills zircons must have been inherited from an older granite and so from older continental crust. So far this is not very exciting – rocks are known that are almost 1 Ga older than the Jack Hills quartzites, so why the excitement?

Detailed isotopic analysis of the Pb isotope composition of these zircons (Figure 2.17) has revealed that they range in ages from 3.05 Ga, which is close to the age of deposition, up to much older ages – a significant number of crystals are in excess of 4 Ga. Indeed, the oldest has an age of 4.4 Ga and formed only 170 Ma after the start of the Solar System. These zircons are the oldest solid material yet found on Earth and, because zircon can only form in high temperature igneous processes in a high-silica magma, they are thought to represent a tiny sample inherited from some of the Earth's original continental crust.

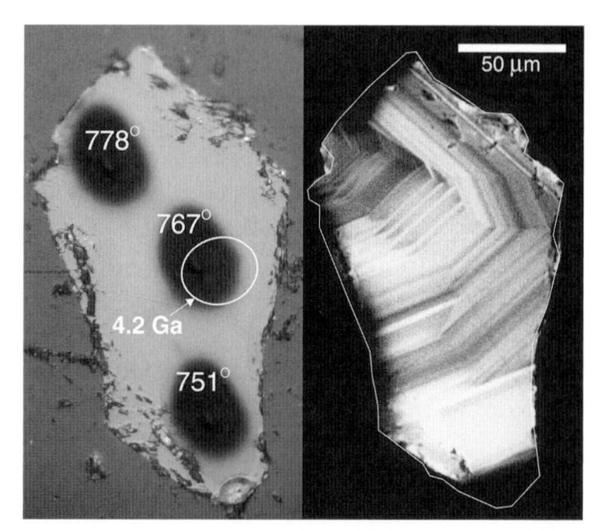

Figure 2.17 Images of a zircon from the Jack Hills quartzites in western Australia: (a) shows the damage produced during ion-microprobe analyses. Temperatures are in degrees Celsius and refer to results from an assessment of their Ti contents. The white ellipse shows the region of the crystal dated at 4.2 Ga; (b) shows the same crystal from cathode-luminescence, revealing compositional banding. (Watson and Harrison, 2005)

That conclusion is in itself exciting – solid material from the Hadean and evidence for continental crust – but apart from their age, the Jack Hills zircons show compositional variations that allow conditions of crystallisation to be inferred. It turns out that the zircons crystallised at a temperature of around 700 °C, a temperature that is no greater than the temperature that granites crystallise at today. As you will see in Chapter 6, granites can only form at this temperature *if water is present*. They form by *dehydration* reactions in which mica breaks down, releasing substantial amounts of water that dissolve in the silicate melt so generated. The presence of granite and its crystallisation at low temperatures is powerful evidence that water was present in the earliest crust of the Earth.

This argument may appear a little convoluted, and the leap from the tiny Jack Hills zircons to a planet with continents and an active low-temperature hydrosphere is large, but the logic is sound. The information gleaned from these zircons is that the surface environment during the Hadean may have been remarkably similar to the Earth today!

2.5.4 The Archaean atmosphere, hydrosphere and biosphere

The nature of the Earth's atmosphere and hydrosphere after the Earth had cooled and the first rocks were preserved at the surface is the subject of considerable debate. It has been suggested that the warm conditions and presence of water may have provided conditions suitable for life to develop. Several geological settings might have hosted such early life. It may have been supported in ocean basins during transient heating after a major meteorite impact, or possibly life could have existed in a hydrothermal system. Submarine systems would have also offered a protective setting against UV radiation.

Although the rocks at Isua show clear evidence that water was present at the Earth's surface, there are no fossils. Again, evidence for life comes from geochemical clues. Carbonate in some of the volcanic rocks contains carbon with a heavy isotope signature, i.e. it shows an enrichment in ^{13}C over ^{12}C. This is complementary to the ubiquitous depletion in ^{13}C relative to ^{12}C, which is characteristic of 'light' carbon that has been involved in biological processes.

However, the carbonate is all non-biological in origin and secondary (i.e. precipitated after the crystallisation of the volcanic rock) and therefore may be much younger than the host volcanic rock. Isotopically light carbon also occurs in graphite flakes in some rocks and this chemical signature is thought by some to be of biological origin. However, this interpretation remains controversial, as the graphite is found in highly metamorphosed and deformed rocks, and is usually associated with secondary carbonate deposition.

The sequence at Isua also contains rocks that are thought by some to provide possible evidence for the changing nature of the Earth's atmosphere as a result of the appearance of life. **Banded iron formations (BIFs)** are characterised by finely banded dark-brown, iron-rich layers alternating with lighter-coloured, iron-poor layers (Figure 2.18). The layers range in thickness from less than a millimetre to about a centimetre. The iron-rich bands contain the highly insoluble iron oxides: hematite (Fe_2O_3), limonite ($Fe_2O_3.3H_2O$) and magnetite (Fe_3O_4). Chert, a rock composed of precipitated silica, occupies the iron-poor bands. Individual bands, often only a few millimetres thick, can extend for several kilometres. How BIFs formed is not entirely clear, as we have no modern analogues to guide us. However, the involvement of iron oxides in their formation suggests that the process that led to the formation of BIFs must have affected the oxidation state of iron and hence such rocks may contain information about the oxidation state of the Earth's ocean and atmosphere at that time.

Figure 2.18 Close-up view of a banded iron formation. The light bands are dominated by silica-rich chert and the darkest bands by red–brown oxides of iron. Each band is ~1 cm thick. (Bill Bachman/SPL)

Whatever the chemistry occurring in the formation of BIFs, large amounts of oxygen were incorporated into BIFs very early in the Earth's history. One interpretation is that this suggests oxygen was available in the shallow seas where most BIFs were formed. Most theories for the origin of BIFs involve a significant role for hydrothermal activity on the early Earth. The seawater flowing through these hydrothermal systems would have dissolved iron-containing minerals, so that iron in a reduced form was subsequently injected into the deep ocean through hydrothermal vents. It is generally accepted that the deep ocean on the early Earth was extremely oxygen deficient or anoxic, so that iron escaped oxidation and precipitation at the vents themselves but was deposited in much shallower, more oxygenated water.

How did the iron get from the deep ocean to shallow water, crossing large expanses of oceans in so doing? One idea is that the iron was actually consumed by bacteria that flourished near the vents and that these bacteria then drifted away in vast colonies into shallow water where they died, depositing a thin film of organic-rich material. After a while the organic material would have been recycled, leaving the iron behind in its highly insoluble oxide form.

■ Hydrothermal vents provide a potential source for the large amounts of iron involved in the formation of BIFs, but where might the large amounts of oxygen come from?

■ Various mechanisms have been proposed to account for the oxidation of BIFs and the precipitation of Fe(III):

– abiotic breakdown of water by UV radiation

– direct oxidation by anaerobic bacteria

– photosynthesis, which generates free oxygen.

Summary of Chapter 2

The sequence of events accompanying the evolution of the early Earth is summarised in Figure 2.19 (overleaf). Taken together, the evidence suggests that the development of the core and magma ocean, and outgassing to form the atmosphere, were contemporaneous and interdependent processes.

The issue of whether accretion of the Earth was homogeneous or heterogeneous has long been debated. To some extent the Earth must have accreted heterogeneously because early planetesimals are likely to have experienced different degrees of differentiation. Furthermore, meteorites with different chemical compositions from Earth continue to accrete right up to the present day. However, if the early Earth were covered by a deep magma ocean, any effect of accreting material with diverse compositions would be erased in the magma ocean. A possible exception to this might be the highly siderophile elements. The abundances of these elements in the silicate mantle might be explained by metal–silicate equilibration in a magma ocean or may have been added as a 'late veneer' after core formation.

Finally, it would appear inescapable that most of the water on Earth arrived as part of the accretion process and the presence of that water influenced primary differentiation. Accretion of hydrous materials and subsequent outgassing led to the formation of a dense 'steam' atmosphere, which in turn served as a thermal blanket allowing the long-term persistence of a terrestrial magma ocean. The higher pressure of water in the atmosphere would have allowed the presence of

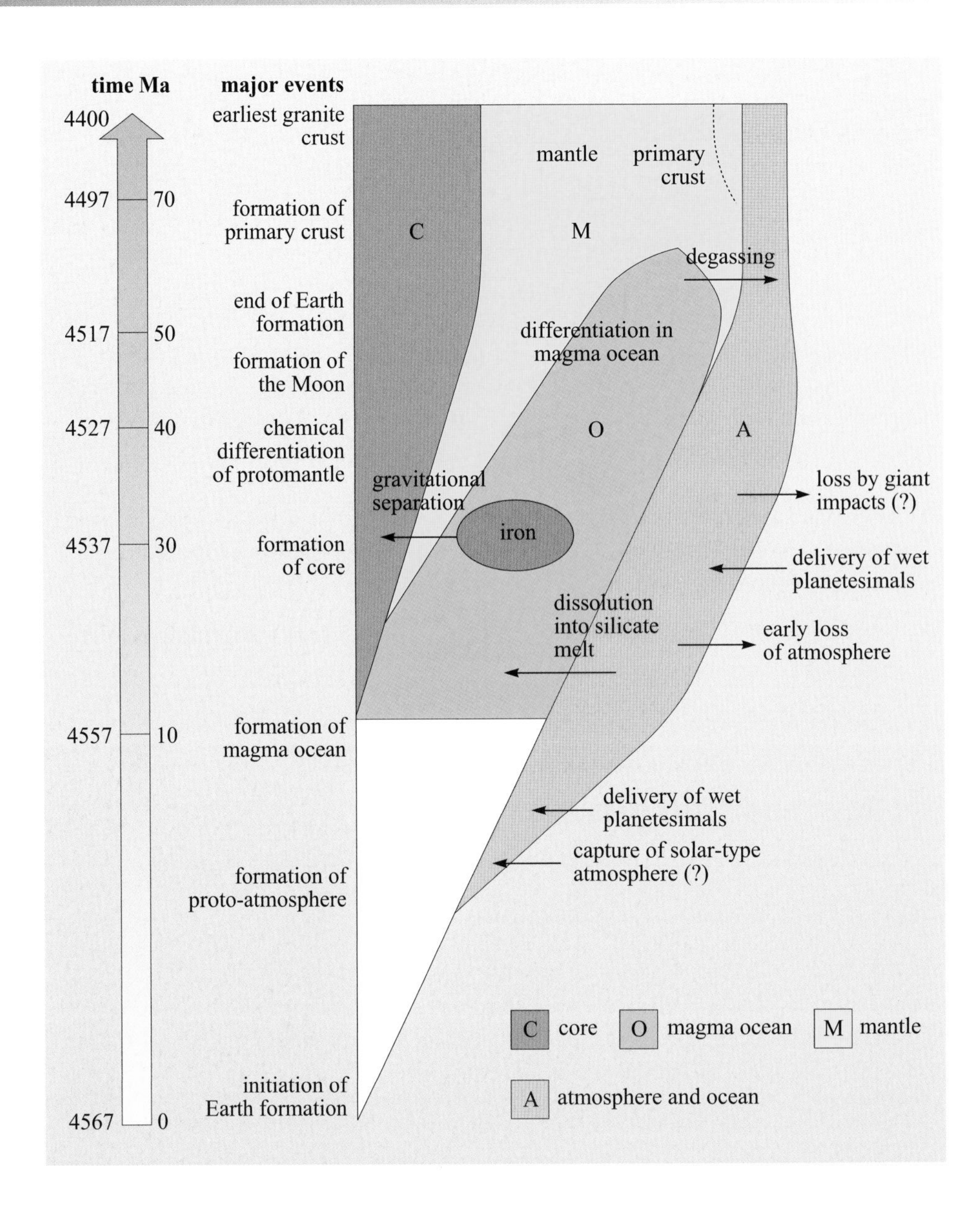

Figure 2.19 Schematic diagram showing the timescale of major events in the evolution of the early Earth. The timescale is indicative rather than accurate. (Adapted from Abe et al., 2000)

liquid water at the Earth's surface, and a significant amount of water would have been able to dissolve in the silicate liquid of the magma ocean. Cooling and crystallisation of the magma ocean would have provided a further source of water for the atmosphere and hydrosphere, and a mantle reservoir of stable hydrous phases.

The formation of the hydrosphere probably occurred soon after the atmosphere because water is also a volatile molecule. The oldest zircons indicate that water must have been present at the Earth's surface 170 Ma after accretion. This is because these ancient zircons crystallised from granite that formed at a temperature only possible if water was present. It is also possible that hydrous komatiites may have been produced either by an Archaean mantle with a high volatile content, or by melting of hydrous crust in an early subduction zone environment. Following the end of heavy bombardment of the Earth around 3.85 Ga, it is possible that early life may have been supported in ocean basins heated by meteorite impact or hydrothermal systems.

Learning outcomes for Chapter 2

You should now be able to demonstrate a knowledge and understanding of:

2.1 The different heat sources and transfer mechanisms involved in driving planetary differentiation and the early chemical and physical evolution of the Earth.

2.2 The current theories associated with the mechanisms and timing of accretion, core formation, and the formation and existence of early magma ocean.

2.3 The evidence for the origin and age of the Moon.

2.4 The evidence for the formation and chemical evolution of the Earth's atmosphere and hydrosphere.

2.5 The nature, composition and age of the earliest continental crust.

2.6 The role of partial melting and the presence of water in generating silicic crustal material from the mantle.

Plate tectonics

The Earth's face is changing all the time, but at barely perceivable rates. It is now known that the Earth is a highly dynamic planet – far more so than the other terrestrial planets (Mercury, Venus and Mars) and the Moon – and one that has altered its outward surface many times over geological time. On Earth, this dynamism is manifest in the opening and closure of ocean basins and the associated movements of continents called **continental drift**. At times, continental drift has resulted in the continents fragmenting into many smaller land masses, while at other times collisions have assembled vast supercontinents with immense mountain chains along their joins (Figure 3.1). Such constant rearrangement has had a profound effect upon the surface geology of our planet. It has also affected the hydrosphere through the changes in the shape and size of the oceans, the ocean–atmosphere circulation, the configuration and extremities of the Earth's climate zones and, perhaps, even the nature of the biosphere and the course of the evolution of life itself.

In Chapters 1 and 2 you discovered how the Earth formed from a primordial nebula and then developed its layered structure (i.e. core, mantle and crust). From your studies of those chapters you will have begun to discover the physical and chemical differences between those three layers, the differences that distinguish the mantle and the crust, the heat that is generated within them, and how this heat then makes its way to the surface to be lost into space. It is the movement of this internal heat that drives the forces that result in the formation and destruction of ocean basins and, ultimately, the movement of the continents.

In this chapter, you will examine how the evidence for the movement of continents was gathered and how this movement relates to, and generates, geological features and phenomena such as ocean basins, mountain ranges, volcanoes and earthquakes. You will learn how and why the continents have moved, and continue to move, and the forces that drive them around our globe.

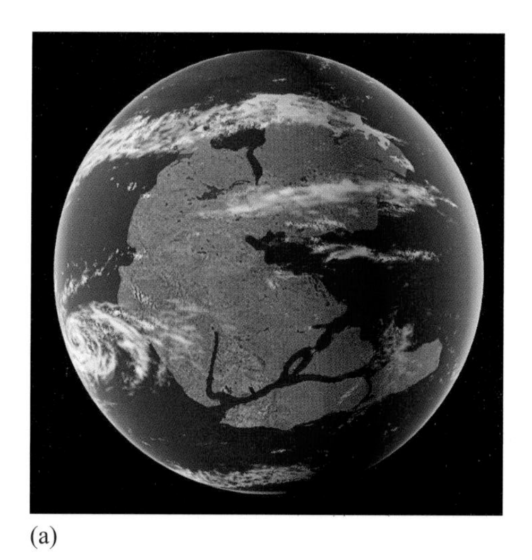

(a)

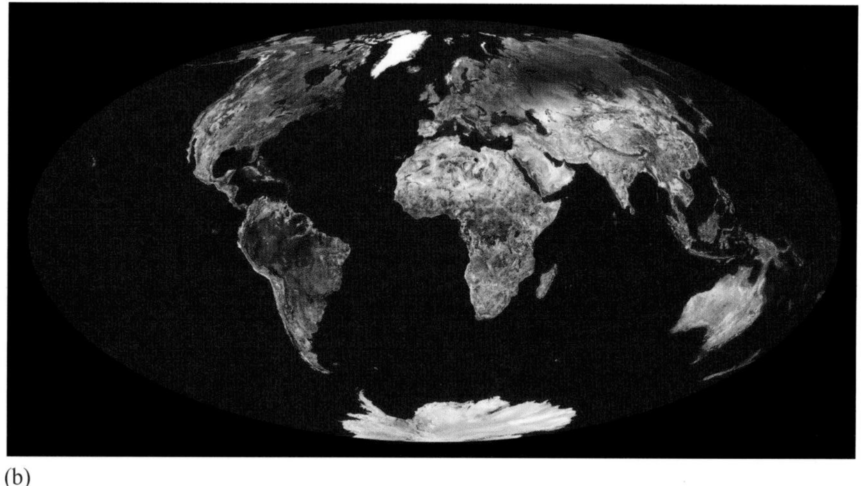

(b)

Figures 3.1 An example of (a) a continental reconstruction for the late Carboniferous showing the supercontinent of Pangaea compared with (b) today's more fragmented arrangement. ((a) Christian Darkin/SPL; (b) M-Sat Ltd/SPL)

3.1 From continental drift to plate tectonics

The remarkable notion that the continents have been constantly broken apart and reassembled throughout Earth's history is now widely accepted. The greatest revolution in 20th century understanding of how our planet works, known as **plate tectonics**, happened in the 1960s, and has been so profound that it can be likened to the huge advances in physics that followed Einstein's theory of relativity. According to the theory of plate tectonics, the Earth's surface is divided into rigid plates of continental and oceanic lithosphere that, through time, move relative to each other, and which increase or decrease in area. The growth, destruction and movement of these lithospheric plates are the major topics of this chapter, but it is first worth considering how the theory actually developed from its beginnings as an earlier idea of 'continental drift'.

3.1.1 Continental drift

The German meteorologist Alfred Wegener (1880–1930) is largely credited with establishing the fundamentals of the theory that we now call plate tectonics. The idea that continents may have originally occupied different positions was not a new one (Box 3.1), but Wegener was the first to present the evidence in a diligent and scientific manner.

Box 3.1 Continental drift to plate tectonics: the evolution of a theory

1620 Francis Bacon commented on the 'conformable instances' along the mapped Atlantic coastlines.

1858 Antonio Snider-Pellegrini suggested that continents were linked during the Carboniferous Period, because plant fossils in coal-bearing strata of that age were so similar in both Europe and North America.

1885 Austrian geologist Edward Seuss identified similarities between plant fossils from South America, India, Australia, Africa and Antarctica. He suggested the name 'Gondwana' (after the indigenous homeland of the Gond people of north-central India), for the ancient **supercontinent** that comprised these land masses.

1910 American physicist and glaciologist Frank Bursley Taylor proposed the concept of 'continental drift' to explain the apparent geological continuity of the American Appalachian mountain belt (extending from Alabama to Newfoundland) with the Caledonian Mountains of northwest Europe (Scotland and Scandinavia), which now occur on opposite sides of the Atlantic Ocean.

1912 Alfred Wegener reproposed the theory of continental drift. He had initially become fascinated by the near-perfect fit between the coastlines of Africa and South America, and by the commonality among their geological features and fossils, and evidence of a glaciation having affected these two separate continents. He compiled a considerable amount of data in a concerted exposition of his theory, and suggested that during the late Permian all the continents were once assembled into a supercontinent that he named **Pangaea**, meaning 'all Earth'. He drew maps showing how the continents have since moved to today's positions. He proposed that Pangaea began to break apart just after the beginning of the Mesozoic Era, about 200 Ma ago, and that the continents then slowly drifted into their current positions.

1920–1960 A range of geophysical arguments was used to contest Wegener's theory. Most importantly, the lack of a mechanism strong enough to 'drive continents across the ocean basins' seriously undermined the credibility of his ideas. The theory of continental drift remained a highly controversial idea.

1937 South African geologist Alexander du Toit provided support through the years of controversy by drawing maps illustrating a northern supercontinent called Laurasia (i.e. the assembled land mass of what was to become North America, Greenland, Europe and Asia). The idea of the Laurasian continent provided an explanation for the distribution of the remains of equatorial, coal-forming plants, and thus the widely scattered coal deposits in the Northern Hemisphere.

1944 Wegener's theory was consistently championed throughout the 1930s and 1940s by Arthur Holmes, an eminent British geologist and geomorphologist. Holmes had performed the first uranium–lead radiometric dating to measure the age of a rock during his graduate studies, and furthered the newly created discipline of geochronology through his renowned book *The Age of the Earth*. Importantly, his second famous book *Principles of Physical Geology* did not follow the traditional viewpoints and concluded with a chapter describing continental drift.

1940–1960 The complexity of ocean floor topography was realised through improvements to sonar equipment during World War II. Accordingly, there was a resurgence of interest in Wegener's theory by a new generation of geophysicists, such as Harry Hess (captain in the US Navy, later professor at Princeton), through their investigations of the magnetic properties of the sea floor. In addition, an increasing body of data concerning the magnetism recorded in ancient continental rocks indicated that the magnetic poles appeared to have moved or 'wandered' over geological time. This **apparent polar wander** was explained by the movement of the continents, and not the magnetic poles.

1961 The American geologists Robert Dietz, Bruce Heezen and Harry Hess proposed that linear volcanic chains (mid-ocean ridges) identified in the ocean basins are sites where new sea floor is erupted. Once formed, this new sea floor moves toward the sides of the ridges and is replaced at the ridge axis by the eruption of even younger material.

1963 Two British geoscientists, Fred Vine and Drummond Matthews, proposed a hypothesis that elegantly explained magnetic reversal stripes on the ocean floor (Section 3.1.2). They suggested that the new oceanic crust, formed by the solidification of basalt magma extruded at mid-ocean ridges, acquired its magnetisation in the same orientation as the prevailing global magnetic field. These palaeomagnetic stripes provide a chronological record of the opening of ocean basins. By linking these observations to Hess's sea-floor spreading model, they laid the foundation for modern plate tectonics.

1965 The Canadian geophysicist J. Tuzo Wilson offered a fundamental reinterpretation of Wegener's continental drift theory and became the first person to use the term 'plates' to describe the division and pattern of relative movement between different regions of the Earth's surface (i.e. plate tectonics). He also proposed a tectonic cycle (the Wilson cycle) to describe the lifespan of an ocean basin: from its initial opening, through its widening, shrinking and final closure through a continent–continent collision (Section 3.5.1).

1960s–present day There was an increasingly wide acceptance of the theory of plate tectonics. A concerted research effort was made into gaining a better understanding of the boundaries and structure of Earth's major lithospheric plates, and the identification of numerous minor plates.

Evidence for continental drift

Wegener's ideas on continental drift based on the following evidence remain the root of modern continental reconstructions.

1 Geometric continental reconstructions

Ever since the first global maps were drawn following the great voyages of discovery of the 15th and 16th centuries, it has been realised that the coastline geography of the continents on either side of the Atlantic Ocean form a pattern that can be fitted back together; in particular, the coastlines of western Africa and eastern South America have a jigsaw-like fit (Box 3.1).

Although some coastline fits are striking, it is important to note that the current coastlines are a result of relative sea level rather than the actual line along which land masses have broken apart. Indeed, coastline-fit is a common misconception – Wegner himself pointed out that it is the edge of the submerged **continental**

shelf, i.e. the boundary between continental and oceanic crust, that actually marks the line along which continents have originally been joined.

It was not until 1965 that the first computer-drawn reassembly of the continents around the Atlantic Ocean was produced by the British geophysicist Edward Bullard and his colleagues at Cambridge University. They used spherical geometry to generate a reconstruction of Africa with South America, and Western Europe with North America, which were all fitted together at the 500 fathom (about 1000 m) contour, which corresponded to the edge of the continental shelf. This method revealed the fit to be excellent (Figure 3.2), with few gaps or overlaps.

- Figure 3.2 shows some overlaps in the way in which the continents fit together around the Atlantic. Why might these exist?

- Most of the overlaps are caused by features that have formed since the continents broke up or rifted apart, such as coral banks (Florida), recent river deltas (Niger) and volcanoes (Iceland).

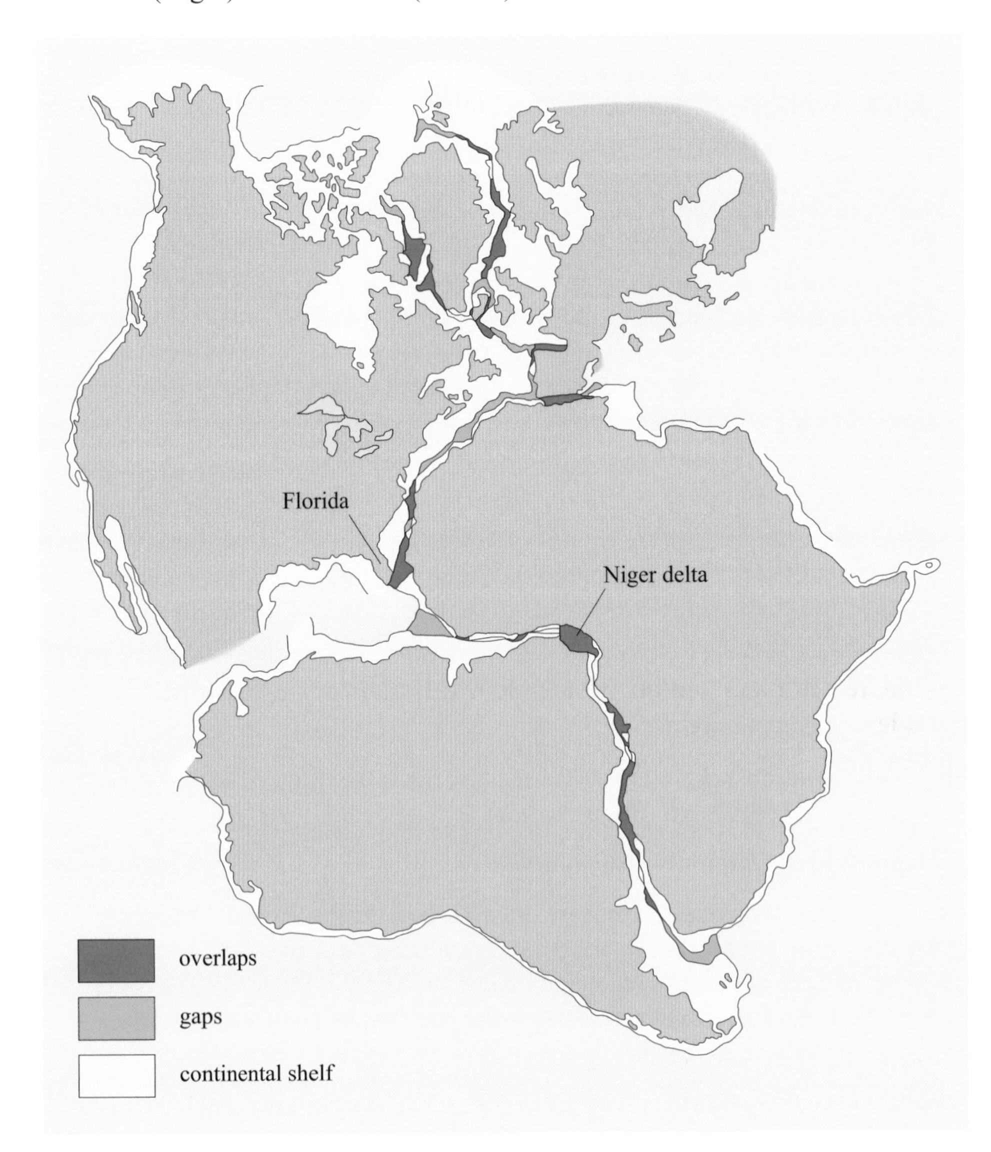

Figure 3.2 A computer-generated spherical fit at the 500 fathom contour (i.e. edge of continental shelf) showing Bullard's fit of the continents surrounding the Atlantic.

2 Geological match and continuity of structure

Previous configurations of continents can also be recognised by the degree of geological continuity between them. These include similar rock types found on either side of an ocean or, more commonly, successions of strata or igneous bodies that have otherwise unique characteristics. Taylor (Box 3.1) was first prompted to consider continental drift by noting the similarity of the rock strata and geological structures of the Appalachian and Caledonian mountain belts of eastern USA and northwestern Europe respectively. Similarly, Wegener investigated the continuity of Precambrian rocks and geological structures between South America and Africa (Figure 3.3).

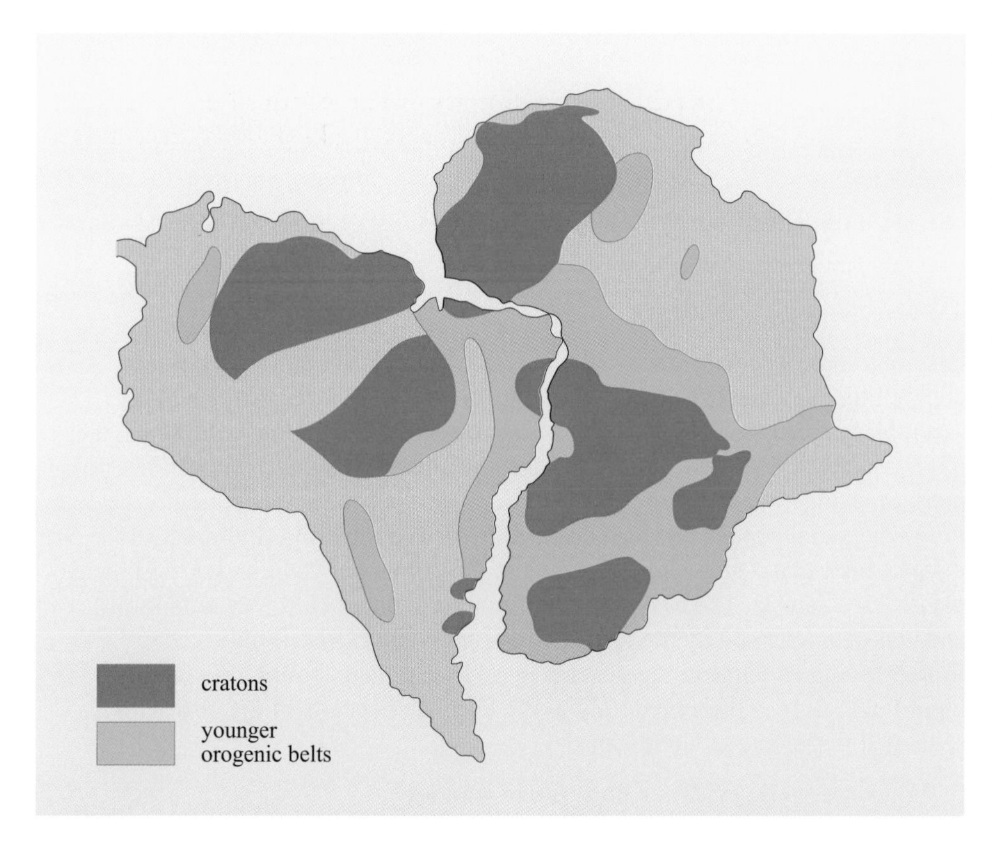

Figure 3.3 Continuity of Precambrian rocks. There is good correlation between these geological units when the continents are fitted along their opposing margins. The immense periods of time over which these Archaean and Precambrian units were formed (>2 Ga) indicate that South America and Africa had together formed a single land mass for a considerable part of the Earth's history. (Adapted from Hallam, 1975)

3 Climate, sediment and the mismatch of sedimentary deposits with latitude

The climate of modern Earth may be divided into different belts that have cold arctic conditions at high latitudes and hot tropical conditions at equatorial and low latitudes. The nature and style of rock weathering and erosion varies according to these climate belts, such that glaciation and freeze–thaw action predominate at present-day high latitudes, while chemical alteration, aeolian and/or fluvial processes are more typical of present-day low latitudes. Once a rock is weathered and eroded, each climatically controlled suite of processes gives rise to its own type of sedimentary succession and landforms:

- sand dunes form in hot, dry deserts
- coal and sandstone successions form in tropical swamps and river deltas
- boulder clay deposits and 'U-shaped' valleys form where there are ice sheets and glaciers.

It has long been recognised that geologically ancient glacial-type features are not just restricted to the present-day, high-latitude locations, but also occur in many warm-climate continents such as Africa, India and South America. Similarly, warm-climate deposits may be found in northern Europe, Canada and even Antarctica. For instance, coal is one of our most familiar geological materials, yet the European and North American coal deposits are derived from plants that grew and decayed in hot, steamy tropical swamps 320–270 Ma ago during the late Carboniferous and early Permian Periods. Reasons for these unusual distributions are often provided by reconstructing the ancient continental areas and determining their original positions when the deposits or landforms were created.

Question 3.1

Late Carboniferous coalfields are found in northern Britain around latitude 55° N. If these coals formed from plants that grew in the tropics between 23° N and 23° S, what is the *minimum* distance Britain has travelled in 300 Ma? At what rate has it travelled (in mm y^{-1})? (Assume the radius of the Earth is 6370 km.)

4 Palaeontological evidence

Palaeontological remains of fossil plants and animals are among the most compelling evidence for continental drift. In many instances, similar fossil assemblages are preserved in rocks of the same age in different continents; the most famous of these assemblages is the so-called *Glossopteris* flora. This flora marks a change in environmental conditions. In the southern continents, the Permian glacial deposits were succeeded by beds containing flora that was distinct from that which had developed in the climatically warm, northern land masses of Laurasia. The new southern flora grew under cold, wet conditions, and was characterised by the ferns *Glossopteris* and *Gangamopteris*, the former giving its name to the general floral assemblage. Today, this readily identifiable flora is preserved only in the Permian deposits of the now widely separated fragments of Gondwana.

5 Palaeomagnetic evidence and 'polar wander'

The Earth has the strongest magnetic field of all the terrestrial planets, with similar properties to a magnetic dipole or bar magnet. As newly erupted volcanic rocks cool, or sediments slowly settle in lakes or deep ocean basins, the magnetic minerals within them become aligned according to the Earth's ambient magnetic field. This magnetic orientation becomes preserved in the rock. The ancient inclination and declination of these rocks can then be measured using sensitive analytical equipment.

As a continent moves over the Earth's surface, successively younger rocks forming on and within that continent will record different palaeomagnetic positions, which will vary according to the location of the continent when the rock was formed. As a result, the position of the poles preserved in rocks of different ages will apparently deviate from the current magnetic pole position (Figure 3.4a). By joining up the apparent positions of these earlier poles, an apparent polar wander (APW) path is generated. It is now known that the Earth's magnetic poles do not really deviate in this manner, and the changes depicted in APW paths are simply a result of the continent moving over time (Figure 3.4b).

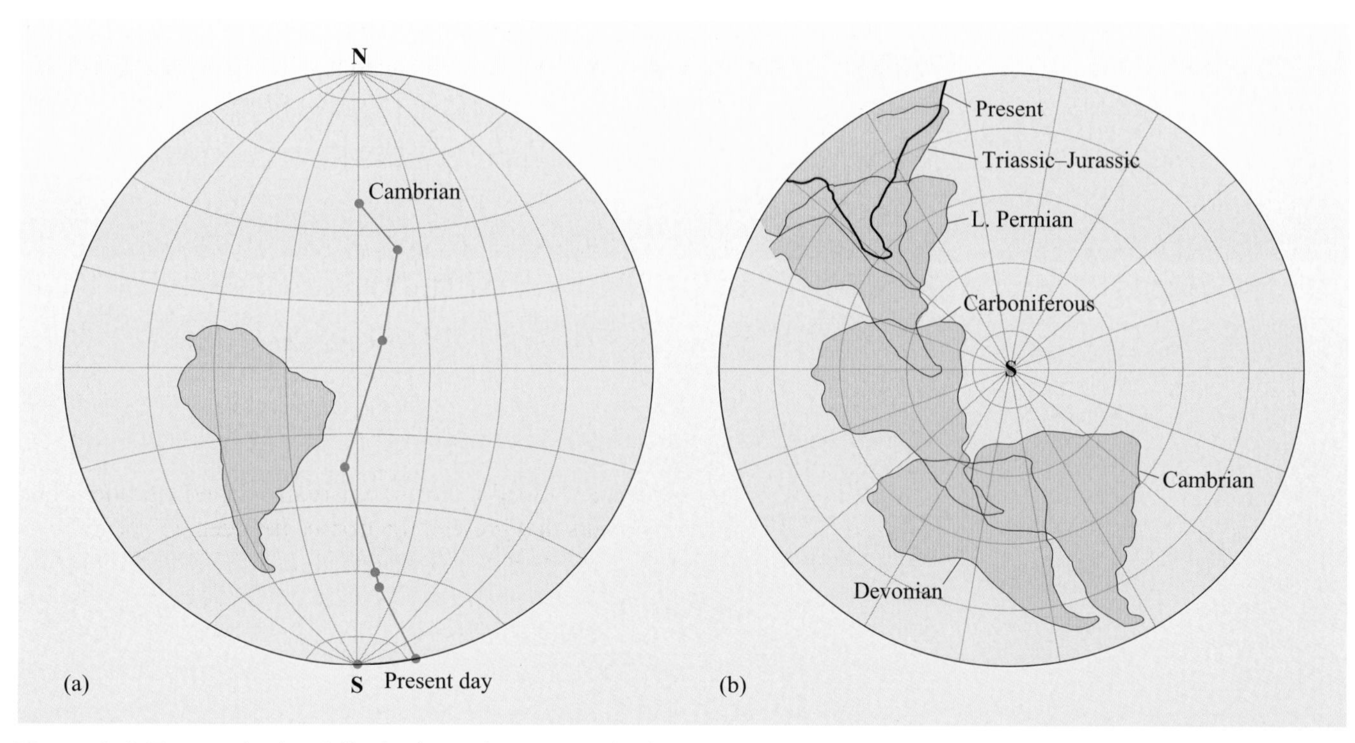

Figure 3.4 Two methods of displaying palaeomagnetic data: (a) assumes that the continent has remained fixed over time, and records the apparent polar wandering path of the South Pole; (b) assumes the magnetic poles are fixed over time, and records the latitude drift of a continent. (Adapted from Creer, 1965)

Nevertheless, APW paths remain a commonly used tool because they provide a useful method of comparing palaeomagnetic data from different locations. They are especially useful in charting the rifting and suturing of continents.

Figure 3.5a shows North America and Europe have individual apparent polar wander paths. However, they are broadly alike in that they have similar changes in direction at the same time. Figure 3.5b shows the APW paths if the Atlantic Ocean is closed by matching the continental shelves.

▼ **Figure 3.5** (a) Apparent polar wander paths for North America and Europe, as measured. (b) Apparent polar wander paths for North America and Europe with the Atlantic closed. Poles for successive geological periods are shown. (c) Apparent polar wander paths for Europe and Siberia. (Adapted from Mussett and Khan, 2000)

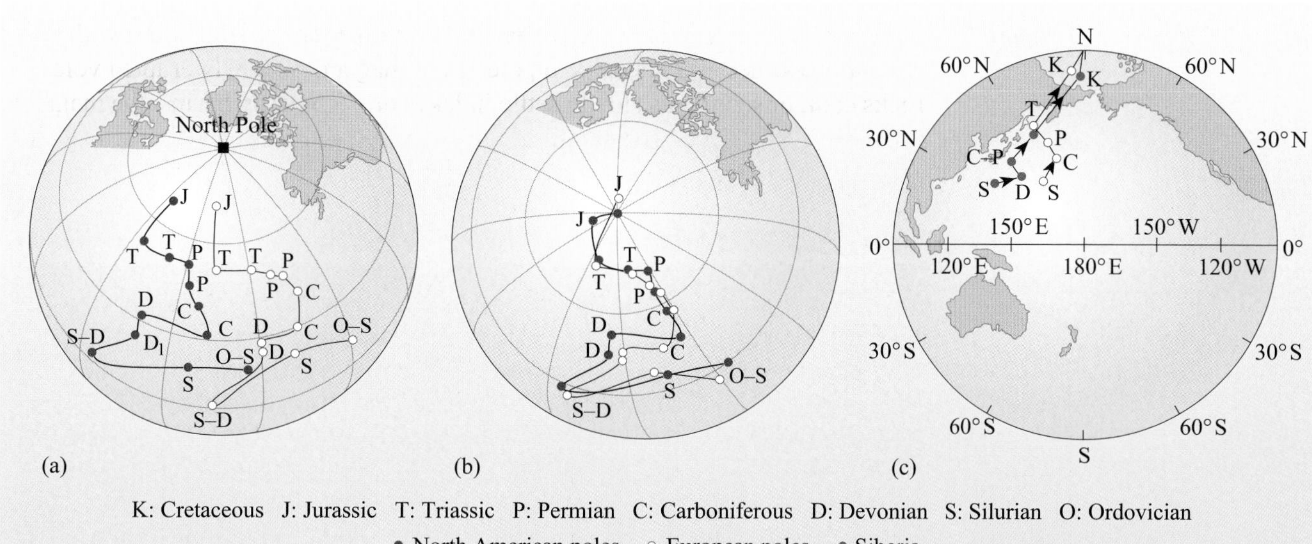

- What does this tell you about the North American and European continental masses during the periods spanned by these palaeomagnetic records?
- The two continents were moving together as one mass from the Ordovician right through to the opening of the Atlantic Ocean during the Jurassic Period.

Conversely, if the APW paths of two regions were different to begin with, but became similar later on, one explanation would be that the two regions were originally on independent land masses that then collided and subsequently began to move together as a single continental unit.

Question 3.2

What do the APW paths in Figure 3.5c tell you about the way in which Europe and Siberia have drifted from the Silurian Period to the present day?

Despite Wegener's amassed evidence and the increasing body of geological, palaeontological and palaeomagnetic information, there remained strong opposition to his theory of continental drift, leaving just a few forward-thinking individuals to continue seeking evidence to support this theory (Box 3.1).

The scientific opposition reasoned that if continents move apart, then surely they must either leave a gap at the site they once occupied or, alternatively, must push through the surrounding sea floor during their movement. The geophysicists of the day quickly presented calculations demonstrating that the continents could not behave in this way and, more importantly, no one could conceive of a physical mechanism for driving the continents in the manner Wegener had proposed. Consequently, the theory of continental drift did not gain scientific popularity at the time and became increasingly neglected for several decades. To gain a wider scientific acceptance, Wegener's ideas had to await a greater understanding of the internal structure of the Earth and the processes controlling the loss of its internal heat.

3.1.2 Sea-floor spreading

During and just after World War II, the technological improvement to instrumentation led to an improvement in underwater navigation and surveying that revealed many intriguing underwater features. The most important of these were immense, continuous chains of volcanic mountains running along the ocean basins. These features are now termed **mid-ocean ridges** or, more accurately, **oceanic ridge systems**.

Using this new information, three American scientists – Hess, Dietz and Heezen (Box 3.1) – proposed that the sea floor was actually spreading apart along the ocean ridges where hot magma was oozing up from volcanic vents. They further suggested that the oceanic ridges were the sites of generation of new ocean lithosphere, formed by partial melting of the underlying mantle followed by magmatic upwelling. They named the process **sea-floor spreading**. Moreover, they proposed that the topographic contrast between the ridges and the oceanic **abyssal plains** was as a consequence of the thermal contraction of the crust as it cooled and spread away from either side of the ridge axis. Most importantly, because new oceanic crust is generated at the ridge, the ocean must grow wider over time and, as a consequence, the continents at its margin move further apart. The evidence to support this model was found, once again, in the magnetic record of the rocks, but this time using rocks from the ocean floor.

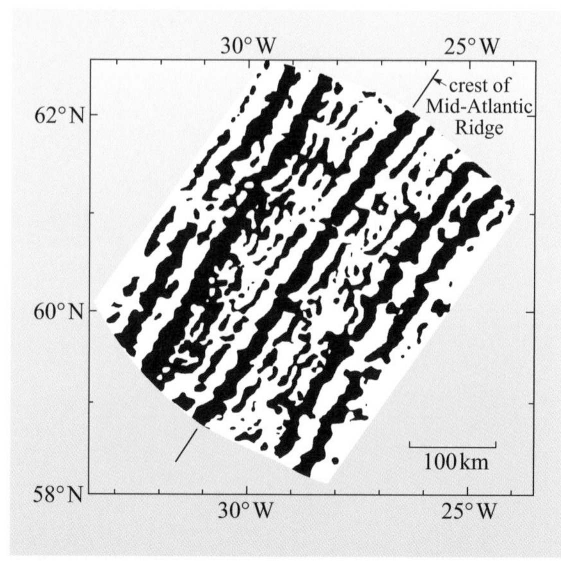

Figure 3.6 A modern map of symmetrical magnetic anomalies about the Atlantic Ridge (the Reykjanes Ridge), south of Iceland. (Adapted from Hiertzler et al., 1966)

Linear magnetic anomalies – a record of tectonic movement

At the time that sea-floor spreading was proposed, it was also known from palaeomagnetic studies of volcanic rocks erupted on land that the Earth's magnetic polarity has reversed numerous times in the geological past. During such **magnetic reversals**, the positions of the north and south magnetic poles exchange places. In the late 1950s, a series of oceanographic expeditions was commissioned to map the magnetic character of the ocean floor, with the expectation that the ocean floors would display largely uniform magnetic properties. Surprisingly, results showed that the basaltic sea floor has a striped magnetic pattern, and that the stripes run essentially parallel to the mid-ocean ridges (Figure 3.6). Moreover, the stripes on one side of a mid-ocean ridge are symmetrically matched to others of similar width and polarity on the opposite side.

In 1963, two British geoscientists, Vine and Matthews (Box 3.1), proposed a hypothesis that elegantly explained how these magnetic reversal stripes formed by linking them to the new idea of sea-floor spreading. They suggested that as new oceanic crust forms by the solidification of basalt magma, it acquires a magnetisation in the same orientation as the prevailing global magnetic field. As sea-floor spreading continues, new oceanic crust is generated along the ridge axis. If the polarity of the magnetic field then reverses, any newly erupted basalt becomes magnetised in the opposite direction to that of the earlier crust and so records the opposite polarity. Since sea-floor spreading is a continuous process on a geological timescale, the process preserves rocks of alternating polarity across the ocean floor (Figure 3.7a). Reading outwards in one direction from the mid-ocean ridge gives a record of reversals over time, and this can be matched with the record read in the opposite direction.

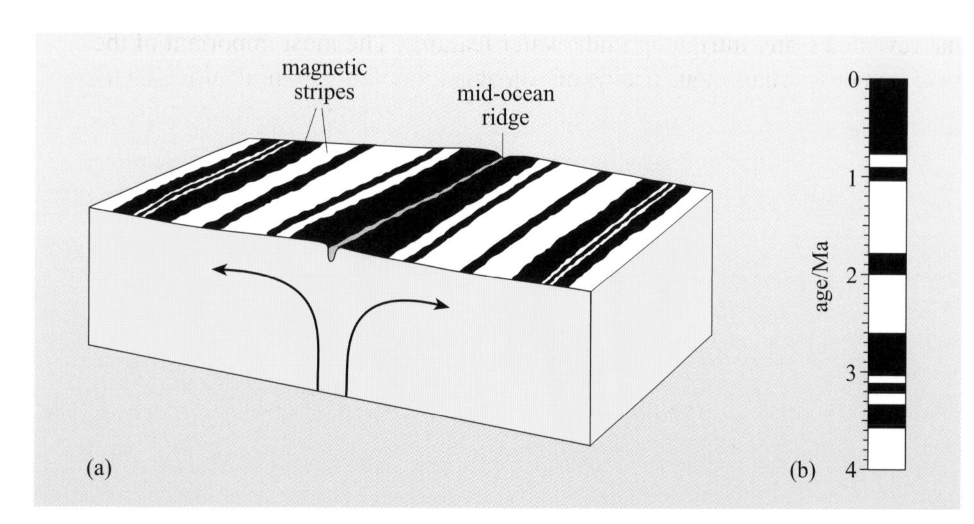

Figure 3.7 (a) Pattern of magnetic reversals on either side of a mid-ocean ridge. Black = normal magnetic field; white = reversed field. When these reversal data are combined with age data (derived by radiometric dating of rocks dredged from the sea floor), a geomagnetic timescale (b) can be produced. Detailed geomagnetic timescales have now been produced for all of the geological time since the Jurassic Period.

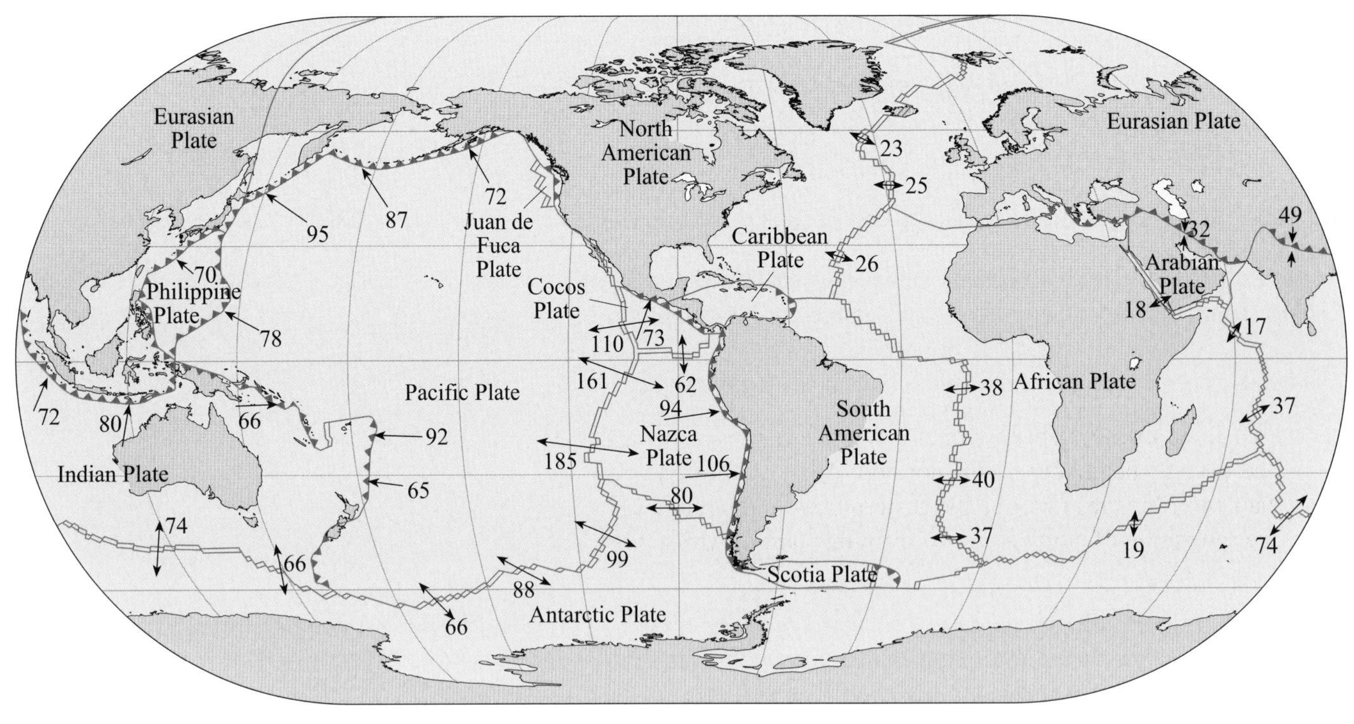

Figure 3.8 Map showing the global distribution of tectonic plates and plate boundaries. The black arrows and numbers give the direction and speed of relative motion between plates. Speed of motion is given in mm y^{-1}. (Adapted from Bott, 1982)

Magnetic and oceanographic surveys of the ocean floor have collected information on both its palaeomagnetic polarity and its absolute age (by radiometric dating of retrieved sea-floor samples). Combining these two records has helped establish a **geomagnetic timescale** (Figure 3.7b) and, by using samples from the oldest sea floor, this timescale has now been extended back into the Jurassic Period, allowing the ages and rates of sea-floor spreading to be established for all the world's oceans, as shown in Figure 3.8. (The different types of plate boundary shown in Figure 3.8 are discussed later in the text.)

Two measures of spreading rate are commonly cited:

- where the rate of spreading is determined on one side (i.e. the rate of movement away from the ridge axis), this is termed the **half spreading rate**
- where the rate is determined on both sides (i.e. the combined rate of divergence), the combined value is termed the **full spreading rate**.

■ Assuming symmetrical spreading rates, use the data given in Figure 3.8 to discover the maximum and minimum spreading rates, and half spreading rates for the ocean ridge system of (i) the Atlantic Ocean and (ii) the Pacific Ocean.

■ (i) The maximum and minimum spreading rates for the Atlantic Ocean are 40 mm y^{-1} and 23 mm y^{-1} (as shown by the double-ended arrows along the central Atlantic ridge); these represent half spreading rates of 20 mm y^{-1} and 11.5 mm y^{-1} respectively. (ii) The maximum and minimum spreading rates for the Pacific are 185 mm y^{-1} and 66 mm y^{-1} (as shown by the double-ended arrows along the ridge). These represent half spreading rates of 92.5 mm y^{-1} and 33 mm y^{-1} respectively.

Question 3.3

The width of ocean floor between the spreading ridge in the South Atlantic Ocean at 30° S and the edge of the continental shelves along the east coast of South America and the west coast of southern Africa at 30° S is approximately 3100 km and 2700 km respectively. Assuming that the spreading rate on this segment of the ridge is 38 mm y^{-1}, estimate the maximum age of the sea floor on either side of the South Atlantic.

Magnetic stripes not only tell us about the age of the oceans, they can also reveal the timing and location of initial continental break-up. The oldest oceanic crust that borders a continent must have formed after the continent broke apart initially, and just as sea-floor spreading began. In effect, it records the age when that continent separated from its neighbour. In the northern Atlantic, for example, oceanic crust older than 140 Ma is restricted to the eastern USA and western Saharan Africa, therefore separation of North America from this part of Africa must have commenced at this time. The oldest oceanic crust that borders South America and sub-equatorial Africa is only about 120 Ma old. Accordingly, it follows that the North Atlantic Ocean started to form before the South Atlantic Ocean.

If new sea floor is being created at spreading centres, then old sea floor must be being destroyed somewhere else. The oldest sea floor lies adjacent to deep ocean trenches, which are major topographic features that partially surround the Pacific Ocean and are found in the peripheral regions of other major ocean basins. The best known example is the Marianas Trench where the sea floor plunges to more than 11 km depth. Importantly, ocean trenches cut across existing magnetic anomalies, showing that they mark the boundary between lithosphere of differing ages. Once this association had been recognised, the fate of old oceanic crust became clear – it is cycled back into the mantle, thus preserving the constant surface area of the Earth.

... and on to plate tectonics

The combination of evidence for continental drift with the increasing evidence in favour of sea-floor spreading finally led to the development of plate tectonics (Box 3.1). Ideas developed in the 1960s and the 1970s have survived largely unaltered to the present day, albeit modified by more sophisticated data and modelling methods.

3.2 The theory of plate tectonics

The surface of the Earth is divided into a number of rigid plates that extend from the surface to the base of the lithosphere. A plate can comprise both oceanic and continental lithosphere. As you already know, continental drift is a consequence of the movement of these plates across the surface of the Earth. Thus the need for the continents to plough through the surrounding oceans is removed, as is the problem of the gap left in the wake of a continent as it drifts – both issues that led to scientific opposition to Wegener's ideas (Section 3.1.1).

The theory of plate tectonics is based on several assumptions, the most important of which are:

1 New plate material is generated at ocean ridges, or **constructive plate boundaries**, by sea-floor spreading.

2 The Earth's surface area is constant, therefore the generation of new plate material must be balanced by the destruction of plate material elsewhere at **destructive plate boundaries**. Such boundaries are marked by the presence of deep ocean trenches and volcanic island arcs in the oceans and, when continental lithosphere is involved, mountain chains.

3 Plates are rigid and can transmit stress over long distances without internal deformation – relative motion between plates is accommodated only at plate boundaries.

As a consequence of these three assumptions, and particularly the third assumption, much of the Earth's geological activity, especially seismic and volcanic, is concentrated at plate boundaries (Figure 3.9). For example, the position of the Earth's constructive, destructive and conservative plate boundaries can be mapped largely on the basis of seismic activity. However, it is not enough just to know where boundaries are. In order to understand the implications that plate tectonics has for Earth evolution and structure, you first need to explore the structure of lithospheric plates, their motion – both relative and real – and the forces that propel the plates across the Earth's surface.

3.2.1 What is a plate?

In order to understand how and why the lithospheric plates move it is first necessary to understand their physical and thermal structure. In Chapter 1 you learned that:

- the Earth can be divided into the core, mantle and crust based on its physical and chemical properties
- the lithosphere comprises the Earth's crust and the upper, brittle part of the mantle.

The thickness of the lithosphere is variable, being up to 120 km thick beneath the oceans; it is considerably thicker beneath ancient continental (cratonic) crust. However, the thermal structure of a plate is best illustrated with reference to the ocean basins and how their thermal characteristics change with time.

3.2.2 Heat flow within plates

As newly formed lithosphere moves away from an oceanic ridge, it gradually cools and heat flow (Box 3.2) decreases away from constructive plate boundaries.

■ If a body cools, what happens to its density?

■ It contracts, so its volume decreases, resulting in an increase in density.

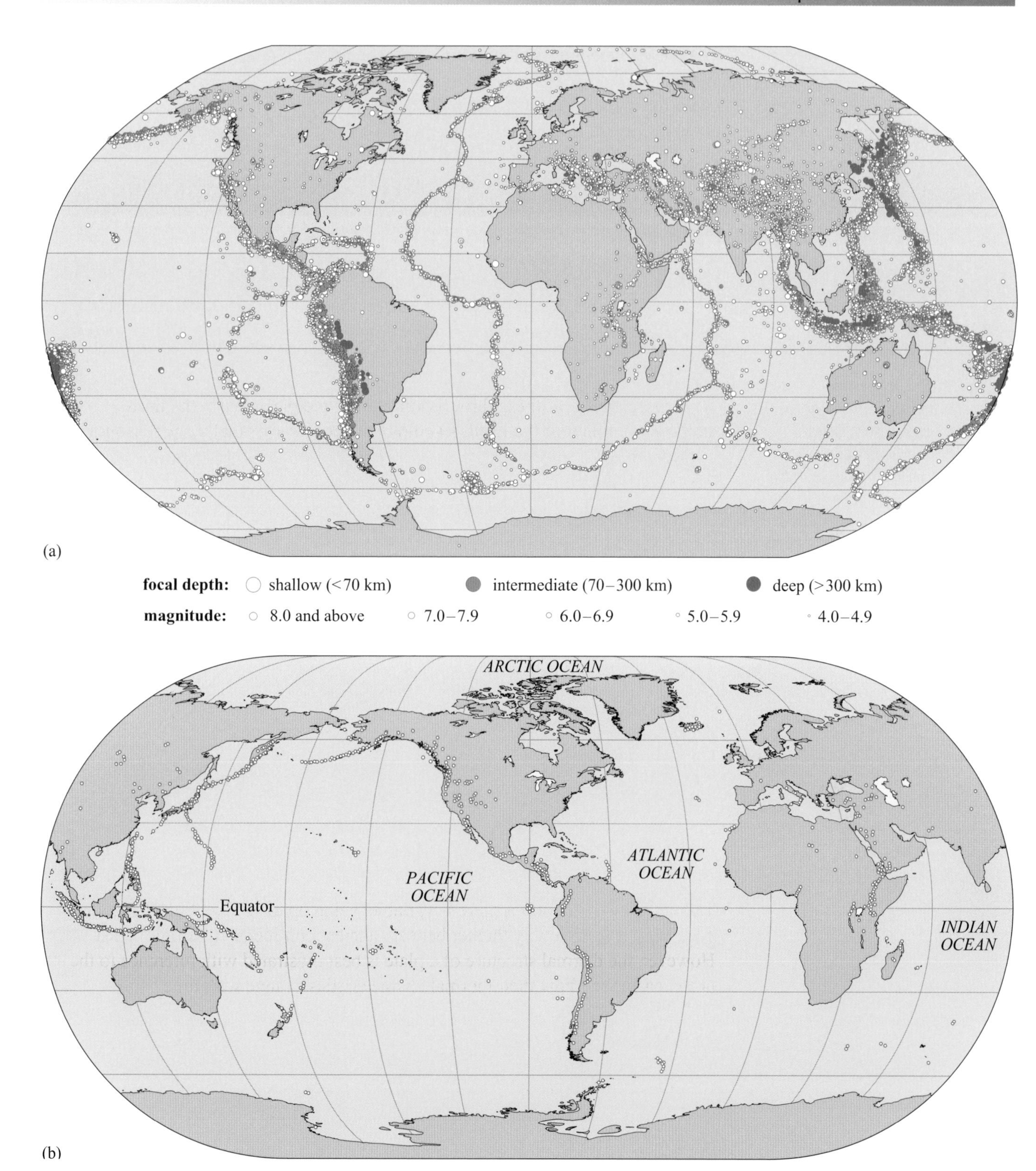

Figure 3.9 (a) Global earthquake epicentres between 1980 and 1996. Only earthquakes of magnitude 4 and above are included. (b) Locations of active, sub-aerial volcanoes. Enlarged versions of Figures 3.9a and b are in Appendix D. ((a) BGS; (b) adapted from Johnson, 1993)

The cooling and the shrinking of the lithosphere result in an increase in its density and so, as a result of isostasy (Section 1.5.1), it subsides into the asthenosphere and ocean depth increases away from the ridge, from about 2–3 km at oceanic ridges to about 5–6 km for abyssal plains. Indeed, one of the more remarkable observations of ocean-floor bathymetry is that ocean floor of similar age always occurs at similar depths beneath sea level (Figure 3.10). The relationship between mean oceanic depth (d in metres) and lithosphere age (t in Ma) can be expressed as:

$$d = 2500 + 350t^{1/2} \quad (3.1)$$

If the depth of the ocean floor can be determined, then the approximate age of the volcanic rocks from which it formed may also be estimated, and vice versa.

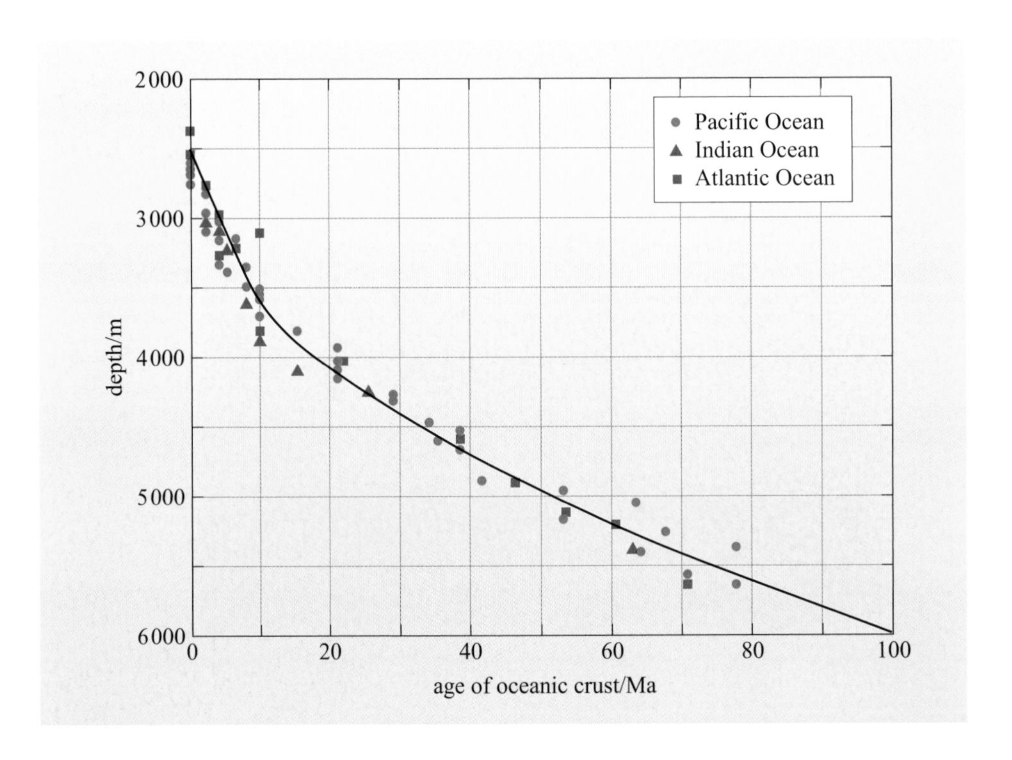

Figure 3.10 Observed relationship between depth of the ocean floor (for both ocean ridges and abyssal plains) and age of formation of the oceanic crust for the Pacific, Indian and Atlantic Oceans. The solid curve shows the relationship between age and ocean depth according to Equation 3.1.

Question 3.4

The ocean depth at a distance of 1600 km from the Mid-Atlantic Ridge is 4700 m.

(a) Calculate:

(i) the age of the crust at this location

(ii) the mean spreading rate represented by this age.

(b) Is this a half or a full spreading rate?

Box 3.2 Earth's heat flow

As you know from Section 2.1, the sources of the Earth's internal heat are:

- heat remaining from the initial accretion of the Earth
- gravitational energy released from the formation of the core
- tidal heating
- radiogenic heating within the mantle and crust.

Although the proportion of each heat source cannot be determined accurately, radiogenic heat is considered to have been the major component for much of the Earth's history. There are three main processes by which this internal heat gets to the Earth's surface, these being conduction, convection and advection (Section 2.1.2).

Heat flow (or heat flux), q, is a measure of the heat energy being transferred through a material (measured in units of watts per square metre; W m^{-2}). It may be determined by taking the difference between two or more temperature readings (ΔT) at different depths down a borehole (d), and then determining the thermal conductivities (k) of the rocks in between. The values of q can then be calculated according to the relationship:

$$q = \frac{k\Delta T}{d} \tag{3.2}$$

Earth scientists are interested in the heat flow measured at the Earth's surface because it reveals important information concerning the nature of the rocks and the processes that affect the lithosphere.

The total annual global heat loss from the Earth's surface is estimated as 4.1–4.3 × 10^{13} W. This yields an average of $q \geq 100$ mW m^{-2} (milliwatts per square metre), although individual measurements may be much higher than this. However, values of q decrease to less than 50 mW m^{-2} for oceanic crust older than 100 Ma (Figure 3.11). In continental areas, the younger crust (i.e. mountain belts that are less than 100 Ma) have relatively high values of q, which are 60–75 mW m^{-2}, while old continental crust and **cratons** have much lower heat flow values, averaging q = 38 mW m^{-2}. Thus, variations in heat flow are closely related to different types of crustal materials and, importantly, different types of tectonic plate boundary.

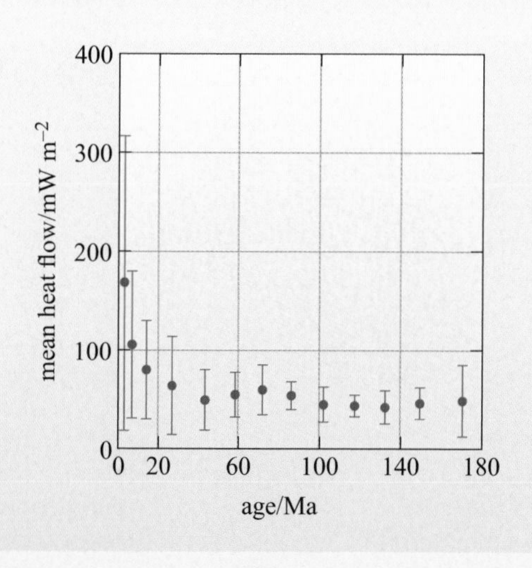

Figure 3.11 Mean heat flow (q) and associated standard deviation (vertical lines) plotted against the age of the oceanic lithosphere for the North Pacific Ocean.

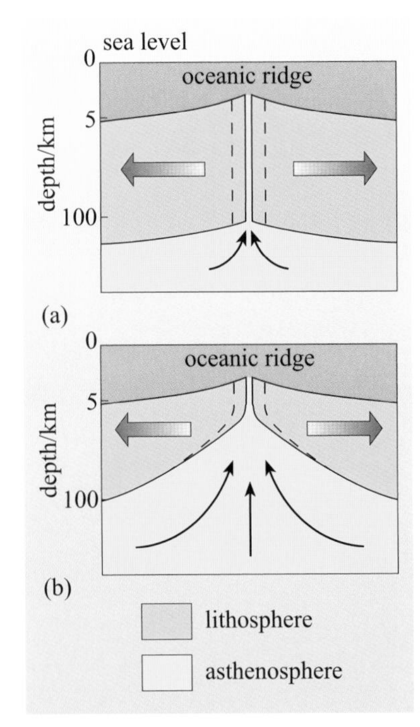

Figure 3.12 Schematic sections through oceanic lithosphere formed at an oceanic ridge: (a) plate model, in which oceanic lithosphere thickness remains constant as the lithosphere moves away from the ridge; (b) boundary-layer model, in which the lithosphere thickens as it ages and cools. The dashed lines show where new lithosphere forms in both models.

There are two general models for the thermal evolution of the oceanic lithosphere: the plate model and the boundary-layer (or half-space) model.

- The **plate model** (Figure 3.12a) assumes that the lithosphere is produced at a mid-ocean ridge with constant thickness and that the temperature at the base of the plate corresponds to its temperature of formation.
- The **boundary-layer model** (Figure 3.12b) assumes that the lithosphere does not have a constant thickness, but thickens and subsides as it cools and moves away from the ridge. This is achieved by loss of heat from the underlying asthenosphere, which progressively cools below the temperature at which it can undergo solid-state creep and is transformed from asthenosphere to lithosphere.

The thermal consequences of these two models can be calculated from a knowledge of the temperature of the mantle at depth (the geotherm) and the thermal conductivity of the rocks in the lithosphere. As it turns out, both models predict similar results for heat flow and ocean depth for young lithosphere but deviate for older lithosphere, as shown in Figure 3.13.

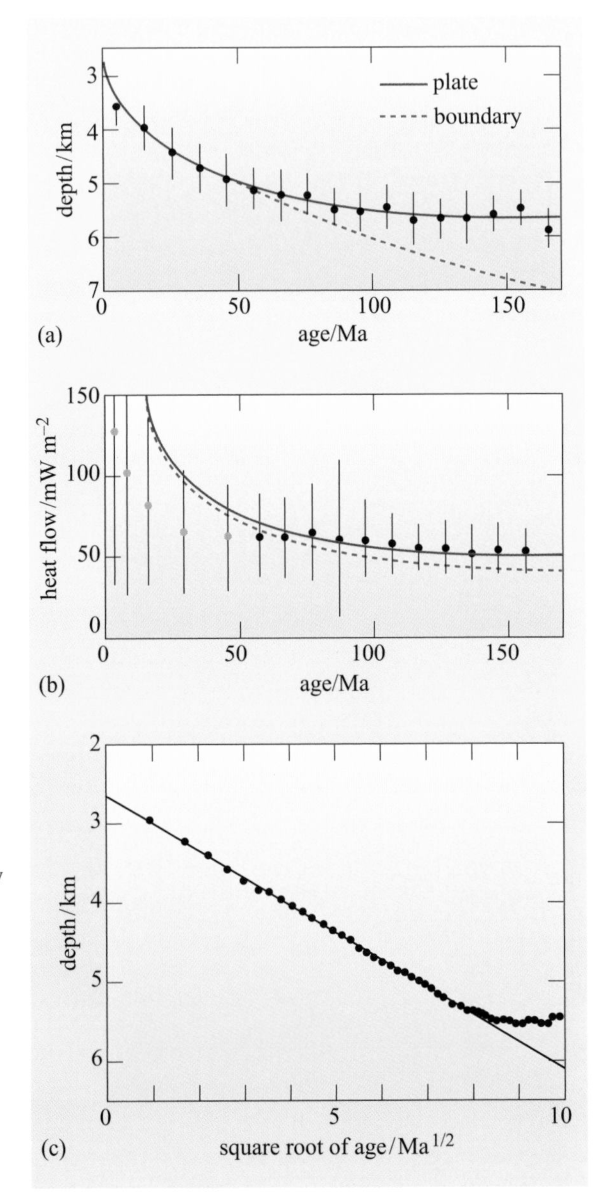

Figure 3.13 Graphical plots of (a) depth and (b) heat flow against age for the Pacific Ocean. The line labelled plate refers to the plate model, whereas the curve labelled boundary refers to the boundary layer model. (c) Graphical plot of ocean bathymetry (depth) against the square root of the age of the lithosphere for the Pacific Ocean. Note the good linear relationship for young lithosphere and the deviation from the simple linear trend for older lithosphere. (Stein and Stein, 1992)

Question 3.5

Study Figure 3.13c and then answer the following questions.

(a) Is the correlation between depth and the square root of the age of the lithosphere positive or negative?

(b) At what age does the relationship depart from a linear correlation?

(c) Does this departure imply that older oceanic lithosphere is warmer or cooler than predicted from the simple, linear boundary-layer model?

Both models predict the observed linear variation between ocean depth and the square root of age of the lithosphere, showing that the oceanic lithosphere cools and subsides as it ages away from a spreading centre. However, the plate model fits the data better for older lithosphere (>60 Ma), suggesting that once lithosphere has cooled to a certain thickness, the thickness remains more or less constant until the plate is subducted.

Both models also predict greater heat flow from young oceanic crust than that observed in the ocean basins, as shown in the example in Figure 3.13b.

- What do you think the cause of this difference might be?
- The thermal models are based on the assumption that heat is lost by conduction only, whereas in reality other mechanisms of heat transfer might be in operation.

The formation of oceanic lithosphere involves contact between hot rocks and cold seawater. As the rocks cool and fracture, they allow seawater to penetrate the young, hot crust to depths of at least a few kilometres. During its passage through the crust the seawater is heated before being cycled back to the oceans. This process is known as **hydrothermal circulation** and the development of submarine hydrothermal vents ('black smokers') close to mid-ocean ridges is the most dramatic expression of this heat transfer mechanism. However, less dramatic but probably equally significant lower temperature circulation continues well beyond the limits of oceanic ridges and contributes to heat loss from the crust up to 60 Ma after formation.

The geophysical evidence from seismology and isostasy suggests that the oceanic lithosphere increases in thickness as it ages until it reaches a maximum of about 100 km. By contrast, bathymetry and heat flow indicate a more constant thickness for older plates. To explain these observations, a plate structure such as that shown in Figure 3.14 (overleaf) has achieved wide acceptance. The plate is divided into two layers: an upper, rigid **mechanical boundary layer** and a lower, viscous **thermal boundary layer**. Both layers thicken progressively with time until the lithosphere is about 80 Ma old, after which the thermal boundary layer becomes unstable and starts to convect. Convection within this layer provides a constant heat flow to the base of the mechanical boundary layer. Thus old lithosphere maintains a constant thickness because the heat flow into its base, and hence the basal temperature, is maintained by convection.

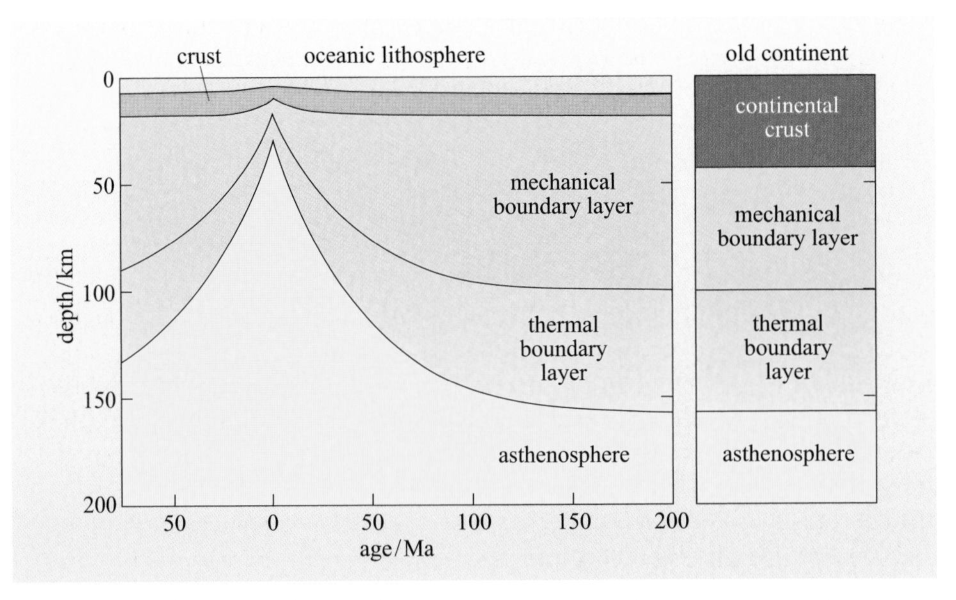

Figure 3.14 Thermal model of lithospheric plates beneath oceans and continents. Note the thicker crust and thinner mantle lithosphere in the continental section. (Fowler, 2005)

3.2.3 Constructive plate boundaries

Constructive plate boundaries or margins are regions where new oceanic crust is being generated. However, in order for the magma to ascend to the surface and build new lithosphere, the earlier formed crust must be pulled apart and fractured to create a new magma pathway. Hence constructive plate boundaries are regions of extensional stresses and **extensional tectonics**. The process of fracturing, injection and eruption is repeated frequently, so that tensional stresses do not have time to accumulate significantly and, as a result, constructive plate boundaries are characterised by frequent, low-magnitude seismicity (typically less than magnitude 5), occurring at shallow crustal depths (<60 km) along the ocean ridge systems.

- Why do you think earthquakes are restricted to shallow depths beneath ocean ridges?

- Earthquakes can only occur in brittle rocks that fail by fracture rather than solid-state creep. As a result of high geothermal gradients beneath oceanic ridges, the brittle lithosphere is thin and so earthquakes are restricted to shallow depths.

Sonar surveying, and direct investigation by sea-floor drilling or deep sea submersibles, has revealed that volcanism along the ridge systems typically consists of a series of individual, active eruption centres. Each eruptive centre is no more than about 2–3 km long, and along the ridge axis they are often separated from each other by an inactive gap of about 1 km. Beneath the spreading ridge the feeder magma chambers that supply the volcanic centres are more continuous, often linking between and across the active segments. This means that magma generation occurs along much of the length of the ridge even though it is erupted via a chain of individual volcanic centres.

Plates move away from constructive boundaries at speeds that can be as low as <10 mm y^{-1} to so-called ultra-fast spreading ridges where half spreading rates can exceed 100 mm y^{-1}. Examples of slow spreading ridges include parts of the mid-Atlantic Ridge and the southwest Indian Ridge. Fast and ultra-fast ridges occur in the East Pacific, along the East Pacific Rise and the Galápagos spreading centres (Figure 3.8).

The depth structure of a constructive plate boundary can be further defined from the variation in the Earth's gravity (Box 3.3). Figure 3.15 shows gravity anomalies across the Mid-Atlantic Ridge. Despite the topographic rise associated with the ridge, the **free-air gravity anomaly** is relatively flat and close to zero across the whole structure. This indicates that there is no mass deficit or excess down to the level of isostatic compensation, i.e. the ridge is in isostatic equilibrium with the lithosphere of the abyssal plains. By contrast, when the free-air anomaly is corrected for the effects of the ridge topography and overall altitude, the resulting **Bouguer gravity anomaly** is strongly positive but with a local dip across the ridge axis. The positive anomaly occurs because of the raised topography of the ridge, but the ridge zone itself is underlain by lower-density material. A possible density model is shown in Figure 3.15b.

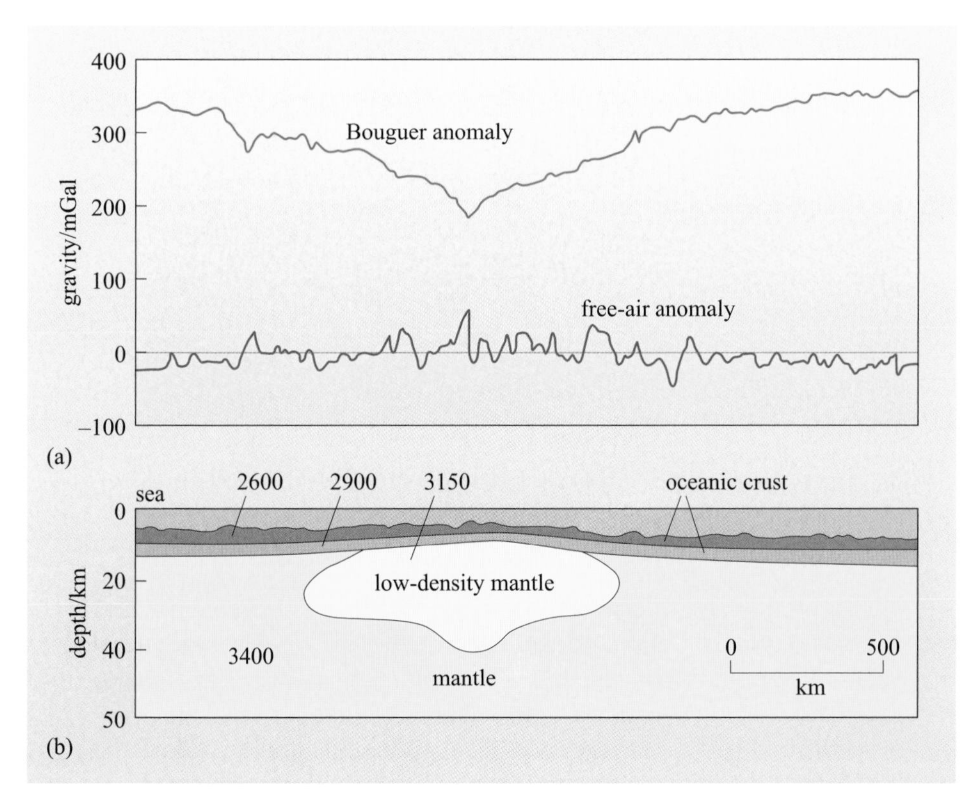

Figure 3.15 (a) Bouguer and free-air gravity anomalies across part of the Mid-Atlantic Ridge. (b) One possible density model that could produce the observed anomalies and satisfies other constraints (e.g. seismic structure). Numbers give the densities of different layers in kg m^{-3}.

■ What do you think the low-density material beneath the ridge might represent?

■ Hot, possibly partially molten mantle that feeds the basaltic volcanism of the ridge axis.

Box 3.3 Gravity and gravity anomalies

Gravity is the attractive force experienced by all objects simply as a consequence of their mass. The magnitude of the attraction is determined from Equation 3.3:

$$F = \frac{Gm_1m_2}{r^2} \quad (3.3)$$

where m_1 and m_2 are the masses (measured in kg) of two objects, r is the distance (measured in m) between them and G is the universal gravitational constant (6.672×10^{-11} N m^2 kg^{-2}).

If one object is the Earth, with mass M, and the other is a much smaller object with mass m, Equation 3.3 can be rewritten:

$$F = \frac{GMm}{d^2} \quad (3.4)$$

where d is the radius of the Earth, i.e. the distance to the Earth's centre of gravity.

However, the force experienced by an object at the surface of the Earth can be measured:

$$F = mg \quad (3.5)$$

where m is its mass and g is the acceleration due to gravity. Hence:

$$F = mg = \frac{GMm}{d^2} \quad (3.6)$$

Cancelling m gives:

$$g = \frac{GM}{d^2} \quad (3.7)$$

Hence g is proportional to the mass of the Earth and inversely proportional to the square of the distance from the centre of the Earth. Note that Equation 3.7 gives a way of measuring the mass of the Earth (M) if g can be measured and G and d are known.

■ Given that the Earth is not truly spherical and that the poles are closer to the centre than the Equator is, will g at the poles be greater than, less than or similar to g at the Equator?

■ Because the poles are closer to the centre of the Earth than the Equator, d^2 in Equation 3.7 will be lower, therefore g will be slightly greater.

Variations in gravity are measured in milligals (mGal) and 1 mGal is equivalent to 10^{-5} m s^{-2}. Since $g = 9.81$ m s^{-2}, this means that 1 mGal ~ 10^{-6} g.

Measured variations in gravity across the surface of the Earth relate to the mass in the vicinity of the point of measurement. Thus a region underlain by dense rocks, such as basalt, will exhibit a slightly stronger gravitational pull than those underlain by less dense rocks, such as granite or sediments. In addition, gravity is affected by the underlying topography and the altitude at which the measurement was made. Thus measurements of the variation in the Earth's gravitational field require numerous corrections for latitude and topography, the details of which are beyond the scope of this book (but see the texts recommended in the Further Reading section). The resultant gravitational anomaly is known as a Bouguer anomaly and this reflects the variations in the Earth's gravity due to the underlying geology. Bouguer gravity is usually calculated over continental regions where the surface topography and the underlying geology are both highly variable. In marine surveys, fewer corrections are applied to the measured value of g, and gravitational anomalies are conventionally referred to as free-air anomalies. Gravity measurements, particularly over the oceans, are now routinely recorded from satellites and have resulted in accurate and detailed maps of the free-air anomaly over most of the Earth.

Hot material is generally less dense than cold material, so the low density of the mantle beneath the ridge is related to the locally high geothermal gradient, as also indicated by the presence of basaltic magma and the restriction of earthquakes to the upper levels of the lithosphere.

Finally, it should be noted that constructive plate boundaries by definition cannot occur within continental lithosphere as they must be bounded by new oceanic lithosphere. There are regions of the Earth's crust where a constructive boundary (or boundaries) can be traced into a continental region, for example at the southern end of the Red Sea and the Gulf of Aden the marine basins join and extend into the Ethiopian segment of the African Rift Valley by way of the Afar Depression. While these three features are all part of an extensional tectonic regime, the African Rift Valley cannot be considered to be a true constructive plate boundary, although, in future, if plate configurations are suitable, it may provide the site for the opening of a new ocean.

3.2.4 Destructive plate boundaries

Destructive plate boundaries are regions where two lithospheric plates converge. This situation provides a more varied range of tectonic settings than do constructive plate boundaries. Firstly, and in contrast to constructive plate boundaries, destructive plate boundaries are asymmetrical with regard to plate speeds, age and large-scale structures. Secondly, whereas true constructive boundaries occur almost invariably in oceanic lithosphere, destructive boundaries also affect continental lithosphere – they can occur entirely within continental lithosphere. Consequently, there are three possible types of destructive plate boundary:

- those involving the convergence of two oceanic plates (ocean–ocean subduction)
- those where an oceanic plate converges with a continental plate (ocean–continent subduction)
- collisions between two continental plates (continent–continent destructive boundaries).

These can be thought of as representing three stages in the evolution of destructive boundaries.

In addition to the disappearance of old lithosphere, destructive boundaries associated with ocean–ocean subduction and ocean–continent subduction are also characterised by:

- ocean trenches, generally 5–8 km deep, but sometimes up to 11 km deep. The sea floor slopes into the trenches from both the landward and the oceanward sides. They are continuous for many hundreds of kilometres, occurring both adjacent to continents and wholly within oceans
- a belt of earthquakes that are shallow-centred closest to the trench and deeper further away. Earthquakes can occur as deep as 600–700 km
- most destructive boundaries are associated with a belt of active volcanoes that, in the case of intra-oceanic boundaries, form chains of islands known as **island arcs**.

1 Ocean–ocean (island-arc) subduction

The convergence of two oceanic plates represents the simplest type of destructive plate boundary and exemplifies most of the features associated with the destruction of oceanic lithosphere. Around the northern and western edges of the Pacific Ocean, many islands are arranged in gently curved archipelagos: anticlockwise these include the Aleutian Islands, the Kuril Islands, Japan, the Mariana Islands, the Solomon–New Hebrides archipelagos and the Tonga–Kermadec Islands north of New Zealand. These all occur some distance off the edge of the continental areas, but lie adjacent to a deep ocean trench. To the oceanward side of the deep trenches the oceanic lithosphere is among the oldest on Earth. For example, the oceanic crust adjacent to the Marianas Trench, the deepest trench on Earth, is Jurassic in age and up to 180 Ma old. The trenches are sites where old oceanic lithosphere is being destroyed, or **subducted**, beneath younger lithosphere. For this reason, destructive boundaries are often referred to by their alternative name of **subduction zones**.

The typical pattern of earthquakes associated with ocean–ocean subduction is well illustrated in Figure 3.16, which shows the distribution of earthquakes associated with the Tonga Trench in the southwest Pacific Ocean. The earthquake data summarised in this diagram clearly define a zone of earthquakes deepening to the west away from the Tonga Trench, beneath the Tonga volcanic islands, and reaching a final depth in excess of 600 km. This inclined plane of earthquakes associated with the Tonga Trench (and every other deep ocean trench) in known as a **Wadati–Benioff zone**, after the first seismologists to recognise its existence.

■ Using the information in Figure 3.16, estimate the angle of the Wadati–Benioff zone beneath the Tonga island arc at ~20° S. (Assume the first occurrence of >400 km earthquakes along this line is representative of earthquakes with a minimum depth of 400 km.)

■ Earthquakes at about 400 km depth are located ~400 km west of the Tonga Trench. Hence the tangent of the angle of subduction is 400/400 = 1, so the angle of subduction is ~45° (tan 45° = 1).

■ Is the angle constant along the whole length of the Tonga Trench?

■ No. To the north at ~16° S the horizontal distance between the trench and the deepest earthquakes is much greater than at 20° S, so the angle of subduction is shallower.

The presence of earthquakes at great depth in Wadati–Benioff zones shows that rocks are undergoing brittle failure throughout this depth range.

■ What does this observation tell you about the thermal state of subducted lithosphere?

■ Subducted lithosphere must remain cool to fail seismically.

Subduction involves the recycling of old, and therefore cold, oceanic lithosphere back into the mantle. A common misconception is that earthquakes represent failure between the subducted lithosphere and the overlying mantle. While this

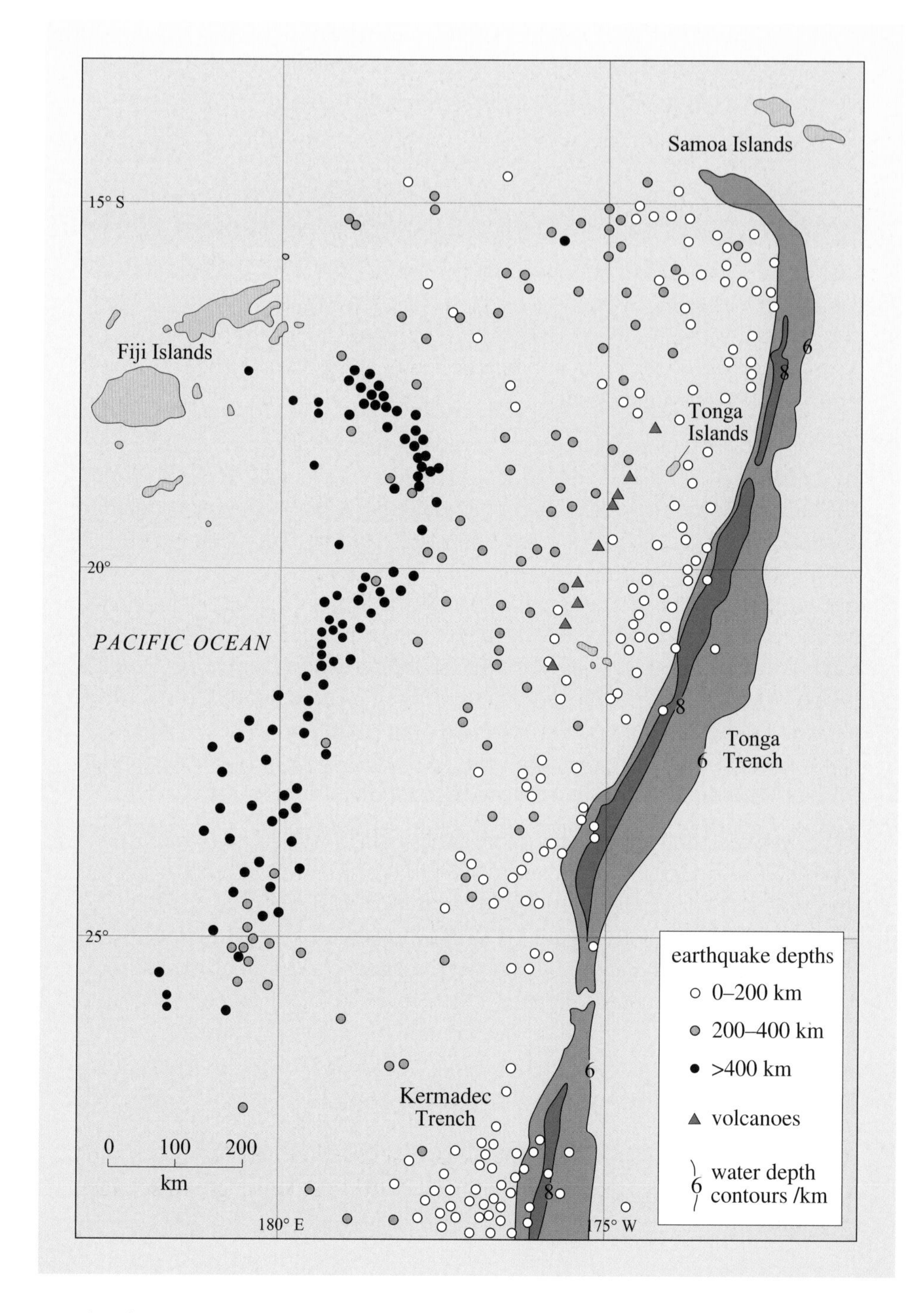

Figure 3.16 Map showing the distribution of earthquake foci associated with the Tonga Trench in the SW Pacific Ocean. (Sykes, 1966)

may be the case for the shallowest levels of subduction, analysis of earthquake waves from deeper seismic events indicates that earthquake foci lie *within* the subducted lithosphere and reflect differential movement between rigid blocks in the subducted lithosphere as it heats up and expands.

The fact that the subducted slab remains colder than ambient mantle at a comparable depth raises a problem for the generation of volcanic activity – a diagnostic feature of subduction. If subduction zones are regions where cold material is being cycled back into the mantle, why are they the site of volcanic activity? Constructive plate boundaries are underlain by regions of hot mantle that rises in response to the separation of the overlying plates and so it is easy to see how magma can be generated from the mantle. Beneath island arcs there is less evidence for hot mantle and so melting must be triggered by some alternative mechanism. Seismic profiles of subduction zones show that melt is generated immediately above the subducted plate, and most volcanic arcs are located approximately 100 km above its surface, so there is clearly a relationship between subducted oceanic lithosphere and the presence of island-arc volcanism.

The link between subduction and volcanism lies in the composition of the subducted lithosphere.

- ■ In addition to basaltic crust and mantle rocks, what other rocks would you expect in the subducted plate?
- ■ As the plate ages it accumulates a veneer of sediments on its surface. Also hydrothermal processes close to the ridge add water to the upper layers of the basaltic crust. So the subducting plate will include sediments and altered basaltic crust.

Subduction provides a mechanism for introducing water-bearing sediments into the mantle, and as the subducted lithosphere heats up the water is gradually released. Water has the effect of reducing the melting temperature of the mantle. It is this process that allows the generation of magma at depth that feeds surface volcanism. As a result, subduction-related magmas are also richer in volatiles than similar rocks from other tectonic environments, such as constructive plate boundaries. This subject will be covered in more detail in Chapter 5.

All of the above characteristics are more or less diagnostic of an oceanic destructive plate boundary. There are, however, a number of other structural features that may or may not be present, but reflect different processes associated with subduction. Figure 3.17b shows the mean ocean depth across the Kuril Trench in the NW Pacific Ocean. On the oceanward side of the Kuril Trench, as with all deep ocean trenches, the ocean depth is between 4 km and 6 km, whereas the trench is 2–4 km deeper still. Note that the vertical scale has been exaggerated fifty times in Figure 3.17b and the actual angles of the sides of the trench are quite shallow, being between 20° and 5°.

- ■ What happens to ocean depth immediately to the ocean side of the trench?
- ■ It decreases slightly.

The decrease in ocean depth towards the trench is characteristic of all island-arc systems and can elevate the ocean floor by as much as 0.5 km. It is caused by the flexure of the lithosphere in response to its entry into the subduction zone and is known as a **flexural bulge**. It is analogous to flexing a ruler over the edge of a table. If you place an ordinary plastic ruler on the edge of a table so that about one-third of it protrudes over the edge and then apply pressure to the extreme tip of the ruler while holding the other end firmly on the table, the ruler will flex and the part that was lying flat on the table will rise slightly. When the pressure is released the ruler will return to its original position because it is a rigid but elastic material. (Imagine trying this experiment with plasticine.)

The flexural bulge is a common feature of ocean-trench systems and is marked by a small increase in free-air gravity, while the trench itself is marked by a large

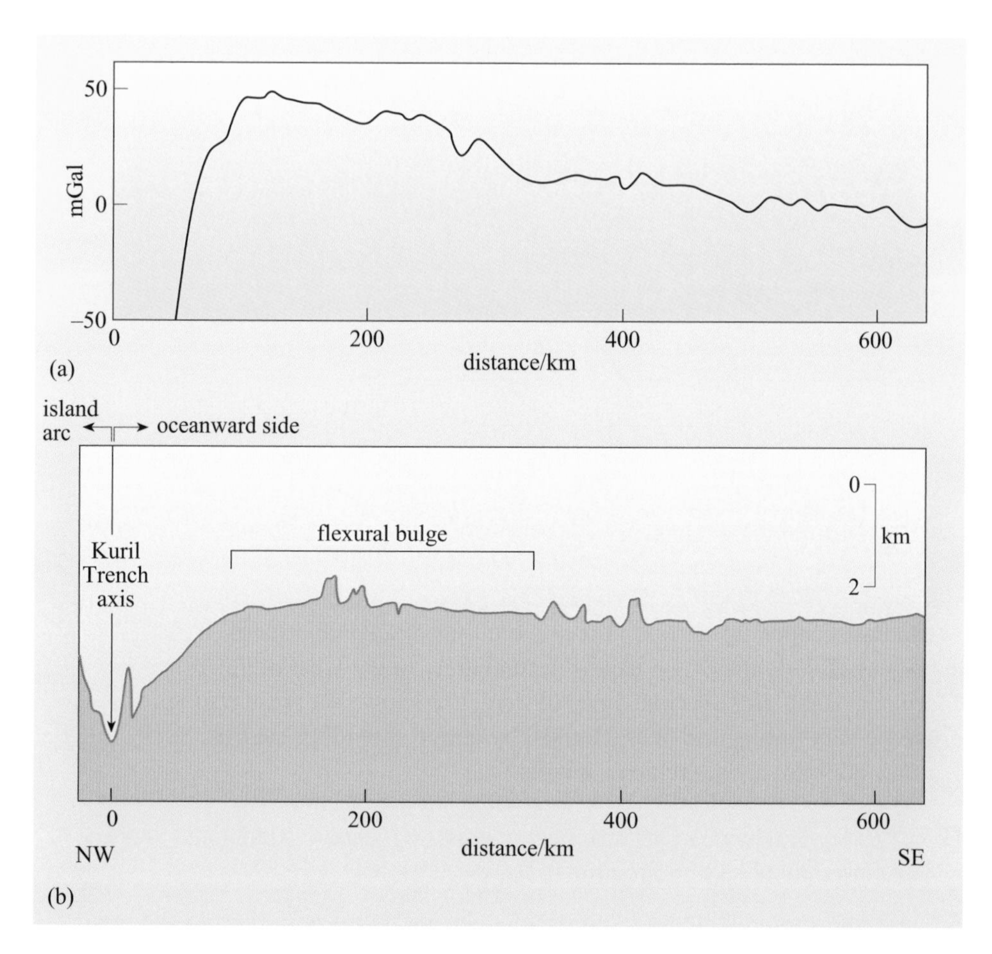

Figure 3.17 (a) Free-air gravity variations and (b) topography across the Kuril Trench. (Adapted from Watts, 2001)

decrease in free-air gravity (Figure 3.17a). Such gravity variations imply that arc systems are out of isostatic equilibrium – the negative anomaly over the trench reflects a mass deficit, meaning that the crust in the trench must be being held down, while the increase over the flexural bulge implies that it is underlain by dense material beneath the plate. The interpretation is that as the plate flexes upwards it 'pulls in' the asthenospheric mantle beneath. The isostatic imbalance, with the trench held down and the bulge supported, is due largely to the rigidity of the subducting plate.

While many ocean trenches are particularly deep, others are not. However, they are still characterised by a strongly negative free-air gravity anomaly, implying that they are filled with low-density material.

■ What do you think this low-density material might be?

■ Sediments.

As a plate ages, it accumulates a veneer of deep-sea sediments made up of clays and the remains of microorganisms in the oceans. At the subduction zone, this sedimentary cover is partly scraped off against the overriding plate to form huge wedges of deformed sediment that can eventually fill the trench system. This material is often known as an **accretionary prism**. Not all the sediment is removed, however, and some remains attached to the descending oceanic plate and may become attached to the base of the overlying plates or even be carried into the upper mantle.

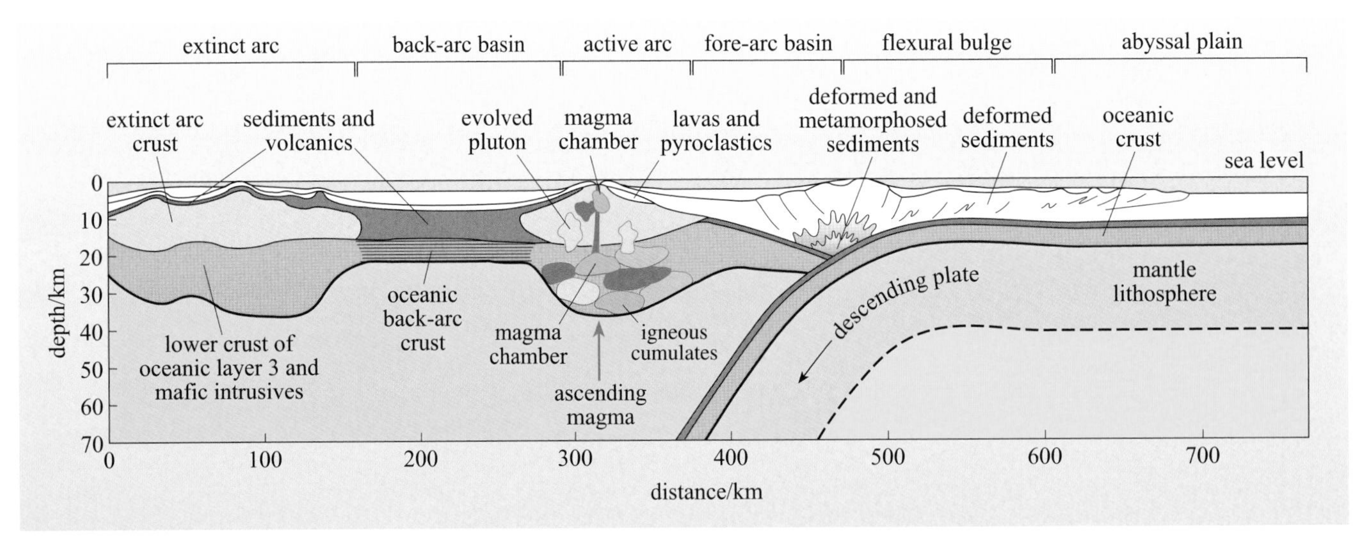

Figure 3.18 Schematic cross-section through an ideal island arc. Note that not all of the features shown here will be present in any one arc system. (*Note*: 2× vertical exaggeration.)

Behind many island-arc systems, especially those in the western Pacific, small ocean basins open up between the arc and the adjacent continent. Typical examples include the ocean basins immediately to the west of the Tonga and Marianas island arcs. Various lines of evidence show that these regions of ocean crust are very young and characterised by active spreading centres. Such features are known as **back-arc basins**.

- Does the existence of young oceanic crust suggest an extensional or a compressional tectonic regime in back-arc basins?

- An extensional tectonic regime. New ocean crust is only produced when lithospheric plates move apart.

The presence of an extensional regime in the back-arc region basin may appear counter-intuitive because where two plates are converging the dominant tectonic regime should be compressional. The mechanisms that give rise to back-arc tension may relate to convection in the asthenosphere underlying the back-arc region. Alternatively, it has been suggested that old, dense slabs may subside into the mantle at a faster rate than the plate is moving, causing the trench to migrate towards the spreading centre (euphemistically called 'slab roll-back'). This gives rise to an extensional regime not only in the back-arc basin but also across the whole arc, even to the extent of suggesting that back-arc basins may originate as arcs that have been split by extension as a consequence of slab roll-back.

All of the major features of an oceanic destructive boundary are included in Figure 3.18, which is an idealised cross-section through an oceanic island-arc system.

2 *Ocean–continent (Andean type) subduction*

When an oceanic plate converges with a continental plate, it is always the ocean plate that subducts beneath the continental plate. You should recall from Chapter 1 that continental lithosphere lies at a higher surface elevation than oceanic lithosphere because of its lower overall density. The resistance shown by continental lithosphere to subduction is simply a further reflection of its lower density.

This type of destructive plate boundary is characterised by the west coast margin of South America. Here, the oceanic lithosphere of the Nazca Plate is being subducted beneath the overriding continental lithosphere that forms the western part of the South American Plate. The overriding continental edge is uplifted to form mountains (the Andes) and the collision zone itself is marked by a deep ocean trench that runs parallel to the continental margin. A chain of active volcanoes runs along much of the length of the South American Andes from Colombia to southern Chile. The ocean trench is similarly characterised by a dipping Wadati–Benioff zone, and is marked by earthquakes reaching depths of several hundred kilometres. The shallowest earthquakes (<60 km) occur at, or near, the trench, with the deepest ones recorded as being up to 700 km deep – the depth before which the plate becomes sufficiently heated to prevent further brittle behaviour. The overall structure of such **Andean margins** is very similar to those described in the previous section, with the added influence of the greater thickness of the overriding continental lithosphere and the probable increased flux of sedimentary material into the system as a result of continental erosion.

3 Continent–continent destructive boundaries

When two continental plates meet at a destructive boundary, the continents themselves collide. These types of continental collision are typically the result of an earlier phase of subduction of intervening oceanic lithosphere that has resulted in the closure of an ocean. Perhaps the best known and most spectacular example is the collision of peninsular India with Asia, which began 50 Ma ago, following the closure of an intervening ocean and produced the Himalayas and the Tibetan Plateau (Chapter 6). Even today, India continues to move northwards, indenting the southern edge of Asia at a rate of 40–50 mm y^{-1}. Such collisions result in intense deformation at the edges of the colliding plates, and those sea-floor sediments that were not subducted become folded and compressed into immense mountain chains or **orogenic belts**. Active mountain belts, such as the Alps and the Himalayas in Eurasia, and the Rocky Mountains in the USA and Canada, are generally much wider than mountain belts associated with Andean-style arc systems, with deformation belts occurring many hundreds of kilometres into continental interiors.

- What does this observation suggest about the strength of continental lithosphere relative to oceanic lithosphere?

- It suggests that the continents are less strong and less rigid than oceanic lithosphere.

The continents are made up of less-dense rock than oceanic lithosphere and are dominated by quartz and feldspars. At elevated temperatures, these minerals are much weaker than the olivine and pyroxene characteristic of the oceanic crust and mantle. Moreover, continental crust contains a higher concentration of the heat-producing elements K, U and Th. The overall higher heat production conspires with the dominance of weaker minerals to make the continental crust much less rigid than the crust beneath the oceans and, therefore, easier to deform. During deformation the continental crust is thickened and this gives rise to the dramatic topography of active mountain ranges. However, once the forces that drive collision are removed, erosion takes over and the high topography is reduced to more modest elevations. Older mountain belts, such as the Appalachian and Caledonian orogenic belts, which are the products of continental collisions that occurred hundreds of millions of years ago, are now supported by a balance between isostatic support of their thickened crustal roots and erosion that is controlled by climate. This topic will be dealt with in more detail in Chapter 6.

3.2.5 Conservative plate boundaries and transform faults

Conservative plate boundaries and transform faults occur when plates slide past each other in opposite directions, but without creating or destroying lithosphere. Transform faults connect the end of one plate boundary to the end of another plate boundary, so there are potentially three types of transform fault:

- those that link two segments of a constructive boundary
- those that link two destructive boundaries
- those that link a destructive boundary with a constructive boundary.

Transform faults linking two constructive boundaries are the most common, and account for the displacements between adjacent segments of mid-ocean ridges. Accordingly, this type of ocean transform fault forms an integral part of constructive plate boundaries, and their position is made obvious by the jagged shape of parts of the ocean-ridge system that are split into several segments by series of so-called **fracture zones**. Examples can be easily seen on the Cocos–Nazca Ridge (also known as the Galápagos Spreading Centre), and the Pacific Ocean spreading ridge (i.e. East Pacific Rise) between 10° N and 10° S, and 40° S and 55° S respectively, or manifest as shorter segments along the Atlantic Ocean spreading ridge between 0° and 40° S. Generally, oceanic transform faults occur at right angles to spreading ridges and, therefore, their orientation is indicative of the direction of plate motion.

Transform faults are seismically active – but only where two different plates are adjacent to one another. In Figure 3.19, the fault trace marks the boundary between plates A and B. Plate A is moving towards the east while plate B is moving towards the west.

■ Describe the sense of relative movement along the length of the fracture zone between W and X, X and Y, and Y and Z.

■ Between W and X, the fracture zone separates different parts of plate B and so there is no differential movement. Between X and Y it separates plate A from plate B, which are moving in opposite directions. Between Y and Z it separates different parts of plate A and there is, again, no differential movement.

■ Which part of the fracture zone between W and Z will be seismically active?

■ Only that segment between X and Y where two different plates are adjacent to one another.

Only those sections of fracture zones between two segments of constructive boundaries (e.g. the segment between X and Y in Figure 3.19) are seismically active and therefore real plate boundaries. Fracture zones continue to exhibit a topographic expression beyond the constructive plate boundaries, even though only the short length occupied by the transform fault is active. This topographic expression is simply a result of the different ages of adjacent oceanic lithosphere: younger lithosphere rests at a higher elevation than older lithosphere – this situation is illustrated schematically in Figure 3.20.

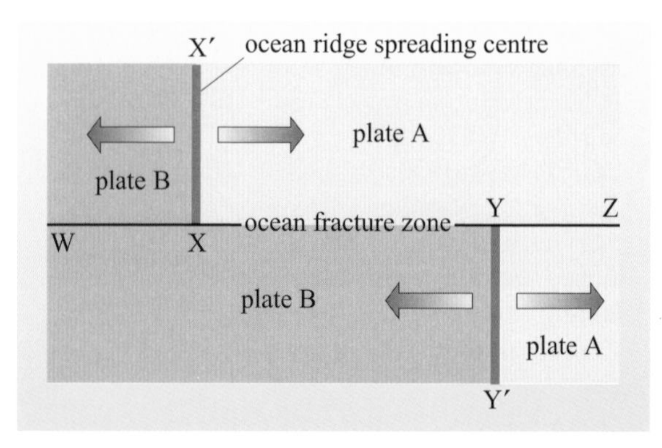

Figure 3.19 Diagram showing relative movements across an oceanic transform fault W, X, Y and Z off-setting a constructive plate boundary. The large arrows indicate the sense of plate motion away from the ridges XX′ and YY′.

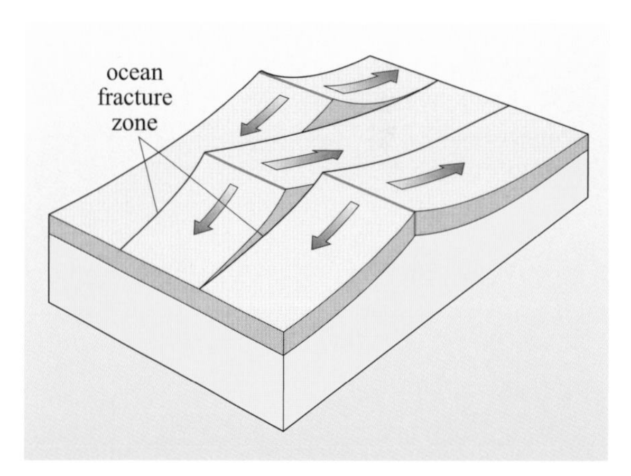

Figure 3.20 Block diagram showing how the topographic contrast across an ocean fracture zone (transform fault) develops as a consequence of lithospheric age as opposed to differential vertical movement.

Transform faults associated with subduction zones are much less common, and destructive plate boundaries do not, in general, show the segmented structure so common in constructive boundaries. An example occurs at the eastern end of the Cocos–Nazca Ridge, where a heavily faulted seismic zone delineates a transform fault (the Panama Fracture Zone), connecting a constructive boundary (the eastern end of the Costa Rica Rift, which is the easternmost part of the Cocos–Nazca Ridge) with the eastern end of a destructive boundary (the Middle America Trench). Similarly, the Scotia arc in the southern Atlantic is terminated in the north by a long transform fault along the North Scotia Ridge that marks the boundary between the South American Plate and the Scotia Plate.

Occasionally, conservative plate boundaries occur in continental plates. The most famous example is the San Andreas Fault of California, which marks a segment of the boundary between the North American and Pacific Plates. Here, Baja and southern California (including Los Angeles) are moving slowly northwards relative to the rest of California. This type of transform boundary produces shallow earthquakes and accompanying ground faulting. The friction between the two plates is often so great that the two sliding margins become 'stuck' together, allowing stresses to build up, which are then relieved by large earthquakes.

3.2.6 Triple junctions

All of the plate boundaries discussed so far have involved junctions between two plates. However, there are some localities where three plates are in contact, and these are termed **triple junctions**. Triple junctions between three ocean ridges, such as that in the South Atlantic between the African, South American and Antarctic Plates, are known as ridge–ridge–ridge, or RRR triple junctions. A similar notation can be used to identify triple junctions involving ocean trenches (T) or transform faults (F). For instance, a ridge–ridge–trench junction would be termed an RRT triple junction. The ordering of the letters is not significant. Considering all the geometric possibilities of fitting together three plate boundaries and their relative motions, there are actually only ten possible triple junctions. Some of these, such as RRR junctions, are termed **stable triple junctions**, which means they maintain their form over time. However, some can only exist briefly before they evolve into another plate configuration and these are termed **unstable triple junctions**. Figure 3.21 shows the evolution of three triple junctions over time.

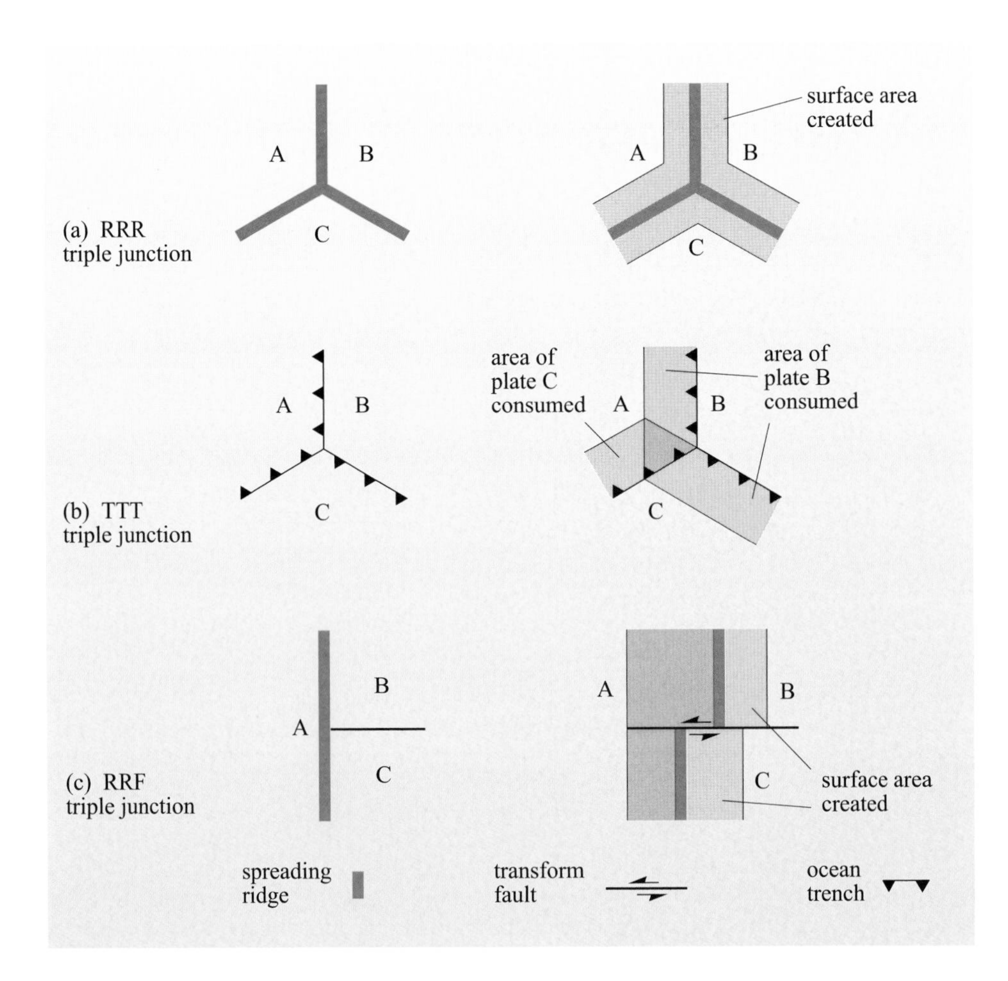

Figure 3.21 The evolution of triple junctions with time. (a) A triple junction involving three ridges (RRR triple junction) is always stable, and the magnetic anomalies within the surface area created have Y-shaped patterns around the spreading ridges. (b) A triple junction between three trenches (TTT) is almost always *unstable* except in the special circumstance shown here when the relative motion of plates A and C is parallel to the plate boundary between B and C. (c) A triple junction between two ridges and a transform fault (RRF) can only exist for a short instant in geological time, and decays immediately to two FFR stable plate junctions. (Dalrymple et al., 1973)

The RRR junction shown in Figure 3.21a is always stable, regardless of the relative rates of spreading at each of the three ridges. The TTT junction in Figure 3.21b is basically unstable except if, by coincidence, the movement rates are the same and if the direction of subduction of plate C below plate A is exactly parallel to the boundary between plates B and C. The triple junction in Figure 3.21c is an RRF junction, and is unstable because there is relative motion between plate B and plate C. The RRF triple junction evolves immediately to form two RFF junctions. FFF and RRF junctions are always unstable.

Seven types of stable triple junction exist in the present plate tectonic configuration. These are:

- RRR (e.g. in the South Atlantic, the Indian Ocean and west of the Galápagos Islands in the Pacific)
- TTT (e.g. central Japan)
- TTF (e.g. off the coast of Chile)
- TTR (e.g. off Moresby Island, western North America)
- FFR, FFT (e.g. junction of the San Andreas Fault and the Mendocino Transform Fault off western USA)
- RTF (e.g. southern end of the Gulf of California).

3.3 Plate tectonic motion

Plates move relative to one another and relative to a fixed reference frame, such as the rotational axis of the Earth. Plates also move across the curved surface of the Earth and so should not be considered as flat sheets on a flat surface but as caps moving across the roughly spherical surface of the Earth. Consequently, plate motion is not as simple as it might at first appear. This section begins with a consideration of relative plate motions and how they can be measured before moving on to the more complex assessment of plate motions across a sphere and how true plate motions may be measured.

3.3.1 Relative plate motions

In previous sections you have already tackled the problem of assessing the full and half spreading rates of ocean ridges.

- Recall two different methods of determining spreading rates.
- The use of dated magnetic anomalies, and the known relationship between ocean depth and age (Sections 3.1 and 3.2).

A simple calculation dividing the distance from the ridge by the age gives the plate speed and, combined with the direction of travel, its velocity.

Question 3.6

Figure 3.22 shows a section through the Earth from the Atlantic to the Indian Ocean, cutting across three different plates and two constructive plate boundaries. The half spreading rates are shown for each plate at each plate boundary. For the situations in Table 3.1, estimate the relative rates and relative directions of motion of the African Plate.

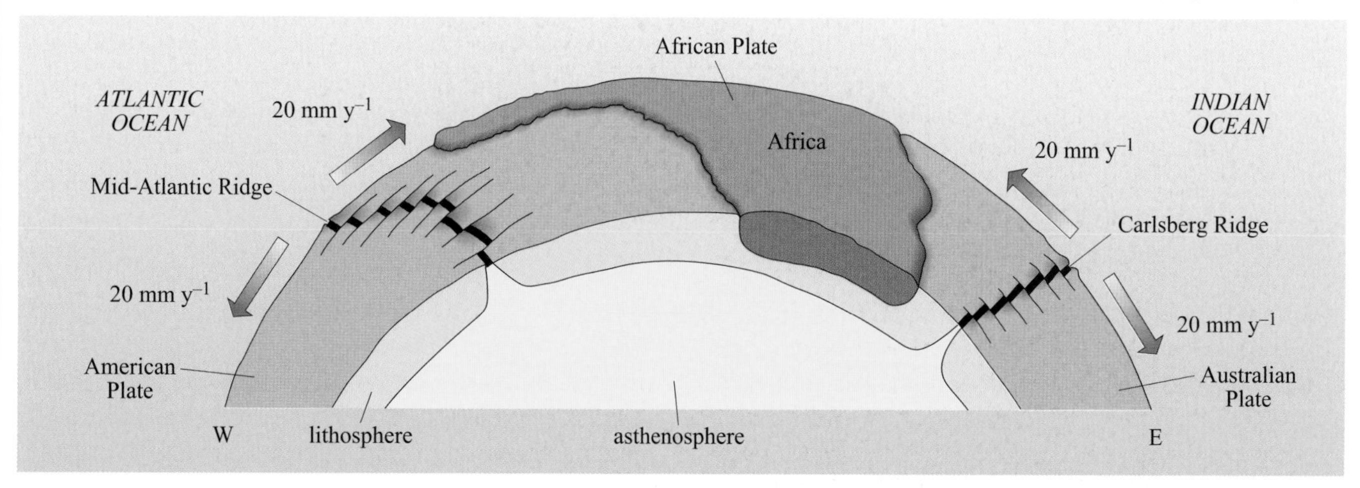

Figure 3.22 Section through the Earth from the Atlantic Ocean to the Indian Ocean showing the relative spreading rates across the Mid-Atlantic and Carlsberg Ridges. (Clague and Dalrymple, 1987)

Table 3.1 For use with Question 3.6.

	Relative plate motion/mm y^{-1}	Direction of plate motion
American Plate fixed		
Australian Plate fixed		

The answer to this question reveals two important points about plate motion:

- measured plate velocities (speed and direction) must be stated relative to one another
- plates and their boundaries cannot be fixed in relation to the mantle.

The second point derives from a consideration of the African Plate – as it grows in size, at least one of its constructive boundaries must have moved. So how can we determine the true movement of a plate against a truly fixed frame of reference? One possibility would be to assume that one plate is stationary and to determine plate movement relative to that 'fixed' reference (as in Question 3.6). However, polar wander curves for all of the continents show that all plates bearing continents have moved relative to the Earth's magnetic pole over periods of tens of millions of years. Another way would be to determine plate motion relative to surface features that might be more firmly rooted in the deeper mantle and for this we turn to volcanism that is not dependent on plate boundary interactions – the so-called within-plate or **hot-spot volcanism** on ocean islands.

3.3.2 Hot-spot trails and true plate motions

In addition to volcanism associated with constructive and destructive plate boundaries there is a third important component to global volcanism. This occurs in the interior of plates and is associated with broad surface up-doming, which is often 1000 km across and hundreds of metres in elevation. Gravity anomalies across these domes show that they are not in isostatic equilibrium, but are supported from sub-lithospheric depths, presumably by upwelling mantle. Perhaps the best-known example is located beneath the active volcanoes of Hawaii (Box 3.4), whose long history of volcanism has been related to a structure in the deep mantle known as a **mantle plume**. Mantle plumes are an important feature of mantle convection and will be discussed in more detail in Chapter 7, but for now it is sufficient to know that they produce surface volcanism that is not necessarily associated with plate boundaries. There are numerous plumes of different sizes recognised around the globe. Some are associated with chains of islands and seamounts, whereas others have produced long ridges in the ocean floor. A good example is the Ninetyeast Ridge in the Indian Ocean. Termed **aseismic ridges** because of a lack of seismicity along their lengths, these ridges are very different structures from the ocean ridges associated with constructive plate boundaries.

Box 3.4 The Hawaiian hot spot

Hawaii is part of an extensive chain of islands and submarine volcanic peaks (called seamounts) stretching almost 6000 km across the floor of the Pacific Ocean. The chain forms an 'L'-shaped chain of volcanic islands and seamounts across the sea floor that increase in age northwards from Hawaii (Figure 3.23).

Hawaii rises from the sea floor some 6 km below the Pacific Ocean to a summit elevation of about 4 km above sea level, making it taller than Mt Everest. Although all of the Hawaiian Islands are volcanic in origin, only Hawaii is currently active and still growing in size. The islands are situated within the Pacific Plate some 4000 km from the nearest plate boundary. The magma that has caused the volcanism is the result of a plume of anomalously hot material rising through the mantle. Where this plume impinges upon the base of the lithosphere, magma finds its way to the surface to produce a so-called **hot spot**. Other island and seamount chains located on the Pacific Plate (Figure 3.23) show a similar pattern of age progression, and are related to different hot spots. In fact, hot-spot volcanoes may be found dotted around the world and most of them are remote from plate boundaries.

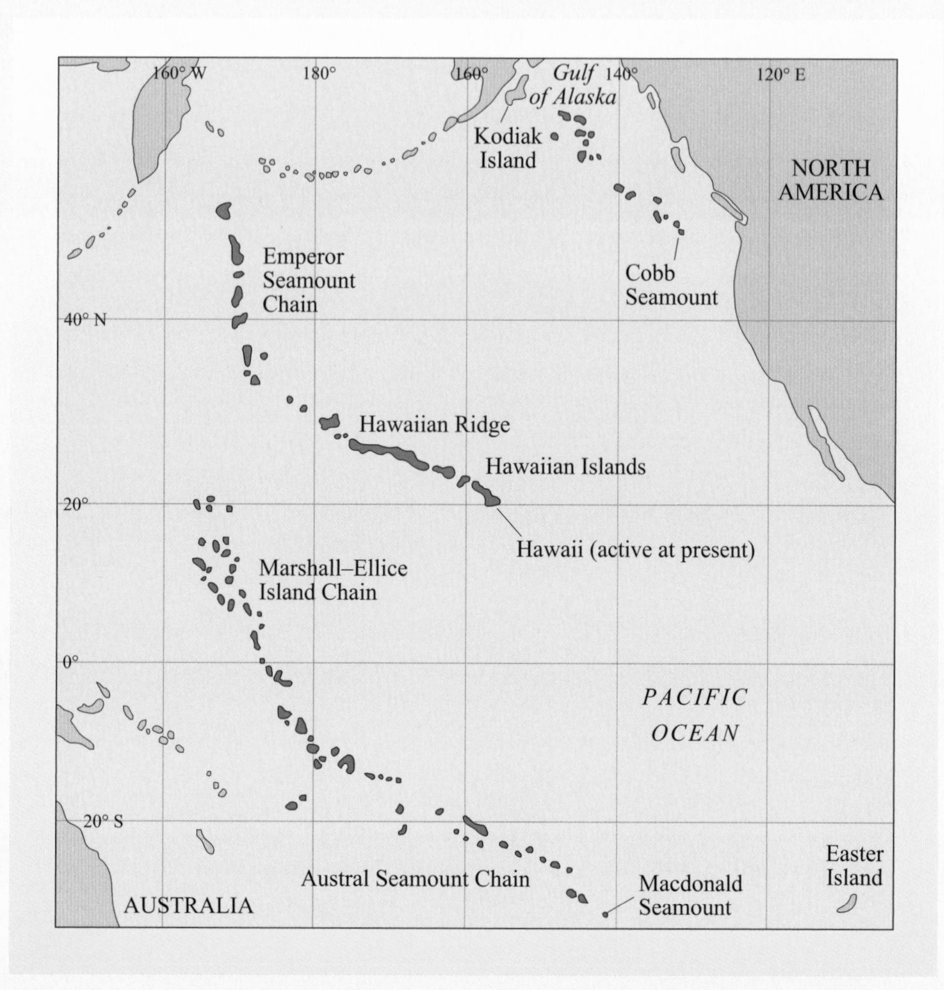

Figure 3.23 Map showing three of the major seamount and island chains in the Pacific Ocean. These are the Kodiak Island–Cobb Seamount Chain in the northeastern Pacific, the Marshall–Ellice Islands–Austral Seamount Chain of the southern Pacific, and the well-known Hawaiian Ridge–Emperor Seamount Chain in the central Pacific. The Hawaiian Islands are located at the southeastern end of the Hawaiian Ridge.

The ages of the islands and seamounts are proportional to their distance away from the currently active site of Hawaii, as shown in the graph in Figure 3.24 (overleaf). The best-fit line through the data points indicates that the site of volcanic activity has apparently migrated at a constant speed along the chain. Each island or seamount has been constructed as the Pacific Plate has moved over the stationary hot spot. If it is assumed that the Hawaiian hot spot has remained stationary with respect to the Earth's axis, then the rate of migration of volcanism along the chain gives the rate of plate movement across the Hawaiian hot spot.

- The Hawaii–Emperor chain of islands and seamounts is not straight, but kinked. What do you think the kink represents?

- If the hot spot is stationary then this must represent a change in direction of the Pacific Plate.

Careful interpretation of the age progression along the Hawaii–Emperor chain suggests that this change in direction occurred 43–50 Ma ago, which ties in with a series of tectonic adjustments around the world, including the start of continental collision between India and Asia.

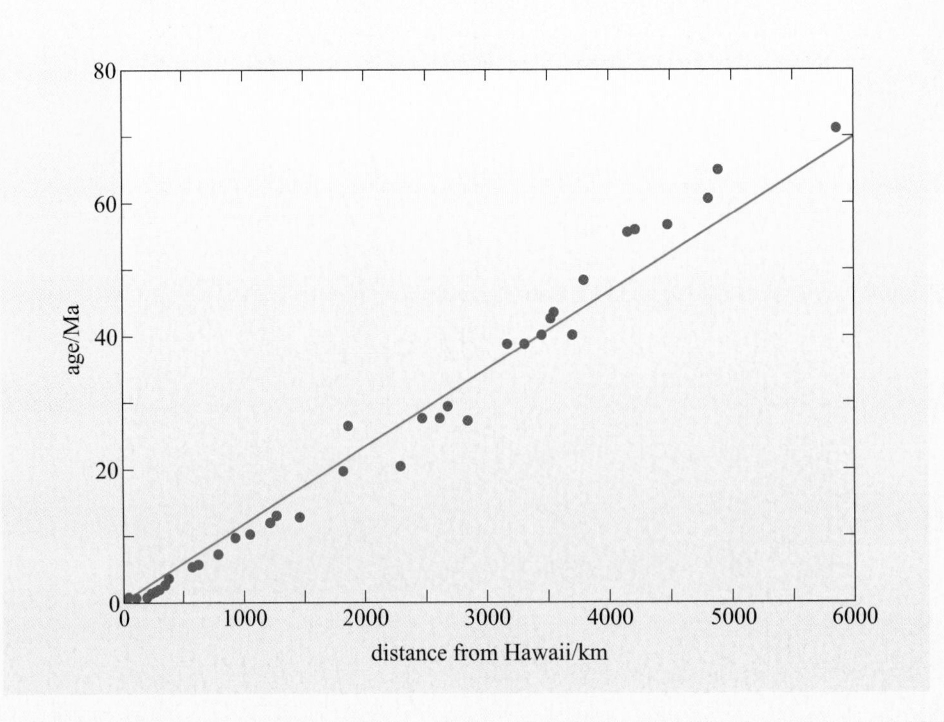

Figure 3.24 Graph of age versus distance from Hawaii measured along the Hawaii–Emperor chain of islands and seamounts. A best-fit line has been drawn through the data.

Question 3.7

Use Figure 3.24 to estimate the average rate at which volcanic activity has appeared to move along the Hawaii–Emperor Seamount Chain. Express your answer in mm y^{-1} to two significant figures. With reference to Figure 3.23, determine in which direction the active volcanism has moved and, from this, the direction the plate has moved. *Note*: the speed at which volcanic activity moves is the inverse of the gradient of the best-fit straight line.

By assuming mantle plumes and hot spots are stationary, the motions derived from age progressions such as those along the Hawaii–Emperor Seamount Chain represent **true plate motions**. However, if the mantle plumes underlying hot spots have also moved with respect to the Earth's axis, but at a rate different from that of plate movement, then the motions defined by a hot-spot trace (see Figure 3.23) may be misleading.

- How do you think plate motions could be verified?
- By measuring true plate motion from another hot-spot trace.

For the Pacific Plate, the Hawaiian hot spot may be the largest hot spot, but it is not the only example of a hot-spot trace on the ocean floor. Two other examples related to the mantle plumes currently beneath the Cobb and the Macdonald Seamounts are shown in Figure 3.23. You should be able to appreciate from this figure that both of these chains of seamounts and islands are broadly parallel with the Hawaii–Emperor Seamount Chain, and the Austral–Marshall Islands Seamount Chain shows a similar bend. Detailed geochronology of these chains reveals that the age progression along them is also consistent with the rates of northwesterly plate motion derived from the Hawaii–Emperor Seamount Chain. Since the chances of three (and more) hot-spot traces moving independently of the plates and one another giving such similar results is remote, these results strongly indicate that hot spots do provide a reference frame within which true plate motions can be measured.

This methodology has been extended to studies of numerous hot-spot trails on other plates. Combined with a knowledge of relative plate motions between adjacent plates, such measurements have allowed the development of a framework of true plate motions around the globe, which are summarised in Figure 3.25.

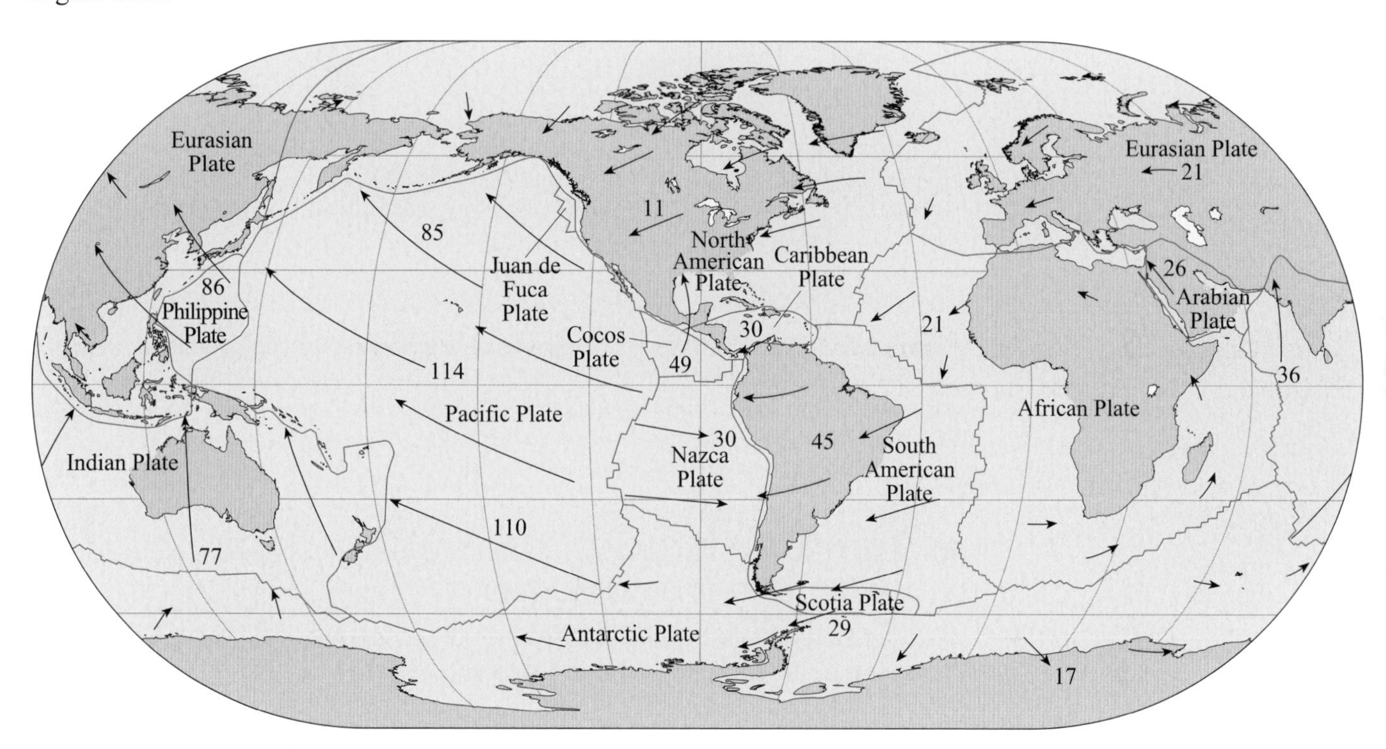

Figure 3.25 Map showing present-day motion of lithospheric plates (indicated by arrows) relative to hot spots (i.e. true plate motion). The lengths of the arrows indicate the amount of movement that would occur over a period of 50 Ma and the numbers are the current mean true plate speed in mm y^{-1}.

3.3.3 Plate motion on a spherical Earth

Earth's tectonic plates are continuously in motion with respect to each other, and together they form the closed surface of a sphere (i.e. the Earth's surface). Understanding the movement of plates, therefore, requires a geometrical analysis of motions over a spherical surface. This is described in the Euler (pronounced 'oiler') geometrical theorem, which shows that every displacement of a plate from one position to another on the Earth's surface can be regarded as a simple rotation of that plate about a suitably chosen axis, known as an Euler pole or **pole of rotation**, which passes through the centre of the Earth.

Plate movement on a spherical Earth is illustrated in Figure 3.26. Two plates, A and B, are separating at a constructive plate boundary and rotating about a pole of rotation (which in this case is the North Pole). The constructive plate boundaries lie along lines of longitude, whereas the transform faults are parallel to lines of latitude. Lines of longitude are great circles with centres that coincide with the centre of the Earth; lines of latitude, which are at right angles to lines of longitude, are small circles with centres displaced from the Earth's centre. When the plates separate, the gap between them increases but, quite obviously, they separate a smaller amount close to the North Pole than at the Equator.

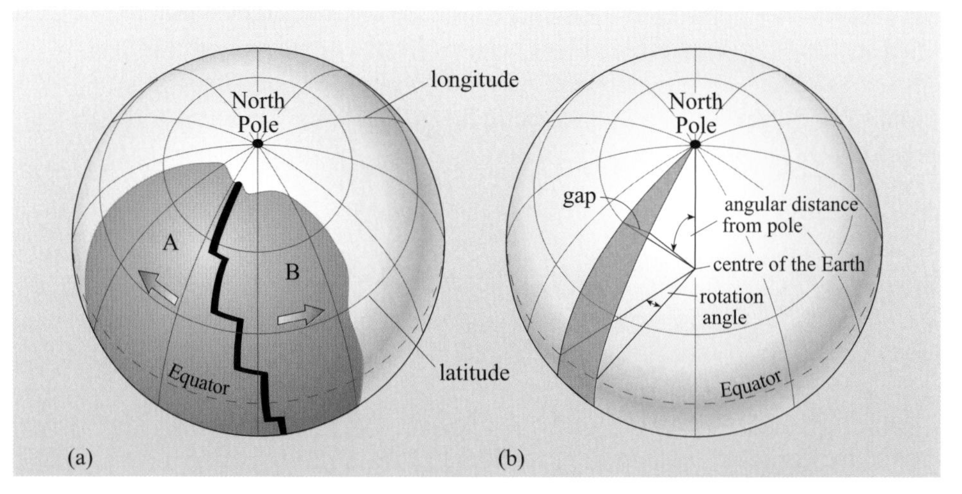

Figure 3.26 (a) The geometry of a constructive boundary in which two plates are separating. (b) The geometrical relationships that may be used to describe plate movement. Note that while the rotation angle remains the same along the length of the split, the width of the gap increases with angular distance from the pole of rotation (in this case the North Pole).

■ What effect does this have on local full and half spreading rates along the length of the constructive plate boundary?

■ Spreading rates measured close to the pole of rotation will be small while those further away will be much greater.

This aspect of spherical geometry explains why relative plate motions expressed simply in terms of mm y^{-1} vary along the length of a plate margin. Of course, poles of rotation are not all located at the geographical poles but can fall anywhere on the Earth's surface. For example, the Mid-Atlantic Ridge has variable spreading rates along its length, with values of 37–40 mm y^{-1} in the South Atlantic, to lower values of 23–26 mm y^{-1} in the North Atlantic (Figure 3.8).

■ What do these variations in spreading rate tell you about the location of the pole of rotation of the South American Plate?

■ Spreading rates are lower in the North Atlantic and so the pole of rotation must be located somewhere in the northernmost Atlantic Ocean.

The location of the pole of rotation can be determined more accurately from the orientation of the transform faults that cut the Mid-Atlantic Ridge. Transform faults lie along small circles and, by definition, lines at right angles to them will pass through the pole of rotation (Figure 3.27a). Using this method, the pole of rotation of the South American Plate can be shown to be located in the North Atlantic, south of Greenland (Figure 3.27b).

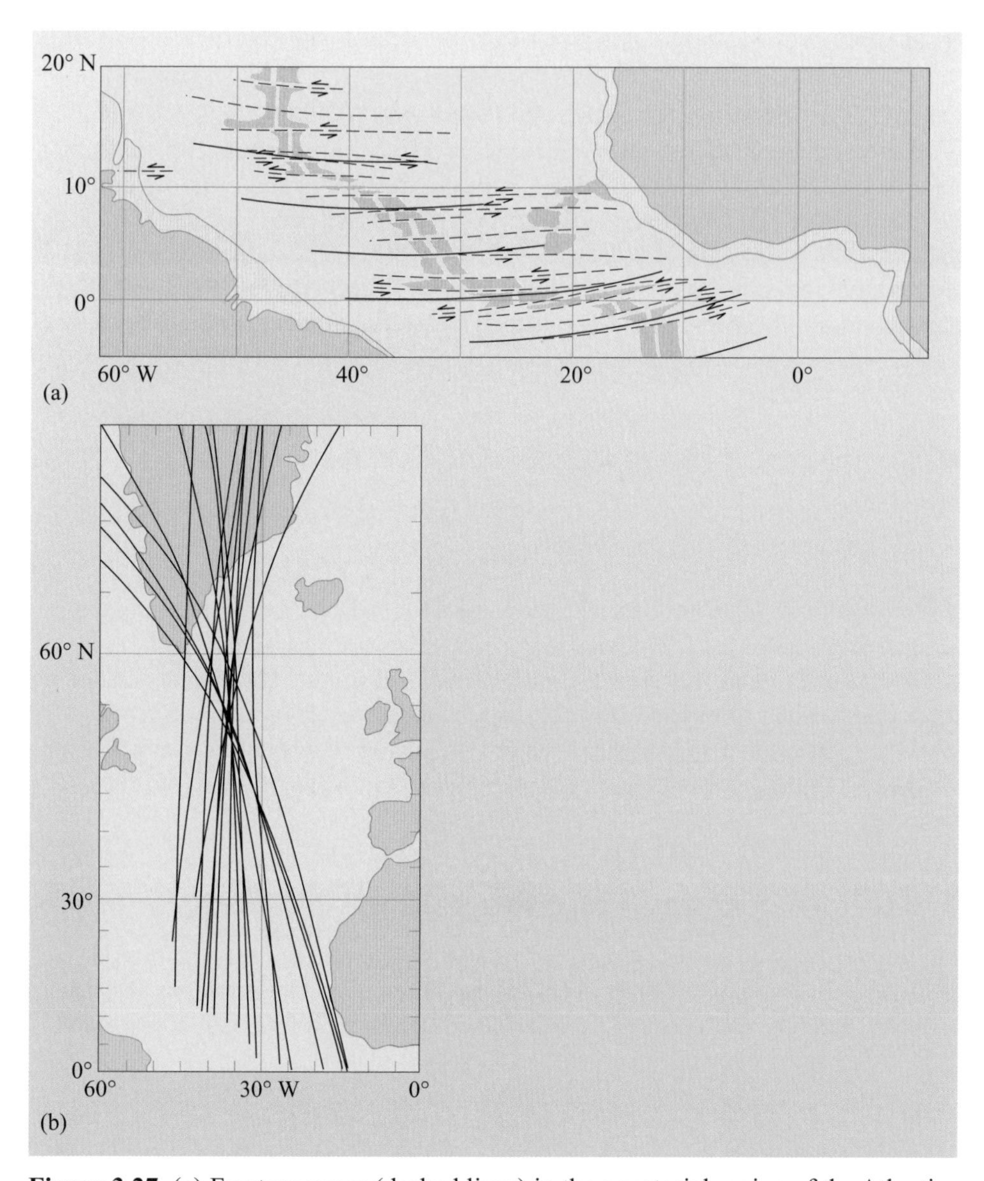

Figure 3.27 (a) Fracture zones (dashed lines) in the equatorial region of the Atlantic Ocean compared with small circles (solid lines) concentric about a rotation pole at 58° N, 36° W. (b) Great circles drawn perpendicular to the transform faults that offset the Atlantic spreading ridge. The great circles converge on an area in the North Atlantic, defining the general position of the pole of rotation of the plate. (After Morgan, 1968)

Finally, some plates have an internal pole of rotation that results in some complex consequences for relative plate motions. The first, and possibly most obvious, is that the real motion of the plate becomes one of rotation. The African Plate (Figure 3.25) is a good example. The southern part of the African Plate is moving to the east while the northern part is moving to the west – the plate is rotating anticlockwise about a pole located somewhere near the Canary Islands, off the northwest coast of Africa. The second consequence is that an internal pole can lead to a given plate boundary transforming from a constructive plate boundary through a conservative plate boundary to a destructive plate boundary along its length, although this is not apparent around the African Plate.

3.4 Plate driving forces

One of the key questions associated with plate tectonics is why plates move and what drives them. Plate tectonics is an expression of the thermal state of the Earth's interior and is the way in which the Earth is currently losing a large proportion of its internal heat. Hot lithosphere generated at constructive plate boundaries loses its heat to the oceans, atmosphere and, ultimately, space by conductive cooling as it ages and spreads away from the ridge before being recycled into the mantle by subduction. It is perhaps, tempting to relate this cycling of material to a convective cycle, with hot material upwelling beneath ocean ridges and cold material sinking in subduction zones.

- Can you think of a reason why this might not be the case?
- Plate boundaries migrate over time and it is unlikely that convection cells within the mantle would migrate with them.

The migration of plate boundaries across the surface of the Earth means that they are not firmly fixed into the underlying convective motions of the mantle – from our analysis of real plate motions, mantle convection is more likely to be related to the location of hot spots and mantle plumes, a subject to which you will return in Chapter 7. So what, then, drives the plates? An answer to this question may lie in an analysis of the forces acting on plates, both on their undersides and at their boundaries.

3.4.1 Forces acting upon lithospheric plates

Figure 3.28 provides a very simplified overview of the forces that are thought to affect the movement of lithospheric plates. The relative contribution of these forces to plate motion needs to be established before their roles can be explored.

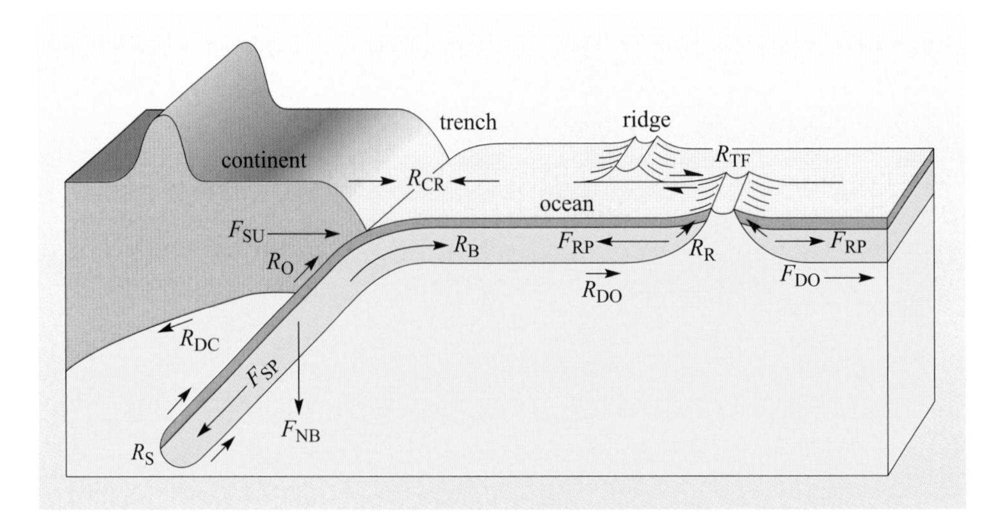

Figure 3.28 Forces acting upon a plate. F denotes a driving force, whereas R denotes a retarding force. See text for explanation. (Forsyth and Uyeda, 1975)

Forces acting on the underside of lithospheric plates

Lithospheric plates are decoupled from the rest of the mantle because the underlying asthenosphere is weak. However, plates may be driven, at least in part, by forces imparted by the convecting mantle. For instance, if a lithospheric plate is being carried along by a faster moving asthenosphere, then the force acting along the bottom surface of the oceanic plate can be considered as an **ocean driving force**, F_{DO}, which helps the plate to move. By contrast, if the asthenosphere is moving slower than the plate in the direction of plate movement, or even in the opposite direction, then the force acting along the bottom surface of the oceanic plate can be considered as an **ocean drag force**, R_{DO}, retarding the movement of the plate.

Continental lithosphere is thicker than oceanic lithosphere, so continents almost always have a 'keel' of lithospheric material projecting downwards. As a result, the resistance to movement might be greater beneath continental plates than oceanic plates. Accordingly, continental plates might be associated with an additional **continental drag force**, R_{DC}, and so the resistive force acting on the base of a continental plate would be the sum of both oceanic and continental drag forces, $R_{DO} + R_{DC}$.

Forces acting at plate margins

Other forces that act on plates must be generated at their boundaries. These forces push from the ridge, drag the plates down at the trenches, or act along the sides of plates at conservative boundaries (Figure 3.28).

At constructive boundaries, the upwelling of hot material at ocean ridges generates a buoyant effect that produces the ocean ridge, which stand some 2–3 km above the surrounding ocean floor. Here, oceanic plates experience a force that acts away from the ridge, the so called **ridge-push force** (F_{RP}), which is a result of gravity acting down the slope of the ridge. The occurrence of shallow earthquakes, resulting from the repeated tearing apart of newly formed oceanic crust, indicates there is also some frictional resistance to this force at ridges. This can be called **ridge resistance**, R_R. Bounding the ridge segments, the oceanic transform faults, where the plate segments slide past each other, encounter resistance to movement, and produce a series of earthquakes: this retarding force is the **transform fault resistance**, R_{TF}.

The situation at destructive plate boundaries is more complex. A major component is the downward gravitational force acting on the cold and dense descending slab as it sinks into the mantle. This gravity-generated force pulls the whole oceanic plate down as a result of the negative buoyancy of the slab: this is the **negative buoyancy force**, F_{NB}. The component of this downward-acting force that is transmitted to the plate is the **slab-pull force**, F_{SP}. The magnitude of the slab pull is related to the angle at which the plate descends, and is greater for steeply dipping plates. However, the sinking slab encounters resistance as it descends, both from the frictional drag on its upper and lower surfaces and from the viscosity of mantle material that is being displaced: this combined resistive force is termed **slab resistance**, R_S. A further complication is that the downwards moving plate must flex at the trench before it begins to slide beneath the opposing plate. This provides a further resistance to the plate motion; this is labelled **bending resistance**, R_B. Due to the pushing of the subducting slab against the overriding plate there is, in addition, frictional resistance that gives rise to shallow and deep earthquakes at subduction zones. These frictional forces can be labelled collectively as **overriding plate resistance**, R_O.

For the overriding plate, another theoretical force analogous to the ocean driving force has been proposed, which is derived from convection induced in the mantle above the subducted plate. Cooling of the mantle wedge against the upper surface of the subducting plate induces convection that sucks more mantle into the wedge. This is the **trench suction force**, F_{SU}, which serves to pull the plate towards the trench. The collision of plates and the associated deformation processes generate a **collisional resistance force**, R_{CR}. This acts in the opposite direction within each converging plate, but it is equal in magnitude in both.

The velocities of present-day plate motions appear to be constant, indicating a state of dynamic equilibrium where a balance exists between the driving and resistive forces. However, each plate moves at its own rate – which suggests that the relative importance of the driving and retarding forces must vary from plate to plate. It seems unlikely that any single force is the sole driving mechanism of plate motions. For example, if the ridge-push force is the only driving force, why does the Philippine Plate, with no ridge on its boundary, move at a similar rate (70 mm y^{-1}) to the Indian Plate, which is bounded by both the Carlsberg and South Indian Ocean spreading ridges? Plate motions must, therefore, be controlled by a combination of forces.

Question 3.8

For each of the forces listed in Table 3.2, indicate whether they are likely to act as a driving force or a resistive force, and place a tick in the appropriate space. Note that some forces can act as either a driving force or resistive force.

Table 3.2 For use with Question 3.8.

Force	**Acts as a driving force**	**Acts as a resistive force**	**Might act as *either* a driving force *or* a resistive force**
oceanic drag			
continental drag			
ridge-push			
transform fault			
slab-pull			
slab resistance			
trench suction			

To investigate which are the most important forces that act on plates, the reasoning process applied in the previous paragraph can be adopted and the plate speed compared quantitatively with factors that relate to the different driving forces. For example, if the dominant driving force is ridge-push (F_{RP}), then the fastest moving plates should be the oceanic plates with the highest ratio of ridge length to surface area. The forces acting at plate boundaries should have magnitudes that are proportional to the length of ocean ridge (in the case of ridge-push), the length of ocean trench (for slab-pull (F_{SP}), trench suction (F_{SU}) and inter-plate resistances) and the length of transform fault (for transform fault resistance (R_{TF})). Oceanic and continental drag act over the lower surface of the plate, and so they should be proportional to the area of the plate. The rates of true plate motion in relation to the relative lengths of ocean ridge, ocean trench, transform fault, and plate area must, therefore, be examined to estimate the effects of each of the plate driving forces. The motions of the twelve plates listed in Table 3.3 are shown graphically in Figure 3.29a–e (overleaf). Some of the important physical properties of each plate are listed in the table, together with the average true velocity of each plate calculated relative to a hot-spot reference frame. To explore whether or not each of these physical attributes of plate configuration are significant in producing plate motion, the degree and character of data correlation need to be examined.

Table 3.3 Dimensions and true velocities of lithospheric plates.

Plate	Total plate area $\times 10^6$ /km^2	Total continental area $\times 10^6$ /km^2	Average true velocity /mm y^{-1}	Circumference $\times 10^2$/km	Length (effective length)* $\times 10^2$/km		
					Ocean ridge	Ocean trench	Transform fault
(a) Eurasian	69	51	20	421	90 (35)	0	56
(b) N. American	60	36	11	388	146 (86)	12 (10)	122
(c) S. American	41	20	45	305	87 (71)	5 (3)	107
(d) Antarctic	59	15	17	356	208 (17)	0	131
(e) African	79	31	10	418	230 (58)	10 (9)	119
(f) Caribbean	4	0	30	88	0	0	44
(g) Arabian	5	4	26	98	30 (27)	0	36
(h) Indo-Australian	60	15	70	420	124 (108)	91 (83)	125
(i) Philippine	5	0	86	103	0	41 (30)	32
(j) Nazca	15	0	30	187	76 (54)	53 (52)	48
(k) Pacific	108	0	100	499	152 (119)	124 (113)	180
(l) Cocos	3	0	49	88	40 (29)	25 (25)	16

*Effective lengths (in brackets) are the lengths of plate boundary that are capable of exerting a net driving or resistive force.

Figure 3.29a is a plot of total plate area against average true plate velocity. The graph reveals that while it is true that the plate with the largest area (k, the Pacific Plate) has one of the highest velocities, it does not always follow that large plates have high velocities – the plate with the second largest area (e, the African Plate) has a relatively low velocity. Moreover, the plate with the highest velocity (l, the Cocos Plate) is one of the smallest. While it would be possible to draw a best-fit line through some of the data points (for instance, g, j, c, h and k) that showed reasonable positive correlation (i.e. that larger plates have faster velocities and smaller plates have slower velocities), there is little obvious pattern if the data are taken as a whole. Accordingly, it is reasonable to conclude that there is no real correlation between these two variables of plate area and true plate velocity. This logic can be extended to examine the role of oceanic drag: if this were a significant motive force then these same data would show a positive correlation; if it were a significant plate-retarding force then the data would show a negative correlation. Since neither is true, the role of oceanic drag appears to have an insignificant effect on plate motion.

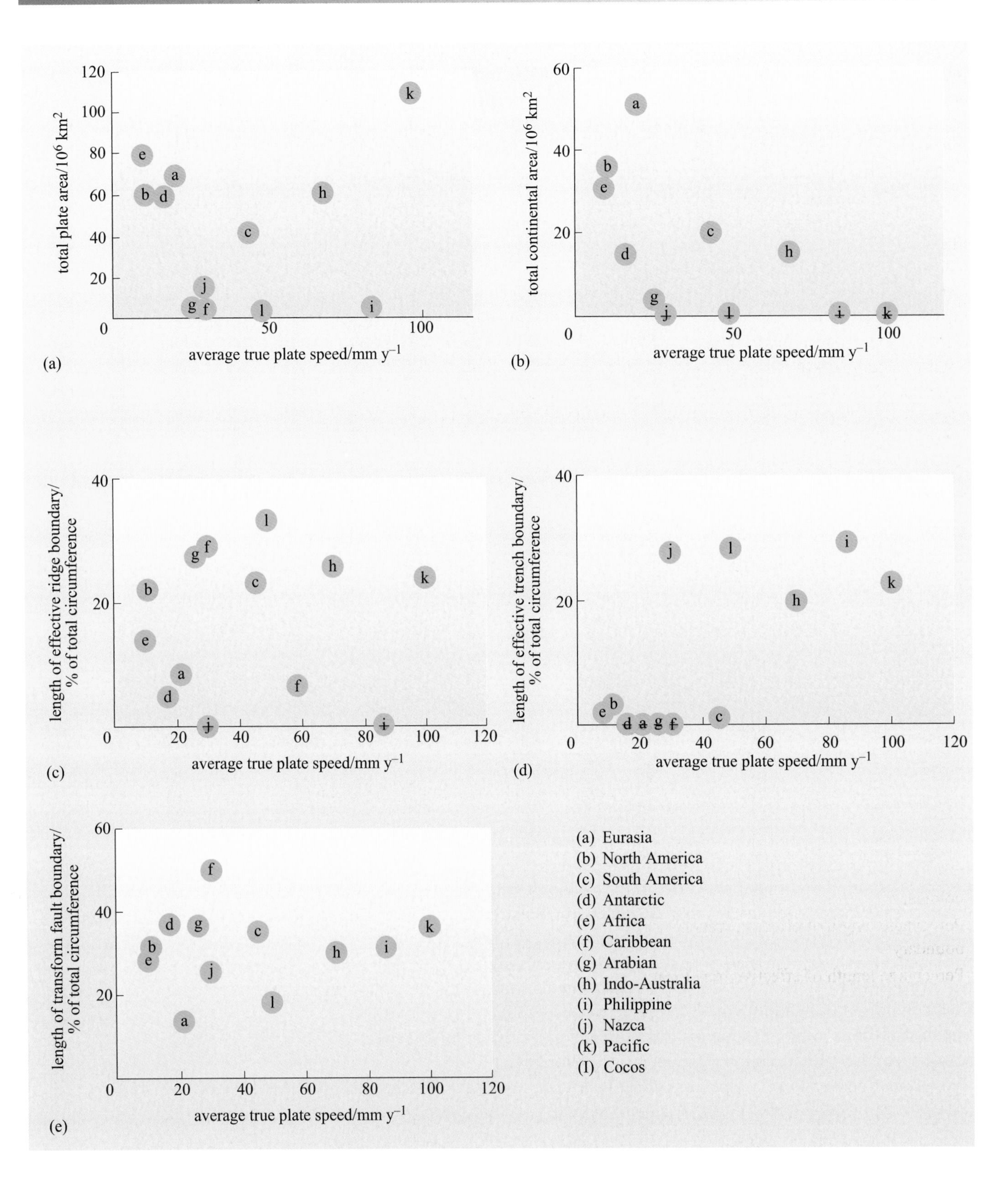
total plate area/10⁶ km²
average true plate speed/mm y⁻¹
(a)
total continental area/10⁶ km²
average true plate speed/mm y⁻¹
(b)
length of effective ridge boundary/ % of total circumference
average true plate speed/mm y⁻¹
(c)
length of effective trench boundary/ % of total circumference
average true plate speed/mm y⁻¹
(d)
length of transform fault boundary/ % of total circumference
average true plate speed/mm y⁻¹
(e)
(a) Eurasia
(b) North America
(c) South America
(d) Antarctic
(e) Africa
(f) Caribbean
(g) Arabian
(h) Indo-Australia
(i) Philippine
(j) Nazca
(k) Pacific
(l) Cocos

◀ **Figure 3.29** (a) Plot of total plate area against average true plate speed. The letters for the plates are as in Table 3.3. (b) Plot of total continental area of each plate against average true plate speed. (c) Plot of length of effective ocean ridge boundary (expressed as a percentage of the total circumference of the plate) against true plate speed. (d) Plot of length of effective ocean trench boundary (expressed as a percentage of the total circumference of the plate) against true plate speed. *Note*: plates a–g have relatively little or no boundary along which the plate is being subducted. (e) Plot of total length of transform fault boundary for each plate (expressed as a percentage of the total circumference of the plate) against true plate speed.

Figure 3.29b is a graph showing the continental area of each plate compared with the average true velocity of that plate. If the continental 'keel' were acting to retard plate motion, those plates with the largest proportion of continental lithosphere might be expected to display the slowest velocities. Interestingly, the plate with the largest *continental* area (a, the Eurasian Plate) has one of the lowest velocities, while the plate with the lowest continental area (k, the Pacific Plate) has the highest velocity; this is the negative correlation predicted above. Clearly, the data points in Figure 3.29b do not fall on a line, but the general trend is for low speed to be associated with a large continental area. Accordingly, these data appear to indicate that continental drag is an effective resistance to plate movement.

Question 3.9

Examine Figure 3.29 and, by ticking the appropriate columns in Table 3.4, indicate the type of association shown by the y-axis variables listed in the left-hand column with the average true plate velocity (x-axis).

From the associations you have identified, what do you conclude about the driving mechanisms of plate motions?

Table 3.4 For use with Question 3.9.

	Positive association	**No clear association**	**Negative association**
Total plate area			
Total continental area			
Percentage length of effective ridge boundary			
Percentage length of effective trench boundary			
Percentage length of effective transform fault boundary			

Figure 3.29 shows that the clearest association is between true plate speed and the length of effective ocean trench boundary, i.e. those destructive boundaries where the plate is attached to a subducting slab (Figure 3.29d). Of the five plates with more than 20% of their boundaries attached to a descending slab, namely the Indo-Australia, Philippine, Nazca, Pacific and Cocos Plates (h, i, j, k and l), four have true plate speeds of 40 mm y^{-1} or more, whereas six of the seven plates with <5% of their boundaries attached to a slab have true plate speeds of 30 mm y^{-1} or less. Of the remaining plates, none of which are attached to large subducting slabs, only the South America Plate has a true plate speed greater than 30 mm y^{-1}. It is reasonable, therefore, to conclude that slab-pull is a dominant plate-driving force.

There is a poorer positive association between true plate speed and length of effective ridge boundary such that, in general, four of the five plates with true speeds above 40 mm y^{-1} have between 20% and 40% of their plate boundaries occupied by constructive margins. From this, it may be concluded that the so-called 'ridge-push' is also an effective plate-driving force.

In summary, the clearest associations in Figure 3.29, as summarised in Table 3.7 (in the answer to Question 3.9), suggest that true plate velocity is largely determined by the slab-pull and ridge-push forces, with continental drag acting as a major retarding force. However, the quality of the correlations in Figure 3.29 is not good and other factors, such as the age and density of the subducting lithosphere and the nature of local tectonics in the overriding plate, can also play a significant role in determining true plate speed. Some of these ideas will be discussed in later chapters.

Finally, it should be noted that slab-pull and ridge-push forces are largely a consequence of density differences and gravity–slab pull relates to the gravitational sinking of a cold slab of lithospheric material into the mantle whereas ridge-push relates to the gravitational potential energy of the ocean ridge standing 2–3 km above the surrounding ocean floor. So the answer to the question, 'What drives plate tectonics?' is not directly related to mantle convection, but to gravity acting on density differences in the lithosphere that have resulted from its own thermal history.

3.5 Implications of plate tectonics

3.5.1 The Wilson cycle

High-quality, palaeomagnetic data are now sufficiently abundant that it is possible to reconstruct the movement of the continents throughout the past 500–600 million years (i.e. the Phanerozoic) and, with increasing uncertainty, back to 750 Ma and possibly earlier. From these reconstructions it became apparent that the continental masses have been assembled previously into supercontinents that have broken apart, dispersed, and then later reassembled in a different configuration to form another supercontinent. This observation was noted by Wilson (Box 3.1) who proposed that an ocean basin has a lifespan with several stages: it begins with the initial opening, and then goes through a widening phase before starting to close and on to its ultimate destruction. This theory accounts for the cycle of continental break up and reassembly, and became known as the **Wilson cycle** in his honour. From the palaeomagnetic reconstructions, it appears that the cycle of supercontinent assembly – break-up and subsequent reassembly – takes about 500 million years to complete. This time period can be further explained by a simple calculation.

Question 3.10

Imagine that a roughly circular supercontinent, 5000 km in radius, and located about the Equator, rifts in two along a north–south line. A new spreading centre between the rifted halves spreads at an average rate of about 3 cm y^{-1}. How long would it take for the two halves to first meet again on the opposite side of the globe? (Assume that the circumference of the Earth is 40 000 km.)

Clearly, this is an average estimate because spreading rates vary, and continental configurations are far more complex than the simple two-continent rifting model outlined in Question 3.10. But if it is correct, and given that Pangaea formed about 300 Ma ago, the next supercontinent is due to begin to assemble in about 200 million years, perhaps once the Pacific Ocean has been closed by the subduction zones that surround it.

Various stages have been identified for the Wilson cycle, and all of these stages can be recognised in different parts of the Earth today.

1. The earliest stage, called the *embryonic stage*, involves uplift and crustal extension of continental areas with the formation of rift valleys (e.g. the East African Rift System).
2. The *young stage* involves the evolution of rift valleys into spreading centres with thin strips of ocean crust between the rifted continental segments. The result is a narrow, parallel-sided sea, for example the Red Sea that is opening between northeast Africa and Arabia.
3. The *mature stage* is exemplified by widening of the growing basin and its continued development into a major ocean flanked by continental shelves and with the continual production of new, hot oceanic crust along the ridge system (e.g. the Atlantic Ocean).
4. Eventually, this expanding system becomes unstable and, away from the ridge, the oldest oceanic lithosphere sinks back into the asthenosphere, forming an oceanic trench subduction system with a Wadati–Benioff zone demarking the descending plate and associated island arcs, such as the situation in the western Pacific Ocean, or Andean-type volcanism. The onset of subduction at the ocean boundary marks the *subduction stage* of the cycle (e.g. the Pacific Ocean).
5. Once subduction outpaces the formation of new crust at the constructive boundary, the ocean begins to contract. Island arc complexes, complete with their inventory of sedimentary and volcanic rocks, collide and create young mountain ranges around the periphery of the contracting ocean. These features mark the *terminal stage* of the cycle (e.g. the Mediterranean).
6. The *end stage* occurs once all the oceanic crust between the continental masses has subducted, and the continents converge along a collision zone characterised by an active fold mountain belt, such as the Himalayas. Finally, the plate boundary becomes inactive, but the site of the join, or suture, between the two continental masses is a zone of weakness in the lithosphere that has the potential to become the site of a new rift and so the cycle continues.

3.5.2 Plate tectonics and climate change

This chapter began by considering the evidence in the Earth's past for the existence of supercontinents and how evidence of past climates recorded in continental rocks can be used to reassemble ancient continental configurations. The evidence was interpreted in such a way that the continents were considered as passive recorders of the surface conditions that they have experienced on their inexorable passage across the Earth's surface. While such an assumption is broadly correct, it does not take more than a momentary glance at a map of the world today to realise that the disposition of the continents has a marked effect on both local and global climate. Not the least of these effects results from the difference in the thermal properties of land versus ocean – a continental region will be colder in winter and warmer in summer than an oceanic region at any given latitude. Moreover, mountain belts formed as a consequence of plate tectonic activity dramatically modify rainfall through the effects of **orography** – the development of a rain shadow on the leeward side of mountain belts.

Global climate is also strongly controlled by ocean currents. For example, northwestern Europe is significantly warmer than other regions at similar latitudes because of the warming effects of the Gulf Stream and North Atlantic Drift. The reversal of oceanic currents in the equatorial Pacific – a phenomenon known as El Niño – has a far-reaching effect on climate around the Pacific. Ocean currents depend on the geometry of the oceans and this is controlled by plate tectonics. Hence, over geological timescales the movement of plates and continents has a profound effect on the distribution of land masses, mountain ranges and the connectivity of the oceans. As a consequence, plate tectonics has a very direct and fundamental influence on global climate.

To illustrate this effect, Box 3.5 briefly describes the opening of a seaway between the southern tip of South America and Antarctica, and how that affected global climate.

Box 3.5 Opening up of the Drake Passage

The climate of modern Antarctica is extreme. Located over the South Pole and in total darkness for six months of the year, the continent is covered by glacial ice to depths in excess of 3 km in places. Yet this has not always been the case. About 50 Ma ago, even though Antarctica was in more or less the same position over the pole, the climate was much more temperate – there were no glaciers and the continent was covered with lush vegetation and forests. So how did this extreme change come about?

The modern climate of Antarctica depends upon its complete isolation from the rest of the planet as a consequence of the Antarctic Circumpolar Current that completely encircles Antarctica and gives rise to the stormy region of the Southern Ocean known as the roaring forties. The onset of this current is related to the opening of seaways between obstructing continents. Antarctica and South America were once joined together as part of Gondwana and were the last parts of this original supercontinent to separate. By reconstructing continental positions from magnetic and other features of the sea floor in this region, geologists have shown that the Drake Passage opened in three phases between 50 Ma and 20 Ma, as illustrated in Figure 3.30. At 50 Ma there was possibly a shallow seaway between Antarctica and South America, but both continents were moving together. At 34 Ma the seaway was still narrow, but differential movement between the Antarctic and South American Plates created a deeper channel between the two continents that began to allow deep ocean water to circulate around the continent. Finally, at 20 Ma there was a major shift in local plate boundaries that allowed the rapid development of a deep-water channel between the two continental masses.

■ What other major change in global plate motions occurred between 43 Ma and 50 Ma?

■ The change of orientation of the Hawaiian hot-spot trace shows that at this time the Pacific Plate changed from a northward velocity direction to a northwestward direction.

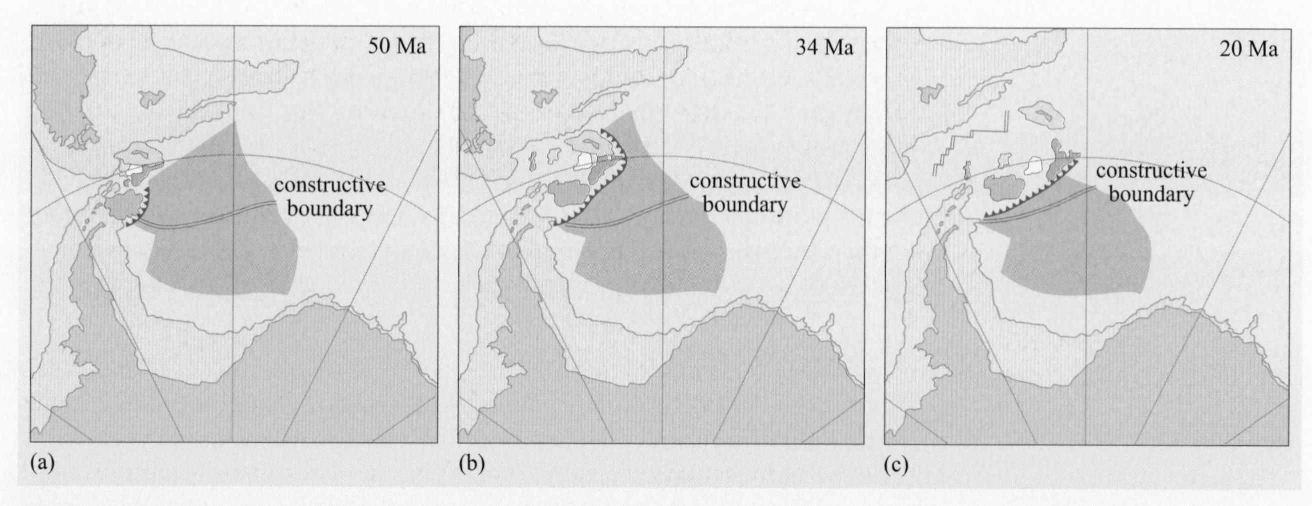

Figure 3.30 Plate reconstruction of the southern tip of South America and the Antarctic Peninsula at (a) 50 Ma, (b) 34 Ma and (c) 20 Ma. The development of a destructive plate boundary at 50 Ma results in the destruction of the ocean lithosphere previously generated at the spreading centre as shown. Continued subduction leads to the development of a back-arc basin and the eastward migration of continental fragments shown in different colours, so that by 20 Ma there is a deep seaway floored by oceanic crust beneath the two continental masses. The darker area represents oceanic lithosphere. (Adapted from Livermore et al., 2005)

The coincidence of the change in motion of the Pacific Plate with changes in plate motions between S. America and Antarctica shows how the motions of all the plates are interconnected – a change in the true motion of one plate leads to changes in the true motions of many others.

While these plate motions were taking place the effect on Antarctica was profound. By 34 Ma the climate cooled from the temperate conditions that previously existed. This was sufficient for glaciers to begin their advance, and was followed by a period of continued cooling until, at about 20 Ma, glaciation was complete. Even though the Drake Passage first opened at 50 Ma it was not until it opened to deep water at 34 Ma that glaciation really took hold.

Today, the Antarctic Circumpolar Current is the strongest deep ocean current and its strength is responsible for the 'icehouse' climate that grips the planet. The opening of the Drake Passage had both a local and a global effect, initially cooling the climate of Antarctica from temperate to cold and ultimately playing an important role in the change from global 'greenhouse' conditions 50 Ma ago to the global 'icehouse' of today.

This example shows how plate tectonics, continental drift and the opening and closing of seaways can have a profound influence on both local and global climate. Throughout the Phanerozoic there were long periods when the Earth was much warmer than today – often called a 'greenhouse' climate – and other times when it was cold – called an 'icehouse' climate. These cycles, like the Wilson cycle, occur over periods of 100 Ma, reflecting the timescale of plate movements and the growth and destruction of oceans. Given the clear link between ocean circulation and climate, and the similar timescales of global climate change and plate motions, it is inescapable that one of the chief controls on long-term changes in the global climate must be plate tectonics.

Summary of Chapter 3

Plate tectonics is the grand, unifying theory of Earth sciences, combining the concepts of continental drift and sea-floor spreading into one holistic theory that explains many of the major structural features of the Earth's surface. It explains why the oceanic lithosphere is never older than about 180 Ma and why only the continents have preserved the Earth's geological record for the past 4000 Ma. It provides the framework to explain the distribution of earthquakes and volcanoes and a mechanism for the slow drift of the continents across the Earth's surface. The theory has now reached such a level of scientific acceptance that the movement of plates, both relative to one another and to the hot-spot reference frame, are being used to infer movement of the hot-spot reference frame with respect to the Earth's rotational axis.

Plate tectonics is an expression of the convective regime in the underlying mantle, but the link between individual convection cells and plate boundaries is not direct because plate boundaries are not fixed and, like the plates, move relative to one another. Plate movements are driven by gravity, largely by cold, dense lithospheric slabs pulling younger lithosphere towards a destructive boundary. A less-powerful driving force is generated by the potential energy of spreading centres, elevated some 2–3 km above the general level of the abyssal plains. You will learn more about mantle convection and how it relates to plate tectonics in Chapter 7.

As ideas concerning plate tectonics have evolved since the 1970s, it has become apparent that while the theory can be applied rigorously to the oceans, the same cannot be said of the continents. Because of the strength and rigidity of oceanic plates, deformation is focused into narrow linear zones along plate margins. By contrast, when continental lithosphere approaches a plate boundary, deformation can extend hundreds of kilometres into the continental interior because continental plates are less strong. Such deformation gives rise to the major mountain belts of the Earth, as exemplified by the Alpine Himalayan Chain, part of which will be described in Chapter 6. Before that, however, you will focus on the processes that occur at constructive (Chapter 4) and destructive plate boundaries (Chapter 5), the major sites of interaction and exchange between the interior of the Earth and the hydrosphere, atmosphere and biosphere.

Learning outcomes for Chapter 3

You should now be able to demonstrate a knowledge and understanding of:

3.1 The theory of tectonic plates and the different forms of evidence (e.g. palaeontology, palaeomagnetism and continuity of structures) that can be used to understand the movement of the lithospheric plates over geological time.

3.2 The mechanisms of crustal growth and transfer of heat at spreading ocean ridges.

3.3 The three main types of plate boundary (constructive, destructive and conservative) and how they interact at triple junctions.

3.4 The difference between relative and true plate motion.

3.5 The driving and retarding forces that influence plate motion at constructive, destructive and conservative plate boundaries.

Processes at constructive plate boundaries

Chapter 4

Two-thirds of the Earth's surface (about 300×10^6 km^2) is made of oceanic lithosphere, and all of that lithosphere has formed during the last 4% of the Earth's history. On average, about 3 km^2 of new oceanic lithosphere is made every year, which is distributed along the constructive plate boundaries that are found at mid-ocean ridges. Although these statistics imply that mid-ocean ridges are important components of the Earth's dynamic systems, direct observations of this prodigious activity have lagged behind observations of geological processes on land, for obvious reasons. It is only since the 1970s that submersible craft have been able to take cameras, other instruments and, importantly, people, to the dark ocean depths of the mid-ocean ridges. There, some 2.5 km below sea level, very young lava flows and hot springs that gush water at up to 400 °C show, together with the longer known background buzz of shallow-focus earthquakes, that mid-ocean ridges are hot-beds of geological activity (Figure 4.1). While Chapter 3 described the tectonics of constructive plate boundaries and their role in driving sea-floor spreading, this chapter describes the processes that generate oceanic lithosphere from the mantle that underlies the mid-ocean ridges. The overall aim of this chapter, therefore, is to discover what happens, why it happens, and where it happens.

(a)

(b)

Figure 4.1 Signs of activity at mid-ocean ridges. (a) The aftermath of a volcanic eruption on the East Pacific Rise (9° 50.6′ N) in 1991 shows the corrugated surface of lava littered with fragments of organisms killed by the eruption. (b) The plume of black 'smoke' issuing from a chimney of metallic sulfide minerals is a cloud of dark metal sulfide particles precipitated when hot (380 °C) mineral-rich water jets out of the sea floor and mixes with cold (2 °C) seawater. ((a) R.M. Haymon; (b) Dr Ken MacDonald/Science Photo Library)

Section 4.1 reviews the structure of the oceanic lithosphere – the finished product of constructive plate boundary processes. The later sections then show how that finished product is generated, beginning in Section 4.2 with the starting material – mantle rock. Subsequent sections deal with the processes that cause the mantle to melt and produce magma (Sections 4.3 to 4.5), the history of that magma as it makes its way to the surface (Sections 4.6 to 4.7), and the interaction between igneous oceanic crust and the oceans (Section 4.8).

4.1 The structure and composition of oceanic lithosphere

From the point of view of plate tectonics, oceanic lithosphere is the relatively cold, strong, outer part of the Earth in the ocean basins. Far from mid-ocean ridges, this layer has a fairly uniform thickness of about 120 km. To understand how the Earth generates new oceanic lithosphere, it is helpful to know its composition and structure. But because the oceanic lithosphere is largely hidden from view and inaccessible, finding out about these rather fundamental aspects of the Earth's outer layer has required ingenuity and the piecing together of different types of information. Apart from the areas at and near mid-ocean ridges, the seabed is covered in very fine mud. Drilling into this sediment layer shows that it can be up to several hundred metres thick and that it is underlain by volcanic rocks of basaltic composition. The crests of mid-ocean ridges, on the other hand, are devoid of any obscuring sediment but display signs of recent volcanic activity (Figure 4.1a). These observations are a key factor in suggesting that the mid-ocean ridges are the youngest parts of the ocean basins and that they have an igneous origin. At some mid-ocean ridge sites, plumes of hot, metal-rich brines rise from the sea floor (Figure 4.1b). Both of these manifestations of heat loss from the Earth's interior signal that, multiplied along the 64 000 km length of the mid-ocean ridge system, constructive plate boundaries play a significant role in the slow but continuous cooling of the Earth.

To see the interior of the oceanic crust it is necessary to visit areas where faults have sliced through the oceanic lithosphere to expose deeper layers, as if cutting through a cake to see what lies beneath the outer layer of icing. One such opportunity is provided by oceanic fracture zones that have cut cross-sections through oceanic crust. These can be investigated from submersible craft or by dredging samples from the rock face. The rocks found in the fracture zones are basalt lava flows, basalt dykes, gabbro and peridotite (Box 4.1) arranged in a layered structure (Figure 4.4).

Box 4.1 Basalt, gabbro and peridotite

The rock types basalt, gabbro and peridotite are defined according to the minerals they contain and/or the rock's texture. In these rocks, the most common minerals are olivine (Figure 4.2a), pyroxene (Figure 4.2b) or plagioclase feldspar (Figure 4.2c). Olivine is a greenish-coloured, magnesium–iron silicate with the chemical composition $(Mg,Fe)_2SiO_4$, where the brackets in the mineral formula signify that for every four oxygen atoms there are a total of two **cations** comprising a mixture of magnesium and iron. Pyroxene is dark-green to black, with the composition $(Ca,Mg,Fe)_2Si_2O_6$. Plagioclase feldspar is a whitish silicate mineral containing aluminium, calcium, sodium and a very a small amount of potassium with the composition $(KSi,NaSi,CaAl)AlSi_2O_8$.

The chemical compositions of basalt, gabbro and peridotite vary within bounds according to the exact composition of the minerals present and their proportions. However, given the general mineralogical make-up of these rocks, peridotite is poorer in Si, Al, Na and Ca, but richer in Mg than either gabbro or basalt. Gabbro and basalt have the same composition (referred to as basaltic); the differences between these two rock types are textural, with basalt having smaller and fewer crystals.

- Basalt (Figure 4.3a) is a fine-grained volcanic rock that typically contains small crystals of pyroxene, plagioclase and, in some cases, olivine.
- Gabbro (Figure 4.3b) is a coarse-grained rock composed of pyroxene and plagioclase and, in some cases, olivine.
- Peridotite (Figure 4.3c) is a coarse-grained rock composed of at least 90% olivine and pyroxene, with at least 40% olivine.

Basalt forms by relatively fast cooling of basaltic magma, either as a volcanic rock or in dykes or other minor intrusions. Gabbro forms when basaltic magma cools so slowly that the entire magma crystallises.

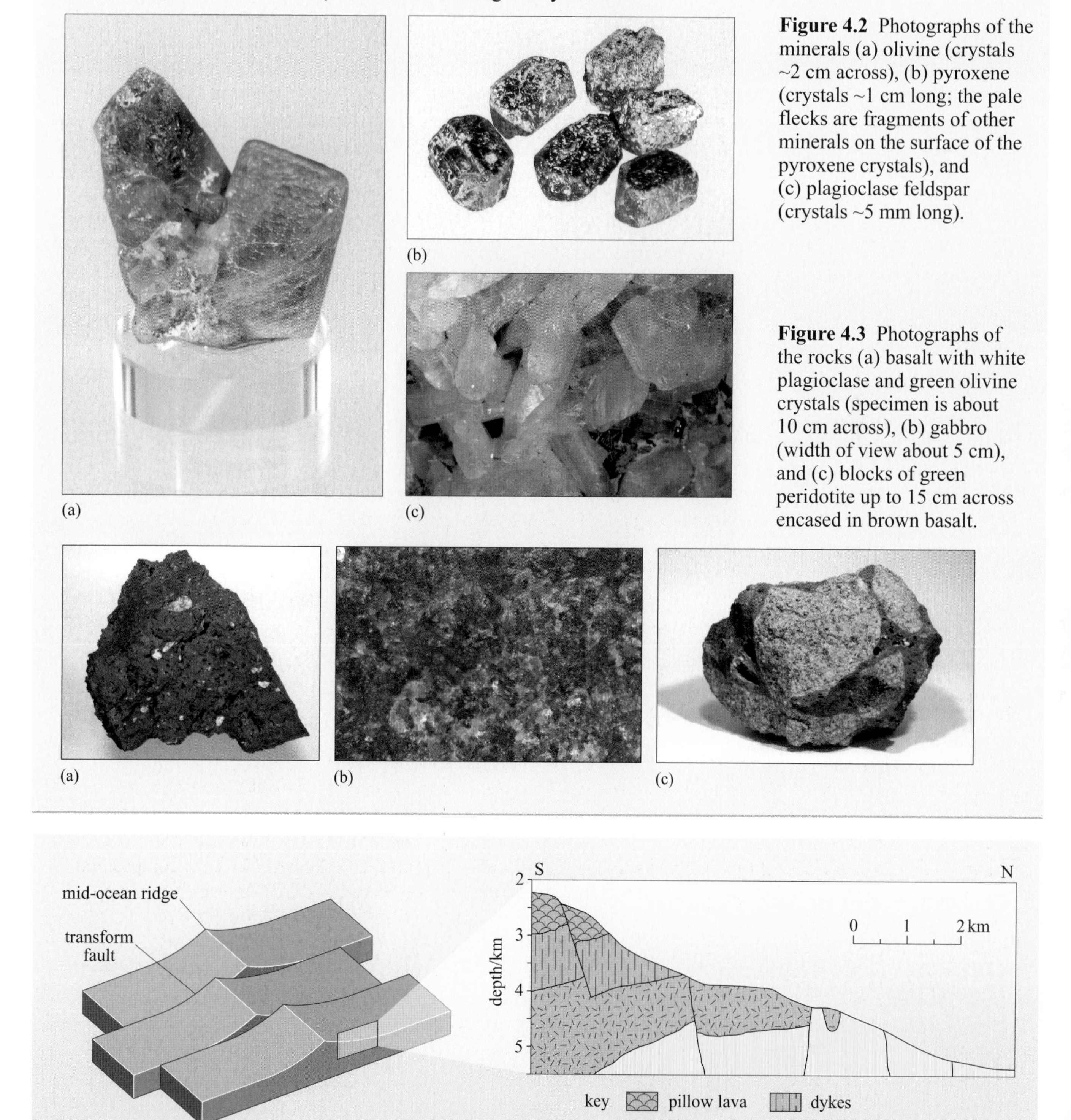

Figure 4.2 Photographs of the minerals (a) olivine (crystals ~2 cm across), (b) pyroxene (crystals ~1 cm long; the pale flecks are fragments of other minerals on the surface of the pyroxene crystals), and (c) plagioclase feldspar (crystals ~5 mm long).

Figure 4.3 Photographs of the rocks (a) basalt with white plagioclase and green olivine crystals (specimen is about 10 cm across), (b) gabbro (width of view about 5 cm), and (c) blocks of green peridotite up to 15 cm across encased in brown basalt.

Figure 4.4 (a) Block diagram showing how transform faults at oceanic fracture zones expose the upper parts of oceanic crust. (b) Geological structure of the Vema Fracture Zone that offsets the Mid-Atlantic Ridge (around 10.75° N, 42° W) shows layers offset by faults and exposed by erosion. ((b) Adapted from Auzende et al., 1989)

Evidence for an igneous, layered oceanic crust with the composition of basalt also comes from, of all places, continental settings where convergent plate motions have caused segments of oceanic lithosphere to be tectonically emplaced onto continental and arc crust in a process called **obduction** (in contrast to subduction, which removes oceanic lithosphere from the Earth's surface down into the mantle). The rocks that make up these obducted fragments are deep-sea sediments, basaltic pillow lavas, basaltic dykes (**sheeted dykes**), gabbro and peridotite (Figure 4.5). This association of rock types was first recognised in the early 19th century, when it was given the name ophiolite. However, it was not until the 1960s and the acceptance of plate tectonics that it was realised that ophiolites were stranded blocks of oceanic lithosphere trapped between colliding continents (Figure 4.6). Examples of ophiolites that teams of geologists have studied in detail occur in the Troodos Mountains of Cyprus, and in Oman, where the barren landscape gives almost complete exposure of rocks that once formed a complete section of oceanic crust (Figure 4.5).

Figure 4.5 (a) Photograph of pillow lavas in the Oman ophiolite formed by the flow of basalt lava under water. (b) Photograph of a cliff face composed entirely of near-vertical dykes of basalt in the Oman ophiolite. Each dyke has intruded rock made entirely of other dykes, resulting in a sheeted dyke complex. (Nigel Harris/OU)

Figure 4.6 Map showing the global occurrence of ophiolites (marked as red dashed lines) illustrating how they often occur in long chains that were formed when they were obducted during mountain-building phases of continental collision. (Gass, 1990)

The geologist's view of oceanic crust, based on studies of ophiolites and deep-sea rocks, can be compared with the results of geophysical investigations of the physical properties of oceanic crust. Foremost among these is the velocity of P-waves in rocks (v_P), because this depends on the mineral composition, making seismology a useful way of probing the geological composition and structure of the subsurface. The results of some seismic refraction experiments at several locations in the Atlantic Ocean off the coast of Florida are shown in Figure 4.7. These indicate that P-wave velocity increases with depth in the oceanic crust. The different profiles from this area share some common features: with increasing depth, v_P increases rather irregularly over the top 2 km. Below this depth, v_P increases much more gradually over a further 6 km until, at a depth of about 8 km, it increases sharply to a value close to 8 km s^{-1}, which is typical of the P-wave velocity in peridotite. The layer of nearly constant v_P above this boundary is interpreted as being made of gabbro, whereas the shallower layer most probably represents a series of pillow lavas that grade downwards into sheeted dykes.

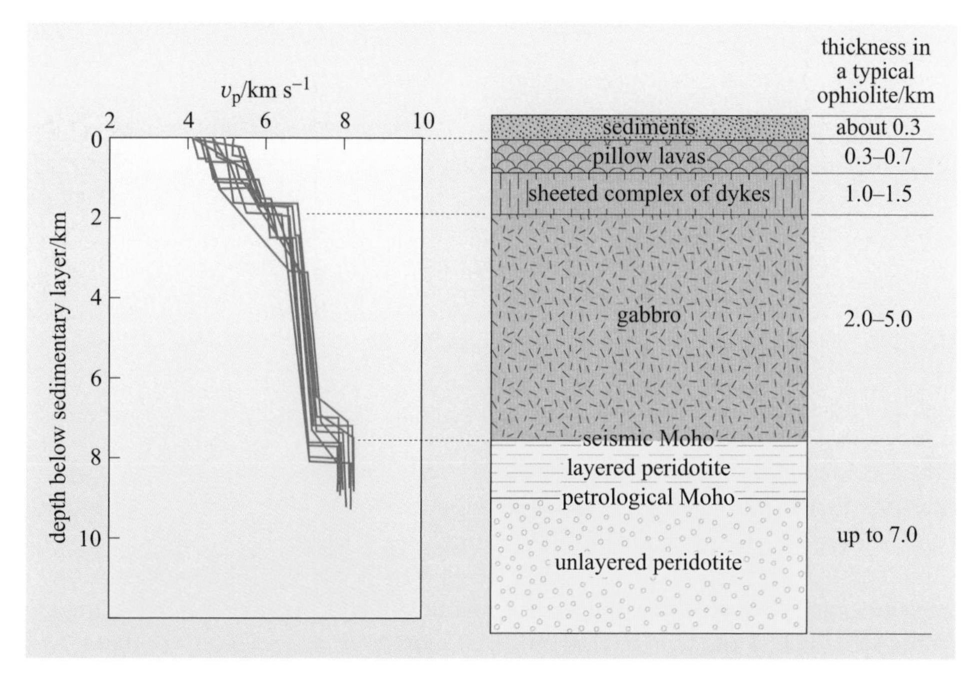

Figure 4.7 Seismic profile of 142 Ma-old oceanic crust in the Atlantic Ocean off the coast of Florida that lies beneath the surface layer of sediment. It shows a layered crustal structure that can be correlated with the layered geology seen in ophiolites. (Seismic profiles from White et al., 1992)

The peridotites found in ophiolites can be classed into two types based on their physical appearance: layered and unlayered. These are distinguished in Figure 4.7, but as they have the same P-wave velocities they cannot be distinguished seismically. The seismically defined Moho (i.e. the **seismic Moho**) is therefore interpreted as marking the shallowest occurrence of peridotite. Layered peridotites are thought to originate by crystallisation of olivine in magma chambers at constructive plate boundaries within oceanic crust, so the top of the mantle is best defined as being the contact between the layered and unlayered peridotite. This interface is known as the **petrological Moho** (Figure 4.7), **petrology** being the study of the composition, texture and structure of rocks. However, as the thickness of the layered peridotites is measured in hundreds of metres and the total thickness of the crust is measured in kilometres, the depth of the seismic Moho is a perfectly good estimate of the thickness of oceanic crust. On average, the igneous part of oceanic crust is 7 km thick. The sedimentary layer is much thinner and variable (between 0 km and 3 km, with an average thickness of 0.4 km).

Nearly the entire thickness of the oceanic crust has an igneous origin and is made from the solidified remains of the stored, transported and eruptive components of a basaltic magma system:

- gabbro that formed by solidification of slowly crystallising basaltic magma
- sheeted dykes that were formed by solidification of basalt magma intruded into crust that was composed of older basalt dykes
- lavas that were emplaced when the dykes broke through onto the sea floor (Figure 4.8).

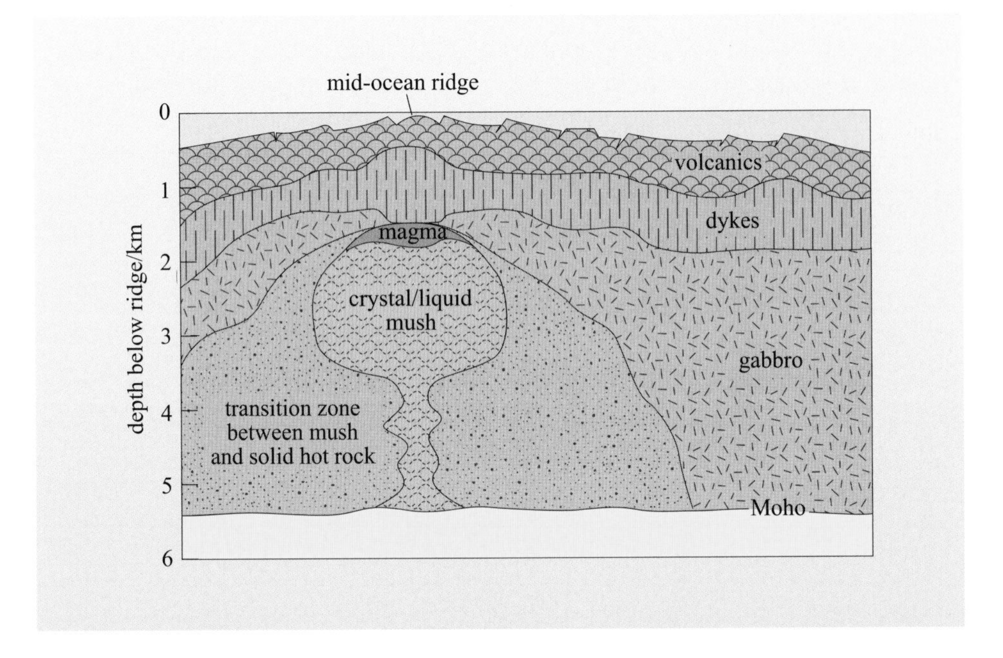

Figure 4.8 Schematic cross-section of a mid-ocean ridge showing how the components of the system that produce the basaltic oceanic crust may be arranged. (Sinton and Detrick, 1992)

Basaltic magma is the raw material from which oceanic crust is constructed by sea-floor spreading, but why, when seismology shows that the mantle is normally solid, should magma be available beneath mid-ocean ridges? In other words, why should parts of the mantle be melting beneath mid-ocean ridges? To answer these questions, you will need to learn more about the material that the mantle is made from (Section 4.2) and then to find out what conditions are required for that material to melt (Section 4.3).

4.2 Rocks and minerals in the upper mantle

Parts of the Earth's mantle can be seen at the surface in ophiolites and oceanic fracture zones, but these are not the only source of mantle samples available to geologists. Lumps of mantle rock are occasionally picked up and carried by rapidly rising magma as it bursts upwards to the surface. These accidental but very special rock fragments, known as mantle xenoliths, can be found at some basaltic volcanoes such as Lanzarote in the Canary Islands, and Kilbourne Hole in New Mexico (Figure 4.3c). A rare type of CO_2-rich magma known as **kimberlite** (named after the Kimberley area of South Africa) also carries a cargo of xenoliths and crystals from the mantle, including diamond, which is the high-pressure form of carbon that forms at depths in excess of 150 km. Before looking more closely at the types of rock found in the upper mantle, the next section gathers together information about the minerals that are found in these (and other) rocks.

4.2.1 Minerals in the upper mantle

There are only a few different minerals that are commonly found in mantle rocks, and these are listed in Table 4.1.

Table 4.1 Chemical formulae of minerals found in peridotite.

Mineral	Formula
olivine	$(Mg,Fe)_2SiO_4$
pyroxene	$(Ca,Mg,Fe)_2Si_2O_6$
garnet	$(Ca,Mg,Fe)_3Al_2Si_3O_{12}$
plagioclase feldspar	$(KSi,NaSi,CaAl)AlSi_2O_8$
spinel	$(Mg,Fe^{2+})(Al,Cr,Fe^{3+})_2O_4$

Question 4.1

The minerals in Table 4.1 can be classified in various ways. Which minerals contain the following?

(a) silicon and oxygen

(b) *no* silicon

(c) aluminium

(d) iron and magnesium.

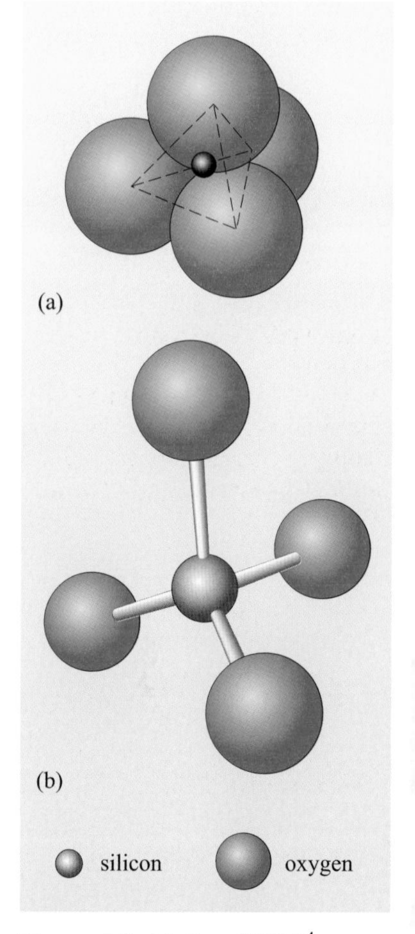

Figure 4.9 (a) One $(SiO_4)^{4-}$ tetrahedral unit (silicate tetrahedron) with the four atoms of oxygen and the one atom of silicon shown to true relative scale. (b) The atomic structure of the $(SiO_4)^{4-}$ unit showing how the silicon and oxygen atoms are linked.

Most mantle minerals are silicates that also contain iron and magnesium. These are examples of **ferromagnesian silicate minerals**. They are also known as **mafic** minerals because of their high content of magnesium and iron (particularly Fe(II) iron, Fe^{2+}).

On the atomic scale, the silicate minerals are constructed from building units that contain one Si^{4+} ion and four O^{2-} ions arranged in a pyramid-shape known as a tetrahedron (Figure 4.9). Each tetrahedron has the Si^{4+} ion in the centre of the pyramid and one O^{2-} at each corner; this is the **silicate tetrahedron**.

Question 4.2

What is the overall electronic charge of a silicate tetrahedron?

Silicate minerals are made up of silicate tetrahedra linked to each other, or occurring separately, with the intervening spaces filled by cations of the appropriate size and charge. This means that the chemical formula and atomic structure of each mineral are related. The main features of the minerals are as follows.

Olivine

The general formula for olivine is $(Mg,Fe)_2SiO_4$. The brackets indicate that a range of compositions is possible. For example, olivine can be Fe_2SiO_4, $Fe_{0.2}Mg_{1.8}SiO_4$, $Fe_{1.1}Mg_{0.9}SiO_4$ or any other combination of Fe and Mg as long as the atomic proportions of Fe and Mg atoms add up to 2 (for every four O atoms).

You would be right in thinking that olivine contains individual $(SiO_4)^{4-}$ tetrahedra because SiO_4 appears in the formula. In the case of pure magnesium-rich olivine, the charges from two Mg^{2+} ions (4+) are required to balance the charge of one silicate tetrahedron (4−), which is why the chemical formula is Mg_2SiO_4. The magnesium cations and silicate tetrahedra are arranged in a particular geometric pattern that allows the sizes and charges of the cations and tetrahedra to stay in stable positions, giving a crystal a specific shape and symmetry (Figure 4.2a).

1 picometre (pm) = 10^{-12} m

But what happens if some of the Mg^{2+} ions are replaced with Fe^{2+} ions? The radius of an Mg^{2+} ion is 66 pm and the ionic radius of an Fe^{2+} ion is 74 pm. Because of their similar size and identical charge, Fe^{2+} ions can take the place of Mg^{2+} ions without distorting the overall mineral structure of olivine. The Mg and Fe atoms can substitute for each other and, in olivine, all compositional variants between Fe_2SiO_4 (**fayalite**) and Mg_2SiO_4 (**forsterite**) are possible because **complete ionic substitution** is allowed because of the similarity in size of the Fe^{2+} and Mg^{2+} ions. These variants are referred to as members of a **solid-solution series**, and the endmembers of the olivine solid-solution series are fayalite and forsterite. Fayalite and forsterite have identical atomic structures and are said to exhibit **isomorphism**, a term which means 'equal form'.

Can Ca^{2+} ions enter the olivine structure? Calcium forms a much larger cation (radius = 99 pm) and cannot be accommodated easily in the structure of olivine. Consequently, only very limited ionic substitution of Ca^{2+} for Fe^{2+} or Mg^{2+} occurs, and it is unusual for olivine to contain much more than 1% Ca.

As a general rule, one ion may completely replace another to form a solid-solution series if the difference in their ionic radii *does not exceed 15%* of the radius of the smaller ion. If the size difference is greater than this, then **partial ionic substitution** will occur, and the compositional variation cannot extend continuously between the endmembers.

Pyroxene

In the pyroxenes, the silicate tetrahedra are linked together by sharing two of their oxygens with two neighbouring tetrahedra, forming a chain of tetrahedra (Figure 4.10). The pyroxene structure has six oxygens for every two silicons, giving units of $(Si_2O_6)^{4-}$ that are charge balanced by two divalent cations – a mixture of Mg^{2+} and Fe^{2+} with some Ca^{2+}, hence the general pyroxene formula $(Ca,Mg,Fe)_2Si_2O_6$, which is sometimes written in simplified form as $(Ca,Mg,Fe)SiO_3$.

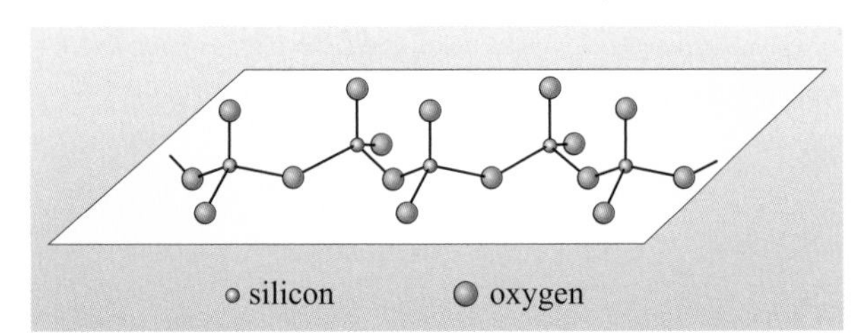

Figure 4.10 A chain of linked silicate tetrahedra, as found in pyroxene.

- Would you expect to find complete ionic substitution between $Mg_2Si_2O_6$ and $Ca_2Si_2O_6$?

- No. The '15% rule' applies and, because the ionic radius of Ca^{2+} is greater than that of Mg^{2+} by 50%, only partial ionic substitution occurs.

Because Ca^{2+} is considerably larger than Mg^{2+} and Fe^{2+}, the atomic structure becomes increasingly distorted as the Ca content increases. When Ca makes up less than about 10% of the cations in pyroxene, the silicate chains and intervening cations form crystals with three symmetry axes at right angles (i.e. orthogonal), which is identical to the symmetry of a cereal box. These are the **orthopyroxenes** (so-called because of their characteristic orthogonal symmetry axes). The simplest orthopyroxenes form a solid-solution series between the endmembers **enstatite** ($Mg_2Si_2O_6$) and **ferrosilite** ($Fe_2Si_2O_6$).

Pyroxenes with more than about 10% Ca have an atomic structure with only two symmetry axes because of the distortions required to fit the large Ca^{2+} ions into the spaces between the chains of silicate tetrahedra. This group of pyroxenes are called **clinopyroxenes**. Their chemical composition can be expressed as mixtures of the endmembers **diopside** ($(Ca,Mg)_2Si_2O_6$), **hedenbergite** ($(Ca,Fe)_2Si_2O_6$), enstatite ($Mg_2Si_2O_6$) and ferrosilite ($Fe_2Si_2O_6$). Note that diopside is a chemical mixture of $Mg_2Si_2O_6$ and $Ca_2Si_2O_6$. Similarly hedenbergite is a chemical mixture of $Fe_2Si_2O_6$ and $Ca_2Si_2O_6$.

The mineral with the formula $Ca_2Si_2O_6$ is **wollastonite**, but it is not a pyroxene because the structural adjustments needed to fit in all the Ca^{2+} ions are so great that it no longer has the same structure of pyroxenes; hence wollastonite is recognised as a different mineral in its own right. Nonetheless, wollastonite is often considered as an endmember when describing the chemical compositions of pyroxenes. For example, diopside can be described as 50% enstatite, 50% wollastonite.

Pyroxene from mantle rocks also contains small amounts of Al, Ti, Cr and Na.

Garnet

Like olivine, garnet contains individual $(SiO_4)^{4-}$ silicate tetrahedra. Three of these are required to account for the Si_3O_{12} part of the mineral's chemical formula ($(Ca,Mg,Fe)_3Al_2Si_3O_{12}$). Unlike olivine, garnet contains trivalent ions, notably Al^{3+} and, in some cases, Fe^{3+}. A comparatively large range of cations can fit into garnet (it might even be called a 'chemical dustbin'), so there are many endmembers to the garnet **isomorphous series**. Here, we mention only pyrope ($Mg_3Al_2Si_3O_{12}$), almandine ($Fe^{2+}{}_3Al_2Si_3O_{12}$) and grossular ($Ca_3Al_2Si_3O_{12}$). Garnet found in mantle peridotites is rich in the pyrope (Mg) component.

Feldspar

As noted earlier, in the pyroxenes the corners of two silicate tetrahedra are shared with neighbours to produce a chain structure (Figure 4.10). In feldspar all four corners are shared, resulting in a three-dimensional framework. The simplest silicate mineral with a framework structure is quartz (which is *not* a feldspar), which has the simple formula SiO_2 because each Si^{4+} ion can lay claim to only half of each of the four O^{2-} ions that surround it. Quartz is a common mineral in the continental crust but it is absent from mantle peridotite. In feldspar, either one or two of the Si^{4+} ions in the framework are replaced by Al^{3+} ions.

Where *one* out of every four silicons is replaced by aluminium, charge balance is re-established by introducing one Na^+ or one K^+, and this yields the **alkali feldspars**: **albite** ($NaAlSi_3O_8$) and **orthoclase** ($KAlSi_3O_8$).

■ Would you expect complete ionic substitution between albite and orthoclase, given ionic radii of 97 pm for Na^+ and 133 pm for K^+?

■ The radius of K^+ is $\frac{(133-97)}{97} \times 100\% = 37\%$ greater than that of Na^+.

Because this is higher than the 15% limit, complete ionic substitution between albite and orthoclase does not occur.

When Al^{3+} replaces Si^{4+} in *two* out of every four tetrahedra, the divalent cation Ca^{2+} restores the neutral charge, giving **anorthite** ($CaAl_2Si_2O_8$).

- Will there be a solid-solution series between anorthite and albite?

- The similarity of ionic radii for Ca^{2+} and Na^+ (99 pm and 97 pm respectively) allows complete ionic substitution to occur between albite and anorthite to form a solid-solution series.

This is known as the plagioclase feldspar series. As Na^+ is replaced by Ca^{2+} in going from albite ($NaAlSi_3O_8$) to anorthite ($CaAl_2Si_2O_8$), so a second Si^{4+} (in every four tetrahedra) is replaced by Al^{3+}. In other words, this solid-solution series involves substitution of CaAl for NaSi to keep the structure electrically balanced, and this is an example of **coupled substitution**. A small amount of K is also present in most plagioclase feldspars because K can substitute for Na to a limited extent.

Spinel group

This is a group of non-silicate minerals loosely described by the chemical formula MOR_2O_3, where M is a divalent cation (e.g. Mg^{2+}) and R is a trivalent cation (e.g. Al^{3+}). The formula of spinels can also be written as MR_2O_4. Important endmembers within this rather diverse group are spinel ($MgAl_2O_4$), magnetite ($Fe^{2+}(Fe^{3+})_2O_4$) and chromite ($Fe^{2+}Cr_2O_4$).

Mineral compositions

Most of the minerals in mantle rocks are ferromagnesian minerals with solid solution between Mg and Fe endmembers, but we have not said which endmembers dominate. In all cases, the mineral composition can be expressed in terms of the proportion of the Mg endmember (relative to the total amount of the Mg and Fe^{2+} endmembers) using the ratio $Mg/(Mg + Fe^{2+})$, where Mg and Fe^{2+} are the number of Mg and Fe^{2+} atoms in the mineral formula. This ratio can be multiplied by 100 to express it as a percentage. The value, either of the ratio or of the percentage, is called the **magnesium number** and is often abbreviated as **Mg-number** or **Mg#**.

Question 4.3

(a) An olivine crystal has the composition $Mg_{1.8}Fe_{0.2}SiO_4$. What is its Mg# as a ratio and as a percentage?

(b) A clinopyroxene crystal has the composition $Mg_{1.2}Fe_{0.4}Ca_{0.4}Si_2O_6$. What is its Mg# as a ratio and as a percentage?

Chemical analysis of the ferromagnesian minerals in mantle rocks shows that they are much richer in the Mg endmember than the Fe endmember. Olivines in peridotite xenoliths have Mg# in the range 88 to 92. In other words, they have between 88% and 92% forsterite, and between 12% and 8% fayalite. Using the conventional abbreviations Fo for forsterite and Fa for fayalite, these compositions are written more concisely as $Fo_{88}Fa_{12}$ to $Fo_{92}Fa_8$. The Mg#s of orthopyroxenes and clinopyroxenes cover a similar range to those of olivine but extend to slightly higher values. Mantle-derived garnets are rich in Cr and Mg.

4.2.2 Rock types in the upper mantle

Rocks from the mantle, whether from the base of ophiolite sequences or xenoliths found in volcanic rocks, are dominated by olivine, orthopyroxene and, to a lesser extent, clinopyroxene. The content of the mafic minerals is so high that the rocks are termed **ultramafic rocks**; these rocks can be subdivided into different types depending on the relative proportions of the three minerals olivine, orthopyroxene and clinopyroxene.

To show the proportions of these three components in any mixture calls for a triangular shaped diagram, known as a **ternary diagram**, as explained in Box 4.2. The diagram used for classifying ultramafic rocks is shown in Figure 4.12.

Box 4.2 Plotting and reading ternary diagrams

A ternary diagram can be used to plot the composition of any substance or mixture in terms of three components. Examples include the chemical compositions of feldspar and pyroxene, and the proportions of different minerals in rocks. Consider a mixture of three components A, B and C. A ternary diagram consisting of a triangle with corners representing the three components (100% A, 100% B, and 100% C) is required to plot the composition of this mixture (Figure 4.11).

- A sample that contained only component A (i.e. 100% A) would plot at the corner labelled A.
- A sample that contained only a mixture of B and C (i.e. 0% A) would plot on the edge of the triangle joining the corners B and C; the greater the proportion of component C, the closer it would be to the C corner.

Figure 4.11a shows how the proportion of A is represented by lines parallel to the edge BC.

Suppose you wanted to plot the composition of a sample that contained 40% A, 30% B and 30% C (a shorthand way of writing this is $A_{40}B_{30}C_{30}$, or $A_{0.40}B_{0.30}C_{0.30}$). This is done as follows.

1. Determine on which 'A' line the composition must lie (as in Figure 4.11a).
2. Find the 'B' line.
3. The point where these two lines intersect marks the composition of the mixture.
4. As a final check, mark the 'C' line, which should pass through the same point (Figure 4.11b).

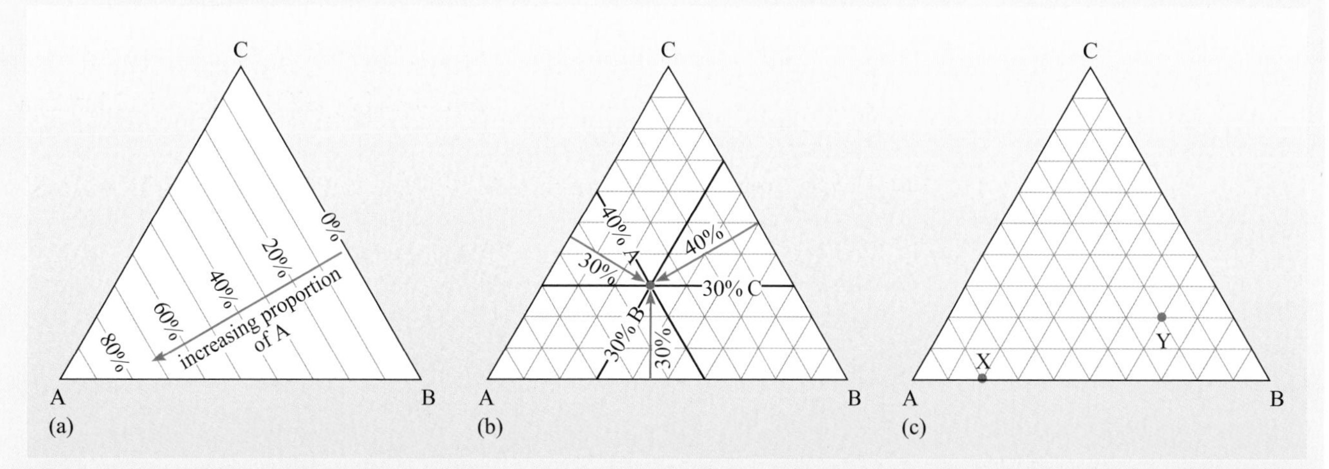

Figure 4.11 (a) A ternary diagram with three components A, B and C. (b) A ternary diagram with the composition $A_{40}B_{30}C_{30}$ plotted. The percentages of each component are indicated by the thick lines, which intersect at a point that shows the overall composition. (c) A ternary diagram for use with Question 4.4.

Question 4.4

Using the ternary diagram in Figure 4.11c, determine the compositions indicated by the two points X and Y.

Figure 4.12 is split into areas, or fields, that define the nomenclature for ultramafic rocks, which are themselves defined as containing more than 90% mafic minerals. Those that contain more than 40% olivine are defined as peridotites. These are further divided according to the proportions of olivine, orthopyroxene and clinopyroxene. The average composition of peridotite xenoliths from basalts and kimberlites is plotted on Figure 4.12 and falls in the lherzolite field. **Lherzolite** samples can be subdivided according to which Al-bearing mineral they contain. Kimberlites have mantle xenoliths that contain garnet, so these are called garnet lherzolite. Spinel lherzolites are found in some basalts from ocean islands (such as Lanzarote) and the continents (e.g. Kilbourne Hole crater, New Mexico) and in oceanic fracture zones. Plagioclase lherzolites are found rarely in oceanic settings.

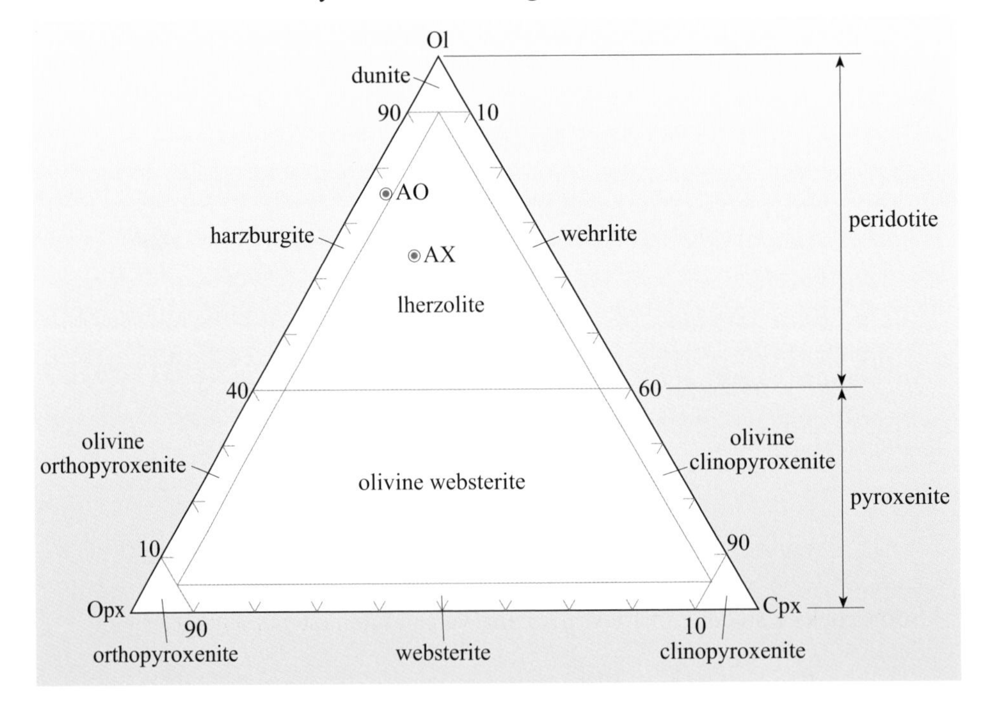

Figure 4.12 Classification of ultramafic rocks based on the relative proportions of olivine (Ol), orthopyroxene (Opx) and clinopyroxene (Cpx). Average compositions of peridotites from ophiolites and oceanic fracture zones (AO) and of xenoliths from basalts and kimberlites (AX). (Dunite, harzburgite, lherzolite, and **websterite** are named after the places where the rock type was first recognised: Dun Mountain in New Zealand, the Harz Mountains in Germany, Etang de Lherz in the French Pyrenees, and Webster in North Carolina, USA; **wehrlite** is named after Adolf Wehrle, a 19th century Austrian Councillor of Mines.)

Question 4.5

(a) What type of ultramafic rock is represented by the average oceanic peridotite composition (labelled AO) in Figure 4.12?

(b) What are the approximate proportions of olivine, orthopyroxene and clinopyroxene in this average composition?

Compared with the average lherzolite, oceanic peridotites have a lower proportion of clinopyroxene and a higher proportion of olivine.

4.2.3 Peridotites and basalts

The chemical compositions of different rocks can be useful in deducing how the rocks formed. Table 4.2 lists typical chemical compositions of a harzburgite from an ophiolite, a lherzolite xenolith, and a mid-ocean ridge basalt. See also Box 4.3.

Table 4.2 Chemical compositions (in weight per cent) of two peridotites and a mid-ocean ridge basalt.

	Harzburgite	Lherzolite	Mid-ocean ridge basalt
SiO_2	44.69	45.35	48.77
TiO_2	0.02	0.16	1.15
Al_2O_3	0.86	4.26	15.90
FeO	8.17	8.24	9.81
MnO	0.12	0.14	0.17
MgO	45.04	38.17	9.67
CaO	1.09	3.39	11.16
Na_2O	0.02	0.29	2.43
K_2O	0.01	0.03	0.08

Box 4.3 The chemical composition of rocks

Although virtually every chemical element can be found in most rocks, oxygen, together with ten other elements (Si, Ti, Al, Fe, Mn, Mg, Ca, Na, K, and P) account for nearly all of the matter in a given rock. These ten elements are therefore known as the **major elements**, as opposed to other elements that occur in much lower abundances, which are known as **trace elements**. Because most minerals contain O, with matching amounts of cations needed to balance the electronic charge, rock analyses are usually reported in terms of the weight percentage of the major element oxides SiO_2, TiO_2, etc., as shown in Table 4.2. Iron can occur in two oxidation states – Fe^{2+} and Fe^{3+} – so some rock analyses give the separate concentrations of their oxides FeO and Fe_2O_3 whereas other analyses report the concentration of all of the iron in terms of either FeO or Fe_2O_3. Some rocks also contain a few per cent water. When the percentages of all the oxides are added up, the total rarely comes to 100%. This is because of (i) analytical errors and (ii) components such as water and/or some trace elements that are actually quite abundant in a given rock were not analysed.

■ In what way do the chemical compositions of lherzolite and harzburgite in Table 4.2 reflect the composition of the minerals found in those rocks?

■ Olivine and pyroxene from peridotites have high Mg#s so, because these minerals are the main constituent of peridotite, the rocks also have much higher MgO contents than FeO contents.

The composition of basalt is starkly different from that of lherzolite or harzburgite, having much less magnesium, slightly higher silica and iron, and much higher concentrations of other oxides. Clearly, basalt is not a molten version of peridotite – the chemical compositions are far too different for them to represent the same material, so basalt must have another origin.

A clue that there is a relationship between the three rock compositions in Table 4.2 comes from noticing that, for every oxide, the composition of the lherzolite falls between that of harzburgite and basalt. It is possible to split lherzolite into two compositionally different portions – a mobile basalt liquid and a harzburgite residue. Basalt, therefore, forms by the partial melting of lherzolite:

$$\text{lherzolite} \rightarrow \text{basalt} + \text{harzburgite}$$

This process is fundamental in processing lherzolite from within the mantle and turning it into oceanic lithosphere comprising basaltic crust that is underlain by harzburgite, as seen in ophiolite sequences (Figure 4.13).

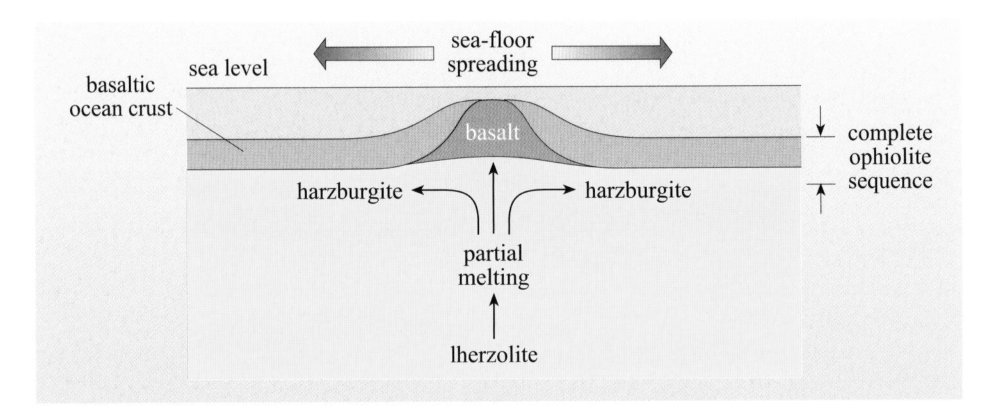

Figure 4.13 Cross-section of a constructive plate boundary, showing partial melting of lherzolite to produce basalt liquid and solid residual harzburgite.

The explanations of why partial melting happens, and exactly where within the mantle it happens, are the topics of the next section.

4.3 Experiments with peridotite

The conditions in the mantle where peridotite begins to melt must involve higher temperature (T) and higher pressure (P) than are encountered on the Earth's surface. But how high must T and P be for melting to happen? Taking a sample of peridotite and subjecting it to high T and P in the laboratory and seeing whether it melts would be an ideal way of answering this question and is, in fact, the approach taken. While the ability to replicate the conditions in the upper mantle in the laboratory is more than fifty years old, the techniques and hardware involved continue to improve. A variety of apparatus exists that has different capabilities in terms of the temperatures and pressures that can be achieved. Whatever the exact design of the equipment, a small sample of powdered rock is subjected to a given pressure and temperature for several hours or even days. Over a long enough time period the sample reaches chemical equilibrium. The temperature and pressure are then quickly returned to room conditions so that the material has no time to change back into its initial state. Instead, the material that was present at high pressure and high temperature becomes quenched, with any liquid that was present being turned to glass. Chemical analysis and microscope studies are then used to identify any minerals present.

To understand the results from these types of experiment you need to understand the terms 'phase' and 'phase diagram'. A **phase** is a substance with a particular chemical composition and molecular structure. Olivine and quartz are examples of two phases because each has its own composition and structure. Basalt liquid and olivine are different phases, as are SiO_2 liquid and quartz (solid SiO_2). Graphite and diamond are both forms of solid carbon but have different structures (it is their structures that make graphite soft and diamond hard). These two phases are both solid. **Phase diagrams** generally illustrate the conditions of temperature, pressure and chemical composition under which different phases exist. Because the processes that form igneous rocks are fundamentally based on reactions between minerals and liquids that either involve melting a crystalline rock to produce a liquid, or crystallising a liquid to produce crystals, phase diagrams that define the relationships between minerals and liquids at different temperatures and pressures are a cornerstone for understanding how igneous processes operate.

4.3.1 The phase diagram of Mg_2SiO_4

The phase diagram of Mg_2SiO_4 (the chemical composition of forsterite, the magnesium endmember of olivine) is shown in Figure 4.14 and introduces some of the main aspects of using phase diagrams. In this example, the chemical composition is fixed (it is Mg_2SiO_4) and the diagram shows the results of experiments in which Mg_2SiO_4 has been held at various high temperatures and high pressures. The outcome at each P and T can be shown using different symbols, in this case a filled square denotes solid crystals of forsterite whereas an open square denotes liquid Mg_2SiO_4. As you might expect, forsterite melts if the temperature is sufficiently high.

The conditions under which forsterite or liquid Mg_2SiO_4 is stable are separated by a line that divides the phase diagram into two areas. Each area is known as a **stability field**, and the boundary between them is a **phase boundary**. Both phases coexist at conditions that plot exactly on the phase boundary, but on either side of it only one phase is stable.

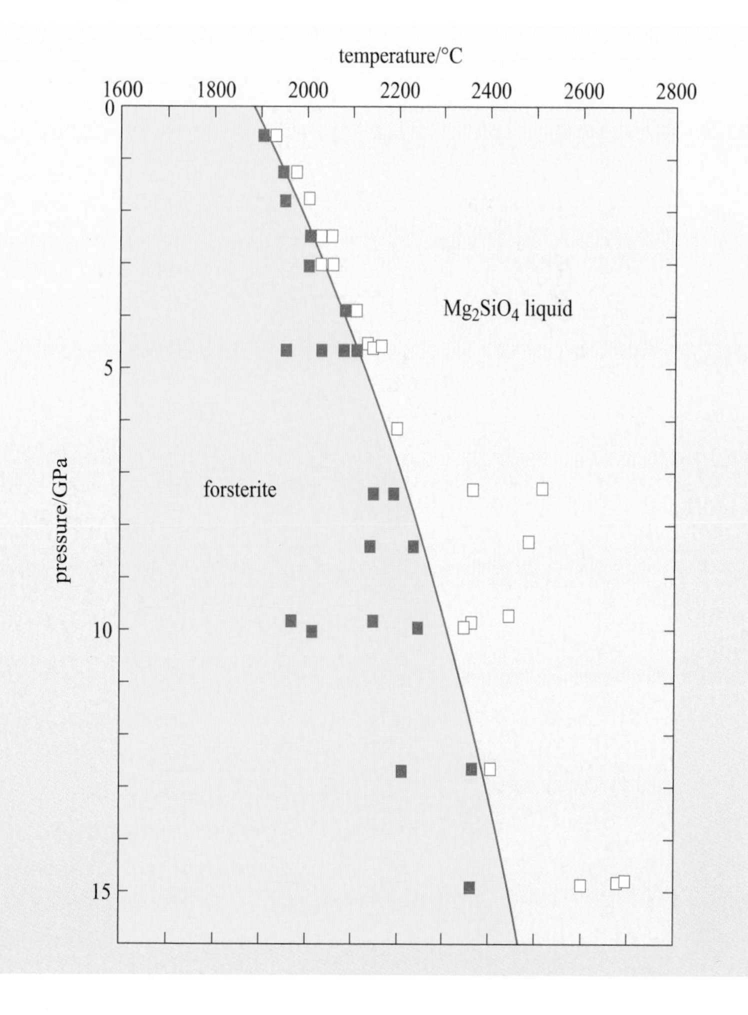

Figure 4.14 Pressure–temperature phase diagram of Mg_2SiO_4. Filled squares represent crystalline Mg_2SiO_4, i.e. forsterite. Open squares represent liquid Mg_2SiO_4. *Note*: pressure increases down the vertical axis. (Adapted from Herzberg, 1987)

■ Does the melting temperature of forsterite depend on pressure?

■ Yes. As pressure increases, the melting temperature increases.

The reason why the melting temperature increases with pressure is that more energy is required to liberate atoms from the ordered atomic structure of a mineral when the mineral's atoms are densely packed together as a result of being compressed under high pressure.

4.3.2 The phase diagram of lherzolite

Although forsterite is the main component of mantle olivine, and olivine is the main component of mantle rocks, the phase diagram of lherzolite is needed to investigate the mineralogy and melting behaviour of the mantle. To this end, the Japanese researcher Takahashi subjected samples of lherzolite xenolith from the Kilbourne Hole crater in New Mexico to pressures as high as 3 GPa, which is equivalent to the pressure at around 100 km below the Earth's surface (Box 4.4), and to high temperatures. The results of his experiments define the phase diagram shown in Figure 4.15, which shows two important features:

- depending on pressure and temperature, the peridotite either remained as a solid, produced a mixture of solid and liquid, or was completely liquid
- depending on pressure, and to a lesser extent temperature, the identity of the mineral that contains most of the aluminium in the rock changes. So, whether plagioclase lherzolite, spinel lherzolite or garnet lherzolite is stable depends on pressure, and hence on depth within the mantle.

Box 4.4 Pressure and depth

The pressure at any point on or inside the Earth depends on the weight of the overlying materials. On the surface, the weight of the atmosphere gives rise to atmospheric pressure. An underwater swimmer experiences a hydrostatic pressure that increases with the depth of water. A rock inside the Earth is subjected to **lithostatic pressure** due to the weight of overlying rock.

The pressure due to an overlying layer of material depends on the density, ρ, and thickness, h, of the layer:

$$P = \rho g h \qquad (4.1)$$

where g is the acceleration due to gravity (approximately 9.8 m s^{-2}) and units are ρ: kg m^{-3}, g: m s^{-2}, h: m, and pressure P: kg m^{-1} s^{-2}, which is equivalent to newtons per square metre (N m^{-2}) or pascals (abbreviation Pa).

$$1 \text{ Pa} = 1 \text{ N m}^{-2} = 1 \text{ kg m}^{-1} \text{ s}^{-2}$$

Atmospheric pressure has a value of 10^5 Pa (which is equal to 0.1 MPa). Lithostatic pressure increases by about 30 MPa for every kilometre depth beneath the surface of the Earth.

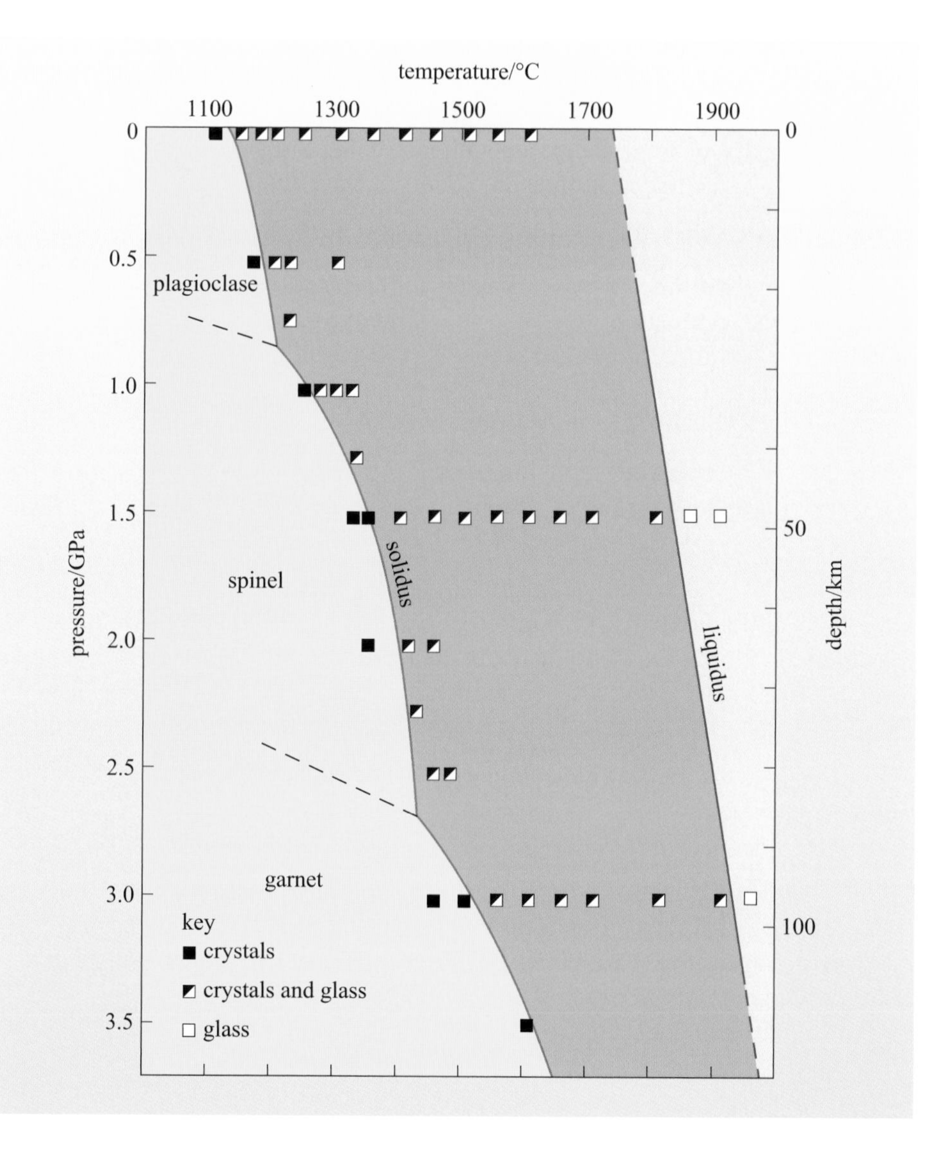

Figure 4.15 Phase diagram of lherzolite sample KLB-1 determined by Takahashi (1986). Open squares represent experiments that produced only glass (quenched liquid), half-filled squares represent a mixture of crystals and glass, and filled squares represent crystals only. Note that pressure increases down the vertical axis and the pressure is proportional to depth (labelled on the right-hand vertical axis). (Adapted from Takahashi, 1986)

Question 4.6

The sample used by Takahashi (Figure 4.15) was a spinel peridotite. From what range of depths could this sample have come from?

One reason for expecting mineralogy to change with pressure is that the structure of a mineral is sensitive to pressure. When pressure is increased, a mineral becomes slightly denser as its atoms are compressed together. At some point, the amount of compression becomes too much for the mineral to sustain its structure, forcing a major readjustment of the positions of the atoms to make a much more tightly packed structure. This is what happens when graphite changes to diamond during the manufacture of industrial diamonds by compressing carbon to very high pressure. Diamond and graphite are **polymorphs** (meaning 'in many forms') of carbon, i.e. they are different forms of the same substance. Diamond has a higher density (3500 kg m^{-3}) than graphite (2000 kg m^{-3}), reflecting the tighter packing of carbon atoms in the high-pressure polymorph.

In other cases, the effect of increasing pressure is to cause two or more minerals to react chemically, forming a new assemblage of minerals with an overall higher density, a process that happens during the metamorphism of rocks in the crust. The reactions that occur in lherzolite with increasing pressure (simplified by referring only to the magnesium endmembers of the minerals) are:

$$\underset{\substack{\text{forsterite}\\\text{(olivine)}}}{2Mg_2SiO_4} + \underset{\substack{\text{anorthite}\\\text{(plagioclase)}}}{CaAl_2Si_2O_8} \longrightarrow \underset{\text{spinel}}{MgAl_2O_4} + \underset{\substack{\text{enstatite}\\\text{(orthopyroxene)}}}{Mg_2Si_2O_6} + \underset{\substack{\text{diopside}\\\text{(clinopyroxene)}}}{CaMgSi_2O_6} \quad (4.2)$$

and

$$\underset{\text{spinel}}{MgAl_2O_4} + \underset{\substack{\text{enstatite}\\\text{(orthopyroxene)}}}{2Mg_2Si_2O_6} \longrightarrow \underset{\substack{\text{forsterite}\\\text{(olivine)}}}{Mg_2SiO_4} + \underset{\substack{\text{pyrope}\\\text{(garnet)}}}{Mg_3Al_2Si_3O_{12}} \quad (4.3)$$

Question 4.7

What does the mineralogy of the different types of lherzolite xenoliths found in basalts and in kimberlites tell you about the relative depths from which these two magma types come? (*Hint*: refer to Section 4.2.2.)

Turning now to the melting behaviour of lherzolite, use Figure 4.15 to answer the following question.

Question 4.8

At a pressure of 1.5 GPa (equivalent to a depth of almost 50 km), what is the maximum temperature at which KLB-1 lherzolite is still solid, and what is the minimum temperature at which it is totally liquid?

Constrained by the experimental data points, the line labelled **solidus** in Figure 4.15 is the curve below which the system is entirely solid (where 'below' means 'at lower temperature'). This area, which is below the solidus, is known as the **subsolidus** field. At higher temperature, the **liquidus** is the curve above which the system is completely liquid. Thus, we can say that at 1.5 GPa the solidus is crossed at a temperature of just above 1350 °C and the liquidus is crossed at about 1830 °C (see answer to Question 4.8). Between the solidus and liquidus lies the zone of partial melting, so it is within this region of pressures and temperature that basalts are generated.

Experiments on KLB-1 and other lherzolite samples reveal that with increasing temperature (at a given pressure) there are systematic changes in the degree of melting, the composition of the liquid produced, the identity of the minerals that are still present, and the chemical composition of those minerals. Some of these results are summarised in Figure 4.16 and Table 4.3, and they are explored in Question 4.9.

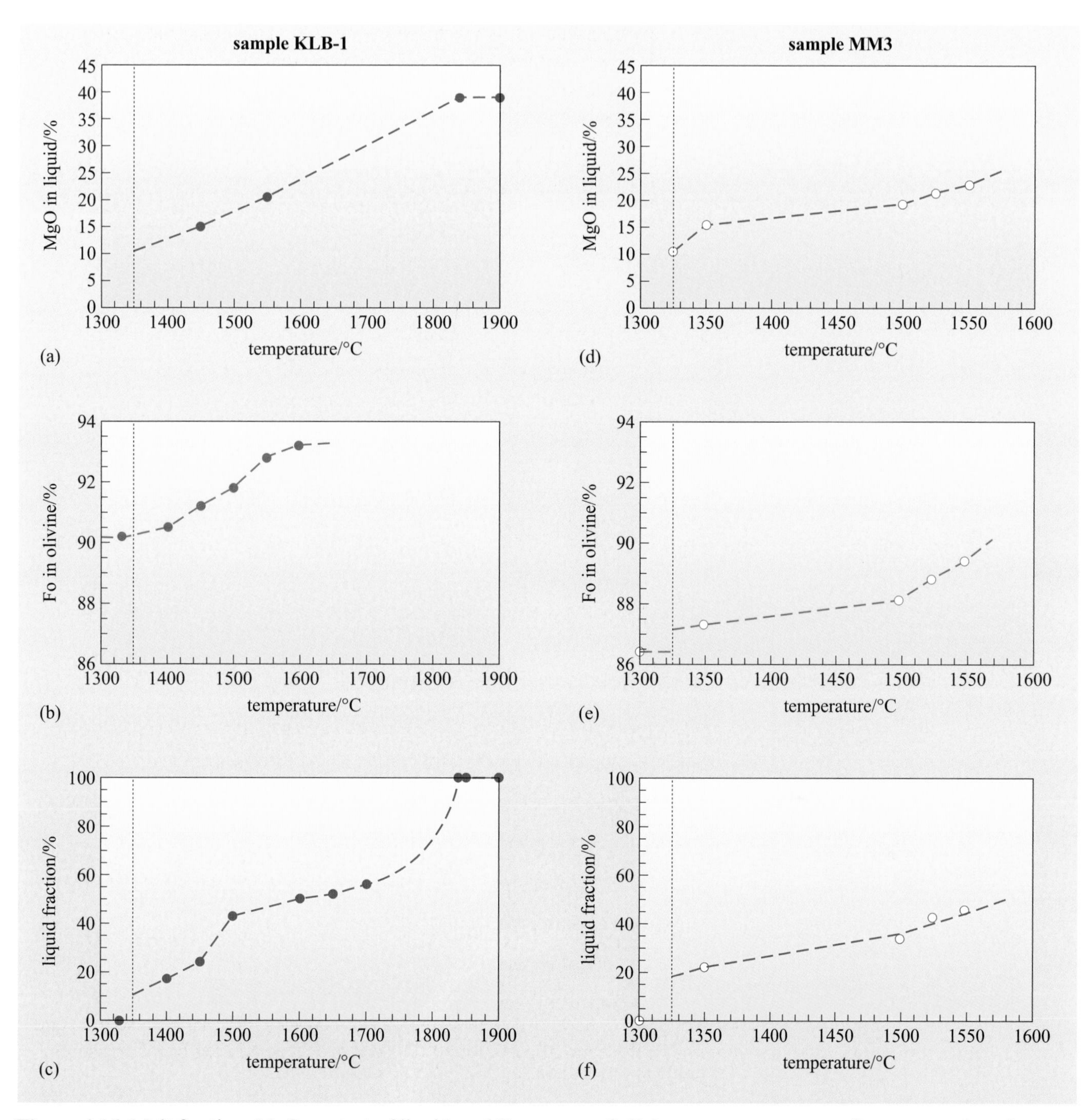

Figure 4.16 Melt fraction, MgO content of liquid, and Fo content of olivine versus temperature for two samples of lherzolite at a pressure of 1.5 GPa: (a–c) KLB-1 and (d–f) MM3. The appropriate solidus temperature is indicated by the vertical dotted line. ((a–c) Adapted from Takahashi et al., 1993; (d–f) adapted from Falloon et al., 1999)

Table 4.3 Phases present in KLB-1 at a pressure of 1.5 GPa (Takahashi, 1986).

Temperature/°C	Olivine	Orthopyroxene	Clinopyroxene	Spinel	Liquid
1325	✓	✓	✓	✓	
1350	✓	✓	✓	✓	
1400	✓	✓	✓		✓
1450	✓	✓			✓
1500	✓				✓
1600	✓				✓
1700	✓				✓
1800	✓				✓
1850					✓
1900					✓

Question 4.9

Use Table 4.3 and Figure 4.16 to decide which of the following statements describe correctly the melting behaviour of spinel lherzolite at 1.5 GPa. For those statements that are wrong, rewrite them so that they are correct.

A The solidus temperature lies between 1300 °C and 1400 °C.

B The first phase to disappear with increasing temperature is orthopyroxene.

C Spinel disappears at a lower temperature than clinopyroxene.

D Orthopyroxene is the mineral phase that survives to the highest temperature.

E Clinopyroxene disappears at a lower temperature than olivine.

F With increasing temperature the amount of partial melting increases.

G The amount of melt produced at the solidus is between 0% and 20%.

H The Fo content of olivine becomes lower with increasing amount of melting.

I The MgO content of the liquid decreases with increasing temperature.

J The MgO content of the liquid produced at temperatures just above the solidus is lower than the MgO content of the peridotite.

The next two sections develop the idea of partial melting by investigating why it is that some materials undergo partial melting rather than melting completely at a single temperature.

4.4 Why does partial melting happen?

To understand the reasons why peridotite does not melt at a single temperature, but undergoes partial melting over a range of temperatures, it is useful to consider some much simpler systems that still have some similarity with peridotite. The aim of this approach is that, by understanding how a simple system behaves, you can gain some insight into the behaviour of the more complex real system. So, because lherzolite is a mixture of olivine, orthopyroxene, clinopyroxene and a small amount of plagioclase, spinel or garnet, and because each of these minerals forms a solid-solution series, two simple versions are considered. These are a single mineral that forms a solid-solution series (forsterite–fayalite), and a mixture of two pure minerals (forsterite–diopside).

4.4.1 The phase diagram of forsterite–fayalite

As there are two components in this system (forsterite and fayalite, abbreviated to Fo and Fa), it is an example of a **binary system**. And as there is a compositional continuum between the two endmember phases, it is a binary system with a solid solution. The most useful phase diagram of such a system, in terms of understanding its melting behaviour, is a diagram that plots chemical composition against temperature (the pressure is fixed) as shown in Figure 4.17.

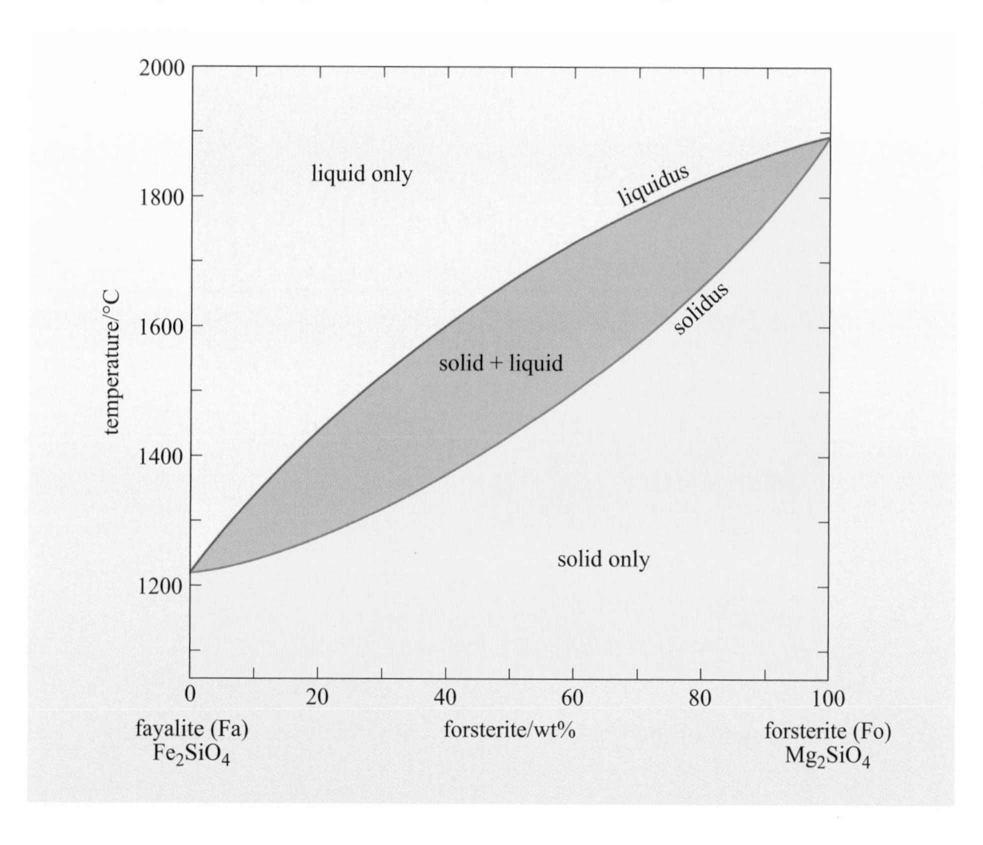

Figure 4.17 Experimentally determined phase diagram of temperature against composition (at atmospheric pressure) for the olivine system.

Composition is plotted on the horizontal axis of Figure 4.17 as the percentage (by weight) of forsterite in olivine. Any olivine composition can be plotted or read off the scale: Fe-rich examples plot to the left and Mg-rich examples plot to the right. The phase diagram is split into three areas by two curves:

- a sample whose composition and temperature plot in the top field (above the liquidus) will be completely molten
- the middle area contains samples in which liquid and solid coexist
- the lowest area (below the solidus) contains samples that are totally solid.

The olivine phase diagram shows how the composition and temperature of liquids and solids are related. For instance, olivine that is 100% forsterite ($Fo_{100}Fa_0$) has a liquidus and solidus that coincide at 1890 °C. This is what you would expect from Figure 4.14, which shows that pure forsterite melts at a single temperature of 1890 °C. Pure fayalite (Fo_0Fa_{100}) also melts at a single temperature, but one that is much less than that of $Fo_{100}Fa_0$.

But more interestingly, imagine a number of experiments in which samples of olivine with the composition $Fo_{50}Fa_{50}$ are heated to different temperatures. At 1150 °C this composition plots in the solid-only field of the phase diagram (Figure 4.18).

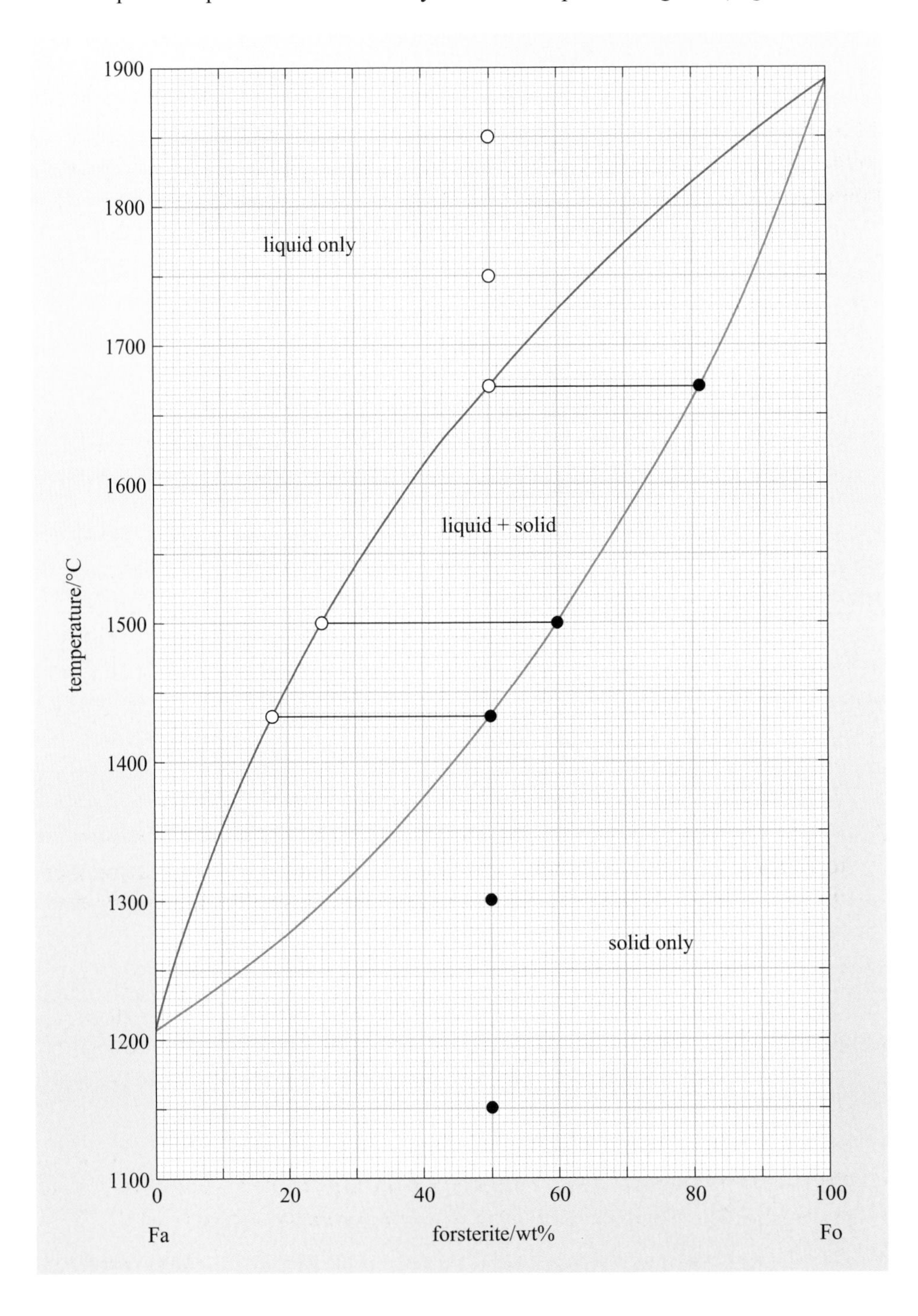

Figure 4.18 Phase diagram of temperature against composition (at atmospheric pressure) for olivine, showing the results of heating $Fo_{50}Fa_{50}$. Filled circles give the composition and temperature of solid olivine. Open circles refer to liquid. Horizontal tie lines join the compositions of solid and liquid phases that coexist at particular temperatures.

Likewise, at 1300 °C it is still solid. As the temperature is increased, the solidus is intercepted at 1433 °C; this is the lowest temperature at which solid and liquid can coexist in a system of composition $Fo_{50}Fa_{50}$.

What will be the composition of the first drop of liquid that appears at the solidus? The liquidus is the curve showing the temperature and composition of liquids that coexist with solid. And the solidus is the curve showing the temperature and composition of solids that coexist with liquid. Thus, when melting starts at 1433 °C, the liquid composition can be read off the phase diagram as $Fo_{19}Fa_{81}$ (to the nearest 1%). A **tie line** has been drawn between phases that coexist at any one time, so the tie line at 1433 °C shows that liquid $Fo_{19}Fa_{81}$ coexists with olivine crystals of $Fo_{50}Fa_{50}$ composition at this temperature. Because the overall composition is still $Fo_{50}Fa_{50}$, the proportion of liquid must be infinitesimally small.

At higher temperatures, the proportion of liquid increases at the expense of solid. At 1500 °C the bulk composition, $Fo_{50}Fa_{50}$, plots between the solidus and liquidus curves and so the sample is partially molten at this temperature. Drawing in the horizontal tie line identifies the liquid composition as $Fo_{25}Fa_{75}$ and this coexists with solid $Fo_{60}Fa_{40}$. In order for the bulk composition of this mixture to be $Fo_{50}Fa_{50}$ there must be more solid than liquid.

At 1670 °C the bulk composition, $Fo_{50}Fa_{50}$, plots on the liquidus.

■ What does this mean?

■ This is the highest temperature at which solid can still exist within a system of bulk composition $Fo_{50}Fa_{50}$.

Here, the liquid has composition $Fo_{50}Fa_{50}$ and, as the position of the tie line indicates, it coexists with an infinitesimally small amount of solid $Fo_{81}Fa_{19}$ olivine. Liquid $Fo_{50}Fa_{50}$ is the only phase present at temperatures above 1670 °C (e.g. the points at 1750 °C and 1850 °C on Figure 4.18 are both in the liquid-only field).

The behaviour of a binary solid-solution system (e.g. olivine) can be summarised as follows.

1. Partial melting occurs in a system showing solid solution.
2. The amount of liquid produced increases continuously from 0% at the solidus to 100% at the liquidus. Strictly speaking, an infinitesimally small amount of liquid is present at the solidus and an infinitesimally small amount of crystals is present at the liquidus.
3. During partial melting the liquid contains a higher concentration, and the solid contains a lower concentration, of the low melting-temperature endmember of the solid-solution series. Thus, liquids formed by partial melting of olivine will always be richer in Fa, and the solids richer in Fo, than the bulk composition.
4. As the temperature increases, both the liquid and the solid become richer in the high melting-temperature component (forsterite in the case of olivine solid solution). This is illustrated in Figure 4.19 for a composition of $Fo_{30}Fa_{70}$.

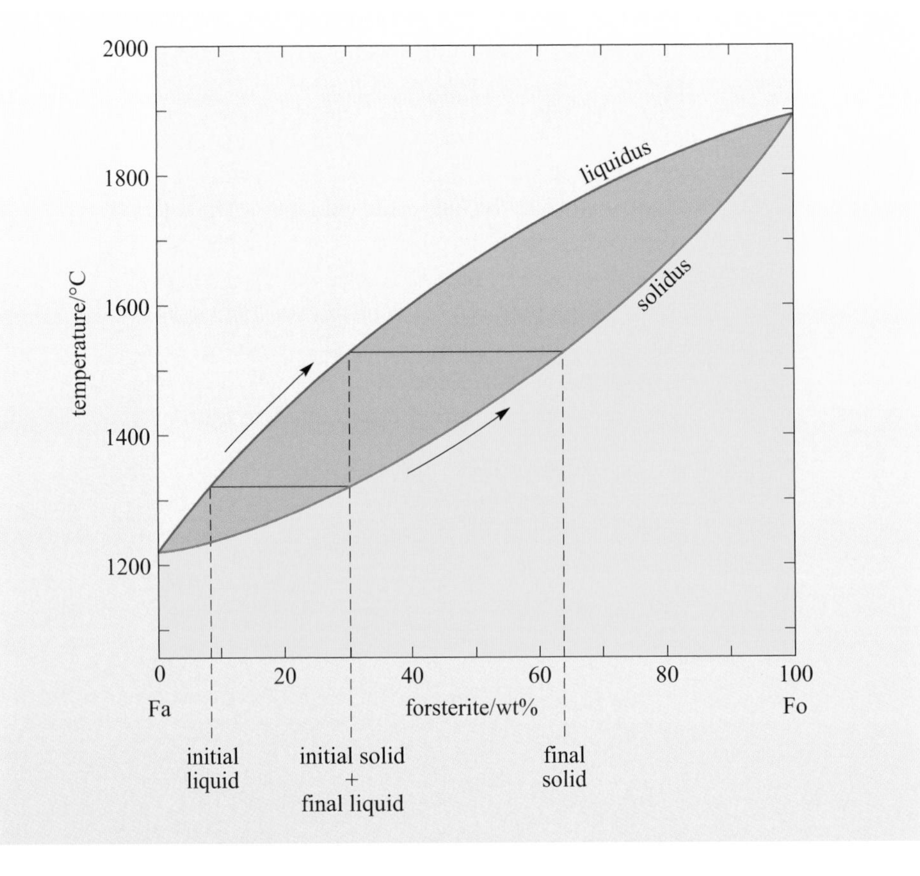

Figure 4.19 The compositions of coexisting solid and liquid evolve along the solidus and liquidus respectively as temperature increases. The example shows the behaviour of a sample of composition $Fo_{30}Fa_{70}$.

Points 3 and 4 are important because they mean that partial melting produces liquids with different chemical compositions from that of the material being melted. Also, partial melting will leave a residual solid whose composition is different from that of the initial material because it will be depleted in the components that preferentially entered the melt.

4.4.2 The phase diagram of forsterite–diopside

Olivine is not the only mineral present in lherzolite, so an alternative to thinking about the mantle in terms of the solid solution Fo–Fa is to consider a mixture of two minerals, such as olivine and orthopyroxene, or olivine and clinopyroxene. To choose an appropriate mixture to help understand partial melting, recall the idea put forward in Section 4.2.3 that harzburgite is the solid residue left behind when lherzolite melts to produce basalt liquid. This involves a decrease in the amount of clinopyroxene as shown, for example, by the shift in composition in Figure 4.12 from AX (lherzolite) to AO (average harzburgite), away from the clinopyroxene corner. So, the melting behaviour of mixtures of olivine (the dominant mineral in peridotite) and clinopyroxene should be a useful model. Furthermore, the Mg endmembers forsterite (Mg_2SiO_4) and diopside ($CaMgSi_2O_6$) are appropriate simplifications for the compositions of these minerals in the mantle.

The phase diagram of the binary system forsterite–diopside (Fo–Di) at a pressure of 1.5 GPa is shown in Figure 4.20. At this pressure pure Fo melts at 1965 °C (Figure 4.14) and pure diopside melts at 1590 °C.

At what temperature will melting start to happen when a mixture of forsterite and diopside crystals is heated?

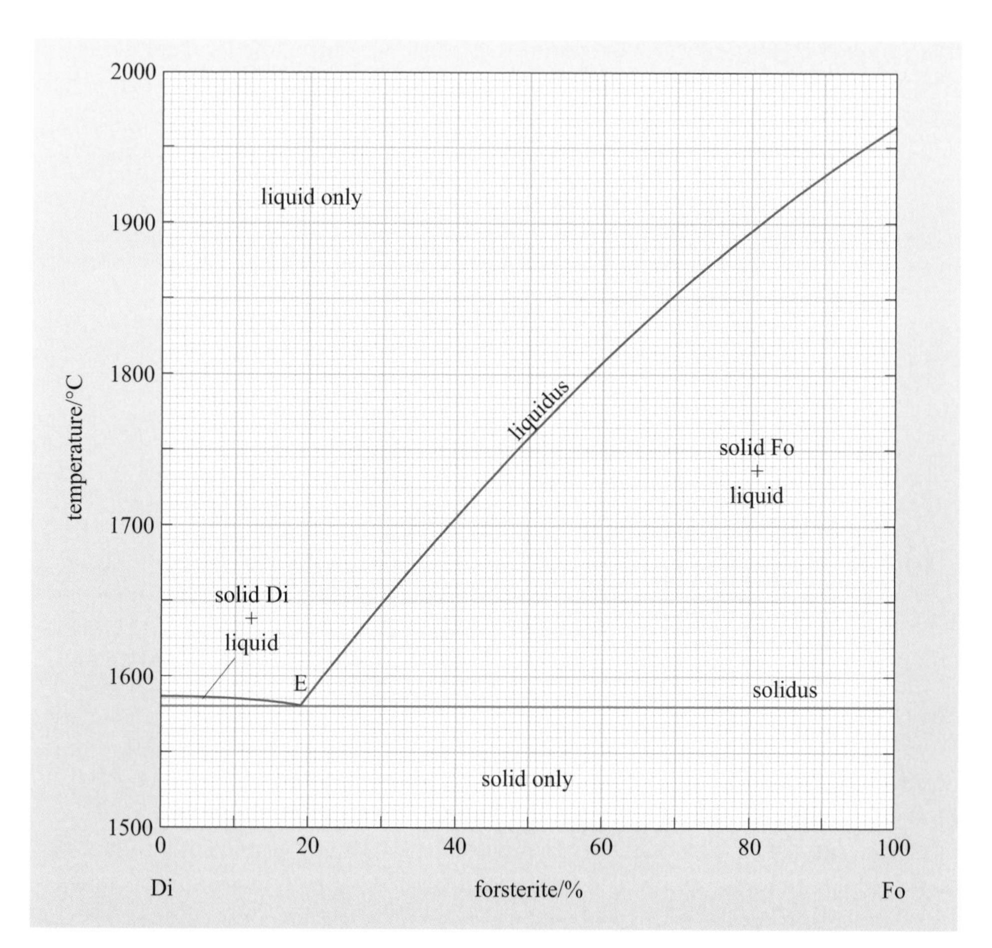

Figure 4.20 Phase diagram of temperature against composition for the system forsterite–diopside at 1.5 GPa pressure. Point E is the eutectic.

On the Fo–Di phase diagram (Figure 4.20), the solidus is a horizontal line at a temperature of 1580 °C, so any mixture of forsterite and diopside in whatever proportions will start to melt at 1580 °C (and 1.5 GPa), which is a lower temperature than the melting temperature of either pure endmember! The coexisting liquid at this temperature lies on the liquidus at 1580 °C, which is at the point labelled E. This point is known as the **eutectic point**. The eutectic point (or just eutectic) is the only location on the phase diagram where the three phases: solid diopside, solid forsterite and liquid coexist. The composition of this liquid is read from the phase diagram as $Fo_{19}Di_{81}$; this is the **eutectic composition**. To recap, on heating any mixture of diopside and forsterite at 1.5 GPa, melting will start at 1580 °C, the liquid will have a composition of $Fo_{19}Di_{81}$ and it will coexist with crystals of Fo_{100} and Di_{100}.

What happens at higher temperatures depends on the bulk composition. Consider a sample with the composition $Fo_{60}Di_{40}$ at a temperature of 1700 °C. This plots in the solid forsterite + liquid field in Figure 4.20, so the sample is now partially molten. The compositions of the two phases present at this temperature are found by drawing the horizontal tie line at 1700 °C, which shows that solid forsterite will coexist with a liquid of composition $Fo_{39}Di_{61}$.

Question 4.10

What are the compositions of the coexisting phases when a sample of $Fo_{60}Di_{40}$ is at 1800 °C?

The sample $Fo_{60}Di_{40}$ plots on the liquidus at 1810 °C. This is the temperature at which the last crystals of forsterite dissolve into the liquid; the sample is completely molten at all temperatures above 1810 °C.

Question 4.11

Consider a sample with the composition $Fo_{10}Di_{90}$ and use the Fo–Di phase diagram (Figure 4.20) to answer the following.

(a) When this sample is heated from 1500 °C to 1700 °C, at what temperature will a liquid first appear?

(b) Which mineral is consumed first?

(c) What is the composition of the first liquid that is generated?

(d) At what temperature does the sample become completely molten?

The answer to Question 4.11b reveals an important aspect of eutectic systems: the first mineral to be completely melted in a eutectic system depends on the relative compositions of the sample and the eutectic. Thus, although pure Di has a much lower melting point than pure Fo, it will be forsterite that disappears first if the sample contains less than 19% Fo.

The amount of liquid that is produced at the eutectic temperature depends on the bulk composition of the sample. In the example of $Fo_{60}Di_{40}$, all but an infinitesimally small amount of the diopside crystals is consumed at the eutectic, so it might be tempting to jump to the conclusion that 40% of the sample becomes liquid at the eutectic. This cannot be the case, however, because the eutectic liquid contains some forsterite (the eutectic composition is $Fo_{19}Di_{81}$), so more than 40% of the sample is molten at 1580 °C. (In fact, in this example, 49% of the sample becomes liquid at the eutectic in order to allow the composition of the eutectic liquid to be derived from the bulk composition of the sample.) This specific example illustrates another important general principle – the amount of liquid produced at the onset of melting in a eutectic system can be large, whereas at the onset of melting in a solid-solution series the initial amount of liquid is infinitesimally small. In both of these model systems the proportion of liquid gradually increases with increasing temperature until it reaches 100% at the liquidus temperature (Figure 4.21).

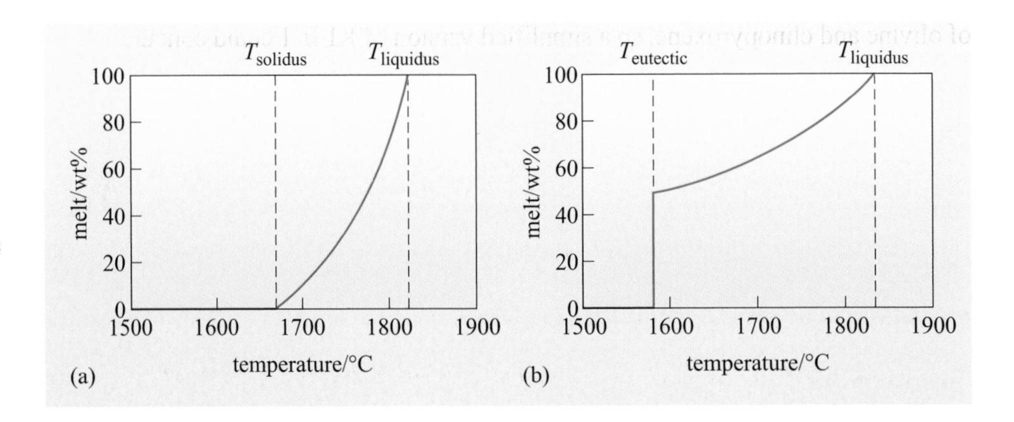

Figure 4.21 Graphs of wt% melt against temperature illustrating the different behaviours of eutectic and solid-solution systems. (a) Binary solid solution. The curve is for $Fo_{80}Fa_{20}$ in the Fo–Fa system at 10^5 Pa (atmospheric pressure). (b) Binary eutectic system. The curve shows the example $Fo_{60}Di_{40}$ in the Fo–Di system at 1.5 GPa discussed in the text.

4.4.3 Partial melting of lherzolite and the generation of basalt

You are now in a position to understand some of the main results that were found in the experiments on peridotite at high pressure and high temperature (Section 4.3.2, Table 4.3, Figure 4.16). Because lherzolite contains different mineral phases, it has a similarity with eutectic systems (e.g. Fo–Di) and, because each of the mineral phases forms a solid-solution series, there are also similarities with systems such as Fo–Fa.

Question 4.12

For each of the following statements describing the partial melting of peridotite, decide whether it is explained by the way solid solutions and/or eutectic systems behave.

A At a given pressure, partial melting occurs over a temperature range of several hundred degrees Celsius.

B The olivine composition of the unmelted residue has a higher Fo content than the olivine in the original, unmelted peridotite.

C The MgO content of the liquid increases as the degree of partial melting increases.

D At least 10% of liquid is produced at temperatures just above the solidus.

E Clinopyroxene disappears at a lower temperature than olivine.

F Spinel disappears at a lower temperature than clinopyroxene.

Lherzolite xenoliths contain a much higher proportion of olivine than clinopyroxene, as seen from the way in which peridotites plot near the olivine corner and near the olivine–orthopyroxene edge of the classification diagram for ultrabasic rocks (Figure 4.12). For instance, the sample KLB-1 whose phase diagram is shown in Figure 4.15 has 58% olivine, 25% orthopyroxene, 15% clinopyroxene, and 2% spinel. In experiments with this sample at 1.5 GPa, no clinopyroxene persists above about 1410 °C, but olivine is always present until the liquidus is reached at 1830 °C. Could this have been predicted from what is known about the Fo–Di phase diagram at 1.5 GPa (Figure 4.20)? In KLB-1, 73% of the rock is composed of olivine and clinopyroxene, so a simplified version of KLB-1 could contain:

$$\frac{58\%}{73\%} \times 100\% = 79\% \text{ olivine and } \frac{15\%}{73\%} \times 100\% = 21\% \text{ clinopyroxene.}$$

Making another approximation by considering that olivine can be represented by forsterite, and clinopyroxene can be represented by diopside, then KLB-1 can be approximated by a mixture with the composition $Fo_{79}Di_{21}$.

- What is the melting behaviour of $Fo_{79}Di_{21}$?

- According to Figure 4.20, this mixture will melt at the eutectic at 1580 °C, with the disappearance of Di and the production of a eutectic liquid. At higher temperatures, as the melt fraction increases, the amount of solid Fo decreases and the liquid becomes richer in the Fo component. At the liquidus temperature of 1890 °C, all of the olivine has melted.

The model composition, therefore, mimics the real system in showing that clinopyroxene will melt before olivine. However, the model predicts a higher solidus temperature and a higher liquidus temperature. The reason for this is that these phase boundaries generally shift to lower temperatures as more chemical components are added. Thus at 1.5 GPa, pure Fo (Figure 4.14) melts at a higher temperature than a Fo–Di mixture (Figure 4.20), which in turn melts at a higher temperature than KLB-1 (Figure 4.15). When lherzolite melts, Al-bearing phases and then clinopyroxene melt completely at lower temperatures than orthopyroxene, while olivine melts at even higher temperatures.

4.4.4 Summary of Section 4.4

The partial melting of lherzolite (and other substances) can be understood using phase diagrams of composition versus temperature for solid-solution series and eutectic systems. Instructive systems that are simplified versions of lherzolite include the forsterite–fayalite (Fo–Fa) and forsterite–diopside (Fo–Di) systems. These show the following features.

The compositional phase diagram of a *binary solid-solution series* is characterised by a loop-shaped liquidus and solidus. This is exemplified by the olivine solid-solution series Fo–Fa (Figure 4.17). Here, partial melting of a given sample produces a liquid that is richer in Fa (the component with the lower melting temperature) than the overall bulk composition. The liquid forms at the expense of solid olivine that, therefore, becomes richer in Fo than the bulk composition. The compositions of coexisting solid and liquid phases plot on the solidus and liquidus curves at either end of a horizontal tie line. As melting progresses (as temperature increases), the compositions of the liquid and coexisting crystals become richer in the high-temperature endmember (Fo). In the melting interval between the solidus and liquidus temperatures, the proportion of melt increases steadily from 0% to 100%.

The partial melting behaviour of lherzolite shows features that are similar to the partial melting behaviour of olivine. There is a continuously smooth increase in melt fraction with temperature, and residual ferromagnesian minerals are richer in Mg (and poorer in Fe) than the same minerals in unmelted samples. The chemical composition of the partial melt also varies with the degree of melting, becoming more Mg rich as the melt fraction increases, analogous to the increase in Fo seen in the Fo–Fa system.

A *binary eutectic system* has a horizontal solidus on a compositional phase diagram (e.g. Figure 4.20). This means that partial melting will start at the same temperature irrespective of the bulk composition. The eutectic temperature is lower than the melting temperature of the system's endmembers. The eutectic point defines the composition and temperature of the first melt that forms. Partial melting at the eutectic produces a significant quantity of liquid, the exact amount of which depends on the composition of the bulk sample and the eutectic composition. This contrasts with the infinitesimally small amount of liquid generated at the solidus temperature in a solid-solution series.

In eutectic systems, the composition of the liquid reflects the amount of melting due to the changing proportions of the different minerals that contribute to the liquid. The phase that is less abundant in the bulk sample compared with the eutectic composition will melt first at the eutectic temperature. This feature is observed when lherzolite melts, with Al-bearing phases and then clinopyroxene melting completely at lower temperatures than orthopyroxene and olivine melting at an even higher temperature. However, melting at the lherzolite solidus does not produce a large quantity of liquid, as in a eutectic system. This is because a solid solution in mantle minerals smoothes out the step-like increases in liquid fraction that characterise melting in eutectic systems.

4.5 Magma generation at constructive plate boundaries

Pressures and temperatures in the mantle beneath mid-ocean ridges are large enough for the mantle to experience partial melting to produce basalt magma. This section starts by outlining the general mechanism for producing magmas at constructive plate boundaries (Section 4.5.1) and then shows how these conditions depend on mantle temperature (Section 4.5.2) so that observations of oceanic crust can be used to infer temperature conditions in the upper mantle (Section 4.5.3).

4.5.1 Crossing the solidus

The temperature structure beneath old oceanic lithosphere (with the exception of hot-spot volcanoes; a separate issue that we return to later in Chapter 7) is not capable of generating magmas. In areas of old oceanic lithosphere, the surface heat flux is <50 mW m^{-2} (Box 3.2) and this heat comes partly from the decay of trace amounts of radioactive U, Th and K and partly from the cooling of the already hot interior. Convection currents in the asthenosphere move heat to the base of the lithosphere, and that heat is then conducted to the surface through the lithosphere, resulting in a geotherm like that shown in Figure 4.22. In the convecting mantle, descending currents of mantle become compressed, causing them to warm slightly as a result of their atoms being squashed together. The reverse happens when rising mantle is decompressed – it cools slightly. (The same effect accounts for the drop in atmospheric temperature with increasing altitude.) In these cases, a rising or falling segment of mantle does not lose or gain heat energy – the heat energy only leaves the convecting system across the steep temperature gradient at the lowest part of the conductive part of the geotherm. The geotherm in the convecting mantle is therefore said to be **adiabatic** (from the Greek for impassable – a reference to the fact that heat passes neither from nor to a substance undergoing an adiabatic process). The temperature gradient in the lithosphere is conductive and, in the underlying mantle where convection occurs, it is adiabatic. Furthermore, the geotherm always lies below the mantle solidus, which explains why the older parts of oceanic basins do not have any volcanic activity.

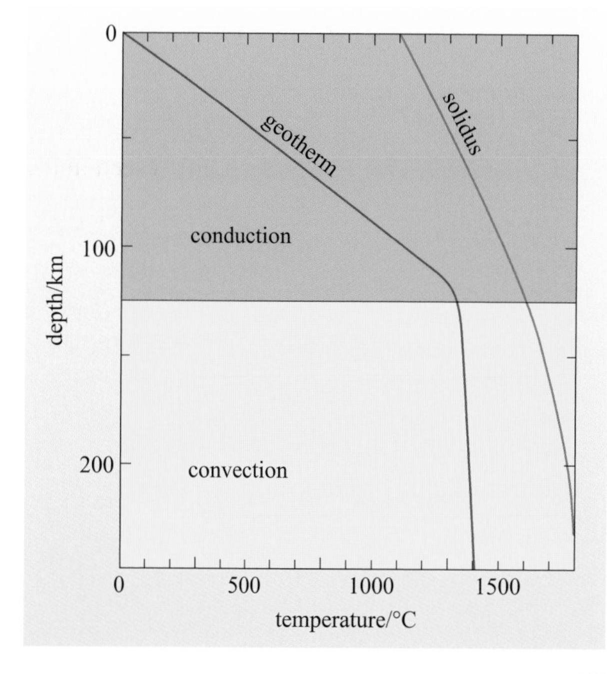

Figure 4.22 The geotherm and mantle solidus in old oceanic lithosphere. The solidus curve is an average taken from experiments on slightly different lherzolite samples.

At mid-ocean ridges the surface heat flux is very high (>100 up to >300 mW m^{-2}), which is an immediate indication that very hot material lies close to the surface. At the ridge, oceanic lithosphere splits apart at rates of a few centimetres per year, so deep material must upwell at the same rate to fill the gap and then flow sideways in a pattern like the one shown in Figure 4.23a (see also Figure 4.13). This flow pattern underneath the plate boundary brings deep, and therefore hot, mantle material towards the surface. Its path is illustrated in Figure 4.23b.

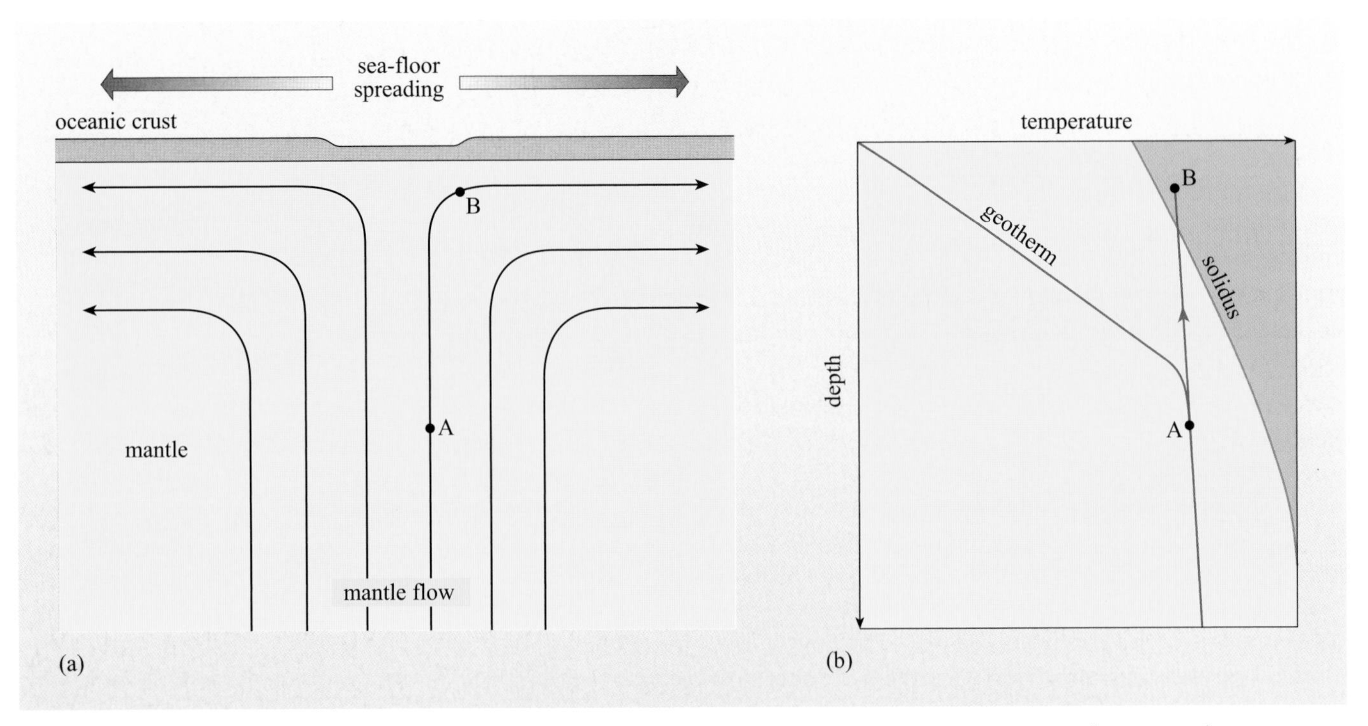

Figure 4.23 (a) Sketch cross-section of a constructive plate boundary. The crust moves apart, opening a gap that becomes filled with basalt escaping from the underlying mantle (see also Figure 4.13). The mantle rises below the plate boundary and then moves sideways to follow the divergent paths of the crust. A parcel of mantle that is originally at point A moves to point B. Its pressure and temperature decrease along the adiabatic gradient as shown in (b) and it crosses the mantle solidus during its journey. ((a) Adapted from Langmuir et al., 1993)

Deep beneath a constructive plate boundary, a region of mantle within the convecting asthenosphere (labelled point A) moves upwards in response to the separation of the plates at the surface. It follows an adiabatic path, so its temperature decreases slightly as it rises along an adiabatic gradient and by the time it has risen to point B it has crossed the solidus and will be partially molten – the basaltic magma needed to make oceanic crust has been created. The change from solid mantle to partially molten mantle has therefore been caused by a decrease in pressure, so it is referred to as **decompression melting**. In this case, partial melting is not caused by heating but by decompression caused by large-scale flow in the mantle that is induced by plate divergence.

Figure 4.23 shows how decompression melting can happen beneath a mid-ocean ridge, and this general concept can be used to link together two aspects of the Earth – the temperature of the upper mantle and the composition and structure of oceanic crust. This link comes about because of a chain of connections:

1 The amount of partial melting will determine how much basalt is generated, and this in turn will determine how thick the oceanic crust is.

2 The amount of partial melting depends on the pressures and temperatures reached within the zone of partial melting where the magma is generated (in other words, how far the decompressed mantle – point B in Figure 4.23 – lies above the solidus).

3 These P and T conditions depend on the initial temperature of the rising mantle and the amount by which it decompresses.

4.5.2 The temperature of the convecting mantle

The temperature in the convecting mantle varies along the adiabatic gradient, decreasing by about 1 °C for every 3 km of ascent. To specify the temperature, a depth, or pressure, needs to be specified. The conventional way of doing this is to extrapolate the **adiabat** to the surface, which is at a pressure of 10^5 Pa, as shown in Figure 4.24a (overleaf). This extrapolated temperature is known as the **potential temperature** (T_p).

If T_p is less than the solidus temperature at atmospheric pressure (about 1120 °C from Figure 4.15), then decompression melting would not happen (Figure 4.24b). This means that T_p for the Earth's mantle must be greater than 1120 °C. When mantle with a higher T_p rises, it will cross the solidus at a pressure, P_s, where the temperature is T_s (Figure 4.24c). For the melting reaction:

solid phases → liquid + residual solid phases

to happen, some energy is needed. This energy (the latent heat of fusion) is supplied by a change in temperature, causing the system to cool as it undergoes more and more melting. Because of the phase changes involved in partial melting, the actual temperature path that the mantle follows once it has crossed the solidus will be similar to the one shown in Figure 4.24c, where T_{act} is the actual temperature that the rising parcel of mantle would have if it rose all the way to the surface, and this is less than T_p.

Sea-floor spreading allows the mantle to come very close to the surface because the oceanic lithosphere at the plate boundary is extremely thin. In detail, the amount of melting achieved will increase as the mantle reaches shallower and shallower levels after it has crossed the solidus because the path it follows moves it further away from the solidus. Figure 4.24d illustrates two hypothetical situations in which mantles with different potential temperatures undergo decompression melting that extends to the surface.

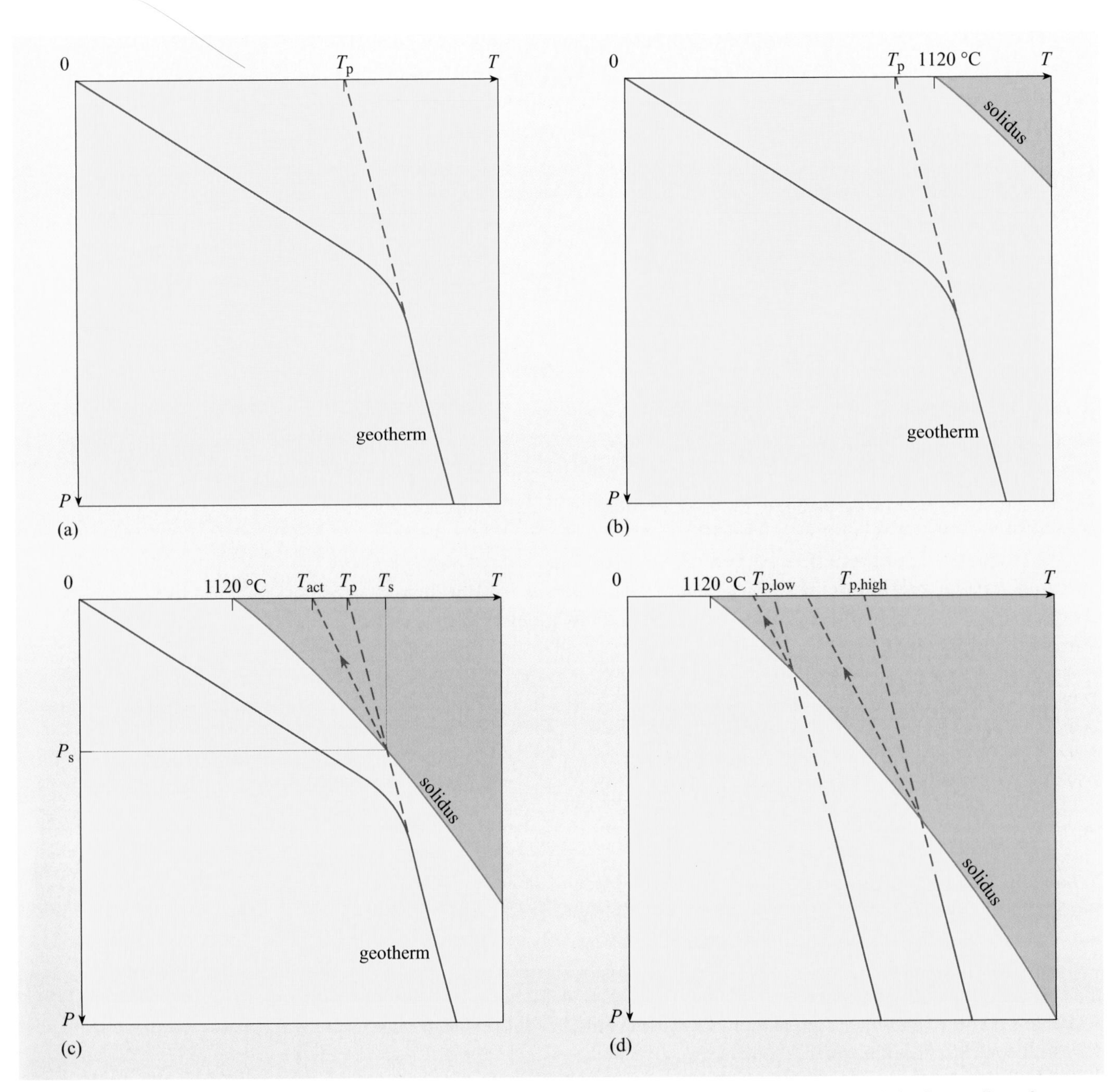

Figure 4.24 (a) Sketch showing how the extrapolation of temperature (dashed line) along the adiabatic gradient from any depth in the convecting mantle to the surface defines the potential temperature (T_p) of the convecting mantle. (b) Mantle with T_p less than about 1120 °C cannot cross its solidus by decompressing. (c) Mantle with T_p greater than 1120 °C crosses the solidus at (T_s, P_s) and continues to decompress along the path leading to (T_{act}, 0). (d) Decompression paths of mantle with potential temperatures of $T_{p,low}$ and $T_{p,high}$ (see Question 4.13).

Question 4.13

Of the two cases ($T_{p,low}$ and $T_{p,high}$) shown in Figure 4.24d, which one will cause partial melting to start at the greater depth, and which will produce the larger amount of liquid?

The answers to Question 4.13 lead to the interesting conclusion that the amount of basaltic magma (and therefore the thickness of basaltic oceanic crust) that can be produced at a mid-ocean ridge will depend on the potential temperature of the mantle. Therefore a measurement of the thickness of oceanic crust can, in principle, be used as a 'thermometer' with which to estimate the temperature of the Earth's upper mantle. This idea is followed to its logical conclusion in the next subsection.

4.5.3 The thickness of oceanic crust and the potential temperature of the mantle

To restate the conclusion of the last section, the amount of liquid produced by decompression melting of the mantle will depend on the potential temperature of the mantle and the lowest pressure reached by the rising mantle. To make this idea quantitative requires measurements of the amount of melting at different pressures and temperatures above the solidus of mantle peridotite. Among the first researchers to take this step were Dan McKenzie and Mike Bickle (Cambridge University, UK), who compiled data from high-pressure, high-temperature experiments and used them to predict the melting behaviour of mantle peridotite. Their study was published in 1988 and, although it continues to be embellished by further experimental data, it sets out the basic quantitative correlation between potential temperature and the extent of decompression melting, under the simplifying assumption of decompression to atmospheric pressure. For a given potential temperature, their model predicts the depth at which partial melting would start, the amount of basalt produced, and the maximum proportion of melting. Their results are summarised in Figure 4.25 (overleaf).

- ■ The average thickness of oceanic crust is about 7 km. What does this imply about the mantle's potential temperature?
- ■ According to the model of Figure 4.25a, this implies a potential temperature of 1280 °C.

As well as helping to explain the basaltic product of partial melting, the model in Figure 4.25 helps to explain why harzburgite is found in the mantle sections of ophiolites and oceanic peridotites. According to Figure 4.25, a potential temperature of 1280 °C should result in up to 25% melting, and this is similar to the amount of melting at which clinopyroxene no longer remains in the mineral assemblage of partially molten peridotite in laboratory experiments. The model also predicts that partial melting will start at a depth of 42 km, which is in the spinel–peridotite field (Figure 4.15).

To summarise, the calculations of partial melting during adiabatic ascent of lherzolite beneath oceanic spreading ridges account for the production of 7 km thick basaltic crust and complementary harzburgite if the mantle has a potential temperature of 1280 °C. Furthermore, the model predicts about 25% melting at the top of a zone of partial melting that extends to a depth of about 40 km.

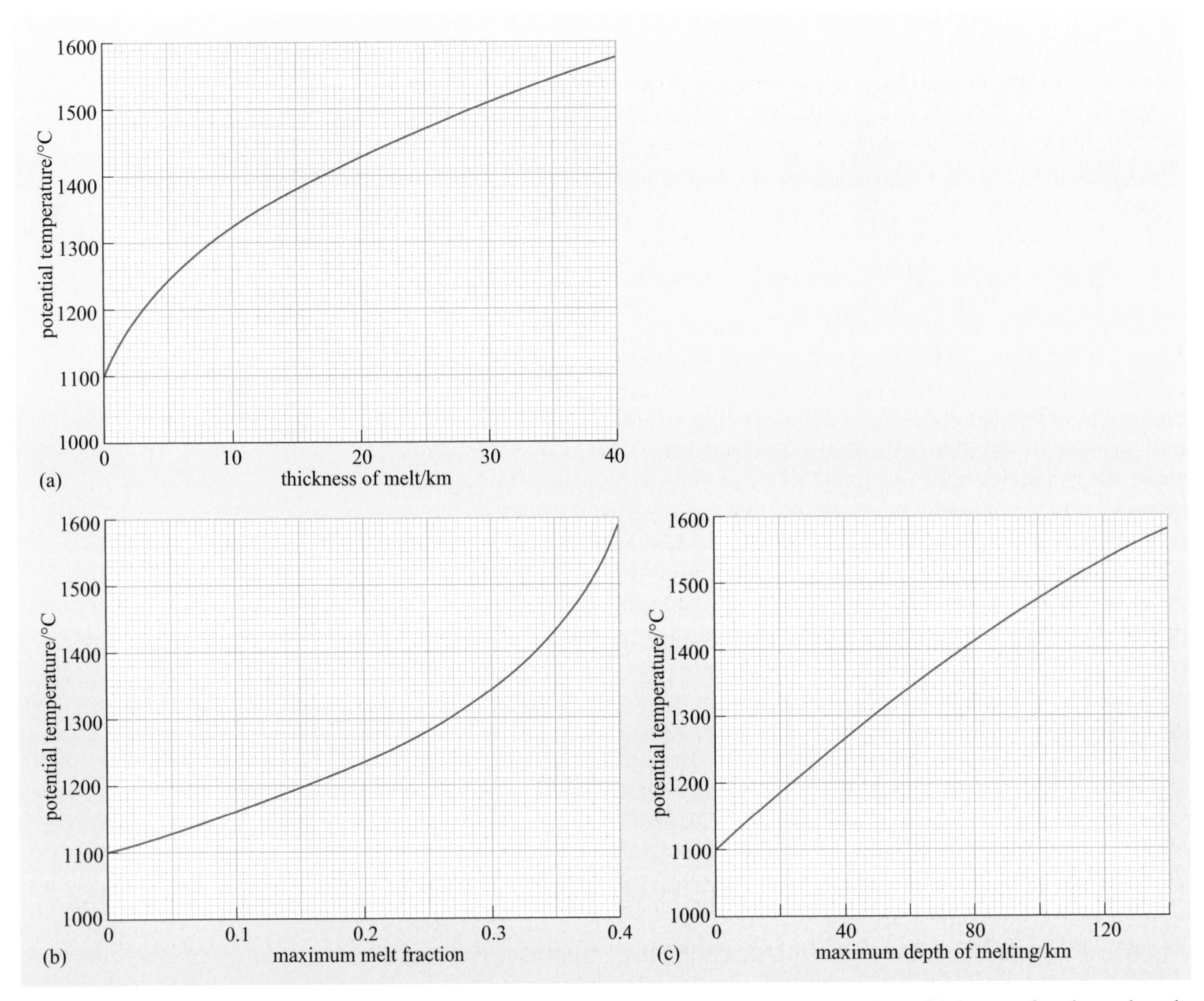

Figure 4.25 Calculated relationships between the potential temperature of the mantle and: (a) thickness of melt produced and extracted; (b) maximum mass fraction of partial melt in the melting zone; and (c) maximum depth at which melting of adiabatically ascending mantle will start to partially melt. The model assumes that decompression continues to atmospheric pressure (10^5 Pa). (Adapted from McKenzie and Bickle, 1988)

Question 4.14

Iceland lies on the Mid-Atlantic Ridge and is part of the constructive plate boundary between the Eurasian and North American Plates. Unusually, this part of the mid-ocean ridge system rises above sea level, and the Icelandic crust has an average thickness of 27 km. What is the potential temperature of the mantle beneath Iceland that is implied by this crustal thickness? Use Figure 4.25 to help you.

4.6 From mantle to crust

Decompression melting produces basaltic magma in the mantle beneath mid-ocean ridges. Being less dense than the residual harzburgite, the basalt liquid is buoyant and therefore migrates towards the surface where it forms oceanic crust, either by erupting onto the sea floor where it forms lava flows, or intruding and crystallising at depth to form the gabbro that makes up the lower oceanic crust. When lava erupts into the cold ocean water at a mid-ocean ridge, the hot lava spreads out from the volcanic vents. Its surface chills to a glassy skin that stretches, breaks and bursts open repeatedly, building up a bewildering accumulation of tubes, channels, ponds and mounds of lava. The chilled rind of these submarine lava flows is composed of dark, glassy rock containing a few crystals of olivine, plagioclase and sometimes clinopyroxene and it represents magma quenched from the state in which it erupted. The larger crystals can reach sizes of up to a few millimetres across and are referred to as **phenocrysts**. Because the glassy sample was quenched by water, the phenocrysts (e.g. the visible plagioclase and olivine crystals in the basalt of Figure 4.3a) must have been present before the lava became solid, and have been carried in the magma as it travelled to the surface. This section shows how the composition of lavas such as these can be used to discover the history of the magma after it has left its source region in the upper mantle. It starts by comparing the magma generated in the mantle with the magma that erupts on the sea floor.

4.6.1 Chemical compositions of partial melts and mid-ocean ridge basalt

If **mid-ocean ridge basalt** (often abbreviated as **MORB**) has come directly from the mantle, then it will be in equilibrium with olivine and the other residual solid phases in its peridotite source region. You can see from the olivine phase diagram (Section 4.4.1) that liquids that are rich in Mg and poor in Fe will be in equilibrium with olivine that is also rich in Mg and poor in Fe. In principle, given the composition of a liquid (such as a basalt lava from a mid-ocean ridge), this can be checked to see whether it is in equilibrium with the olivine found in the mantle source region of basalts, which is taken to be Fo_{88-92} (Section 4.2.1). If the liquid lava is in equilibrium with mantle olivine, then it can be assumed that the magma has come directly from the mantle.

The Canadian geologists Peter Roeder and Ron Emslie were the first to demonstrate experimentally the systematic correlation between the compositions of coexisting basaltic liquids and olivine (Figure 4.26). Their results, which have been corroborated by several hundred more experiments carried out over a wide range of temperatures, pressures and liquid compositions, show that the ratio of the number of moles of Fe^{2+} atoms to Mg atoms in the olivine and in the liquid are related by the equation:

$$\left(\frac{Fe^{2+}}{Mg}\right)_{\text{olivine}} = 0.3\left(\frac{Fe^{2+}}{Mg}\right)_{\text{liquid}} \tag{4.4}$$

An important point here is that the iron in a magma is present as a mixture of Fe^{2+} and Fe^{3+}. In most basalts between about 5% and 20% of the total iron is present as Fe^{3+}, whereas the iron in olivine is exclusively Fe^{2+}. This is why care has been taken in specifying in Equation 4.4 that the iron that is relevant is Fe^{2+}.

Equation 4.4 can also be expressed in terms of the distribution coefficient (denoted K_D), which is defined by the gradient of the straight line passing through the data in Figure 4.26:

$$K_D = \frac{\left(\frac{Fe^{2+}}{Mg}\right)_{olivine}}{\left(\frac{Fe^{2+}}{Mg}\right)_{liquid}} = 0.3 \quad (4.5)$$

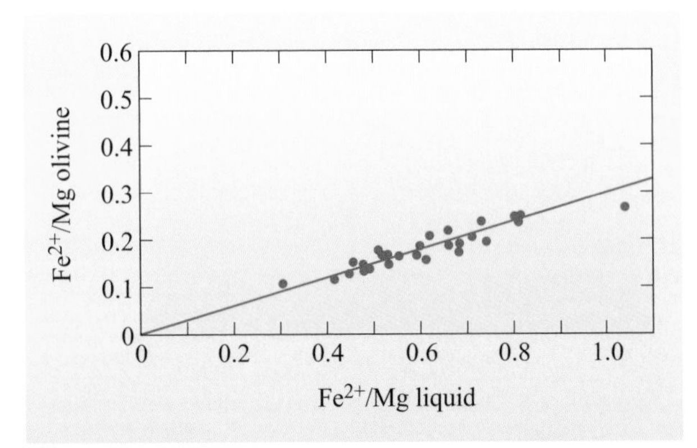

Figure 4.26 The results of experiments by Roeder and Emslie (1970), demonstrating that the ratio of the number of Fe^{2+} to Mg atoms in olivine is proportional to the Fe^{2+}/Mg ratio in the liquid with which it is in equilibrium. (Adapted from Roeder and Emslie, 1970)

This is a very useful result because it means that, without having to know the temperature or pressure, it is possible to test whether a given olivine composition and a given liquid composition are in chemical equilibrium. The practicalities of this approach are demonstrated by considering a specific example.

First, note that the chemical equilibrium between olivine and liquid is based on the ratio of the numbers of Fe^{2+} to Mg atoms in the olivine and in the liquid. However, conventional chemical analyses of rocks and minerals (for example those shown in Table 4.2) are given in terms of the weight per cent oxides and not the proportions of atoms of different elements. You therefore need to convert from wt% FeO and wt% MgO to the atomic proportions of Fe^{2+} and Mg. Consider a basalt containing 8.0 wt% FeO and 12 wt% MgO. From the relative atomic masses of Fe (atomic mass 56) and O (atomic mass 16), one mole of FeO has a mass of 72 g. That means that 100 g of a rock containing 8.0 wt% FeO will contain:

$$8.0\ \text{g} \times \frac{56\ \text{g}}{72\ \text{g}} = 6.22\ \text{g of Fe atoms.}$$

So the number of moles of Fe atoms in 100 g of rock is:

$$\frac{6.22\ \text{g}}{56\ \text{g}} = 0.11$$

Looking at this calculation as a whole, 8.0 wt% FeO is seen to be:

$$8.0\ \text{g} \times \frac{56\ \text{g}}{72\ \text{g}} \times \frac{1}{56\ \text{g}}\ \text{of a mole of Fe atoms.}$$

Because the two 56 grams cancel out, this can be simplified, and shows that 8.0 wt% is equivalent to:

$$\frac{8.0\ \text{g}}{72\ \text{g}} = 0.11\ \text{of a mole of Fe atoms.}$$

Likewise, given that the molar mass of MgO is 40 g, then 12 wt% MgO is:

$$\frac{12\ \text{g}}{40\ \text{g}} = 0.30\ \text{moles of Mg atoms.}$$

The atomic Fe^{2+}/Mg ratio of the rock, therefore, is:

$$\frac{Fe^{2+}}{Mg} = \frac{\left(\frac{8.0\ g}{72\ g}\right)}{\left(\frac{12\ g}{40\ g}\right)} = 0.37 \text{ (to 2 sig. figs).}$$

■ What is the atomic Fe^{2+}/Mg ratio of the olivine in equilibrium with this basalt?

■ Applying Equation 4.4, this liquid should be in equilibrium with an olivine with $Fe^{2+}/Mg = 0.3 \times 0.37 = 0.11$.

Until now, we have considered olivine compositions in terms of the percentages of forsterite (Fo) and fayalite (Fa). The percentage Fo is simply the percentage of Mg in the $(Mg,Fe)_2$ part of the olivine formula:

$$\%Fo = 100\% \times \frac{Mg}{Mg + Fe^{2+}}$$

This ratio was introduced towards the end of Section 4.2.1 as the Mg-number, often abbreviated as Mg#

$$Mg\# = 100\% \times \frac{Mg}{Mg + Fe^{2+}} \tag{4.6}$$

Therefore, an olivine with composition $Fo_{80}Fa_{20}$ has an Mg# of 80 (see also Section 4.2.1). But how does this relate to the Fe^{2+}/Mg ratio of the olivine?

To convert from Mg# to the atomic ratio Fe^{2+}/Mg or vice versa, Equation 4.6 can be rewritten by dividing the top and bottom lines by Mg:

$$Mg\# = 100\% \times \frac{\left(\frac{Mg}{Mg}\right)}{\left[\left(\frac{Mg}{Mg}\right) + \left(\frac{Fe^{2+}}{Mg}\right)\right]} = 100\% \times \frac{1}{1 + \left(\frac{Fe^{2+}}{Mg}\right)} \tag{4.7}$$

and this can be rearranged to make Fe^{2+}/Mg the subject:

$$\frac{Fe^{2+}}{Mg} = \frac{100 - Mg\#}{Mg\#} \tag{4.8}$$

■ What is the Fe^{2+}/Mg ratio of $Fo_{80}Fa_{20}$?

■ Applying Equation 4.8 and substituting an Mg# of 80:

$$\frac{Fe^{2+}}{Mg} = \frac{100 - 80}{80} = 0.25.$$

■ An earlier example established that basalt with $Fe^{2+}/Mg = 0.37$ would be in equilibrium with olivine with $Fe^{2+}/Mg = 0.11$. What are their Mg#s?

■ Applying Equation 4.7 in each case gives a liquid Mg# of 73 and olivine Mg# of 90.

The equilibrium relationship between the atomic Fe^{2+}/Mg ratios of olivine and liquid (Equation 4.4) can now be rewritten in terms of the Mg#s of olivine and liquid:

$$Mg\#_{liquid} = \frac{30\ Mg\#_{olivine}}{100 - 0.7\ Mg\#_{olivine}} \tag{4.9}$$

and this equation is shown graphically in Figure 4.27. It illustrates what we found out from the Fo–Fa phase diagram, namely that olivine has a higher Mg# than the coexisting liquid.

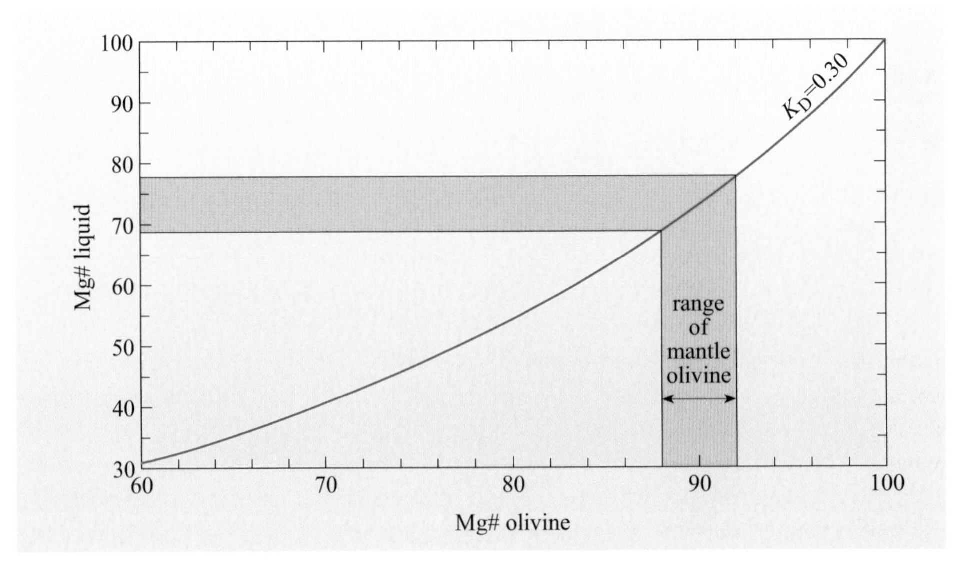

Figure 4.27 Equilibrium relationship between the Mg#s of olivine and liquid for K_D= 0.3. The heavily shaded bands show the ranges in composition of mantle olivines and the liquids with which they are in equilibrium.

The Mg#s of mantle olivine are in the range 88 to 92 so, using Equation 4.9 or Figure 4.27, these olivines will be in equilibrium with basaltic liquids with Mg#s in the range 69 to 78. We now need to compare the actual compositions of basalts erupted at mid-ocean ridges with the compositions produced in the mantle.

The Mg#s of more than 2000 samples of MORB are summarised in Figure 4.28, from which two conclusions can be drawn:

- very few mid-ocean ridge basalts have Mg#s of 69 to 78, therefore very few mid-ocean ridge basalts are in equilibrium with mantle olivine
- the erupted lavas have a wide range of compositions, with most lavas having Mg#s in the range 45 to 65.

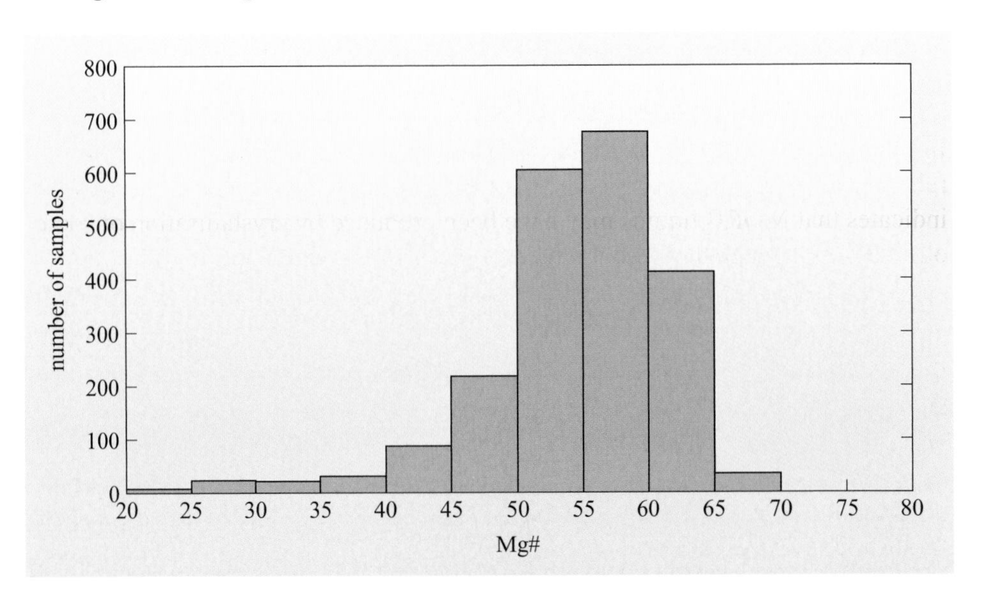

Figure 4.28 Histogram of Mg#s of mid-ocean ridge basalts. (Adapted from Sinton and Detrick, 1992)

It is very rare, therefore, for magma erupted at mid-ocean ridges to have come straight from its mantle source region unchanged. Between leaving the mantle and arriving at the surface, the magma has changed composition. The liquid produced by partial melting is referred to as a **primary magma**, and will have a suitably high Mg# reflecting equilibrium with mantle olivine. The origin of oceanic crust, therefore, must start with a narrow range of primary magmas (based on the narrow range of $Mg\#_{olivine}$) and end with a wide range of magmas with lower Mg#s (see Figure 4.28). The transformations that take place between leaving the mantle and arriving at the sea floor are the subject of the next sections.

4.6.2 The liquidus temperatures of mid-ocean ridge basalt

Sections 4.4 and 4.5 showed how the origin of basalt by partial melting of peridotite could be understood in terms of the pressure–temperature phase diagram of peridotite and the *P–T* path followed by hot mantle as it decompressed. Can phase diagrams be used to help us think about how hot basalt liquid will behave as it moves through cold crust to the surface? We can make an initial stab at what happens by revisiting the Fo–Fa phase diagram. According to Figure 4.18, what will happen to a liquid of composition $Fo_{70}Fa_{30}$ as it cools from 1800 °C?

The liquidus curve is intercepted at 1775 °C, so at this temperature olivine crystals start to grow. The composition of this olivine, read from the solidus at 1775 °C, will be $Fo_{91}Fa_{9}$. With further cooling, the liquid tracks down the liquidus curve and the solid tracks down the solidus curve. So, as this magma cools, Mg-rich olivine crystals grow from the liquid, causing the concentration of Mg in the liquid to fall. As the temperature continues to decrease, more olivine crystallises, and the liquid becomes poorer in the Fo component, as do the coexisting olivine crystals. This is the reverse of the situation during partial melting that we looked at in Figure 4.19. The phase diagram shows that crystallisation is caused by cooling, and that there are correlations between the composition of the liquid, the composition of the crystals and the temperature.

■ In this model of crystallisation during cooling, does the Mg# of the liquid increase, decrease, or stay the same?

■ With decreasing temperature, the liquid becomes poorer in the Mg endmember, and richer in the Fe endmember, so the Mg# of the liquid decreases.

In principle, the relatively low Mg# of mid-ocean ridge basalt glasses (i.e. low relative to the values expected for liquids in equilibrium with mantle olivine) indicates that MORB liquids may have been produced by crystallisation of some olivine from a magma of initially higher Mg# and higher temperature.

Turning to laboratory experiments with real mid-ocean ridge basalts, samples with different compositions (and therefore different Mg#s) are found to have different liquidus temperatures. If a single sample is taken and held at different temperatures, mixtures of crystals and liquids of different composition are produced. The correlation between the compositions of liquids and their liquidus temperatures is shown in Figure 4.29 and it mimics the pattern anticipated from the Fo–Fa phase diagram: high Mg# compositions have higher liquidus temperatures than low Mg# compositions.

The temperature at which the global population of mid-ocean ridge basalts is erupted can be estimated from their chemical compositions, specifically their Mg#s, using Figures 4.28 and 4.29.

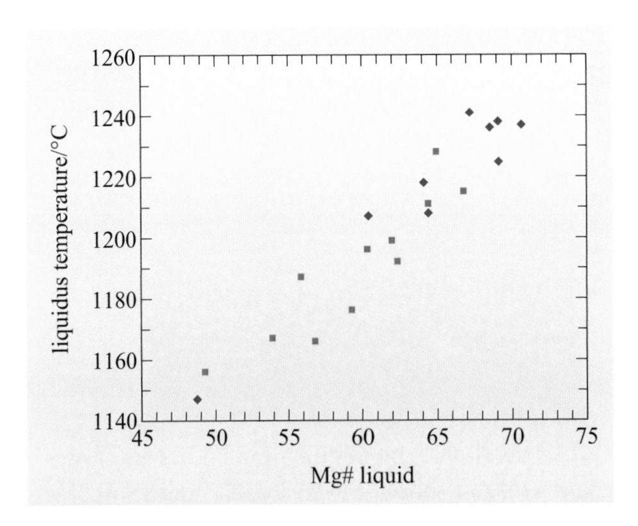

Figure 4.29 Results of melting experiments on mid-ocean ridge basalts done at atmospheric pressure. Diamonds represent the liquidus temperatures of different samples. The squares represent the temperatures and liquid compositions of liquids produced in different experiments on the same sample. Both sets of data overlap on this diagram. (Data from Grove and Bryan, 1983; Juster et al., 1989; Tormey et al., 1987; Grove et al., 1990, 1992; Yang et al., 1996;)

Question 4.15

The quenched liquids (glass) in many mid-ocean ridge lavas have Mg#s in the range 50–65%. What range of liquidus temperatures do these liquids have?

The answer to Question 4.15 is important because it shows that mid-ocean ridge magmas cool by up to 100 °C or more between leaving their mantle source and reaching the sea floor.

4.6.3 Crystallisation of mid-ocean ridge basalt

When basalt magma cools, crystallisation occurs and drives the composition of the remaining liquid to a lower Mg#. But because the erupted lavas have lower Mg#s than the primary magmas, the majority of the crystals that formed after the primary magma had left the region of partial melting in the mantle must have been removed from the magma before it erupted. It is likely that this is achieved where magma accumulates at depth in the crust and slowly cools. It is within such magma chambers that crystals can start to grow. Being denser than the liquid, crystals are likely to sink through the magma, accumulating on the floor of the chamber in layers. The continuous physical separation of the crystals as they grow from the liquid is known as **fractional crystallisation**. This term is used because the sample becomes physically divided into separate fractions, i.e. liquid and crystals. The accumulated crystals are referred to as **cumulate crystals**, or just **cumulates**.

A second way in which the magma can behave is where the crystals remain suspended within the magma, forming phenocrysts. In this case, there are a further two possibilities: either the growing crystals continuously re-equilibrate with the magma, or previously formed crystals become isolated from the liquid by becoming coated by new layers of crystal growth. In the first of these two cases, the phenocrysts will have constant composition, whereas in the second case only the edges of the phenocrysts are in equilibrium with the liquid and each crystal will be compositionally zoned with, for example, the cores of olivine crystals being richer in Fo than the edges of the crystals.

Fractional crystallisation produces a series of liquids whose compositions reflect the amount and composition of removed (cumulate) crystals, but what determines which minerals will crystallise from the cooling magma?

Question 4.16

Results of crystallisation experiments on a Mid-Atlantic Ridge basalt are shown in Table 4.4. What was the sequence of mineral growth as lower temperatures were reached? (You can ignore the small amount of spinel and just concentrate on the silicate minerals.)

Table 4.4 The results of experiments with a sample of basalt when held at different temperatures, showing whether liquid, olivine, plagioclase or clinopyroxene were found. (✓* indicates that a small amount of spinel was also present.) (Data from Grove and Bryan, 1983)

Temperature/°C	Liquid	Olivine	Plagioclase	Clinopyroxene
1245	✓			
1228	✓	✓*		
1215	✓	✓*		
1211	✓	✓*		
1199	✓	✓*	✓	
1196	✓	✓*	✓	
1192	✓	✓*	✓	
1187	✓	✓*	✓	✓
1176	✓	✓	✓	✓
1167	✓	✓	✓	✓
1166	✓	✓	✓	✓
1156	✓	✓	✓	✓

As well as using experiments to discover the order in which different minerals crystallise during cooling, samples of rock brought up from the sea floor can also reveal the order of crystallisation. Each rock is a mixture of glass together with a few phenocrysts, and the identity of the phenocrysts depends on the composition of the glass (i.e. the composition of the liquid in equilibrium with those crystals). Glasses with lower Mg#s represent liquids that have been generated at lower temperatures, and therefore by greater amounts of fractional crystallisation, than glasses with higher Mg#s. The sequence in which different phenocrysts appear with decreasing Mg# (and decreasing MgO content) of the glass thus represents the crystallisation sequence of the minerals in mid-ocean ridge basalt magma. The commonest sequence is olivine, then olivine + plagioclase, and then olivine + plagioclase + clinopyroxene (see Question 4.16).

To summarise:

- during partial melting of peridotite it is found that, with increasing temperature, some minerals are consumed in the melting reactions (plagioclase, spinel, garnet and clinopyroxene) whereas others persist to higher temperatures (orthopyroxene and olivine)
- in the case of progressive crystallisation due to cooling, the reverse phenomenon occurs: certain minerals do not start to crystallise until the temperature has become sufficiently low (and the liquid composition has sufficiently low Mg#).

4.6.4 Variation diagrams

Mid-ocean ridge glasses show a range in composition reflecting the progressive removal of olivine, plagioclase and clinopyroxene. These different minerals have different compositions. For example, relative to basalt:

- olivine is rich in MgO and poor in CaO
- plagioclase is poor in MgO but rich in CaO
- clinopyroxene is rich in both MgO and CaO.

■ When olivine is the only crystallising mineral, what will happen to the MgO and CaO concentrations in the liquid?

■ MgO will decrease and CaO will increase.

This effect can be seen on a plot known as a **chemical variation diagram** (or just **variation diagram**). Take a look at Figure 4.30, which plots the CaO and MgO wt% contents of liquids and olivine. Starting with an initial liquid (L_1), crystallising some olivine causes the liquid composition to move further away from the composition of the olivine (Figure 4.30a) until L_2 is reached. Crystallising still more olivine changes the liquid composition to L_3. Given the array of liquids L_1 to L_3, their alignment on this diagram links them to fractional crystallisation involving olivine removal. In other words, fractional crystallisation causes the composition of the liquid to change, or evolve. Magmas which have undergone only a small amount of fractional crystallisation and are still fairly close in composition to that of a primary magma are said to be **primitive magmas**, whereas those that are further removed are said to be **evolved magmas** (but note these terms are relative). Evolved magmas may also be described as **derivative magmas** or **fractionated magmas**. They can also be called **differentiated magmas** because the production of evolved magmas from a single starting magma is called differentiation.

If at some point plagioclase starts to crystallise alongside olivine, then the liquid composition will evolve in a direction that moves it away from the average composition of the crystallising mixture of olivine and plagioclase. This average composition lies on the straight line joining the compositions of the two minerals and depends on the proportions of each mineral. This is shown schematically in Figure 4.30b.

If a mixture of olivine, plagioclase and clinopyroxene is crystallising, then the compositional path followed by the liquid will move away from the composition of the three-phase mixture (Figure 4.30c). In this case, the overall composition of the crystals plots within the triangle defined by drawing straight lines between each of the three minerals.

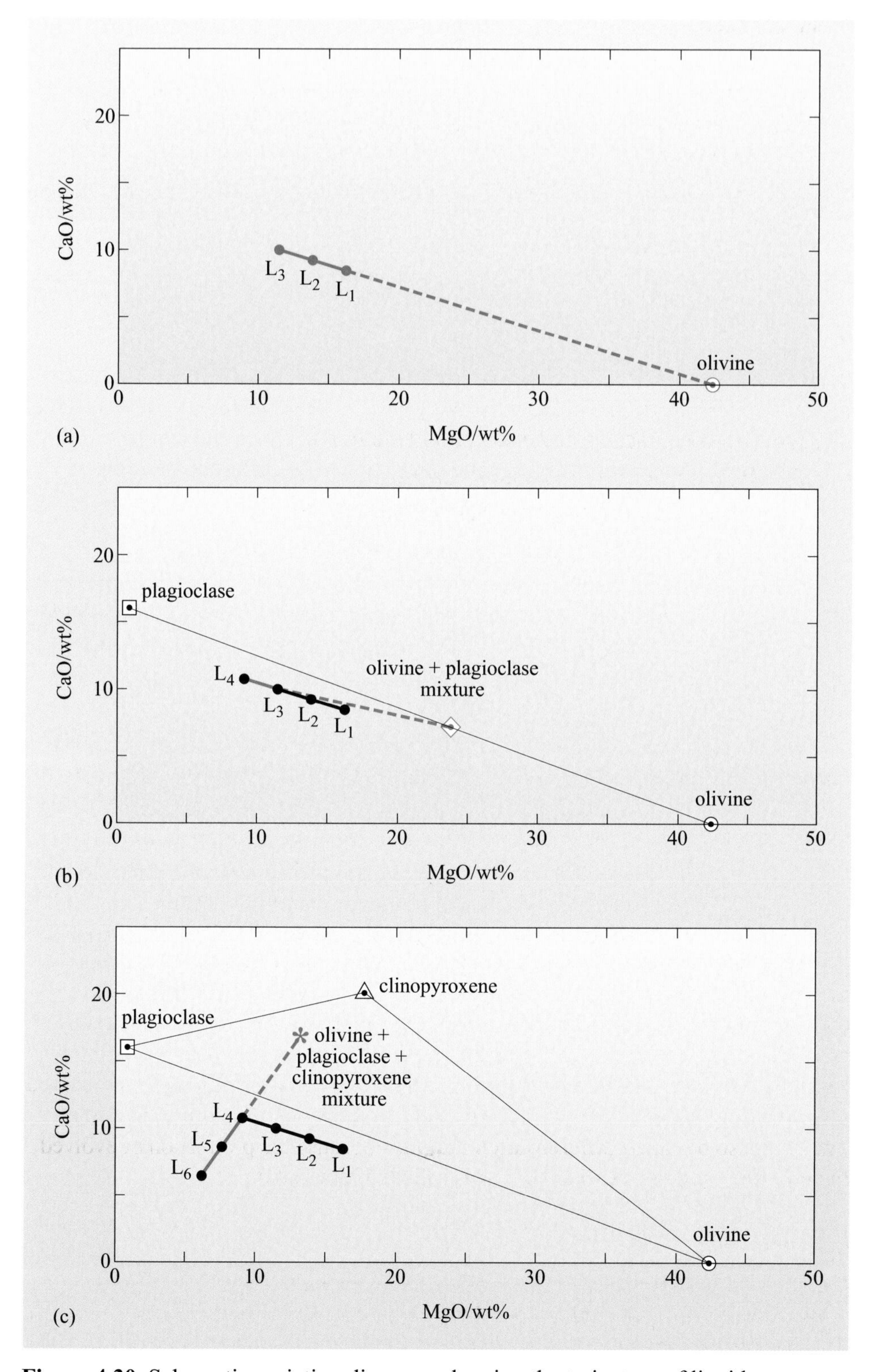

Figure 4.30 Schematic variation diagrams showing the trajectory of liquid composition of basalt undergoing fractional crystallisation of (a) olivine (L_1 to L_3), (b) a mixture of plagioclase and olivine (L_3 to L_4), and (c) a mixture of olivine, plagioclase and clinopyroxene (L_4 to L_6). For simplicity, mineral compositions are assumed to remain constant, whereas solid-solution effects will cause them to change as a result of changing liquid composition.

Variation diagrams such as those in Figure 4.30 are useful for a number of reasons. First, they allow the compositions of many samples to be compared without having to scrutinise tables of data. Second, and more importantly, the shapes of the trends can be used to infer the processes that have controlled the evolution of the magmas. For instance, a trend of decreasing MgO and slightly increasing CaO can be due to fractional crystallisation involving the removal of olivine or a mixture of olivine and plagioclase. A trend of decreasing MgO and decreasing CaO reflects fractional crystallisation of a mixture of olivine, plagioclase and clinopyroxene.

■ Figure 4.31 shows the chemical compositions of glasses from pillow lavas collected from the East Pacific Rise. Describe the trend and suggest how it was formed.

■ The correlation shows decreasing MgO and CaO. This trend can be explained if different samples have been formed by different amounts of fractional crystallisation of olivine, plagioclase and clinopyroxene (as in the trend from L_4 to L_6 in Figure 4.30c).

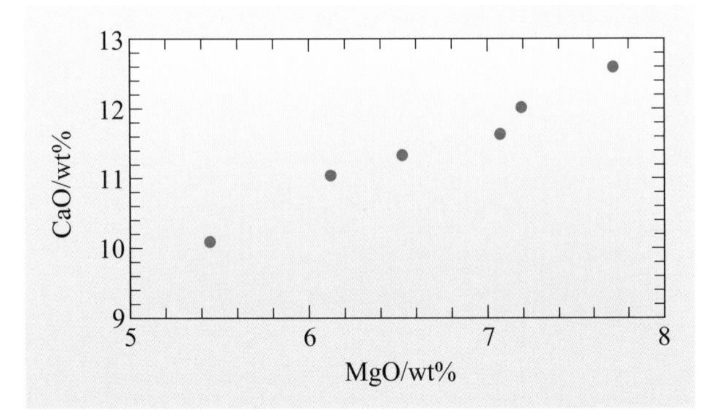

Figure 4.31 Variation diagram of wt% CaO against MgO for basalt glasses from 9° 30′ N on the East Pacific Rise. (Data from Pan and Batiza, 2003)

The interpretation of Figure 4.31 is that magma became lodged in the magma chamber(s) beneath the East Pacific Rise, and cooled sufficiently for olivine, plagioclase and clinopyroxene to crystallise, and that these crystals became separated from the magma, forming cumulates. By this process, fractional crystallisation produced a range of liquid compositions, dependent on the amount of crystallisation: the larger the amount of crystals removed, the lower the MgO content of the liquid.

Depending on the stage of magma evolution, the removed cumulate crystals will be composed of olivine and a mixture of olivine and plagioclase, or a mixture of olivine, plagioclase and clinopyroxene. Rocks made of these minerals are dunites (olivine only) or gabbros. So, dunite and gabbro cumulates are the expected by-products of the fractional crystallisation of basalt, as well as the slowly crystallised equivalents of basalt. Thus, the igneous oceanic crust is composed of evolved liquids quenched as sea-floor lavas or dykes, the cumulate complement of those liquids, and slowly cooled basalt magmas.

4.6.5 The depths of magma chambers beneath mid-ocean ridges

Between the depths where magmas are generated and the sea floor where they erupt, mid-ocean ridge magmas undergo fractional crystallisation to yield the evolved liquids that erupt and the cumulate gabbros that remain in the crust. But what about the depths at which fractional crystallisation occurs?

Studies of the passage of seismic waves through the crust beneath mid-ocean ridges have been successful in locating the tops of magma chambers at between 1.2 km and 2.4 km below the East Pacific Rise (Detrick et al., 1987) and about 3 km below the Valu Fa Ridge (situated in the Lau Basin, North of New Zealand).

But in other cases, no magma chambers have been detected by this method. This could be because either the chambers simply are not there or the chambers are too small or too deep to be detected amongst the noise in the seismic data. Seismology has been only partly successful in detecting magma chambers because of limitations with the method, so other ways of pinpointing the depths at which magmas may be stored or may fractionate need to be used.

Because pressure can have an influence on which minerals crystallise from a magma, there is the potential to use the compositions of the liquids or the crystals in the erupted lavas to infer the pressure (and therefore depth) at which fractional crystallisation occurred. This can be illustrated using a simple model system involving the first mineral to crystallise from primitive mid-ocean ridge basalt (i.e. olivine) and the last mineral to crystallise (i.e. clinopyroxene).

Both of these minerals form solid-solution series, so the first step is to focus on the pure endmembers. The Mg endmembers are usually chosen because the olivine and clinopyroxene phenocrysts in basalts are richer in Mg than in Fe.

- What are the pure Mg endmembers of olivine and clinopyroxene called and what are their chemical compositions?
- Forsterite (Mg_2SiO_4) and diopside ($CaMgSi_2O_6$) respectively.

Figure 4.32 shows the phase diagram of Fo–Di at two pressures: 10^5 Pa (atmospheric pressure) and 2 GPa. In this system, for a starting composition that is rich in Fo, forsterite will be the first mineral to crystallise (which is analogous to the initial crystallisation of olivine when primitive basalt magma cools).

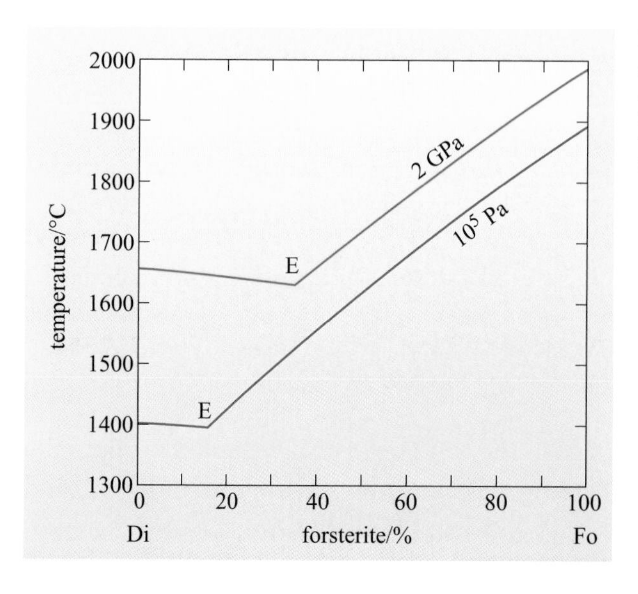

Figure 4.32 Liquidus curves in the system Fo–Di phase at pressures of 2 GPa and 10^5 Pa. E is the eutectic point.

- According to Figure 4.32, what is the composition of the liquid when Di starts to crystallise at 2 GPa, after fractional crystallisation of Fo?
- The composition of the liquid follows the liquidus curve with falling temperature until it reaches the eutectic, at which point Di starts to crystallise. At 2 GPa the eutectic liquid has a composition of about $Fo_{35}Di_{65}$.

■ What is the composition of the liquid that first crystallises Di at 10^5 Pa, and how does this differ from the composition at 2 GPa?

■ At 10^5 Pa, the eutectic composition is at about $Fo_{16}Di_{84}$. This is closer to the composition of pure Di.

The Fo–Di phase diagram illustrates that the amount of Fo that must crystallise out of the magma before Di starts to crystallise depends on the pressure. The lower the pressure, the greater the amount of Fo needed to crystallise before Di appears. Fo has a higher MgO content than Di (57.2 wt% versus 18.6 wt% MgO) so, in terms of the MgO content of the evolving liquid, low-pressure fractional crystallisation of Fo will drive the liquid to a lower MgO content before Di starts to crystallise than will fractional crystallisation at high pressure.

The idea that reducing the pressure delays the start of Di crystallisation can now be applied to the chemically more complex mixture of natural basalt, and a variation diagram can be used to contrast the liquid compositions caused by fractional crystallisation of primitive MORB at high and low pressures. The plot of MgO wt% versus CaO wt% in Figure 4.33 illustrates the compositional evolution of the magma at different pressures:

- at high pressure, fractional crystallisation of a primitive magma with composition A would first involve olivine and then olivine + plagioclase until reaching point B, when clinopyroxene starts to crystallise and compositions evolve along the path of decreasing CaO wt% from B to C. The magmas along the line BC crystallise olivine + plagioclase + clinopyroxene
- at low pressure, the first appearance of clinopyroxene is delayed and does not appear until the magma has evolved from A to D, after which point olivine + plagioclase + clinopyroxene crystallisation drives the composition along the path D to E.

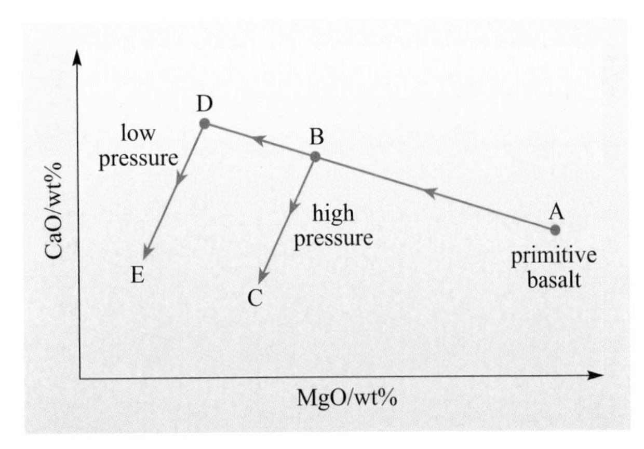

Figure 4.33 Hypothetical variation diagram showing how the compositional evolution of a primitive basalt (A) depends on the point at which clinopyroxene starts to crystallise, and how that point depends on pressure. See text for explanation.

According to Figure 4.33, liquids that are in equilibrium with crystals of olivine, plagioclase and clinopyroxene at low pressure will have different compositions from those at high pressure. For a given MgO content, liquids produced at lower pressure will have a higher CaO content than those produced by fractional crystallisation at higher pressure. Experiments on actual MORB samples bear this out, as seen in Figure 4.34a, which contrasts the compositions of liquids produced in experiments done at atmospheric pressure (10^5 Pa) and 800 MPa. To put these pressures into context, answer the following question.

Question 4.17

What is the pressure (a) on the sea floor below 3 km of seawater (density 1.0×10^3 kg m^{-3}) and (b) at 30 km below the sea floor (assuming an average rock density of 3×10^3 kg m^{-3}) and a water depth of 3 km? In both cases, assume that $g = 9.8$ m s^{-2} and give your answer to the correct number of significant figures. (Use Equation 4.1 in Box 4.4.)

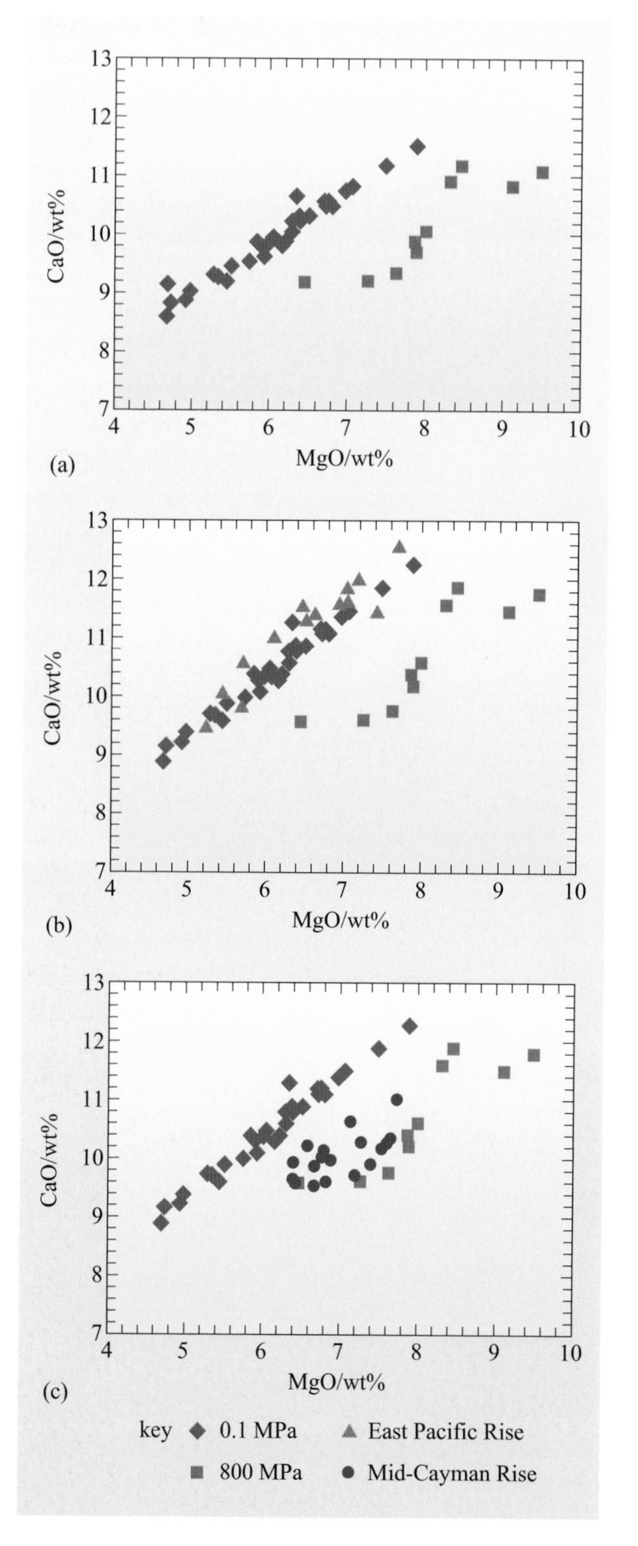

Figure 4.34 (a) Compositions of liquids in equilibrium with olivine + plagioclase + clinopyroxene produced in experiments on mid-ocean ridge basalts at 10^5 Pa and 800 MPa. (b) Compositions of glasses from the East Pacific Rise at 9° 30′ N, in comparison with experimental glasses from (a). (c) Compositions of glasses from the Mid-Cayman Rise, in comparison with experimental glasses from (a). (Data for (a) from Tormey et al., 1987; Juster et al., 1989; Grove et al., 1990, 1992; Yang et al., 1996; for (b) from Pan and Batiza, 2003; for (c) from Elthon et al., 1995.)

In Figure 4.34, liquids at atmospheric pressure fall on a narrow, well-defined trend, whereas the data at 800 MPa are much more scattered, but the points are still clearly separated from those at atmospheric pressure. The experimental compositions produced under known pressure can be compared with the compositions of natural lavas produced by fractional crystallisation at some unknown pressure, in order to estimate the pressures at which the mid-ocean ridge magmas evolved.

The East Pacific Rise and the Mid-Cayman Rise in the northern Caribbean Sea represent a fast-spreading ridge (11–12 cm y^{-1} full spreading rate) with seismic evidence for the roof of a magma chamber at between 1.6 km and 2.2 km depth, and a slow spreading ridge (2.0 cm y^{-1} full spreading rate) respectively. Lavas from these two ridges show trends of decreasing CaO with decreasing MgO (Figure 4.34b–c), indicative of fractional crystallisation of olivine + plagioclase + clinopyroxene (see Figure 4.30c).

The East Pacific Rise lavas plot very closely to the liquids from experiments at atmospheric pressure, so we infer that the East Pacific Rise magmas were produced by olivine + plagioclase + clinopyroxene fractional crystallisation in a magma chamber situated at shallow depths in the oceanic crust. On the other hand, the Mid-Cayman Rise lavas plot in a broader area to the right of the atmospheric pressure experiments but to the left of, and overlapping with, the area occupied by the liquids produced at 800 MPa. The crystallisation pressure of these magmas has been calculated to be about 600 MPa, and this pressure is found at a depth of about 20 km below the sea floor.

This indicates that the fast-spreading East Pacific Rise has a shallower magma chamber than the slow-spreading Mid-Cayman Rise. A possible reason for this difference is that slowly spreading ridges do not allow magmas to escape from depth as easily as at fast-spreading ridges, so the magma becomes trapped at greater depth.

4.7 Interactions with the oceans

So far in this chapter, the main themes have been the processes of partial melting and fractional crystallisation that generate oceanic crust at constructive plate boundaries. The material that forms the Earth's oceanic crust originates in the high-*P*, high-*T* environment of the upper mantle, but it ends up as part of the low-*P*, low-*T* environment of the Earth's surface where it is laid bare to the chemical, physical and biological processes of the hydrosphere and biosphere. Once formed, the oceanic crust continues to evolve, and this is the topic of the final section of this chapter.

Sea-floor spreading and volcanism at mid-ocean ridges are driven by the rise of hot mantle towards the surface. As a result, mid-ocean ridges sustain the highest surface heat fluxes on Earth, with half of the Earth's heat output coming from ocean lithosphere that is less than 65 million years old. Given this high heat flux and the fact that the upper ocean crust is a permeable rock mass soaking in seawater, it is perhaps not surprising that hot springs akin to those found at volcanic regions on land are found at mid-ocean ridges (Figure 4.1b). The first inkling that mid-ocean ridges sustained hydrothermal activity and vigorous submarine hot springs was the discovery, in the mid-1960s, of brine solutions at 40 °C to 60 °C and metal-rich sediment at the constructive plate boundary separating Arabia and Africa on the floor of the Red Sea. Scientists became eager to find more evidence of deep-sea hydrothermal systems, so an expedition was proposed to the Galápagos Ridge in the eastern Pacific Ocean at 86° W, where small thermal anomalies in the otherwise uniformly cold sea-floor ocean water had already been detected by remotely towed instruments. Using the submersible craft *Alvin*, which had been purpose built for deep-sea research, the first views of monotonous, barren, black basalt lava and the occasional clam shell at the mid-ocean ridge were seen in February and March 1977. As the expedition progressed, sensors detected temperatures above the normal 2.0 °C. *Alvin*'s lights lit up the otherwise total darkness of the deep to reveal shimmering warm water rising out of cracks in the sea-floor basalt lava; large mussels, clams, red shrimps, white crabs, limpets and anemones populated the 17 °C water (Figure 4.35 and Box 4.5). Samples of water returned to the surface gave off the rotten-egg stench of hydrogen sulfide (H_2S).

Figure 4.35 A view of the East Pacific Rise, 2600 m below sea level, showing a community of crabs, bivalves, tube worms (the two white 'stalks' attached to the rocky sea floor left of lower centre) and fish. (Dr Ken MacDonald/Science Photo Library)

Box 4.5 Life in the dark

Dwelling on the surface of the Earth, we take it for granted that life's food chain is based on plants and micro-organisms that get their energy from sunlight. These are the photosynthetic organisms. However, no sunlight penetrates to the bottom of the ocean, so what sustains the communities of organisms that are found around mid-ocean ridge vents? At the bottom of the food chain are colonies of micro-organisms that rely on chemical energy rather than energy from sunlight. The main metabolic process in many of these micro-organisms is the oxidation of reduced sulfur. For example, reduced sulfur in H_2S gas is used to take CO_2 gas dissolved in seawater and generate organic carbon by the chemical reaction:

$$S^{2-} + CO_2 + O_2 + H_2O \rightarrow SO_4^{2-} + [CH_2O] \quad (4.10)$$

where $[CH_2O]$ denotes organic carbon. Organisms that live in this way are called **chemosynthetic**, in contrast to photosynthetic organisms. These chemosynthetic organisms tolerate high temperatures – and some require temperatures well above 100 °C. The chemosynthetic organisms and the macrofauna that consume them form deep-sea vent communities that are alien to our normal concept of life, but their existence provides a new impetus to understand the intertwined geological and biological processes on Earth, and even on other planets.

Exploration gathered pace and, in the spring of 1979, diving on the East Pacific Rise at 21° N, *Alvin*'s crew encountered two types of hot-water vent. Not only did they find hot water seeping from cracks on a biologically infested sea floor (similar to the Galápagos Ridge sites found in 1977), they also discovered water at 380 °C spouting from chimneys made of copper and iron sulfides. As the hot fluids rose through the water column they mixed with the ambient seawater, precipitating clouds of billowing black 'smoke' (Figure 4.1b). Here, in these spectacular 'black smoker' vents, was yet more evidence of sulfur-rich hot fluids gushing from the sea floor at mid-ocean ridges. At least fifty such areas of hydrothermal activity are now known from most of the mid-ocean ridge system, as a result of either planned exploration or serendipitous discovery of thermal or chemical anomalies in the oceans.

The underlying processes at hydrothermal areas are clear: cold seawater is drawn into the permeable oceanic crust, becomes heated and reacts with the rocks through which it flows before hot mineral-rich solutions rise to the surface, mixing and cooling to various degrees before flowing back out of the system into the ocean. As a result, the chemical composition of the oceanic crust becomes modified, as does the chemical composition of the ocean, and chemicals are supplied to mid-ocean ridge ecosystems. Two lines of evidence that give clues as to what happens in these hydrothermal systems are:

- the differences between pristine lava erupted on the ocean floor and rocks that have been chemically altered as a result of chemical reactions with hot seawater
- the differences between normal cold seawater that enters hydrothermal systems and the hot effluent that comes out of the hydrothermal systems.

4.7.1 Evidence from rocks

Section 4.6 dealt with the chemical composition of fresh lava that had been quenched on eruption and sampled before any later alteration processes had occurred. The chemical variations were due to the igneous processes that generated the magma and led to its differentiation. Samples that have undergone various degrees of chemical and mineralogical alteration after the magma solidified are needed to investigate hydrothermal alteration processes. Samples of altered ocean crust come from tectonically disrupted sections of sea floor (such as fracture zones), cores recovered by deep sea drilling, and dissected oceanic crust exposed in ophiolites.

These altered rocks are typically pale or dull green, in contrast to the black colour of freshly erupted glassy basalt, so are generally called **greenstones**. The familiar igneous ferromagnesian minerals olivine and clinopyroxene are absent, but minerals such as chlorite (a green mineral with a layered structure similar to mica), epidote (a pale-green mineral) and actinolite (a green Ca-bearing amphibole rich in iron) are present. Another contrast with fresh basalt is that the plagioclase feldspar found in most mid-ocean ridge basalts is relatively rich in the Ca endmember anorthite (abbreviated as An) and poor in the Na endmember albite (Ab), ranging from $An_{85}Ab_{15}$ to $An_{60}Ab_{40}$, whereas in the altered rocks, plagioclase is much richer in albite, ranging from $An_{35}Ab_{65}$ to An_5Ab_{95}. All of these minerals (i.e. chlorite, actinolite, epidote and plagioclase) form solid-solution series so their compositions are variable; examples of their compositions are given in Table 4.5. Chemical analyses of a relatively unaltered sample from the centre of a pillow lava and two more altered samples of greenstones from the edge of the same pillow are given in Table 4.6.

Table 4.5 Chemical analyses (reported as weight per cent oxides) of minerals from altered basalts recovered from the Mid-Atlantic Ridge. Analyses give the total iron (Fe^{2+} and Fe^{3+}) in terms of FeO and is denoted FeO_t. The shortfall between the reported total oxides and the ideal total of 100% reflects analytical uncertainties and components that were not analysed for, particularly water in the case of chlorite, epidote and actinolite. (Data from Humphris and Thompson, 1978)

	chlorite	chlorite	epidote	epidote	actinolite	actinolite	feldspar	feldspar
SiO_2	28.34	28.42	37.50	37.75	49.75	51.01	67.68	63.50
Al_2O_3	17.00	17.54	24.83	24.36	6.30	2.43	20.46	22.18
FeO_t	21.30	23.33	10.45	11.75	17.40	13.32	0.04	0.00
MnO	0.16	0.14	0.03	0.16	0.12	0.14	0.00	0.00
MgO	20.19	17.41	0.03	0.27	12.66	15.16	0.00	0.00
CaO	0.07	0.13	23.4	22.83	11.78	12.63	1.03	2.87
Na_2O	0.07	0.01	0.01	0.04	0.55	0.33	11.39	10.64
K_2O	0.03	0.03	0.03	0.00	0.06	0.07	0.00	0.00
Total	87.16	87.01	96.28	97.16	98.62	95.09	100.60	99.19

Question 4.18

(a) Which of the following statements are true?

(i) Chlorite, epidote and actinolite are silicates.

(ii) Chlorite, epidote and actinolite are hydrous minerals.

(iii) Plagioclase, olivine and clinopyroxene are hydrous minerals.

(b) Apart from silica, what are the main constituent oxides of (i) albite, (ii) chlorite, (iii) epidote, and (iv) actinolite?

(c) What is the range of water content of the greenstones in Table 4.6?

Table 4.6 Chemical analyses (wt%) of greenstones from parts of a pillow lava from the Mid-Atlantic Ridge. Samples A and C are from the outermost and innermost parts of a single pillow respectively. Sample B is from a position between A and C. (Data from Humphris and Thompson, 1978)

	A strongly altered	B moderately altered	C slightly altered
SiO_2	39.31	47.70	52.30
TiO_2	2.25	1.75	1.63
Al_2O_3	17.59	14.15	14.79
FeO_t	12.71	9.50	8.17
MnO	0.11	0.10	0.14
MgO	12.67	9.70	6.03
CaO	3.35	8.40	9.83
Na_2O	2.98	2.97	3.22
K_2O	0.14	0.13	0.15
P_2O_5	0.15	0.14	0.16
H_2O	6.53	3.26	1.85
Total	97.79	97.80	98.27

One of the main chemical differences between greenstones and fresh mid-ocean ridge basalt is that basalt has less than 0.1% H_2O, whereas greenstones can have more than 6% water. This is reflected in the abundance of hydrous minerals in greenstones and is a clear indication that seawater is incorporated into basalt during hydrothermal alteration. Simply adding water will dilute the concentrations of all other elements, in which case the chemical compositions of the greenstones would be intermediate between those of fresh basalt and those of seawater. On a variation diagram they would plot on a straight line joining the basalt and seawater compositions. But is this what the chemical compositions of the rocks show?

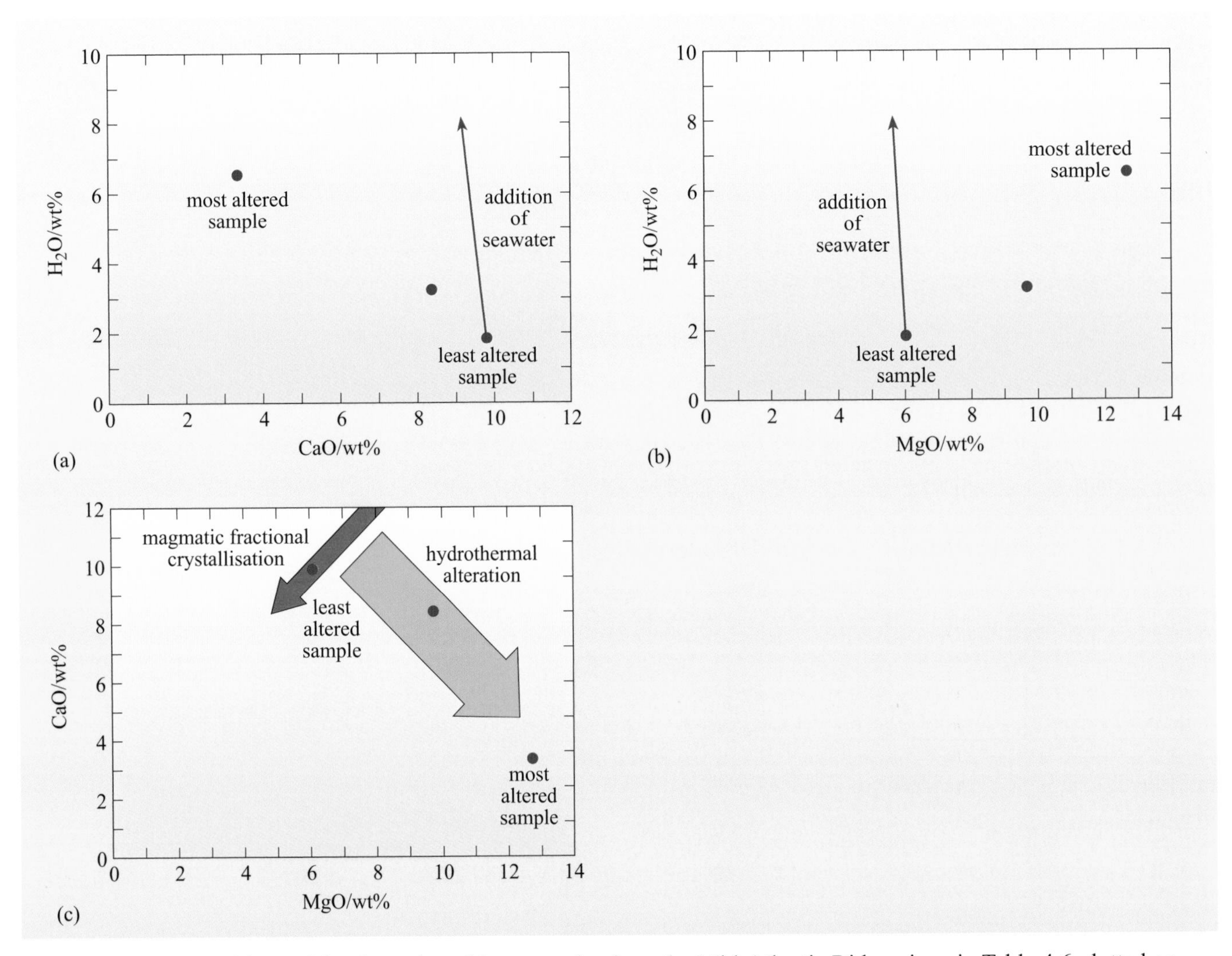

Figure 4.36 Compositions of the three altered lava samples from the Mid-Atlantic Ridge given in Table 4.6 plotted on variation diagrams of wt%: (a) CaO versus H_2O; (b) MgO versus H_2O; and (c) MgO versus CaO. The arrows in (a) and (b) show the trend that would be produced by adding seawater to the least-altered sample. The broad arrows in (c) indicate the general effects that fractional crystallisation and subsequent hydrothermal alteration will have.

- Compare the array of compositions in Figure 4.36 with the trend expected from just adding water to the least-altered sample. How do you account for the composition of these rock samples?

- Because CaO decreases with increasing H_2O more sharply than by dilution, CaO must have been removed from the original rocks. MgO does not follow the expected trend of dilution by seawater either. Instead, it increases in concentration as the percentage of water increases, indicating that water and MgO have both been added to the rock.

Although the greenstones have formed by incorporation of water, their origin also involves the removal of some elements (notably Ca and Si) and the addition of others (notably Mg) and this gives them compositions that are distinct from those produced by igneous processes (Figure 4.36c). Pristine basalts show a trend of decreasing MgO and CaO due to the removal of olivine, plagioclase and clinopyroxene crystals during fractional crystallisation in a cooling magma chamber (Section 4.6.4). The greenstones, on the other hand, have an array of compositions that reflect the addition of MgO and the removal of CaO to the original igneous composition.

Where does the added MgO come from, and where does the CaO that has been removed end up? The answer to this question comes from comparing the composition of seawater with the composition of the water that comes out of mid-ocean ridge hydrothermal systems. Hydrothermal water contains no Mg, whereas normal seawater has about 1290 ppm; hydrothermal waters have up to five times more Ca than seawater does (400 ppm). In overall terms, mid-ocean ridge hydrothermal systems remove Mg from the oceans (adding it to the crust) and add Ca to the oceans (removing it from the crust).

4.7.2 Mid-ocean ridge hydrothermal systems

Hydrothermal systems occupy large volumes of the crust at mid-ocean ridges, and can be divided into different parts according to the processes occurring (Figure 4.37). The broad area on either side of the plate boundary is where seawater penetrates the fractured oceanic crust and is drawn down into deeper, hotter crust; this is known as the **recharge zone** of the hydrothermal system. The hottest, deepest part of the system is where most of the chemical changes that determine the composition of the hydrothermal effluent occur, and this is known as the **reaction zone**. Further reactions occur as the hot fluid rises from the reaction zone to the surface, and this part of the system is known as the **discharge zone**.

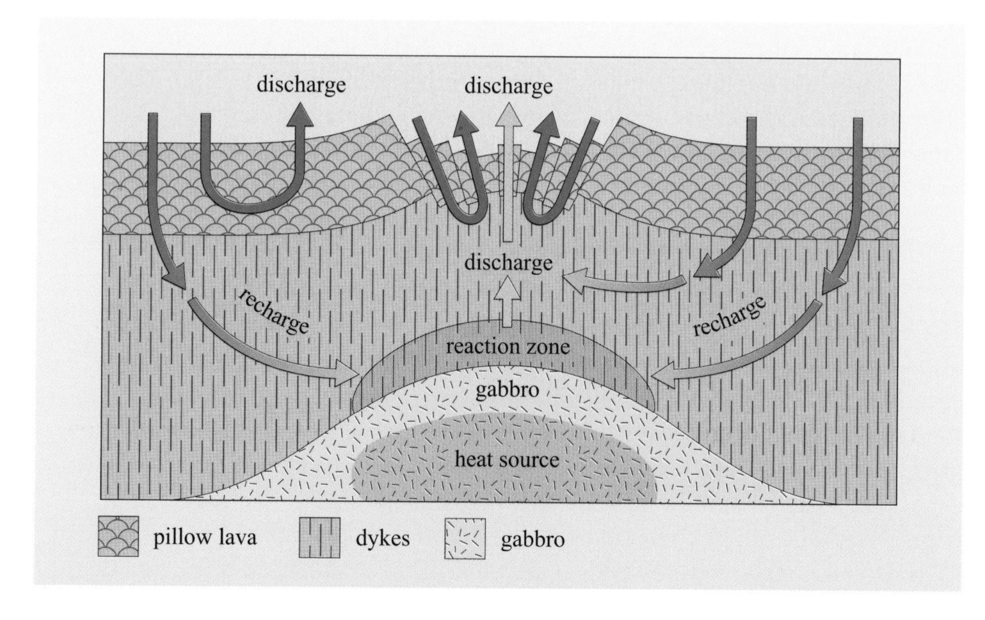

Figure 4.37 Schematic cross-section of oceanic crust at a mid-ocean ridge showing the flow paths of water through the recharge, reaction and discharge zones of the hydrothermal system. (Adapted from Alt, 1995)

Processes in the hydrothermal system have to be inferred from the composition of hydrothermal waters, investigations of hydrothermally altered rocks from ophiolites and ocean drilling, and theoretical modelling of chemical reactions at high temperature and pressure. Combinations of these approaches have led to an overall understanding of what happens in each of the zones.

Heating of seawater in the recharge zone

Direct sampling of water from recharge zones has not yet been achieved, although fluids from shallow off-axis regions have been obtained on the flanks of the Juan de Fuca Ridge (in the NE Pacific Ocean) and may be analogous to those from low-temperature regions of the recharge zone. These water samples, together with consideration of the behaviour of seawater when it becomes heated at crustal pressure, provide some ideas of the processes occurring in the recharge zone. The sampled fluids have less than 10% of the Mg contained in seawater, indicating uptake of Mg by the warm rocks through which the water passes. This happens by incorporation of Mg and water into basaltic rocks and minerals to form new, hydrous, Mg-rich minerals such as clay and chlorite. Other changes come about as a result of the change in the solubility of the dissolved components in seawater when it is heated. Once seawater is above about 150–200 °C (depending on pressure), it becomes saturated in $CaSO_4$ (the mineral anhydrite) such that all of the Ca and about two-thirds of the SO_4^{2-} are removed from the water by the crystallisation of anhydrite in pore spaces and veins within the rock. As well as removing Ca from the solutions that penetrate into deeper regions of the hydrothermal system, removal of SO_4^{2-} in the recharge zone limits the amount of sulfur taken to deeper levels.

Reaction of water with hot rock in the reaction zone

Fluids reaching the reaction zone have compositions that have been changed considerably by their passage through the recharge zone: most Mg has been removed, the remaining sulfur is largely in the form of sulfide ions rather than sulfate ions, and the pH has decreased. High-temperature reactions between this fluid and the hot rock (such as recently solidified intrusive rock or the rocks immediately above a magma chamber) yield new minerals and modify the water composition further. Fe, Mn, Cu, Zn and other metals are soluble in such hot acidic fluids and are therefore leached from the rock along with sulfur. The original igneous minerals (olivine, plagioclase and clinopyroxene) and glass are consumed in reactions that yield mineral assemblages involving albite and hydrous minerals that include chlorite, actinolite and epidote found in greenstones.

Precipitation of sulfides and sulfates, and mixing with seawater in the discharge zone

The hot mineralised solutions within the reaction zone are extremely buoyant with respect to cooler waters and therefore rise rapidly to the sea floor, undergoing decompression, cooling, and mixing with cooler compositionally different solutions. Phase changes that occur can involve the switch from a liquid into a mixture of vapour and a concentrated brine solution. Cooling can lead to deposition of metal sulfides in veins that give rise to mineral deposits. These mineral deposits can be found in ophiolites, notably in Cyprus and Oman, and have also been discovered by deep research drilling in the ocean crust. For instance, chemical analyses of a 1300 m long core that passes through pillow lavas and the underlying sheeted dyke complex of 5.9 million year old crust in the Nazca Plate (in the eastern Pacific Ocean) show the concentration of metal sulfide mineralisation is preferentially concentrated in the region where the dyke complex grades into the lavas. This is probably the location of heterogeneities in the rock structure that favoured mineral deposition.

Black and white smokers and hydrothermal plumes in the ocean

The most dramatic manifestation of hydrothermal systems at mid-ocean ridges are the billowing plumes of hot fluid and particles known as black smokers that flow from tall chimneys and mounds on the sea floor (Figure 4.1b). The black 'smoke' particles and the chimney are formed when hot, H_2S-rich, O_2-poor, low-pH (acidic) hydrothermal fluid mixes with cold, sulfate-rich, O_2-rich seawater. Within the first second of meeting seawater, mixing of the hottest fluids (between 290 °C and 400 °C) precipitates tiny particles of Cu, Fe and Zn sulfides and oxides, and these are carried up in the buoyant plume of water some 100–200 m before spreading out as a broad umbrella cloud. Some particles sink to the sea floor and become hydrothermal sediment that is rich in Fe and Mn.

- As well as the effect of seawater being mixed into hydrothermal fluid, the hydrothermal fluid also heats seawater. Can this lead to any minerals being precipitated?
- Yes. Heating seawater leads to precipitation of anhydrite.

Hot seawater, as well as mixtures of seawater and hydrothermal fluid, precipitates anhydrite where the hydrothermal fluid enters the sea, building a pipe that channels the hydrothermal fluid upwards. This pipe grows in size through a combination of sulfide and sulfate (anhydrite) precipitation, and matures in shape to form a chimney. Temperature gradients across the chimney wall, from the hot hydrothermal fluid flowing up the chimney to the cold seawater outside it, and mixing of fluids in the porous walls of the chimney, generate a compositional structure within the chimney as it grows, with the outside being composed of anhydrite and the inner walls of Cu–Fe sulfide (chalcopyrite). Individual chimneys can have an internal diameter of about 10 cm and can grow to 45 m in height, although most are much shorter than this.

Hydrothermal discharges between 150 °C and 290 °C produce white smokers, the white colour being due to pale-coloured minerals and precipitates, mainly of anhydrite, barite (barium sulfate), and amorphous (non-crystalline) silica.

Over time, mineral growth and dissolution in hydrothermal systems can clog the pathways used by the fluids and discharge may wane or cease, only to start up elsewhere. Once discharge has ceased, a further set of reactions takes place between seawater and the now cold mineral-rich deposits (the sulfide and sulfate-rich chimneys and sediment). Most important is the gradual dissolution of anhydrite into cold seawater. Because anhydrite cements parts of the chimneys together, this process leads to the weakening and collapse of chimneys, eventually reducing them to piles of sulfide-rich rubble.

4.7.3 Hydrothermal systems and the composition of seawater

Hydrothermal systems alter cold seawater by heating it, causing it to react with hot basalt and gabbro of the oceanic crust, and then rise to the surface. Because normal seawater enters the hydrothermal system but hot, mineral-rich water exits the system, the overall process results in a change to the original igneous composition of the crust. Equally, the input of hydrothermal fluid from black smokers and other discharging vents adds components to the ocean water. Through these effects, the oceanic crust becomes hydrated whereas the ocean becomes depleted in some elements (particularly Mg) and enriched in others (such as Mn and Fe). Exact estimates of the strength of these sinks and sources are difficult to make, principally because hydrothermal fluids are very variable in composition and the rate of discharge is known in only a few isolated cases and for a very short period of time. A further complication is that Mn and Fe in hydrothermal fluids are removed from the ocean by precipitation of sulfides. So, the slow evolution of the chemical composition of ocean water by inputs from rivers and interactions with oceanic crust and hydrothermal processes is easier to conceptualise than it is to quantify.

Summary of Chapter 4

This chapter has taken a path that started in the upper mantle and followed the fate of hot peridotite as it rises towards the surface beneath a mid-ocean ridge as sea-floor spreading pulls the lithosphere apart. The geological processes that progressively change the composition of the mantle and crust (partial melting, fractional crystallisation, and alteration) have been studied largely in terms of the chemical changes that are caused when Earth materials experience changes in pressure, temperature or the chemical environment.

Decompression partial melting of peridotite produces basaltic magma that then cools and undergoes fractional crystallisation at shallow (crustal to uppermost mantle) levels to produce the range of volcanic and intrusive rocks that comprise oceanic crust. High-temperature hydrothermal systems at the mid-ocean ridge then modify large sections of the igneous crust, generating a hydrous greenstone crust and sulfide mineral deposits, and moderate the chemical composition of the oceans. High heat and chemical fluxes sustain sulfur-oxidising bacteria that are the primary producers in a food chain of deep-sea biota.

Learning outcomes for Chapter 4

You should now be able to demonstrate a knowledge and understanding of:

4.1 The layered structure of the oceanic lithosphere.

4.2 The different rocks and minerals that make up the mantle and oceanic crustal rocks.

4.3 Variations in the mineralogy and melting temperature of lherzolite with increasing pressure.

4.4 How the magmatic processes of partial melting and fractional crystallisation account for differences in the composition and evolution of mantle and oceanic crustal rocks.

4.5 The processes involved in decompression melting beneath oceanic ridges and how the percentage of melt produced is directly related to the potential temperature of the upper mantle, the depth of melting and the thickness of the oceanic crust.

4.6 The interactions and reactions that occur between seawater and the oceanic crust within hydrothermal systems beneath oceanic ridges.

Processes at destructive plate boundaries

Chapter 5

Destructive plate boundaries are the sites where oceanic plates are recycled back into the mantle and, in terms of understanding the Earth as a system, they are the key locations for material flux back into the mantle. The modification of the subducted plate in the shallow mantle beneath a volcanic arc and the ultimate fate of a plate as it plunges into the deeper mantle play an important role in both the short-term and the long-term recycling of chemical elements within the Earth. In addition to volcanism, the majority of major earthquakes, particularly deep earthquakes, are located at destructive plate boundaries (Figure 3.9), a fact brought into stark reality in 2004 with the 26 December (Boxing Day) tsunami, which was triggered by a shallow earthquake located on the Sumatran plate boundary where the Indo-Australian Plate is being subducted beneath the Eurasian Plate.

In detail, subduction zones are complicated by a wide variety of factors that can influence the style of the plate boundary (Figure 5.1). In this chapter, the key parameters are described and their influence on the structure, tectonics, seismicity and magmatism of subduction zones is discussed.

You should recall from Chapter 3 that there are three major types of destructive plate boundary:

- ocean–ocean boundaries, where oceanic lithosphere is subducted beneath oceanic lithosphere
- ocean–continent boundaries, where oceanic lithosphere is subducted beneath continental lithosphere
- continent–continent boundaries, where two continental plates collide.

The first two plate boundaries are the focus of this chapter and continent–continent collision is discussed in Chapter 6.

It is clear that the dominant material that is subducted is the oceanic lithosphere.

■ What are the main components of oceanic lithosphere?

■ You should recall from Chapter 4 that oceanic lithosphere is made up of oceanic crust, consisting of pillow basalts, sheeted dykes and gabbros, and the underlying residual mantle comprising harzburgite.

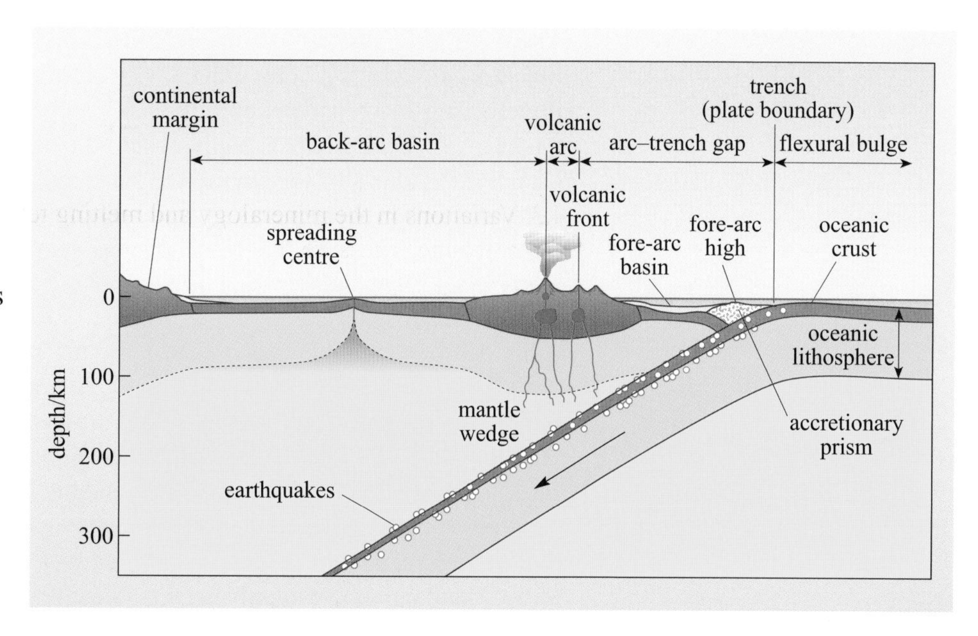

Figure 5.1 Generalised cross-section through oceanic and continental subduction zones. The terms used in this figure are explained in Section 5.1.

Importantly, hydrothermal fluids at the ridge crest chemically modify the oceanic crust. In addition, hydrothermal activity can extend deep within the upper part of the residual mantle to produce **serpentinites**. The oceanic plate, as it moves away from the ridge crest, cools and experiences low-temperature interaction with seawater; sediments gradually accumulate and, in some cases, further volcanic activity related to hot spots, known as intra-plate volcanism, may occur, producing seamounts and ocean islands. Sediment composition can vary widely depending on the latitude and proximity to continents and ocean islands. All of these effects contribute to the variable chemical composition of the oceanic plate, but one key factor is common to all plates: the addition of relatively large amounts of H_2O, both as pore water and locked up in hydrous minerals. The recycling of H_2O and other volatiles through subduction zones is one of the most important aspects of destructive plate boundaries and will be discussed with respect to volcanism in this chapter and to the inner workings of the planet in Chapter 7.

Although mid-ocean ridges, on average, produce more magma per year than volcanism associated with destructive plate boundaries, the majority of the world's explosive volcanism is found at destructive plate boundaries. This volcanism, the most hazardous and destructive, is consequently the one that captures the public's imagination. Mt St Helens, Pinatubo, Krakatau, Montserrat and Fuji are volcanoes that many people could name because of their histories of violent eruptions and all of these volcanoes are located at destructive plate boundaries.

From a geological point of view, destructive plate boundaries play a major role in the evolution of the planet, as subduction zones are thought to be the location of modern-day crustal growth. Furthermore, it is generally agreed that ancient subduction zones were the sites where a significant proportion of the continental crust was produced in the geological past. An understanding of the physical and chemical behaviour of modern-day subduction zones provides an essential framework for unravelling ancient crustal growth mechanisms.

5.1 Anatomy of a subduction zone

There is a large amount of nomenclature associated with subduction zones, much of which is illustrated in Figure 5.1. The **subducted plate** is often known as the **down-going plate**, and at depth is usually called the **subducted slab**. This mechanically strong slab is equivalent to the oceanic lithosphere discussed in Chapters 3 and 4. It is the location of most major earthquakes associated with destructive plate boundaries and the zone of earthquakes is called the Wadati–Benioff zone. The plate that lies above the subduction zone is termed the **overriding plate**, below which is a triangle of asthenosphere known as the **mantle wedge** overlain by the lithosphere of the volcanic arc itself. The **arc lithosphere**, which includes the crust and the lithospheric mantle, can be of variable thickness and composition and plays an important role in controlling the composition of the erupted lavas. In the oceans, arc crust is very thin (7 km) and the lithosphere thickness depends on the age of the plate. With time, the arc crust may thicken because of continued volcanism. The overriding plate at continental subduction zones often comprises pre-existing continental crust (30–50 km thick) and a much thicker lithosphere. The volcanoes form a **volcanic arc**, although the term 'island' arc is commonly used in oceanic arcs because the volcanic arc is defined by discrete islands (e.g. Guam and Krakatau). On a map, the line of the volcanic islands defines the **volcanic front**. Between the volcanic front and the plate boundary, which is usually defined by a **trench**, is the **fore-arc region**. This zone can be very complicated because it can be a region undergoing either extension or compression, depending of the geometry of the plates involved in the subduction zone. The fore-arc usually comprises a low-lying **fore-arc basin** that is filled with sediment and an outer **fore-arc high**. Depending on the tectonics, the fore-arc may also contain an accretionary prism, which is a wedge of sediment scraped off the down-going plate, or faulted regions containing serpentinised peridotites. In continental arcs, the fore-arc may also contain remnants of older arcs, continental crust and material accreted from the down-going plate. Some subduction zones have a second line of volcanoes, usually submarine, where a new spreading centre is underlain by the subducting slab. This region is known as a back-arc basin, or sometimes as a **marginal basin**, from which magmas are erupted that are affected by the subducting slab. This may seem like a lot of names, but it reflects the complexity of destructive plate boundaries.

A rapid review of the global map of plate boundaries (Figures 3.8 and 5.2) shows that the majority of destructive boundaries are found around the periphery of the Pacific Ocean. Consequently, the majority of the world's volcanic arcs are located around the Pacific, and this is commonly referred to as the **Pacific Ring of Fire**.

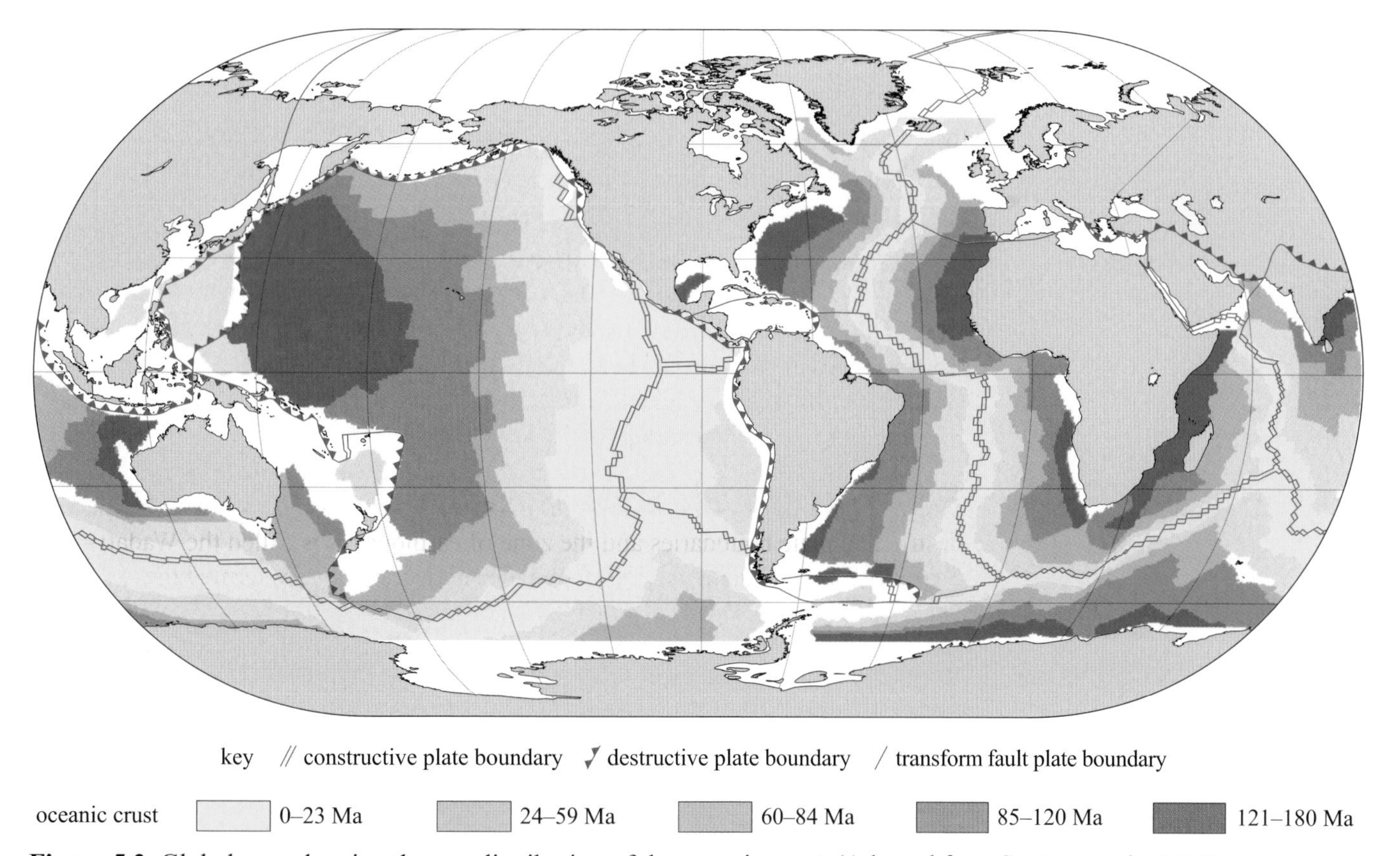

Figure 5.2 Global map showing the age distribution of the oceanic crust. (Adapted from Scotese et al., 1998)

- Using Figure 5.2, what can you say about the age of the subducting plate around the Pacific Ocean?

- Put simply, it varies hugely. The oceanic plate gets older the further away from a spreading centre. In the Pacific Ocean, the crust being subducted beneath the Mariana arc–trench system is the oldest because it is very distant from the East Pacific Rise where it formed. In places the crust is Jurassic in age. By contrast the Tonga–Kermadec system, being further east, has Cretaceous crust being subducted. On the eastern margin of the Pacific, the average age of the subducting plate is younger. The Nazca Plate is <25 Ma in age, and even younger where it is being subducted under southern Chile. The Cocos Plate and particularly the Juan de Fuca Plate being subducted beneath the Cascades are both very young (<10 Ma) because of their proximity to a spreading centre (Figure 3.25).

- Can you recall another property of the subducting slab that varies between different destructive plate boundaries?

- Its convergence rate.

You discovered in Chapter 3 that subduction is driven by the gravitational descent of old, cold lithosphere into the mantle, and once a plate starts subducting it pulls younger lithosphere behind it. Thus, with time, lithosphere of almost any age can be subducted. Moreover, since the plate is being driven by slab-pull, which is the most effective of the plate driving forces, all plates converge on subduction zones at relatively high speeds that vary between 65 and 106 mm y^{-1}.

- What will be the effect of subducting old versus young lithosphere on the composition and physical properties of the subducting slab?

- Old lithosphere is colder than young lithosphere and is more likely to have accumulated a thick layer of marine sediments.

Thus, although the subducting slab is invariably oceanic, its composition varies, as does its thermal characteristics, both of which contribute to the development of the complexity of arc structure.

5.2 Geophysical observations

Whereas Figure 5.1 gives an overall impression of the general structure of a subduction zone, more detailed images of the seismic activity across specific arcs provide greater insights into that structure. In Chapter 3 you were introduced to the classification of shallow, intermediate and deep earthquakes in relation to the Tonga arc. Figure 5.3 is a cross-section of the upper part of the Japanese arc and so all of the earthquakes in this diagram could be described as shallow (0–70 km) or intermediate (70–300 km).

It is easy to see from Figure 5.3 that there are abundant shallow earthquakes located in the volcanic arc itself that may relate to magma movement and storage in the arc crust, or possibly extension in the back-arc region. There is also a zone of earthquake foci that marks out the upper boundary of the descending slab, i.e. the Wadati–Benioff zone. In addition, there are other earthquakes deeper within the slab that may reflect the way in which the slab is deforming as it descends into the mantle. These intermediate earthquakes are more often compressional and relate to the resistance that the slab experiences as it pushes through the mantle. Deep earthquakes are also located within the descending slab, and are now considered to be generated by mineral transformations as the slab slowly heats.

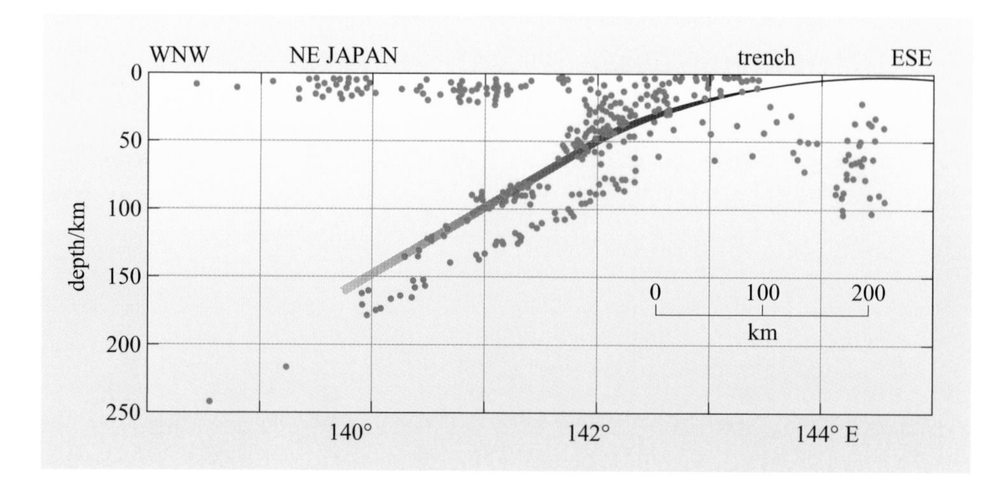

Figure 5.3 Simplified cross-section of the NE Japan subduction zone with the distribution of earthquake foci plotted as dots. The top of the subducting slab is marked with a black line. (Hasagawa et al., 1978)

Yet more information on the structure of subduction zones is derived from a technique known as seismic tomography (see Section 7.3). This technique compares the travel time of millions of individual seismic ray paths and compares them with the expected speed derived from an Earth reference model of the type described in Section 1.2.3. In essence, cold, dense material, such as subducted slab, allows seismic waves to travel faster, whereas warm, less-dense material makes them travel slower than the Earth reference model. For S-waves, the presence of partial melt has an additional slowing effect on the seismic waves. The results of seismic tomography are similar to an X-ray tomographic image of the human body but reveal regions of the Earth's interior where seismic velocities are either slower or faster than expected. To make the images easy to understand, seismically fast regions are coloured blue and seismically slow regions are coloured red. A collection of seismic tomographic images of some of the world's subduction zones is illustrated in Figure 5.4 (overleaf). From these diagrams the cold, dense subducting slab is well resolved, showing that the angle of subduction varies from arc to arc. A compilation of subduction angles is given in Table 5.1.

Table 5.1 Geometrical characteristics of some arc–trench systems.

Arc–trench system	Subduction angle	Curvature
Mariana	80–90°	highly arcuate
Lesser Antilles	45°	moderately arcuate
Tonga	45°	low curvature
Kurile	35–50°	moderately arcuate
Northern Chile	10–30°	low curvature

- Can you think of a reason why there is such a difference between the subduction angles beneath the Mariana and Chile arc–trench systems (Table 5.1)?

- You have already discovered that the age of the subducting crust is different for these two arcs. The steepness of the slab beneath the Marianas Islands is related to its dense, cold nature, which causes it to sink rapidly. This is the slab-pull force in action. Younger crust beneath Chile is more buoyant, and so will not sink into the mantle as easily.

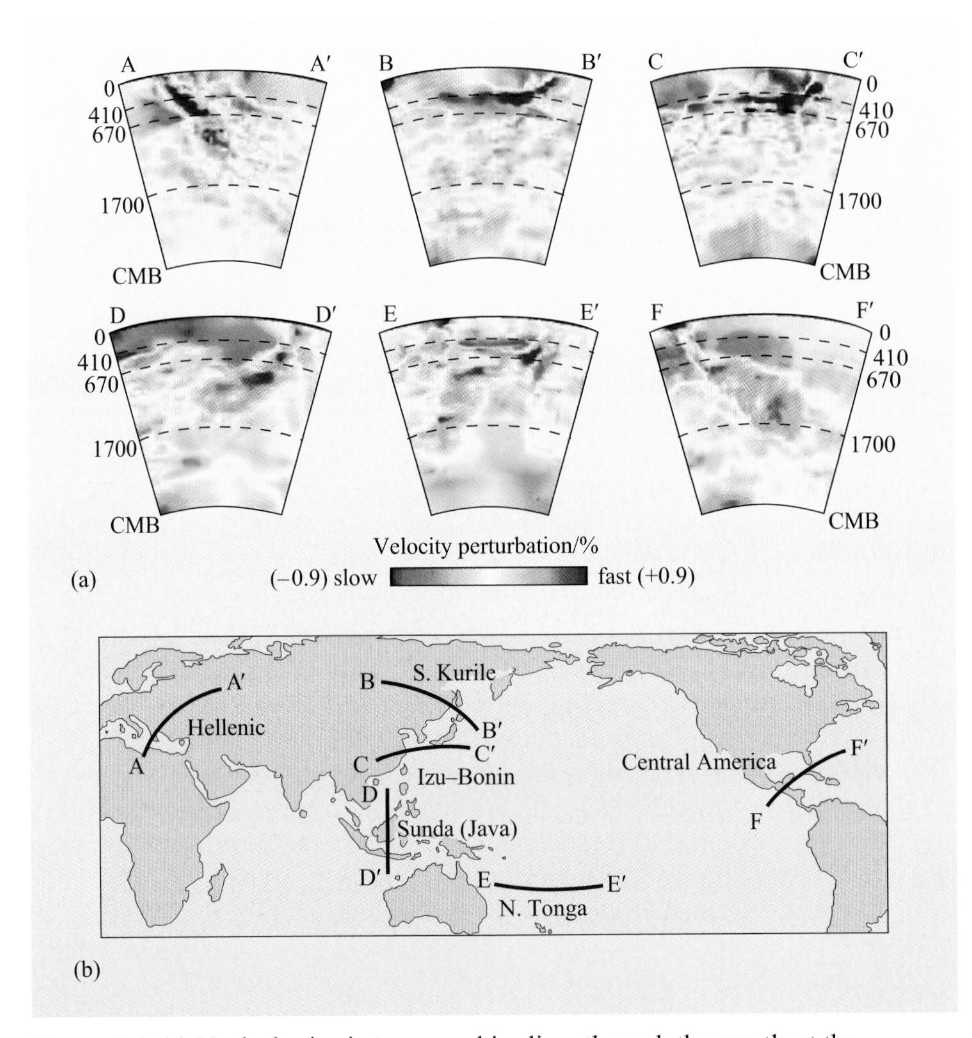

Figure 5.4 (a) Vertical seismic tomographic slices through the mantle at the localities shown in (b). Fast velocities indicate cold, dense anomalies, generally interpreted as subducting slabs or their remnants. Slow velocities imply warm, less-dense material. *Note*: CMB is the core–mantle boundary; depths are in kilometres. (van der Hilst)

The seismic tomographic images also reveal that the slab does not break up and is not mixed into the upper mantle during its voyage down to 670 km into the mantle. It also remains cooler than the surrounding mantle. This is not a surprise, as a simple thermal calculation indicates that it would take several hundred million years for the slab to equilibrate thermally with the mantle at depth, whereas it takes a few tens of millions of years to subduct to 670 km depth. The fate of subduction zones at depths greater than 670 km will be discussed in Chapter 7.

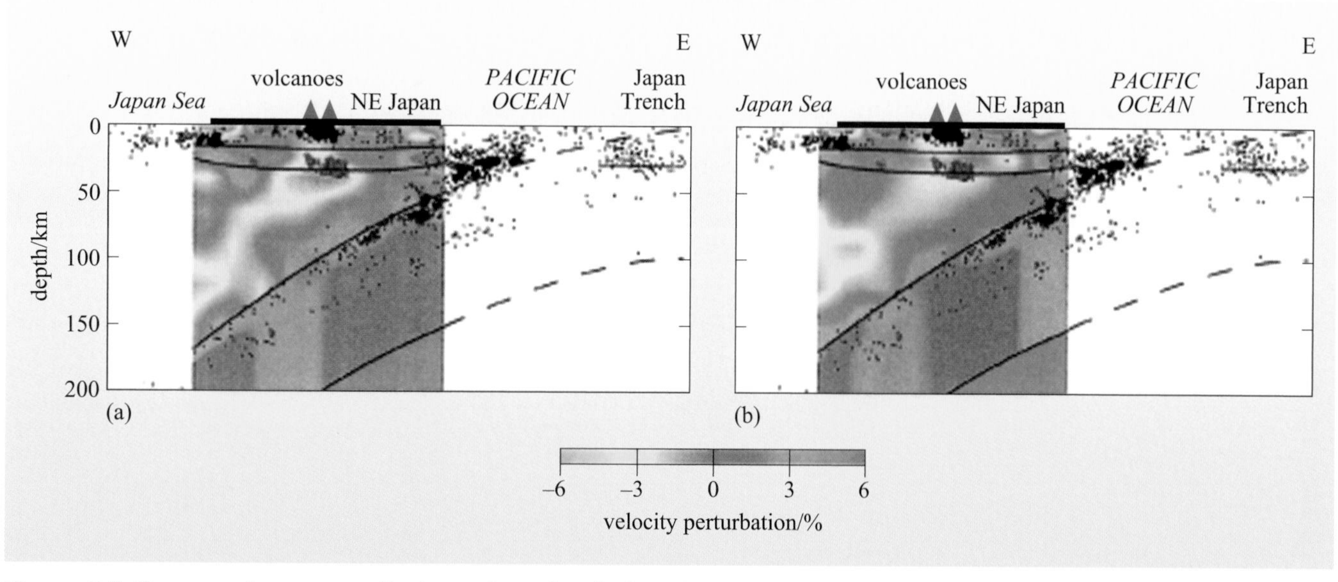

Figure 5.5 Cross-sections perpendicular to the volcanic front in NE Japan, at about 39° 50′ N running westwards from the Japan Trench, showing (a) P-wave and (b) S-wave velocity anomalies. Dots represent earthquake foci. Black lines beneath Japan represent seismic discontinuities. In order of increasing depth these are the Conrad discontinuity (upper/lower crust boundary), the Moho and the top and base of the subducting slab.

Figure 5.5 presents two detailed seismic tomographic images of the NE Japanese arc. Figure 5.5a uses only P-waves while Figure 5.5b uses only S-waves.

- What features can you observe in Figure 5.5 that are similar to and different from the less-detailed images in Figure 5.4?

- You should notice that the slab in Figure 5.5 is very well defined in blue tones (+3% to +6% faster), similar to the images of all the subducting slabs in Figure 5.4. The greater resolution of Figure 5.5 reveals a zone of seismically slow material that crosses the mantle wedge from 150 km depth (just above the slab–mantle wedge boundary) to beneath the volcanic-arc crust. A zone of particularly slow velocity is located beneath the crust in Figure 5.5b.

- What do you think the zones of slower S-wave velocities might represent?

- One possibility is that they represent zones of hotter material. However, you should recall from Section 1.2.1 that S-waves do not propagate through liquids, so the S-wave signature both in the mantle wedge and beneath the arc crust may also indicate the presence of a small amount of melt.

Just how much melt this slow zone represents can be estimated by the decrease in S-wave velocity, and calculations indicate that a few per cent partial melt or fluid must be present. Therefore, the seismic tomography suggests that partial melting occurs within the mantle wedge and that this melt then traverses the wedge and perhaps ponds beneath the arc crust. As you will see in Section 5.4, these observations place major constraints on how melts can be produced within subduction zones.

Other geophysical observations shed further light on the tectonic processes that occur in subduction zones. Figure 5.6 illustrates the free-air gravity and heat flow profile across a simple oceanic subduction zone. There is a small gravity high over the flexural bulge (where the down-going plate bulges just before the trench), a gravity low over the trench and a large gravity high over the fore-arc and volcanic arc. These data indicate that subduction zones are very dynamic and are not in isostatic equilibrium.

- Can you recall from Chapter 3 what these anomalies reflect in the underlying structure?

- The gravity low above the trench indicates that there must be either a downward force or a mass deficit in this region, and this may well be related to the subduction of H_2O-rich sediments. By contrast, the gravity high over the fore-arc and volcanic arc implies a mass excess at depth or, more likely, an upward acting force that could be related to the introduction of magma beneath the arc.

Turning to the heat flow data (Figure 5.6b), the fore-arc and the trench are characterised by low heat flow, whereas the volcanic-arc has high heat flow.

- What do you think are the causes of these variations in heat flow?

- Low heat flow over the fore-arc and the trench corresponds to the subduction of the cold slab, whereas high heat flow through the arc is consistent with the presence of magma at depth.

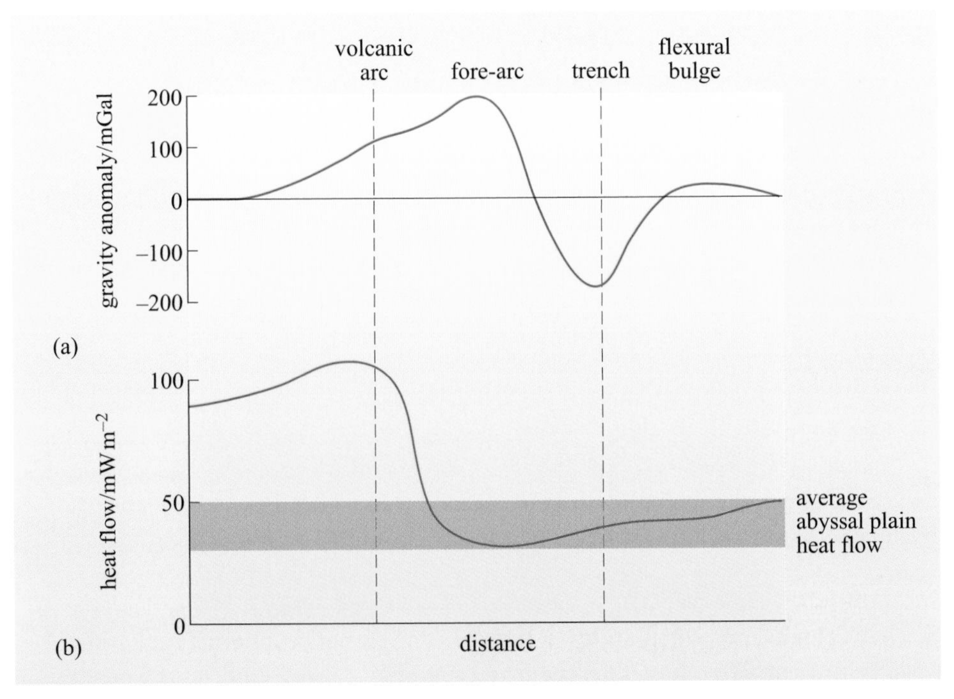

Figure 5.6 Sketch of (a) free-air gravity anomaly and (b) surface heat flow across a destructive plate boundary.

5.2.1 Summary of Section 5.2

You are now at a point where you have enough information to try to think quantitatively about melt generation in subduction zones.

- What are the most likely components in a subduction zone that may contribute to melt production?
- There are four components that could melt:
  (i) the mantle wedge, which is made of peridotite
  (ii) altered oceanic crust
  (iii) a wide variety of hydrated sediments
  (iv) serpentinised peridotites in the subducting slab.

Geophysical information further reveals that melt or fluid is present as deep as 150 km in the mantle wedge. These relatively simple observations allow some important conclusions to be drawn and a debate to be resolved that prevailed during the initial years of the plate tectonic revolution in the late 1960s and early 1970s. At that time scientists turned their attention to magma generation in subduction zones, and particularly to the problem of how subduction of cold material could produce hot magmas. Two competing models existed, one in which the oceanic crust of the subducted slab was heated up and melted and one that involved the melting of hydrous ('wet') peridotite in the mantle wedge. Initially, the slab-melting model was favoured, in part because many workers thought that andesites (which are more silica rich than basalts) were the primary melts found in most arcs, and because slab melts are also rich in silica. This perhaps reflects the fact that many experimental petrologists (people who perform high-pressure and/or high-temperature experiments on igneous rocks) had not visited many island arcs. There is still a perception that subduction zones dominantly generate andesites, whereas in reality modern oceanic arcs produce, on average, basaltic andesites and many of these arcs erupt primitive basaltic lavas (e.g. Vanuatu in the southwest Pacific Ocean, and Grenada in the Lesser Antilles). Therefore, it is now widely accepted that magmatism in the great majority of subduction zones is due to partial melting of the mantle wedge. However, there may be special circumstances where the slab does melt, and some arcs where melts are a mixture of slab melts and melting of the mantle wedge.

5.3 Thermal structure of subduction zones

The thermal structure of the Earth away from destructive plate boundaries is relatively simple, with the temperature increasing with depth along a geotherm.

- Would you expect the geotherm beneath a destructive plate boundary to be simple?
- No. The subduction of cold oceanic lithosphere into the hot mantle should modify the thermal structure of the subduction zone, producing a more complex geotherm.

The geotherm across the mantle wedge first increases with depth to a maximum and then cools again as it approaches the subducted slab. This is known as an **inverse geotherm**.

You have already seen that seismic tomography can image differences in P- and S-wave velocities of up to 12% beneath active arcs that are primarily related to temperature differences in the down-going plate and the overlying mantle wedge. Specifically, the down-going plate remains relatively cold compared with the mantle wedge. Seismic tomography, therefore, provides a qualitative guide to the relative temperature differences in the subduction zone. But can it produce a more quantitative picture of the thermal structure of a subduction zone? It would be particularly useful to have a thermal map of a subduction zone in which key input parameters, such as the rate of subduction, the age of the subducted plate and the angle of subduction, could be varied. Such a model could be tested against modern-day subduction zones where the values for the input parameters are known, but could also be used to make predictions about ancient subduction. Moreover, predicted pressure and temperature conditions could be compared with phase diagrams that describe mineralogical and melting reactions to make predictions about the location and type of melt generation in a given subduction system.

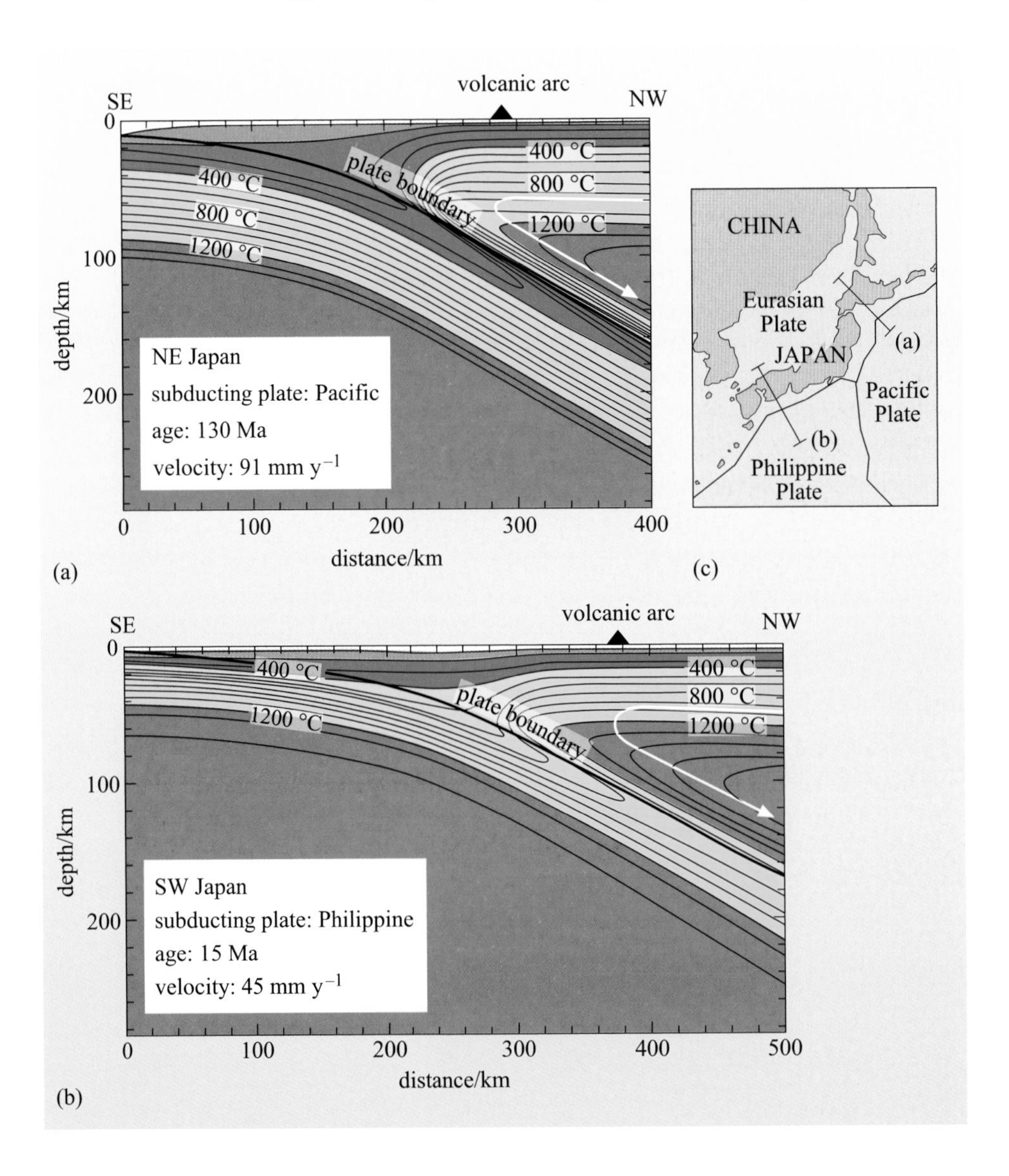

Figure 5.7 Models of the thermal structure of two different subduction zones: (a) an old slab being subducted rapidly; (b) a young slab being subducted relatively slowly. The isotherms are colour coded and the direction of corner flow is illustrated by the white arrow. (Peacock, 2003)

Mathematical models and computer simulations of the thermal structure of subduction zones have existed for many years, but modern computation power allows models of considerable sophistication, particularly with regard to how material flows within the mantle wedge. Some simple thermal models for a subduction zone are illustrated in Figure 5.7. This is a steady-state model because all of the input parameters have been held constant for long enough for the thermal structure to become stable. In all of the models, lines of equal temperature (**isotherms**) are depressed to greater depths in the vicinity of the down-going plate. This thermal perturbation produces a very different heat distribution from that found in a normal plate (as seen at the left-hand edges of all the models). Therefore two key thermal consequences of continuous subduction of cold oceanic lithosphere can be deduced:

- the Wadati–Benioff zone is cooler than the surrounding mantle
- melts generated in the mantle wedge migrate into, and thus heat, the arc lithosphere, particularly the arc crust.

The elevation of the isotherms in the arc lithosphere is hard to see in Figure 5.7 because of the scale, but is obvious from the surface heat flow (Figure 5.6b).

You might expect that continuous subduction would progressively cool the overlying mantle wedge, and this would be true were it not for an additional control on the thermal structure of the subduction zone. As the slab is subducted, it cools the immediately overlying mantle wedge. The resultant viscous drag causes the subducting plate to couple with the overlying mantle wedge, pulling hot mantle down parallel to the convergence direction. This results in **corner flow**, in which material dragged down by the slab is replaced at higher levels by asthenosphere from the back-arc region. The direction of corner flow is indicated on the thermal models in Figure 5.7 by the white arrow. Alternatively, **arc-parallel flow** in which asthenosphere flows parallel to the length of the arc, can also form as a consequence of variations in geometry along the length of the arc. This allows material to be sucked from the edges of the arc to its centre (e.g. the Aleutian and Kurile arcs). This effect extends out of the plane of the thermal models illustrated in Figure 5.7.

For all models, it is important to assess how the different input parameters affect the thermal structure of the subduction zones.

■ What is the effect of the age of the subducting slab on the thermal structure of a subduction zone?

■ As the plate ages it becomes colder and denser and the lithosphere becomes thicker. Therefore, an old, cold slab will take longer to heat up and will depress the isotherms to a greater degree. By contrast, a very young slab will be relatively warm and will heat up more rapidly and depress the isotherms to a lesser degree.

■ What is the effect of the rate of subduction on the thermal structure of a subduction zone?

■ The rate of subduction is directly proportional to the depression of the isotherms. Rapid subduction will allow the cold slab to traverse to greater depths, thus depressing the isotherms to a greater extent. Slow subduction limits the depth to which the cold slab will go before it is heated up, which in turn reduces the extent to which it depresses the isotherms.

These simplified thermal models are a useful starting point, but the most up-to-date models are best illustrated with a specific example that is based on a real subduction zone. You have already encountered seismic tomographic images (Figure 5.5) of Honshu in NE Japan. This is a subduction zone where old oceanic crust is being subducted rapidly. Until recently, most models assumed the viscosity of the mantle was the same throughout the mantle wedge (called **isoviscous**). This was convenient as it simplified much of the mathematics of the modelling that readily produced corner flow. However, models that take into account the effect of H_2O (from the subducting slab) on the viscosity of the mantle wedge produce some rather different results. H_2O has the effect of reducing the viscosity of the mantle, and this allows the mantle to flow more easily. You will see later (Chapter 7) that viscosity is a fundamental parameter in determining how the Earth works, and H_2O has a very strong effect on modifying this property.

In a subduction zone, therefore, there is a strong **positive feedback** between subducting H_2O from the Earth's surface and the thermal and volcanic evolution of the subduction zone. In the Honshu model (Figure 5.8), the reduction in viscosity above the slab–wedge interface has the effect of increasing the flow of hot asthenosphere into the wedge that pulls down hotter isotherms closer to the slab. This produces a very strong thermal gradient across the top of the slab in this model. Previously, the only way to produce this effect in thermal models was to include an arbitrary amount of frictional heating between the slab and wedge that was not well constrained. Reassuringly, the more sophisticated models confirm the effects previously seen in the simple thermal models. Fast subduction of an old slab (Honshu) tends to keep the slab cold, and slow subduction of a young, hot slab keeps the slab hot.

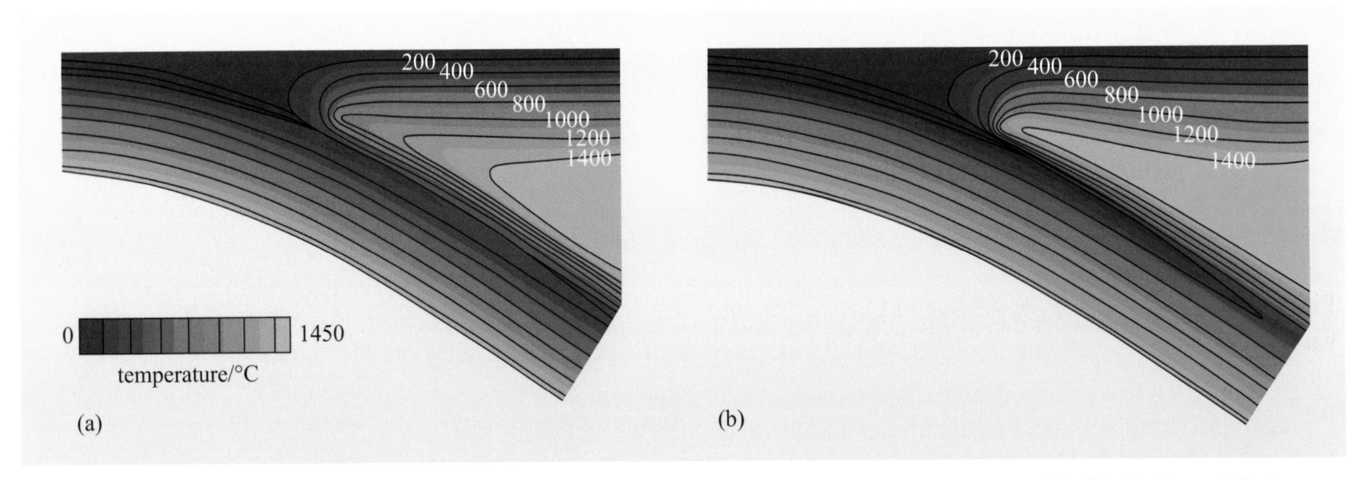

Figure 5.8 Models of the thermal structure of the Honshu subduction zone: (a) isoviscous model; (b) H_2O-modified viscosity. (van Keken et al., 2002)

5.4 Applying the thermal models to melt production

To understand whether any of the components that make up the source of arc lavas could actually melt or influence melt production, the information from the thermal models needs to be translated on to appropriate phase diagrams for each of the possible source compositions. You should now realise that evidence for the mantle wedge melting can be deduced from the observation that primitive basaltic melts are found in many arcs. H_2O influences this melting as indicated by the elevated H_2O contents of many arc lavas. Before considering what happens in the mantle wedge, the next section reviews phase diagrams for each of the three slab components, namely sediment, hydrated oceanic crust and the underlying hydrated peridotite. To help interpret what happens in the slab, a variety of pressure–temperature paths have been added to the phase diagrams. These correspond to thermal models where the age of the slab and the rate of subduction are varied to represent the range of probable conditions found in different subduction zones.

5.4.1 Basalt phase diagram

When geologists started to accept sea-floor spreading and plate tectonics it became obvious that subduction of oceanic basalts was a critical aspect of plate tectonics.

- What types of experiment do you think experimental petrologists attempted?
- The obvious experiment to attempt was to derive a phase diagram for the system MORB + H_2O because hydrothermally altered oceanic basalt is an important component of the subducted slab.

Figure 5.9 is a simplified phase diagram for a H_2O-saturated basalt.

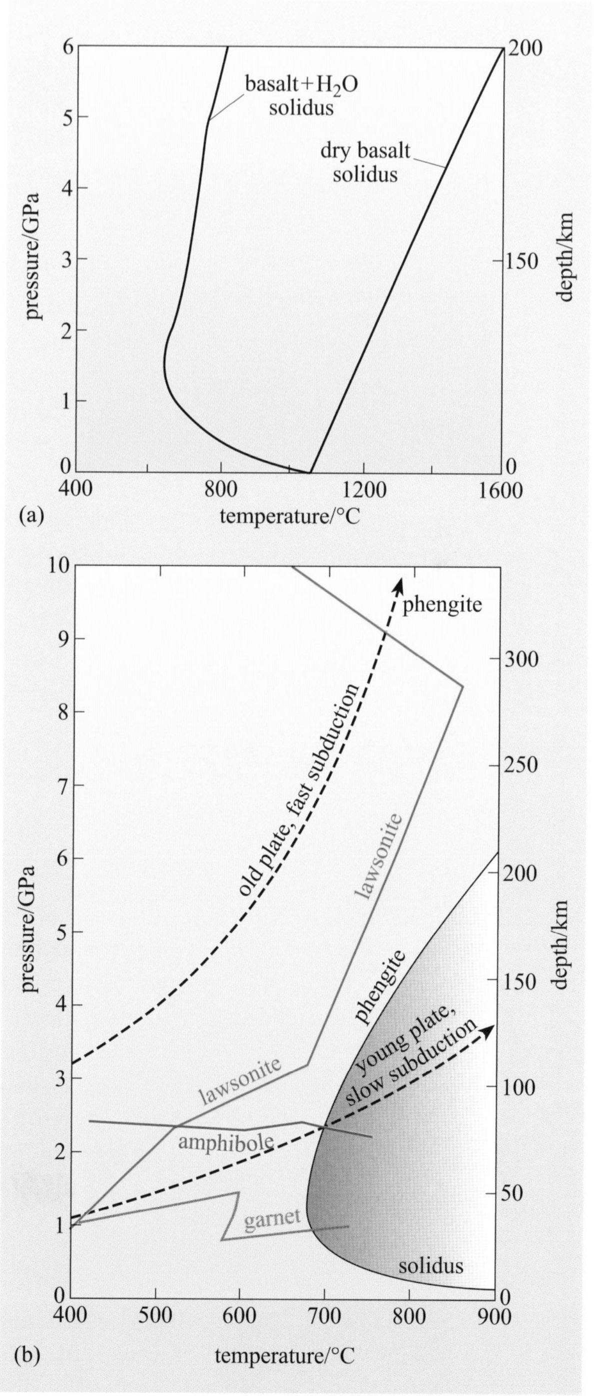

Figure 5.9 (a) Simplified phase diagram for a basalt illustrating the effects of H_2O on the melting temperature of the subducted basaltic slab. (b) Simplified phase diagram for H_2O-saturated basalt with *P–T* paths for the slab for two different subduction zone thermal regimes and the stable sub-solidus phases. Garnet is stable at all pressures above the lower brown line, whereas amphibole is only stable up to pressures below the blue line labelled amphibole. Lawsonite is stable to the left of the labelled red line whereas phengite is stable to the right of the red line labelled lawsonite up to high temperatures and pressures. Several key phases such as clinopyroxene have been excluded from the diagram for clarity reasons and/or because they have a wide stability range. *Note*: the depth increases upwards on this figure. Melt can be present in the shaded area to the right of the solidus. (Adapted from Schmidt, 2002)

- What is the most obvious effect that H_2O has on the basalt phase diagram?
- It reduces the solidus temperature over a range of pressures by as much as 800 °C relative to the dry solidus.

Below the solidus, however, H_2O also interacts with a basaltic rock to produce a great variety of hydrated or hydrous minerals. For example, as the slab gets deeper H_2O reacts with the original igneous minerals so that at depths of a few tens of kilometres the dominant minerals are amphibole, chlorite, clinopyroxene, **lawsonite** (a Ca–Al silicate mineral that contains up to 11 wt% H_2O) and a variety of mica minerals. These phases are not all on Figure 5.9 for reasons of clarity. The amphiboles are sodium rich and have a characteristic blue colour, which explains the name given to **blueschist facies metamorphic rocks**. As pressure and temperature increase, further reactions take place and new minerals appear. These reactions all involve the release of H_2O in what is called a **dehydration reaction**. In particular **garnet** becomes a progressively more important mineral and, by 70 km depth, the blue amphibole dehydrates and the rock is made up of dominantly a sodium-rich clinopyroxene (not shown on Figure 5.9) and garnet. Lawsonite and a type of mica called **phengite** are the remaining H_2O-bearing phases. This pyroxene–garnet bearing rock is known as an **eclogite** and is a characteristic low-temperature, high-pressure metamorphic rock. Overall, the reactions involve the hydrated basalt progressively losing H_2O until a depth of about 200 km, when only about 1 wt% H_2O is present in the rock.

- Study Figure 5.9. Do either of the thermal models predict that basaltic crust will melt during subduction?
- Basalt only melts for young plates subducted slowly.

For most thermal models, for example those illustrated in Figure 5.9, basalt does not melt, although for slow subduction of a young plate it is possible to produce some melt.

Melt produced from hydrated basalt is enriched in SiO_2 and sodium and broadly dacitic in composition, and because it is produced in the presence of garnet it has a characteristic trace-element composition. Some arcs do erupt lavas with compositions that are consistent with a component of slab melt, and such melts are found in arcs that subduct young slabs (e.g. the Aleutian Islands in the North Pacific Ocean, and South Chile). Overall, the role of the altered oceanic crust in most arcs is to give up H_2O progressively as the slab dehydrates under the fore-arc and volcanic front.

5.4.2 Sediment phase diagram

There are a wide range of sediments that could be subducted, which makes using a single phase diagram for a sediment slightly simplistic. However, experiments have been performed on various deep-sea clay compositions that are reasonably representative of pelagic sediment. A simplified phase diagram for such a composition is shown in Figure 5.10. At low pressures, sediments undergo a series of dehydration reactions that drive off some chemically bound H_2O in addition to any free pore fluids, and a large amount of H_2O is lost into the fore-arc region. This is most often expressed on the sea floor as mud volcanoes and the intrusion of masses of serpentinite.

The critical aspects of these experiments are:

1 The solidi temperatures of sediments vary from below that of the solidus of a 'wet' basalt, for clays that contain some quartz, to higher than the 'wet' basalt solidus for quartz-poor clays. This implies that, under certain conditions, sediment can melt without the underlying oceanic crust melting. The most recent thermal models (e.g. Figures 5.7 and 5.8) have a very high thermal gradient close to the sediment–mantle wedge interface, and this makes the melting of sediment more likely without recourse to melting the rest of the slab.

2 At temperatures above the solidus sediment a variety of hydrous phases break down as melting progresses. These are phengite, amphibole and biotite, which break down respectively with increasing temperature. The stability of these high-temperature hydrous phases means that subducted sediment behaves in a similar manner to the altered oceanic crust, in that it continuously dehydrates as the slab descends. The melt produced from sediment melting is essentially rhyolitic in composition, but it also contains high alkalis (Na_2O and K_2O) and in excess of 10 wt% H_2O.

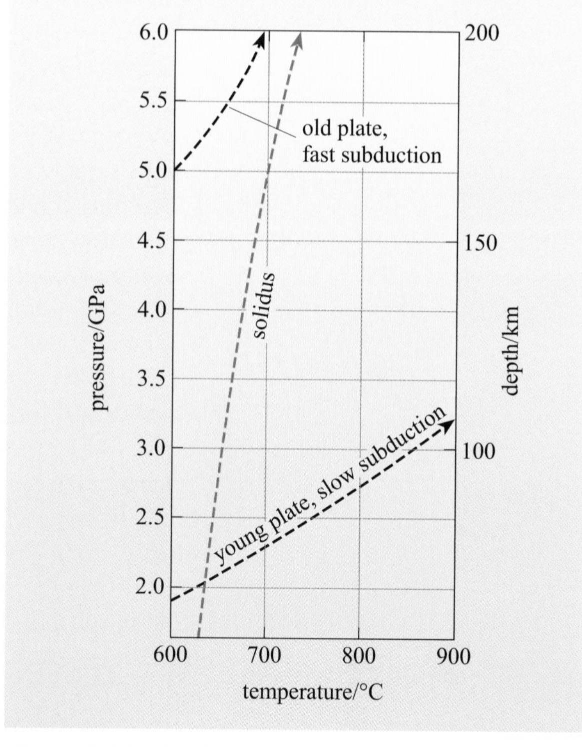

Figure 5.10 Simplified phase diagram for H_2O-saturated red clay with *P–T* paths for the slab from a variety of subduction thermal models.

5.4.3 Peridotite phase diagram

The peridotite phase diagram in Figure 5.11 (overleaf) is for an H_2O-saturated peridotite at low temperatures. At low pressures a whole host of hydrous minerals are present, including serpentine minerals, chlorite, talc and amphibole. As the temperature rises, the number of stable hydrous phases decreases, as does their potential for including structurally bound H_2O. Amphibole is stable to a pressure of ~3 GPa, but at higher pressures only chlorite and then a serpentine mineral called antigorite are present. Thermal models indicate that it is highly unlikely that the hydrous mantle in the slab ever melts, but it potentially will dehydrate and flush H_2O through the overlying basalt and sediment. Moreover, the possibility of stabilising hydrous phases at high pressures in a peridotite composition may allow the transport of H_2O into the deep mantle, a possibility explored in Chapter 7.

■ Is the peridotite–H_2O phase diagram appropriate only for the subducted slab?

■ No, because dehydration reactions in the subducted sediment and oceanic crust release H_2O into the overlying mantle wedge. The inverse geothermal gradient in the mantle wedge means that the mantle just above the slab is also quite cool, and so the same peridotite phase diagram is also useful for understanding the base of the mantle wedge.

Consider what happens to the base of the mantle wedge, which is viscously coupled to the slab. It will receive large amounts of H_2O and possibly some melt from the slab. This hydrated peridotite is then dragged downwards and at some point may release its H_2O to the mantle wedge and initiate melting. The movement of this hydrated amphibole-bearing peridotite in the mantle wedge is an important constraint on the location of partial melting in the mantle wedge. Therefore, it has been the focus of several models for melt generation in subduction zones, which usually involve a series of hydration and dehydration reactions within the peridotite that allow the transport of H_2O from the slab to a hotter part of the mantle wedge before eventually encountering conditions where the mantle melts. Regardless of the actual mechanism, melting in the mantle wedge above a subduction zone is most likely related to the stability of amphibole or other hydrous minerals.

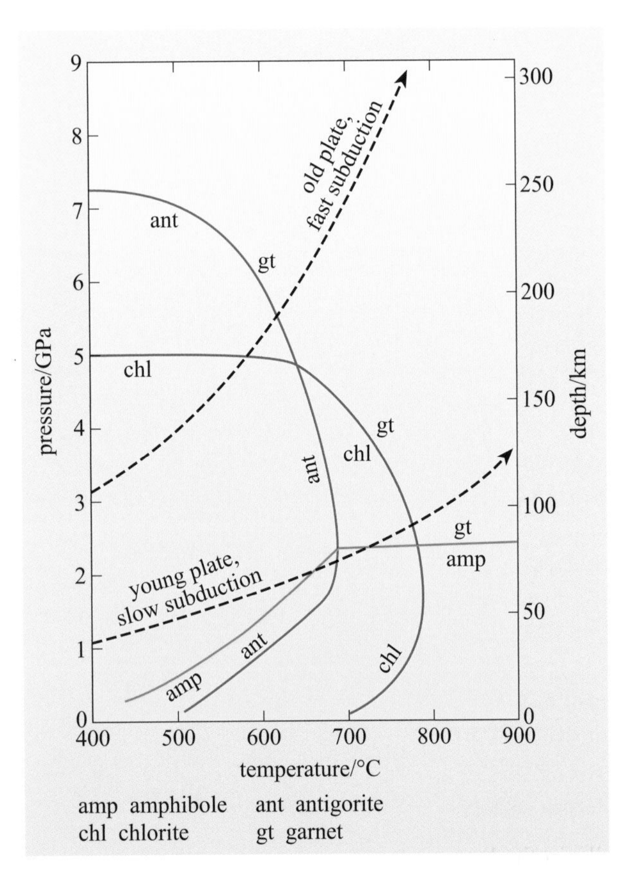

Figure 5.11 Simplified phase diagram for H_2O-saturated peridotite with *P–T* paths for the slab for two different subduction zone thermal regimes and the stable sub-solidus phases. Peridotites stabilise many phases at high temperatures and pressures and on this figure they are simplified into regions that contain the key phases, amphibole (amp), chlorite (chl), the serpentine mineral antigorite (ant) and garnet (gt). Garnet is stable to the right or above any line it is labelled against, meaning it is a high-pressure phase. By contrast, amphibole is stable at low pressures below and to the right of its line. Chlorite and antigorite are stable to the left of their respective stability lines, with antigorite stable to higher pressures. There are regions where more than one of these key phases is stable (e.g. chlorite and antigorite). (Poli and Schmidt, 2002)

- Can you recall any other evidence that might provide more information about the distribution of melt in the mantle wedge?
- The seismic tomography beneath Japan (Figure 5.5) suggests that melt or fluid is present close to the slab at 150 km depth and that melt is distributed along a diagonal path across the mantle wedge.

5.4.4 Melting the mantle wedge

Figure 5.12 is a phase diagram for both a dry (anhydrous) and wet (hydrous) peridotite with two calculated geotherms for the mantle wedge. One geotherm assumes that the viscosity does not change and the other assumes a more realistic mantle wedge viscosity that takes into account the effects of H_2O on mantle viscosity. The geotherm has a distinct but expected shape, with its highest temperature located in the centre of the mantle wedge, away from the cooling influence of the Earth's surface and the subducting slab.

The hydrous peridotite phase diagram is relevant to understanding what happens in the mantle wedge (Figure 5.12). If enough H_2O is added to a mantle peridotite then there is a major change in the shape of the peridotite solidus. The curved shape reflects the stability of amphibole, which is stable in hydrous peridotites below 3 GPa. The key observation is that the temperature at which a hydrous peridotite will start to melt is several hundred degrees lower than for a dry peridotite. However, even small amounts of H_2O decrease the temperature of the solidus, without forming amphibole, as shown by the series of lines roughly parallel to the dry solidus in Figure 5.12.

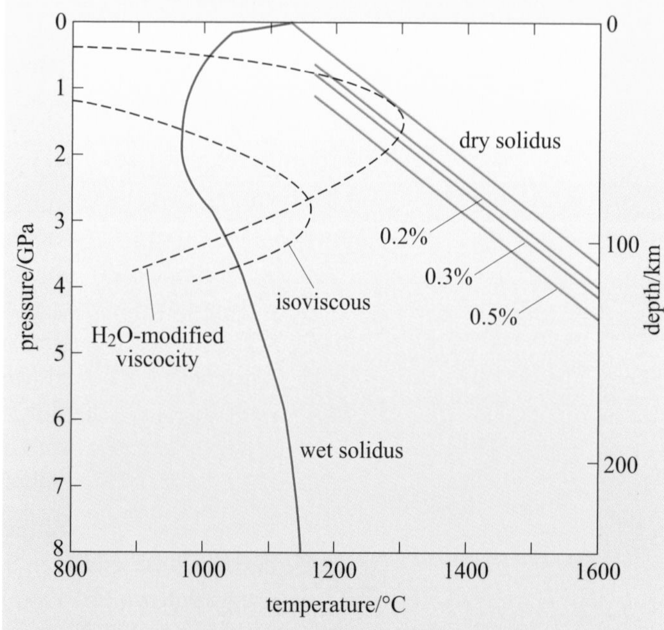

Figure 5.12 Simplified phase diagrams for anhydrous (dry) and H_2O-saturated (wet) peridotite. Also plotted are solidi curves for peridotite plus 0.2, 0.3 and 0.5 wt% H_2O. Also plotted are geotherms for the mantle wedge from two subduction thermal models (see Figure 5.8).

- What key difference is there between the two mantle wedge geotherms?

- The isoviscous geotherm only crosses the H_2O-saturated mantle peridotite solidus, whereas the geotherms using H_2O-dependent viscosities cross both the anhydrous and all of the solidi with as little as 0.2% water. The geotherm also shows that there are parts of the mantle wedge where amphibole would not be stable.

How do these model results compare with real observations on the composition of arc lavas? You should recall that one of the assumptions for earlier thermal models and experimental studies of magma generation in arcs was that andesites were the primary melts in arcs, but that more recent studies have discovered that basalts are very common in many oceanic arcs. Moreover, these basalts have high MgO contents (~10 wt%) and magnesium numbers consistent with being in equilibrium with the mantle (see Section 4.2.1). Experimental studies have also found that these arc basalts were produced at 1300–1400 °C at 2 GPa. However, these same melts have high H_2O contents, and this presents a dilemma. If these magnesian arc basalts are representative of primary magmas generated within the mantle wedge then these temperatures are too high for them to have been produced at the H_2O-saturated solidus (the wet solidus on Figure 5.12). These observations, therefore, imply strongly that, even though hydrous melting is an important aspect of magma generation beneath arcs, melting in the mantle wedge is influenced by an additional process or processes.

Ideas about melting in the mantle wedge have changed substantially in the last 10 years. The observation that many volcanic arcs are located about 110 km above the subducting slab was used to argue that melting was associated with a pressure-related dehydration reaction, usually ascribed to the breakdown of amphibole in the slab, which triggered melting in the wedge. You should now realise that there are many dehydration reactions that occur in the slab, and these all have the potential to add H_2O to the overlying wedge over a range of pressures. Experiments in which hydrous peridotite is melted produce about 8% partial melt when amphibole breaks down. However, most arc lavas are thought to have been generated by at least as much partial melting as that which produces MORBs and possibly by as much as 30% (see Section 5.8).

Models for the generation of hot, H_2O- and MgO-rich subduction zone magmas have to reconcile several observations. They are generally produced by high (~30%) degrees of partial melting, so they cannot be produced simply by amphibole breakdown. They are H_2O-rich, but are too hot to have been produced at the H_2O-saturated solidus. However, because H_2O behaves as an incompatible element during partial melting of the mantle, it will be lost rapidly from the mantle into the melt phase and mantle melting will revert to the anhydrous solidus. Current models favour what is termed a 'flux melting' model. In this model, H_2O is continuously supplied to the hot mantle wedge, which melts one of the lower temperature solidi (equivalent to one of the 0.2%, 0.3%, 0.5% H_2O plus peridotite solidi plotted in Figure 5.12). Partial melting will extract the H_2O into the melt phase, but continued fluxing of water into the mantle wedge will keep the mantle melting, thus generating a high melt fraction. Geochemical evidence not presented in this chapter has found that the degree of partial melting is proportional to the H_2O content of the mantle source, supporting the flux melting model.

Returning to some of the geophysical evidence presented earlier in the chapter, the deep melt generated in the Honshu arc could be related to hydrous peridotite being dragged down to 150 km and eventually releasing H_2O and generating some melt. The melt and H_2O then migrate through the wedge and induce high-temperature melting in the centre of the mantle wedge. Clearly, there is more research needed to understand melting in subduction zones, but hydrous melting of the mantle wedge at high temperatures is a model that currently fits most observations.

To illustrate the effects of varying the rate of subduction and the age of the subducting plate, representative (endmember) pressure–temperature paths have been added to the various slab phase diagrams (Figures 5.9–5.11). These paths are labelled 'old plate, fast subduction' and 'young plate, slow subduction'. The first of these pressure–temperature paths is representative of, for example, NE Japan (see Figure 5.7a), whereas the second is similar to the thermal structure of SW Japan (Figure 5.7b) or the Cascades in western North America. The key is to note whether the pressure–temperature path crosses the solidus leading to melting or whether it just crosses boundaries between mineral reactions.

■ What happens to the various components in the subducted slab (i.e. sediment, altered basaltic crust and hydrated peridotite) during subduction for (a) the old plate, fast subduction and (b) the young plate, slow subduction thermal models?

■ In both models the slab is hot enough to melt the subducted pelagic sediment, although this will happen at great depth in model (a) at pressures greater than those illustrated in Figure 5.10. Modelling using H_2O-modified viscosities produces steeper thermal gradients at the top of the slab and this will make melting of the subducted sediment more likely. By contrast, for the old plate, fast subduction model (a), the basaltic portion of the slab remains cold enough not to melt but hot enough to lose H_2O by dehydration reactions. In the young plate, slow subduction model (b), the basaltic crust will dehydrate and the hydrous parts may also melt. Finally, the upper part of the mantle lithosphere will dehydrate in both models, but it will happen at shallower depths in model (b) below the volcanic front.

Slab melts have been discovered in volcanic arcs such as the Cascades and are known as **adakites**. They have chemical compositions consistent with melts produced in hydrous basalt melting experiments.

Summary

It can be concluded from this discussion of the thermal structure of subduction zones that a series of complicated melting and dehydration reactions can occur both within the slab and the mantle wedge. These are controlled by the rate of subduction and the age of the subducted slab. Moreover, recent modelling strongly suggests that the very process of subduction provides a mechanism for lowering the viscosity of the mantle wedge by adding H_2O, and this may help to enhance the flow of hot (1300–1400 °C) mantle below the arc. Sediment melting may occur in most arcs because there is a very high thermal gradient across the top of the slab due to the enhanced flow of the mantle wedge. The basaltic slab will only melt when young plates (<25 Ma) are subducted slowly, but this is a rare occurrence (e.g. the Aleutian Islands and the Cascades). More commonly, the slab will dehydrate by a series of progressive dehydration reactions. The actual method of melting the mantle wedge is still controversial, but hydrous fluxing of the wedge is a model that currently fits most observations.

5.5 The composition of island-arc magmas

Subduction-zone magmas represent one of the three important broad tectono-magmatic associations, with magmas produced at mid-ocean ridges and intra-plate magmas making up the other two. Given the differences in tectonic setting for these different associations you might expect significant differences in their chemical composition. Figure 5.13 is a plot of magmas from these three tectonic settings on a total alkali ($Na_2O + K_2O$) versus SiO_2 diagram, commonly known as a **TAS diagram**. This type of diagram is used to classify volcanic rocks using simple chemical criteria. Each suite of lavas illustrates the range of lava compositions found in each tectonic setting and, although only a representative fraction of the geochemical analyses available are plotted, some clear distinctions between the magma types can be seen. Mid-ocean ridge samples have a restricted compositional range and are predominantly basaltic in composition with only rare intermediate compositions. Intra-plate magmas span a much

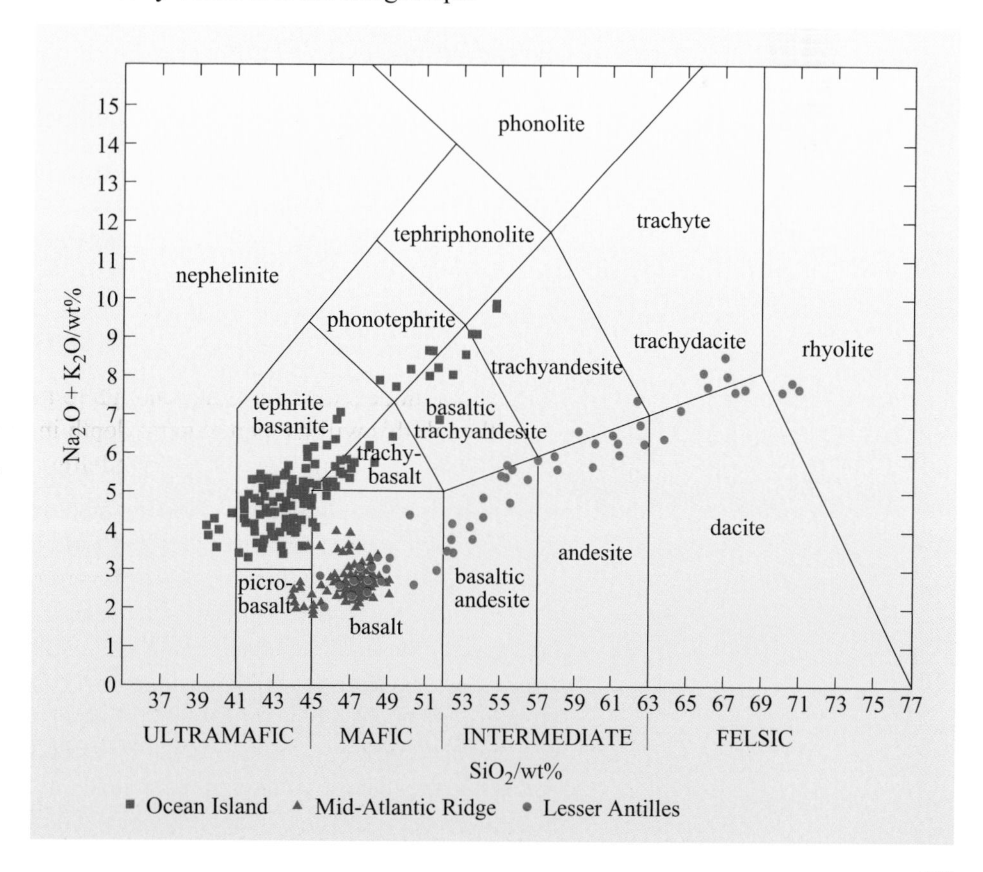

Figure 5.13 Plot of total alkalis ($Na_2O + K_2O$) versus SiO_2 showing fields of different rock names. Representative suites of data from an island-arc volcano (from the Lesser Antilles), an Ocean Island volcano and MORBs from the Mid-Atlantic Ridge are plotted. (Adapted from Le Bas et al., 1986)

wider SiO_2 range, and generally are enriched in the alkali elements, sodium and potassium. Subduction-zone magmas have the greatest range in SiO_2 contents and generally have lower total alkali contents for a given SiO_2 than intra-plate magmas. The predominant rock association found in subduction-zone settings is a **basalt–andesite–dacite–rhyolite association**.

The TAS diagram provides two important insights into magma generation in subduction zones. Magmas from all three tectonic settings include some basaltic compositions, and you should recall from Section 4.6 that such compositions, are generated by partial melting of peridotite in the mantle. Fractional crystallisation, a process that is discussed in the next section, involves the removal of crystals from a magma and a corresponding change in the magma chemistry. Increasing SiO_2 contents in lavas are the result of fractional crystallisation, and the wide range and high SiO_2 contents of arc lavas strongly imply that fractional crystallisation can readily take place in the arc crust.

You will consider the difference in composition between MORBs and arc lavas in more detail later in this chapter, but there is one final difference between these two magmas worthy of mention. Magmas erupted subaerially (or in shallow water depths) tend to lose all or some of their volatiles, such as H_2O, CO_2 and SO_2 to the atmosphere. By contrast, those lavas that erupt beneath significant ocean depths, under pressures of a few hundred times that of the atmosphere, tend to retain volatile components. When MORBs and arc lavas are collected from the deep ocean floor, the glass in these submarine basalts will quench all of the volatiles that were originally dissolved in the magma. Comparison between primitive quenched glasses reveals that subduction-zone basalts contain significantly higher H_2O contents than MORBs and basalts from other tectonic settings. Primitive arc lavas can contain 3–6 wt% H_2O compared with generally less than 1 wt% in MORBs, and they are also enriched in the element chlorine (Cl).

- Where do you think the H_2O and Cl dissolved in arc lavas originated?
- The logical source of these volatiles is from the subducted slab, either from altered oceanic crust or from sediments. Ultimately, the H_2O and Cl come from seawater, providing a direct link between the hydrosphere and the mantle.

5.6 Evolution of arc magmas

The observation that arc lavas extend to higher SiO_2 compositions (Figure 5.13) is not just of academic interest, because high-SiO_2 lavas have some interesting but potentially hazardous characteristics. Increasing the SiO_2 content of a melt increases what is called the **polymerisation** of the melt. In effect, this means that atoms are more strongly bonded to each other (they form the equivalent of bigger molecules) and this makes the melt more viscous. The viscosity of a liquid reflects the ability of the atoms to flow past each other. As an analogy, you can imagine the viscosity of a basalt to be like water and the viscosity of a SiO_2-rich melt, such as a dacite or a rhyolite, to be like very thick porridge. The combination of high viscosity and volatile content is a recipe for potential disaster. Consider the effects of heating a pan of water and a pan of porridge. As the water boils, bubbles of H_2O vapour form and rise easily through the water and burst gently at the surface. By contrast, any gas in the porridge forms much larger bubbles which tend to burst more violently at the surface and spatter hot porridge over the cooker! The analogy for what happens during an eruption is not perfect, but it serves to illustrate that more viscous magmas tend to build up high gas pressures and then erupt explosively. The eruptions of Mt St Helens, Krakatau and Mt Pinatubo are examples of the effects of catastrophic degassing of magma, and contrast with a less viscous and less volatile-rich basaltic lava flow erupting on Hawaii. The eruption of Krakatau obliterated most of the island and ejected volcanic ash and gases high into the stratosphere. This affected atmospheric conditions around the globe and reduced global temperatures, whereas a Hawaiian eruption causes local difficulties related to acidic smog and the disruption of local infrastructure. Clearly, it is useful to be able to understand how and why SiO_2-rich melts are produced in subduction zones.

- Can you think of any reasons why basaltic arc lavas might evolve into such SiO_2-rich compositions?
- There are three possible reasons why arc lavas behave differently from those in other tectonic settings: (i) arc lavas have elevated H_2O contents and this may influence the crystallisation history of the magma; (ii) some volcanic arcs, particularly continental arcs, have thicker lithosphere and this may act to slow down the transport of magma and allow for more extensive crystallisation; (iii) magmas in continental arcs rise from their source through high-SiO_2 crust, which may contaminate them in some way.

Crystallisation has an important role to play, and it is necessary to define what happens when a magma crystallises. In Section 4.6 you used experimental data and phase diagrams to illustrate how a MORB crystallises. Crystallisation usually happens when a magma cools. In the case of a simple experiment where the crystals are continuously in contact with the melt, some simple rules are observed. This type of crystallisation is called **equilibrium crystallisation**, and because the crystals are not separated from the melt the bulk composition of the crystals and liquid added together does not change. Also, because chemical equilibrium is maintained at all times, the crystals are homogeneous (i.e. they do not vary in composition from the core to the rim of the grain). For example in the binary olivine system, although the forsterite composition of the olivine crystals changes continuously during crystallisation it is always homogeneous at any point in the crystallisation history. The simple effect of equilibrium crystallisation is that, if you start with a basalt, although the liquid may change its composition during cooling, the bulk composition remains basaltic, which is not a particularly good start for making a wide range of magma compositions.

There is another type of crystallisation called fractional crystallisation, which involves the continuous or episodic removal of crystals from a magma system. This has a rather dramatic effect on the composition of magmas and can be illustrated using a simple binary phase diagram.

5.6.1 The plagioclase binary phase diagram

Plagioclase is an important mineral in the crystallisation of most lava associations because it has a wide range of compositions and is stable over a wide range of temperature. The plagioclase feldspar system involves a solid solution between anorthite ($CaAl_2Si_2O_8$) and albite ($NaAlSi_3O_8$). The solid solution in the plagioclase system is not as simple as in the olivine system because it involves the coupled substitution of $Ca^{2+}Al^{3+}$ for $Na^{+}Si^{4+}$ (note that the charge is balanced), but the shape of the binary diagram is very similar (Figure 5.14). Anorthite is the higher temperature feldspar and the mass proportion of anorthite is given as the An content (e.g. 50% anorthite is An_{50}). During equilibrium crystallisation, if a liquid of An_{50} composition cools it will start to crystallise An_{82} crystals at 1445 °C (points a and b in Figure 5.14). If the liquid continues to cool to 1385 °C, when it is 50% crystallised (line c to d), it will consist of 50% An_{68} crystals and 50% An_{32} liquid. Below 1290 °C it will be completely crystallised and the crystals will have the same initial bulk composition of An_{50}. The same diagram can be used to illustrate the effect of fractional crystallisation, where some crystals are separated at a point during the cooling process. Starting with the same bulk composition and allowing it to undergo equilibrium crystallisation to 1385 °C, 50% An_{68} crystals are in equilibrium with liquid with a composition of 50% An_{32} (point c). At this point, if all of the crystals are removed then this leaves a liquid with a bulk composition of An_{32}. This composition (point c) can be treated as a new liquid and upon cooling to 1290 °C will be composed of 50% An_{49} crystals and 50% An_{15} liquid (points m and n) and, ultimately, will become completely crystallised with a bulk composition of An_{32} below 1205 °C (point p). This simplified version of fractional crystallisation illustrates that more sodium-rich compositions can be reached at lower temperatures from the same initial bulk composition by this mechanism than is possible by equilibrium crystallisation.

It is possible to envisage a mechanism that removes crystals more than once during crystallisation. This would drive the liquid composition to be very sodium rich and produce a set of crystals with different anorthite compositions. The process described

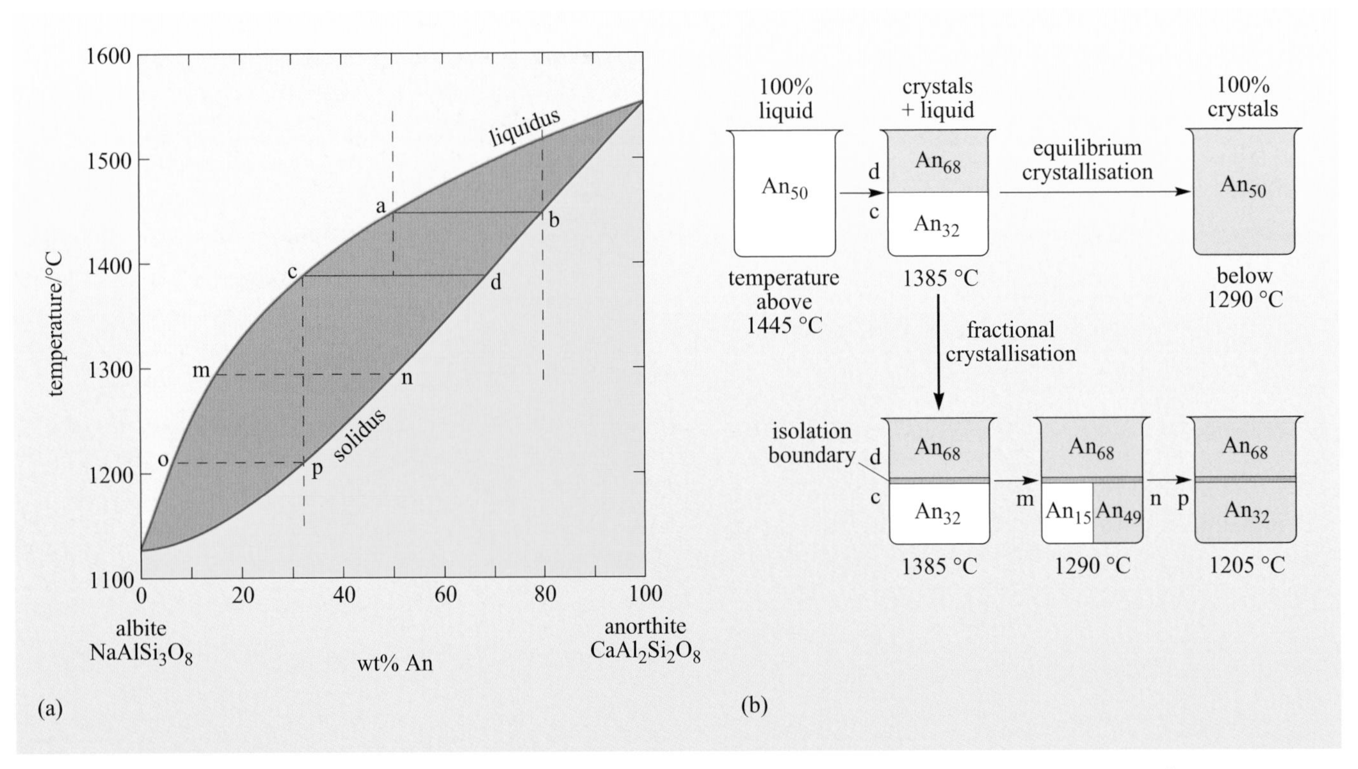

Figure 5.14 (a) Phase diagram of temperature against composition for the binary plagioclase system at 10^5 Pa. A sample of An_{50} crystallises under equilibrium conditions until 1385 °C, whereupon the crystals (point d) become isolated from the liquid (point c) and cool separately. The horizontal lines are tie-lines between coexisting crystals and liquid, before (solid) and after (dashed) the crystals were separated at 1385 °C. (b) Sketch illustrating the crystallisation of the sample of An_{50}.

is still an equilibrium process, in that each crystal that separated out has, up to that point, kept in equilibrium with the liquid and remains homogeneous. But what happens in nature? One of the most striking features of many arc lavas is that they contain plagioclase phenocrysts with complicated zoning. Such a **zoned crystal** is illustrated in Figure 5.15. Each of the concentric rings represents a growth of the crystal with a distinct anorthite content. Clearly, this crystal is not homogeneous and so could not have been formed by simple equilibrium crystallisation.

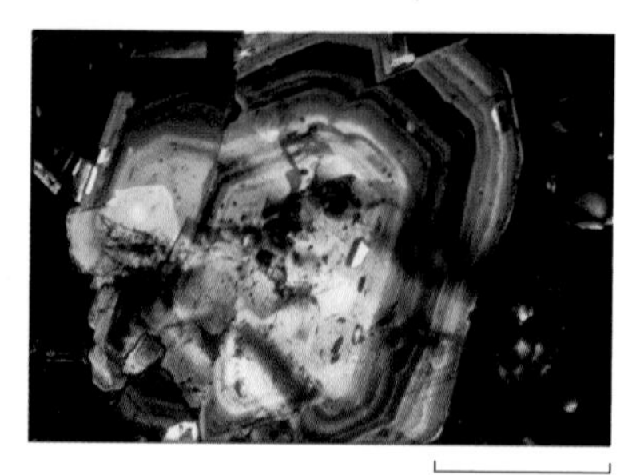

Figure 5.15 A zoned plagioclase phenocryst. The concentric lines represent boundaries between regions of different compositions.

- How might such a zoned crystal be formed?

- There are two possible explanations for generating such a crystal. First, each zone might represent a period of equilibrium crystal growth, and the crystal then separated out. In this case, each zone is like a tree ring that records some chemical information about the magma the crystal was growing in at that time. Second, only the outer layer (rim) kept in equilibrium with the melt. This indicates that there was not enough time for the ions to move (diffuse) across the crystal and to re-equilibrate the inner zone (core) with the magma before the crystal was physically separated from the melt.

Zoned crystals provide good evidence for fractional crystallisation and they potentially record information about changes in magma composition prior to eruption. But before fractional crystallisation can have an effect on magma composition the crystals need to separate from the melt. Crystals that do separate from the melt will accumulate on the sides and floors of magma chambers and conduits to form cumulate rocks, similar to those found in the lower part of the oceanic crust (Chapter 4). Arc cumulates have a distinct mineralogy that provides further information about the evolution of magmas in subduction zones.

5.6.2 The Lesser Antilles: fractional crystallisation in an arc

The Lesser Antilles islands in the eastern Caribbean are a well-studied example of an island arc associated with the westward subduction of the South American Plate beneath the Caribbean Plate. This volcanic arc contains several active volcanoes, including Montserrat, which erupted in the late 1990s, Mt Soufriere on the island of St Vincent, which erupted in 1979, and Mt Pelée on Martinique, which killed 30 000 people in 1902.

Two volcanoes, one on the island of Dominica and one on St Lucia, can be used to illustrate the chemical evolution of arc magma. These two volcanoes have been active in the last few million years and are located close to each other. However, one has erupted mainly basaltic rocks and the other produced andesites and dacites, and by studying these lavas it should be possible to discover whether the higher SiO_2-content andesites and dacites could be produced by fractional crystallisation of the basalts.

Major elements

Figures 5.16 and 5.17 show SiO_2 wt% plotted against CaO and MgO wt% respectively for a range of volcanic rocks from Dominica. The average compositions of four minerals, i.e. olivine, plagioclase, clinopyroxene and orthopyroxene commonly found as phenocrysts in the lavas, are also plotted on the figures. In both Figures 5.16 and 5.17, the volcanic rocks define smooth trends and the shape of the trends can be used to infer which minerals or groups of minerals have fractionated from these magmas. You may remember from Chapter 4 that if a mineral is removed from a magma, then the liquid moves away from the composition of the mineral on a chemical variation diagram. If two minerals are crystallising, then a point along the line connecting the two minerals defines the composition of the phases being extracted, conventionally referred to as the **extract**. If three or four minerals are crystallising, then the composition of the extract lies within a triangle or quadrilateral on the variation diagram defined by the three or four mineral compositions, respectively.

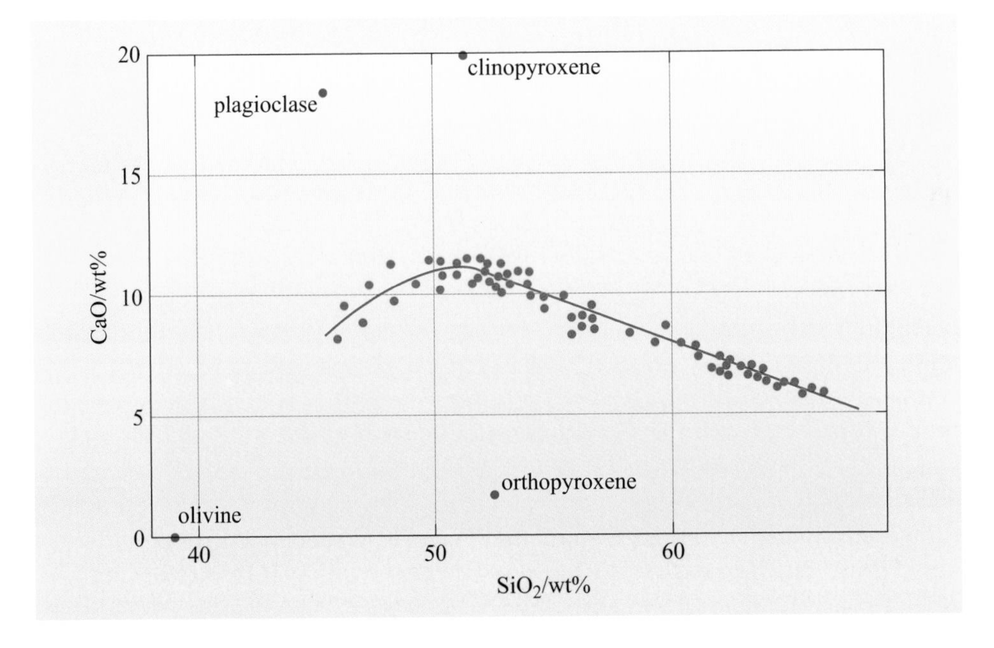

Figure 5.16 CaO versus SiO_2 concentrations from a suite of lavas and phenocrysts from two volcanoes on the island of Dominica. The blue line represents a smoothed compositional trend.

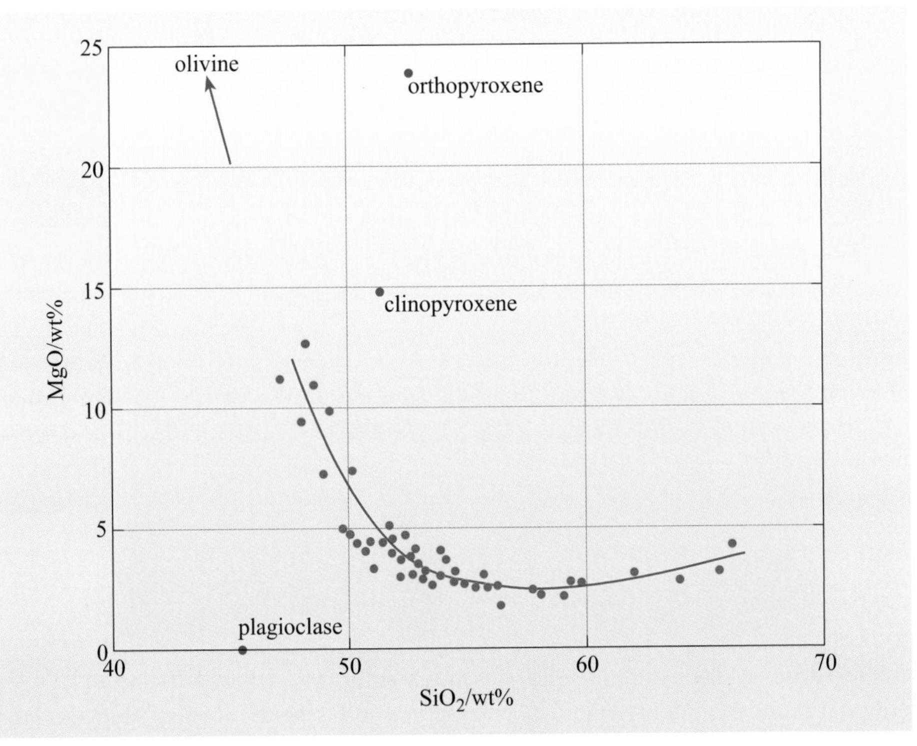

Figure 5.17 MgO versus SiO_2 concentrations from a suite of lavas and phenocrysts from two volcanoes on the island of Dominica. The blue line represents a smoothed compositional trend.

- Are there any kinks in the trends defined by the data in Figures 5.16 and 5.17?

- Yes, there is a strong kink in the trends at about 50–52 wt% SiO_2, particularly in Figure 5.16.

- If the observed variations in SiO_2, CaO and MgO were controlled by fractional crystallisation, what could you say about the composition of the crystals extracted (the extract) compared with that of the liquid? For samples with less than 50 wt% SiO_2, would the extract contain more or less of the following constituents than the liquid: (a) SiO_2, (b) CaO, and (c) MgO?

- The trends in the data for rocks with less than 50 wt% SiO_2 have an increase in CaO and SiO_2 and a decrease in MgO. The extract must, therefore, contain: (a) less SiO_2, (b) less CaO and (c) more MgO than these liquids.

- What about the extract controlling liquids with 55–60 wt% SiO_2?

- The extract contains less SiO_2, but more CaO and just slightly more MgO than those liquids.

- Look at the trend in the data for rocks with less than 50 wt% SiO_2. Does the composition of any one phenocryst plot in such a way that crystallisation of that mineral alone could drive the composition of the liquid along the trend?

- Yes, the removal of just olivine would drive the composition of the liquids along a line of increasing CaO until it starts to flatten out at about 50 wt% SiO_2. Similarly, olivine has a high MgO content and, therefore, removal of olivine reduces the MgO contents of the liquids with less than 50 wt% SiO_2.

- For rocks with 52–60 wt% SiO_2, is it possible that just one of the phenocryst minerals could produce the trends observed in the data?

- The answer is no, because no one mineral plots along the trends of data for rocks with more than 52 wt% SiO_2. If this trend reflects fractional crystallisation, then it must be due to more than one phenocryst phase crystallising. The extract has to be some combination of a high-calcium phase (plagioclase and/or clinopyroxene) and a low-calcium phase (olivine and/or orthopyroxene).

The chemical variation diagrams give a guide to which minerals crystallise and their proportions but cannot give a unique answer. To find out which minerals actually crystallised together in the Dominica lavas involves an examination of the rocks themselves. Phenocrysts are the minerals that were crystallising at the time the liquid was quenched during eruption on the Earth's surface. Thus, identification of the phenocryst minerals and their proportions provides an estimate of the bulk composition of the aggregate minerals (the extract) that were crystallising together shortly before they were erupted. However, the phenocryst assemblage may not be completely representative of the mineral proportions that were extracted during a long period of fractional crystallisation.

By definition, fractional crystallisation involves the separation of crystals, and so earlier phenocryst minerals will now comprise the cumulates that reside at the base and sides of the underlying magma chamber.

A typical basalt from Dominica contains 10.5% olivine, 27.3% plagioclase and 6.2% clinopyroxene. The abundances represent the percentage the phenocrysts make up of the total rock, and so the relative proportion of the phenocryst assemblage, in this case, consists of 24% olivine, 62% plagioclase and 14% clinopyroxene.

- The aggregate phenocryst assemblage discussed above has a bulk composition of 14.2 wt% CaO and 44.5 wt% SiO_2. Where does this composition plot in relation to the compositional trend between 52–60 wt% in Figure 5.16? Could fractional crystallisation of this phenocryst assemblage be responsible for the variations in the lavas with 52–60 wt% SiO_2?

- The bulk composition of the aggregate phenocryst assemblage plots on an extension of the CaO–SiO_2 compositional trend, which could, therefore, be explained by extracting the phenocryst assemblage in the proportions found in the Dominica basalts. The bulk composition of this phenocryst assemblage plots in a position that could explain the trend not only between 52–60 wt% SiO_2, but also all the way up to 66 wt% SiO_2.

This is where other geochemical and geological information needs to be considered. Just using the SiO_2 versus CaO diagram could be misleading, as relying on one variation diagram limits the possible number of geometrical constraints. This is illustrated by the presence of an upturn in the SiO_2 versus MgO plot at about 60 wt% SiO_2 (Figure 5.17), suggesting a change in the phenocryst assemblage at this composition.

The phenocryst assemblage found in the basalts is not going to be the same one that is present in higher SiO_2 melts. High-SiO_2 rocks, such as rhyolites (the volcanic equivalent of a granite), consist largely of different minerals (quartz and feldspars) from those found in basalts (olivine, clinopyroxene and plagioclase). Magnesium-rich olivine and quartz are not stable together in the same rock, reacting together to produce orthopyroxene. Moreover, it has already been shown that during fractional crystallisation the composition of minerals that have a solid solution change their composition as the melt composition evolves. The plagioclase compositions plotted in Figures 5.16 and 5.17 are just average values. Analyses of plagioclase from a rock of higher SiO_2 content on Dominica gives 8.2 wt% CaO and 58.6 wt% SiO_2. This is consistent with the chemical formula of anorthite-poor plagioclase, because the decrease in Ca^{2+} is coupled with the substitution of Si^{4+}. If you plot this composition in Figure 5.16 you will find that it lies on the trend of the rock analyses, which suggests that the separation of anorthite-poor plagioclase alone could control the compositions of rocks with the highest SiO_2 content. The slight rise in MgO contents in the SiO_2-rich samples (Figure 5.17) is consistent with this interpretation, as plagioclase contains virtually zero MgO.

- Would you expect olivine to be crystallising from a melt with 3 wt% MgO and 60 wt% SiO_2 and if so what would be its Mg#?

- The experimental data for crystallising MORB (Section 4.6.3) indicate that olivine becomes a less important phase during extensive crystallisation and, if it did crystallise, it would produce a low Mg# olivine.

In Dominica, olivine, which is abundant in low-SiO_2 rocks, is rarely observed in rocks with more than 60 wt% SiO_2, indicating that the liquids have moved out of the olivine stability field. Instead, orthopyroxene starts to crystallise; and since, like olivine, it contains little CaO and significant amounts of MgO, the transition from olivine to orthopyroxene is not marked by any kink in the trends in Figures 5.16 and 5.17.

The gradual changes in the composition of the lavas from the two volcanoes on Dominica, therefore, indicate that they may be related by fractional crystallisation. From the phenocryst assemblages observed in the rocks and their compositions, a fractional crystallisation path capable of generating the chemical variations observed in the lavas can be deduced. This can be summarised as follows:

1 At less than 50 wt% SiO_2, only olivine crystallises.
2 At 50–60 wt% SiO_2, plagioclase and clinopyroxene crystallise together with olivine (at lower SiO_2) and orthopyroxene (at higher SiO_2).
3 At more than 60 wt% SiO_2, the amount of pyroxene crystallising decreases until the chemical variation in very SiO_2-rich liquids appears to be controlled by crystallisation of anorthite-poor plagioclase alone.

Thus, it can be concluded that the andesites on Dominica appear to have been derived from the fractional crystallisation of a basaltic magma that was produced by partial melting in the mantle wedge.

This section demonstrates that, as the liquid evolves, the phenocryst assemblage changes. Plagioclase feldspar is present through most of the crystallisation history and will produce a wide range of plagioclase compositions, often resulting in zoned crystals. Figure 5.18 presents the composition of plagioclase from a suite of lavas from St Lucia as a function of the composition of the lavas. The lavas vary from basaltic andesites (56 wt% SiO_2) to silicic andesites (61 wt% SiO_2). Plagioclase is zoned in all of the rocks. However, the plagioclase composition varies systematically with the bulk composition of the lava, with a general decrease in the anorthite content of both the rim and the core compositions. The rim composition changes from An_{43} to An_{20} and tracks the cooling liquid composition expected from the binary phase diagram (Figure 5.14). By contrast, the core composition reflects the composition of a crystal that was crystallised from earlier magmas and has not separated completely from the evolving liquid. Where zoning is less marked (i.e. the core and rim compositions are more similar), the plagioclase phenocrysts have reacted more extensively with the magma during cooling.

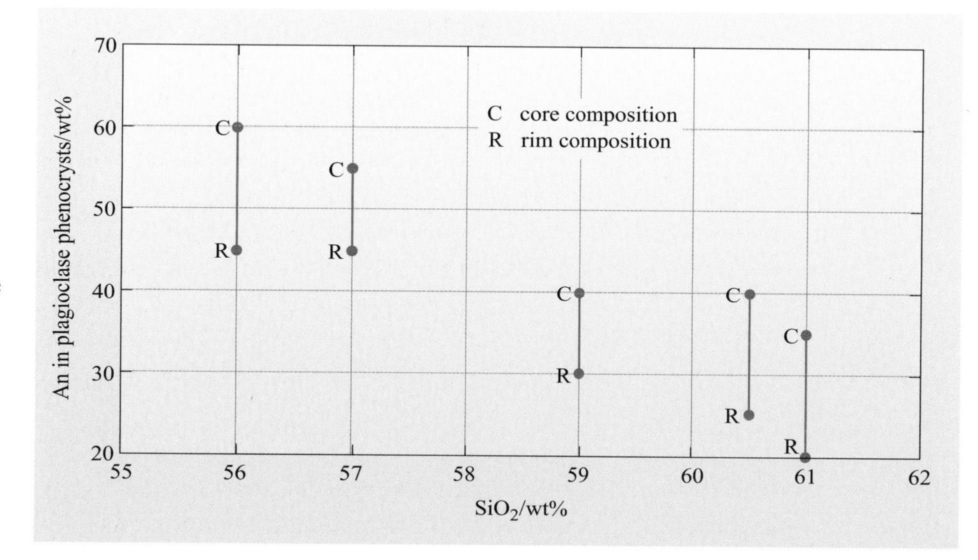

Figure 5.18 Variation of the composition of plagioclase phenocrysts with SiO_2 content of bulk sample from a suite of lavas from St Lucia.

One final comment needs to be made about the crystallisation history of these Lesser Antilles lavas. Plagioclase is an important phase throughout the crystallisation, but it is only stable up to pressures of 0.8 GPa, which equates to 30 km depth. In the Lesser Antilles, the crust is at least 30 km thick, so the presence of plagioclase confirms that crystallisation must take place within the crust rather than within the upper mantle.

Trace elements

Returning to the Dominica lavas, it has already been demonstrated that the major element geochemistry is consistent with crystallisation of olivine, clinopyroxene and plagioclase in varying proportions. Are the trace element variations in these lavas consistent with fractional crystallisation of these phases?

- Look at the partition coefficients for the elements yttrium (Y) and zirconium (Zr) in Table 5.2. What would you predict would happen to the Y and Zr concentrations during fractional crystallisation of olivine, clinopyroxene and plagioclase?
- The partition coefficients for these elements are less than one (except for Y in clinopyroxene). Y and Zr are therefore incompatible elements in olivine plus plagioclase ± clinopyroxene and their concentration should increase with increasing crystal fractionation.

Figure 5.19 is a plot of the Y and Zr variations in the Dominica lavas. Fractionation of basalts to basaltic andesites does indeed increase the concentration of these two elements, but there is a kink in the fractional crystallisation path. Something happens within the andesite field that produces a decrease in the Y content.

- What do you think happens at this point (the decrease in Y content) in the crystallisation history of the Dominica lavas?
- This must indicate where a new mineral crystallises which removes Y relative to Zr (i.e. Y behaves as a compatible element, $D > 1$).

- From the partition coefficients listed in Table 5.2, which new minerals could cause the change in the Zr–Y trend?
- Only garnet and amphibole have Y partition coefficients high enough to produce the decrease in Y.

Table 5.2 Selected trace-element partition coefficients for minerals in a magma of andesitic composition.

	Rb	Sr	Ba	Zr	Y
olivine	0.001	0.001	0.001	0.01	0.01
orthopyroxene	0.003	0.05	0.001	0.08	0.1
clinopyroxene	0.001	0.07	0.001	0.25	1.0
garnet	0.01	0.02	0.02	0.5	11.0
plagioclase	0.04	4.4	0.3	0.03	0.06
amphibole	0.01	0.02	0.04	1.4	2.5

While garnet is rarely found in andesitic lavas, amphibole is common in Dominica andesites and dacites (but not in the basalts). Amphibole is a hydrous mineral that is found in some evolved island-arc lavas, so it is not surprising that it crystallises from the more evolved magma compositions given the higher H_2O content of arc magmas relative to MORBs. However, amphibole stability also requires a reasonably high amount of Na_2O in the magma, which is why it is not usually found in arc basalts. The amphibole in the Dominica lavas has a major element composition very close to the clinopyroxene, and so its effect on the variation of CaO, MgO and SiO_2 contents is very similar. The Y–Zr plot (Figure 5.19) however demonstrates that, in addition to clinopyroxene, amphibole appears as a crystallising phase about halfway through the andesitic compositions causing Y to decrease. Meanwhile Y flattens off as Zr continues to increase in the dacite samples, indicating the percentage of amphibole crystallising decreases in these higher SiO_2 rocks. This has been confirmed by petrographic observation.

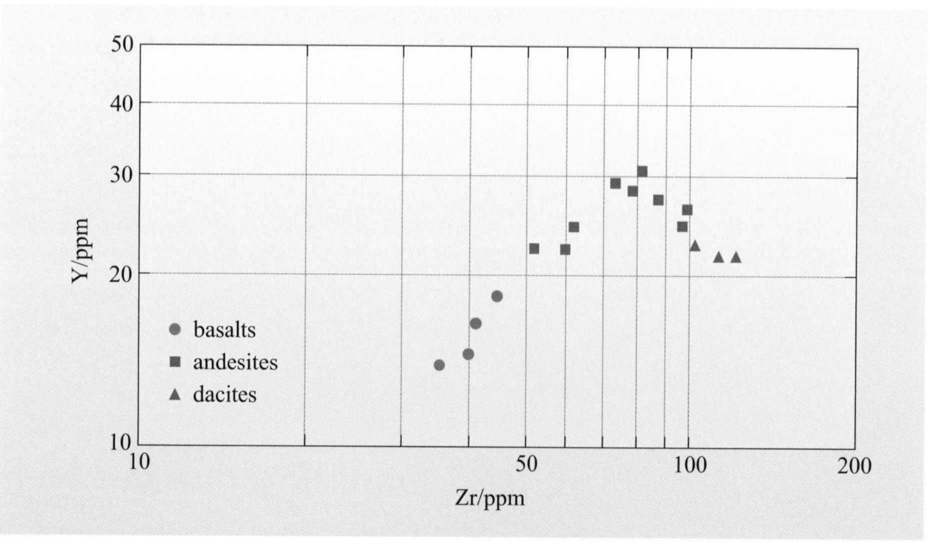

Figure 5.19 Plot of Zr versus Y for lavas from Dominica. Note scales of both axes are logarithmic.

5.6.3 Summary of arc magma evolution

To understand the evolution of an evolving magma, information from the mineralogy of the rocks and their major and trace elements composition is essential. In the case of the lavas from Dominica, fractional crystallisation of olivine, plagioclase and clinopyroxene produced andesitic lavas from a primary basaltic magma, while further crystallisation of plagioclase ± amphibole produced more SiO_2-rich compositions ranging from dacite to rhyolite. There are several factors governing why magmas at destructive plate boundaries evolve to more SiO_2-rich compositions than mid-ocean-ridge lavas. The rate of melt production from the mantle is generally less in arc systems than at mid-ocean ridges, so magmas can be stored in the crust for longer. High melt production island-arc volcanoes, such as Mt Fuji in Japan and Mt Klyuchevskaya in Russia, rarely produce magmas more evolved than basaltic andesites. Magma storage in the crust is also aided by the slightly thicker crust of many island arcs, particularly in continental arcs compared with that at mid-ocean ridges. Continental arcs are dominated by andesite, dacite and rhyolite with very few erupted basalts.

- Why is it difficult to erupt a basaltic lava in a continental arc?

- Basaltic lavas contain dense ferromagnesian minerals and the melt itself is denser than the surrounding continental crust, even if it is considerably hotter. Therefore, it will not easily erupt. As the magmas fractionate, the liquids evolve towards less-dense compositions and these are easier to erupt in terms of buoyancy, if not viscosity.

It is no surprise, therefore, that andesitic lavas are common in the Andes, because the thick continental crust allows for extensive fractionation at depth. Moreover, the thickened crust in the Andes prevents some of the magma from erupting. As a result, rocks of andesitic to rhyolitic composition become trapped in the crust where they cool and crystallise more slowly, producing the coarse-grained, intrusive equivalents of dacite and rhyolite, which are granodiorite and granite respectively. If the new material added to the crust in this manner is derived solely by fractional crystallisation of a basaltic parent, then this represents a process whereby the crust grows, and most geoscientists are of the opinion that active crustal growth today occurs mostly above subduction zones. The mean composition of the continental crust is andesitic, and many modern arcs produce a significant quantity of basaltic andesitic to andesitic crust.

5.6.4 Some final thoughts on H_2O

There is one final aspect of magma generation in arcs to consider. H_2O is the component that makes subduction zones function, but what happens to H_2O and other volatile elements in arc magmas as they evolve? H_2O has a bulk partition coefficient similar to the highly incompatible rare earth element cerium (Ce).

- During fractional crystallisation, what will happen to the H_2O content of a melt?
- The H_2O content should increase exponentially as the melt evolves because it behaves like an incompatible element.

This is true in principle, although the ability of a melt to dissolve H_2O is a function of pressure. As long as the melt is stored at a depth of a few kilometres in the crust the H_2O will stay dissolved in the melt. Volcanic rocks from Dominica stabilise the hydrous mineral amphibole when they evolve to an andesitic composition, indicating that magmas do retain their H_2O, and some evolved arc lavas can retain up to 10 wt% H_2O both dissolved in the melt and bound in hydrous minerals. It was noted earlier in this chapter that SiO_2-rich melts become more viscous and that this stops volatiles, such as H_2O, degassing from magmas as they rise to shallower depths in the crust. When the confining pressure is released the vapour and silicate phases separate rapidly and generate an explosive eruption. This can have devastating effects, and the explosive eruptions of Mt Krakatau, Mt Pinatubo and Mt St Helens all resulted from the eruption of evolved volatile-charged magmas. These eruptions are so destructive because they inject ash and other debris into the atmosphere, but locally they also generated deadly ash flows. The eruption of Mt Pelée on Martinique killed many people because a turbulent flow of hot ash and gases at temperatures of 800 °C flowed at high speeds (~160 km h^{-1}) down a valley and flattened the town of St Pierre. Ash from such eruptions can be distributed around the globe and has the effect of cooling the Earth and disrupting air travel.

Other volatile elements also behave as incompatible elements, and sulfur is particularly concentrated in arc magmas because the melts are oxidised and, therefore, do not stabilise sulfide minerals that can crystallise and reduce the sulfur content of a melt. Sulfur is dissolved as a sulfate species in arc magmas and is readily partitioned into H_2O vapour as the melt degasses. This mechanism efficiently transfers sulfur into the atmosphere. Sulfur from the Mt Pinatubo eruption in 1991 produced a global cooling of about 0.6 °C over the next year, and there are many instances in the geological record of such global cooling events. Sulfur concentration spikes in ice core records correlate with known eruptions (not always from arcs) and these correspond to historical records of global food shortages and cool summers. This is a good example where recycling of an element through a subduction zone has a profound effect on human society.

The longer term aspects of H_2O recycling will be considered further in Chapter 7, but a simplified diagram of the H_2O cycle for a subduction system is shown in Figure 5.20. Estimates of the H_2O fluxes are difficult to determine, but recent calculations suggest that although some H_2O is clearly taken deeper into the mantle, mainly in minerals in altered oceanic mantle, a significant amount is returned through the subduction zone system. This is good news, as it means the oceans are not going to disappear into the mantle!

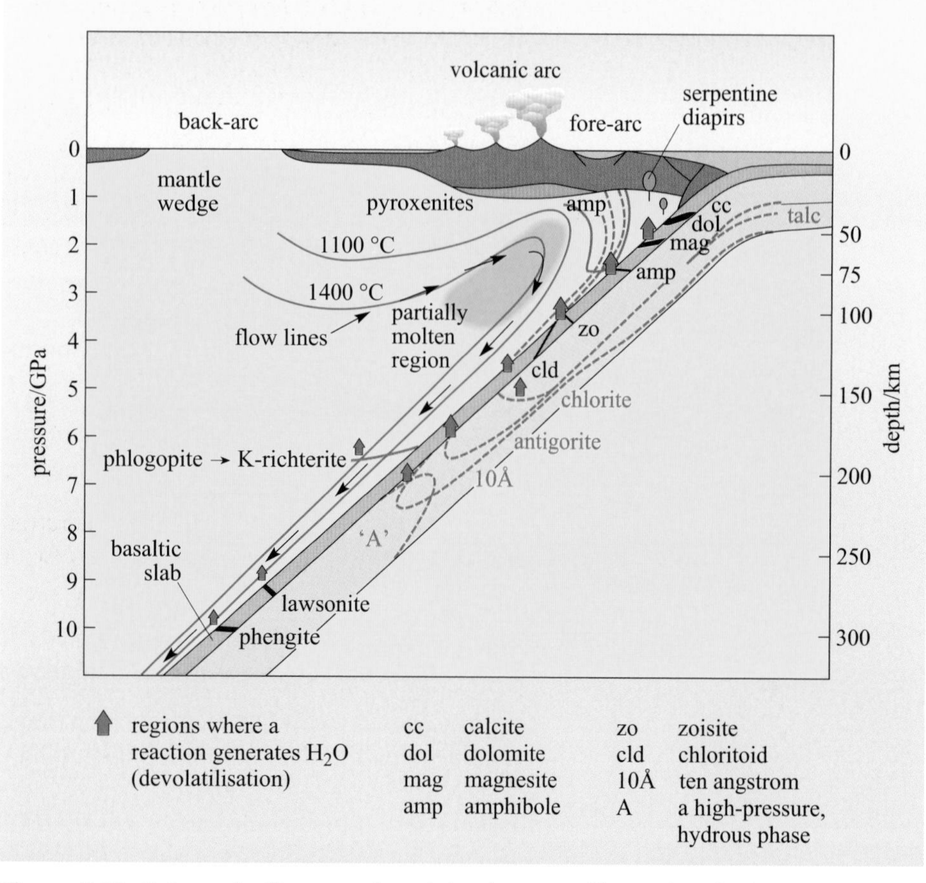

Figure 5.20 Schematic diagram of a subduction zone illustrating the key dehydration reactions that generate H_2O. Solid lines in the mantle wedge represent isotherms whereas the thin arrows are mantle flow lines. Phase boundaries in the basaltic slab are represented by thick lines and in the peridotite section of the slab by dotted lines. It is not important to know the exact position of all the phase boundaries, but that there are series of reactions that generate volatiles at increasing pressures (from calcite to phengite). The ten angstrom phase is an intermediate high-pressure hydrous phase. The 'A' phase in the peridotite section of the slab carries water to depths much greater than the bottom of the figure (>300 km). Large arrows represent regions where a reaction generates H_2O. Mineral names involved in the reactions are labelled on the figure. (Poli and Schmidt, 2002)

5.7 Trace elements and magmagenesis

Before studying the trace element composition of subduction zone magmas in detail, it is important to consider how trace elements can be understood quantitatively. One of the reasons geochemists use trace elements is that they have a wide range of geochemical properties and there are some simple equations that can be used to model their behaviour during partial melting and crystallisation.

5.7.1 A quantitative approach

The distribution of a trace element between a mineral and a melt is described quantitatively by the partition coefficient D_{min}, which is defined as:

$$D_{min} = \frac{C_{min}}{C_{melt}} \qquad (5.1)$$

where C_{min} is the concentration of an element in a mineral and C_{melt} is the concentration of the same element in a coexisting melt.

■ Can you recall from Box 1.3 what is meant by the terms compatible and incompatible elements?

■ A compatible element has a D_{min} value greater than one, and so preferentially partitions into a mineral phase. An incompatible element has a D_{min} of less than one, and so preferentially partitions into the melt.

The main controls on element substitution in crystal structures were discussed in Section 4.2.

■ What are the two major factors that allow element substitution in minerals?

■ You should recall that elements with similar ionic radii and charge tend to be able to partition into the same site within a mineral structure.

Nickel (Ni^{2+}), for example, easily substitutes for Fe^{2+} and Mg^{2+} in the olivine structure because of the similarity in their ionic radii, and it is a compatible element in olivine. However, Th^{4+} has a large ionic radius and is highly charged and, as a result, is highly incompatible ($D_{ol} \approx 0.000\,01$). In a system in which more than one mineral is crystallising or melting it is possible to define a bulk partition coefficient, which is the sum of the proportion of each mineral multiplied by its respective partition coefficient. For example, if a melt is crystallising 20% olivine and 80% clinopyroxene and the partition for the element Y is 0.01 for olivine (ol) and 0.5 for clinopyroxene (cpx) then this mineral assemblage would have the following bulk partition coefficient:

$$D_{bulk} = 0.2D_{ol} + 0.8D_{cpx}$$

$$D_{bulk} = (0.2 \times 0.01) + (0.8 \times 0.5) = 0.002 + 0.4$$

$$D_{bulk} = 0.402$$

This can be written out in a more general form, such that if a mineral assemblage consists of phases A, B, C, D, … in mass proportions X_A, X_B, X_C, X_D, …, where

$$X_A + X_B + X_C + X_D \ldots = 1 \qquad (5.2)$$

and the minerals have partition coefficients D_A, D_B, D_C, D_D, etc., then the bulk partition coefficient is:

$$D_{bulk} = X_A D_A + X_B D_B + X_C D_C + X_D D_D \ldots \qquad (5.3)$$

Having defined how elements are partitioned between minerals and melts their behaviour during melting and crystallisation can be quantified (Box 5.1).

Box 5.1 Equations that describe trace-element behaviour during partial melting and fractional crystallisation

During melting or crystallisation, trace elements partition between the melt and minerals in a predictable way. Consider a source rock with a concentration of a given element C_0 that melts to produce a liquid with a concentration C_L and some residual solid C_S. If the melt fraction is F, then the concentrations of the trace element in the source, melt and residue are related:

$$C_0 = FC_L + (1 - F)C_S \quad (5.4)$$

This is known as a mass balance equation.

With some rearranging the partial melting equation can be derived:

$$C_L = \frac{C_0}{F + (1 - F)D} \quad (5.5)$$

This can sometimes be found written in the alternative form:

$$C_L = \frac{C_0}{D + (1 - D)F} \quad (5.6)$$

Equations 5.5 and 5.6 relate the liquid composition to that of the source, the partition coefficient of the element of interest and the melt fraction. Equation 5.6 shows the partial melting equation in its simplest form because it assumes that the minerals on the solidus enter the melt phase in the same proportion that they are found in the rock. This is known as **modal melting**. You already know from melting of the mantle that a lherzolite containing clinopyroxene melts to produce a basalt and leaves a residual harzburgite (with essentially no clinopyroxene). Therefore in the mantle clinopyroxene must enter the melt at a much higher proportion than it is found in the original rock. When the phases that enter the melt do so in a different proportion from that in the original rock it is known as **non-modal melting**. The equation that takes this effect into account is not presented here as it is more complex, and for most elements the two equations give similar results for the composition of the liquid.

Equation 5.6 assumes that the melt does not separate from the mantle until melting is complete. In nature this is unlikely to happen, and melts can efficiently segregate from the mantle when the melt fraction is less than 1%. These small melts eventually aggregate and form the final erupted lava. Again, the equations are not described here, but the composition of all of the aggregated small melt fractions is similar in composition to the liquid composition derived in Equation 5.6. However, the residual mantle is more affected by the differences in the melting model.

Equilibrium crystallisation can be described by Equations 5.5 or 5.6. In this case F represents the amount of liquid remaining, so the amount of crystallisation is $(1 - F)$.

Fractional crystallisation is described by a different equation. This assumes that each infinitesimally small crystal that is in equilibrium with the melt is then separated from the melt. This yields an equation containing an exponent:

$$C_L = C_0 F^{(D-1)} \quad (5.7)$$

Again, F represents the amount of liquid remaining, so the amount of crystallisation is $(1 - F)$.

Having defined the basic equations, it is useful to plot some simple results using these equations to illustrate how they control the concentrations of trace elements as a function of F and D. Figure 5.21 is a plot of the concentration of an element in a liquid relative to the starting composition. This ratio simply helps us understand whether a liquid is depleted or enriched in an element relative to the original composition. Curves are plotted for various bulk partition coefficients. For an element with $D = 1$, the melt does not change with degree of melting (changing F). For incompatible elements, the initial melt has a very high concentration and this concentration decreases as partial melting increases (increasing F). The concentration is the highest for the most incompatible element and approaches $\frac{1}{F}$ for highly incompatible elements. For compatible elements, the concentration of an element is always lower than the starting composition. At low degrees of melting the concentration is lowest and it increases as the rock progressively melts. The concentrations are lowest for elements with the highest D values.

■ What is the concentration of Y in a melt if the mantle has 4 ppm Y, the bulk partition coefficient is 0.3 and 5% of the mantle has melted?

■ $C_0 = 4$, $D = 0.3$ and $F = 0.05$ therefore using Equation 5.6:

$$C_L = \frac{4}{0.3 + (1 - 0.3) \times 0.05} = 11.94 = 12 \text{ ppm to two significant figures.}$$

For crystallisation, Figure 5.21 is read in reverse, starting with 100% melt at $F = 1$. With increasing crystallisation (decreasing F), the concentrations of the incompatible elements increase and the concentrations of the compatible elements decrease in the liquid:

- for incompatible elements, the increase in concentration is greatest for the elements with the lowest D value
- for compatible elements, the decrease in concentration is greatest for elements with the highest D value.

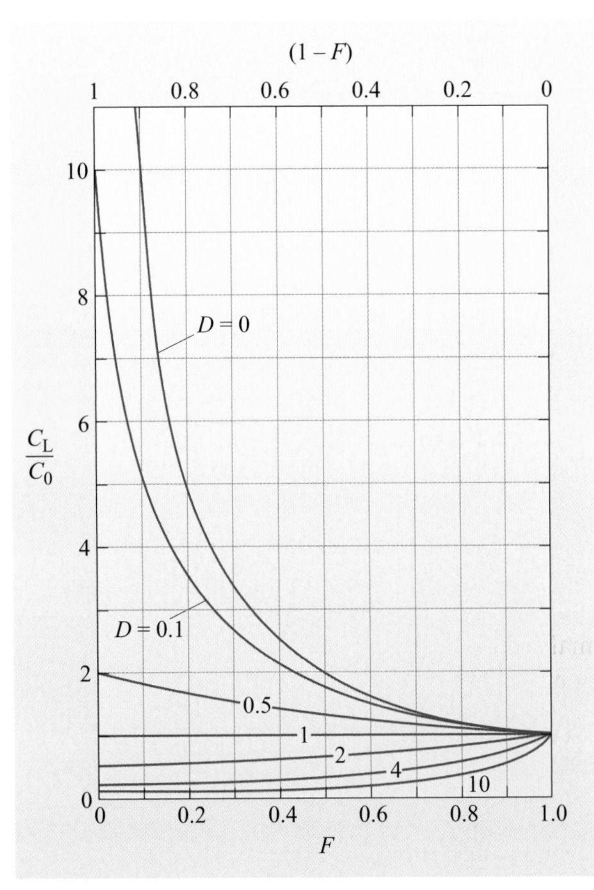

Figure 5.21 Plot of the ratio of the concentration in a liquid relative to the starting composition (C_L/C_0), as a function of the amount of melting F and the amount of crystallisation $(1 - F)$, for different values of the partition coefficient D using the partial melting Equations 5.5 or 5.6. This diagram can also be used for equilibrium crystallisation. (Adapted from Cox et al., 1979)

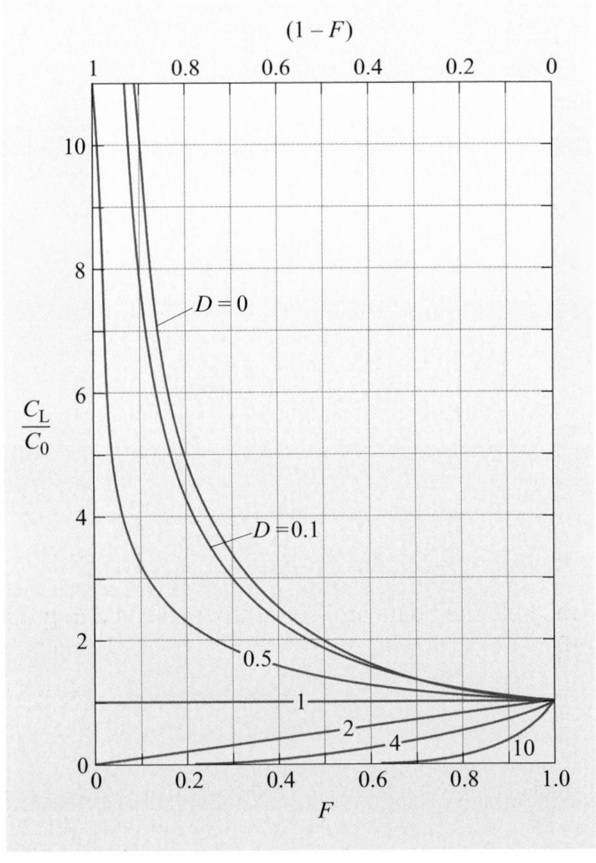

Figure 5.22 Plot of the ratio of the concentration in a liquid relative to the starting composition (C_L/C_0) as a function of F and the amount of crystallisation $(1 - F)$, for different values of the partition coefficient D using the fractional crystallisation Equation 5.7. (Adapted from Cox et al., 1979)

- Sr has a partition coefficient of 4.4 in plagioclase. If plagioclase were the only crystallising phase, what would happen to the Sr concentration in the melt as plagioclase crystallised?
- As the melt crystallises, the Sr concentration will drop because it is concentrated into the plagioclase crystals.

Results using the fractionation crystallisation equation are illustrated in Figure 5.22. The curves are similar to those from the equilibrium crystallisation equation but the effects on concentration are more dramatic. Incompatible elements are more enriched and compatible elements are more depleted in the melts for a given F and D than for equilibrium crystallisation.

Question 5.1

Assuming the mantle contains 60% olivine, 20% orthopyroxene, 10% clinopyroxene and 10% garnet, what would be the bulk partition coefficient for Rb during partial melting of the mantle? Use the partition coefficients provided in Table 5.2.

Question 5.2

Calculate the concentration of Y in a melt, 80% of which has crystallised, for equilibrium and fractional crystallisation. Assume the original melt had 25 ppm Y and the bulk partition coefficient is 0.402.

5.8 The trace-element composition of island-arc magmas

In comparing the composition of **island-arc basalts (IABs)** with MORBs it has been established that primitive lavas from the two tectonic settings have similar major element compositions. One key geochemical contrast is that IABs contain higher H_2O contents, due to partial melting of the hydrous mantle wedge, and the H_2O is ultimately derived from the subducting slab. Given that the slab comprises altered oceanic crust and hydrated depleted peridotite, and may carry an array of exotic sediments into the mantle, it would be surprising if there were not more differences between IABs and MORBs.

Trace-element data are frequently presented in graphical form normalised to a reference composition. This is usually an estimate of the composition of either the Earth's primitive mantle or chondritic meteorites. Elements are ordered from left to right in terms of decreasing incompatibility during partial melting of the mantle. Thus, if a basalt is simply derived by partial melting of the mantle with no added components, then the resultant trace-element pattern will be a smooth curve. Figure 5.23 is a primitive mantle-normalised plot of a representative arc basalt and MORB that have roughly the same MgO contents.

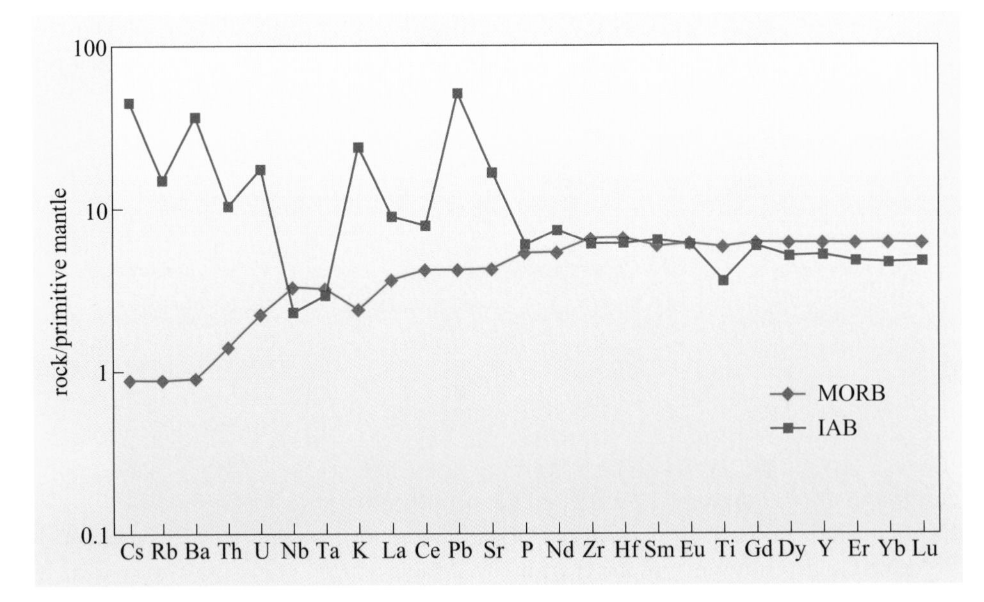

Figure 5.23 Primitive mantle-normalised multi-element patterns of a MORB and an IAB with similar MgO contents. Element names and symbols are given in Appendix A.

- Describe the differences between the mantle-normalised trace-element plots in Figure 5.23.

- There are obviously big differences between the two trace-element patterns. The MORB pattern is very smooth with a steady decrease in the highly incompatible elements to the left of the figure.

This is because MORB represents moderate degrees of partial melting of a mantle that is slightly depleted in highly incompatible elements. The IAB plot has a very spiky pattern. The right-hand side of the pattern is broadly similar in shape to the MORB, although the IAB has slightly lower heavy rare earth elements (HREE) (Dy–Lu) contents. The IAB has lower Zr and Hf contents and large enrichments in K, Pb, Sr, U, Th, Ba, Rb and Cs and smaller enrichments of La and Ce. Of the highly incompatible elements to the left of the diagram, only Nb and Ta are depleted relative to MORB.

By choosing to compare two lavas with the same MgO content, the difference in the trace-element patterns is unlikely to result from differing degrees of crystallisation in a magma chamber. The elements to the right of the pattern such as the HREE and the element Y vary due to the amount of partial melting.

- Y has a bulk partition value of 0.3; what does the lower Y content of the IAB relative to MORB mean in terms of partial melting?

- The partial melting equation dictates that the concentration of an incompatible element decreases with increasing degree of partial melting. Therefore, a first-order interpretation of the right-hand side of the normalised trace-element patterns is that IAB lavas are produced by higher degrees of partial melting. Alternatively, the source region of the IAB may have a lower concentration of Y than that of the MORB.

- Th has a bulk partition coefficient of 0.0001; are the Th concentrations in the IAB and the MORB consistent with the partial melting interpretation put forward above?

- Clearly, the Th concentration is much higher in the IAB than in the MORB, as are most of the other highly incompatible elements. This is not consistent with IABs being generated by higher degrees of partial melting than MORBs, if they come from the same mantle source.

To explain the difference between the two trace-element patterns, there must be a difference in the composition of the material that is melted to produce an IAB. To understand a bit more about how to interpret these diagrams you need to know about the geochemical behaviour of various elements. Trace elements can be subdivided into those that are controlled by partitioning between a melt and a mineral and those that may also be controlled by partitioning between an aqueous fluid and a mineral. The elevated H_2O content of an IAB means that it is important to understand how some elements are partitioned into aqueous fluids. Elements with a low valency and large ionic radii tend to be soluble in aqueous fluids and are commonly known as **large ion lithophile elements (LILEs)**. Cs^+, Rb^+, K^+, Ba^{2+} and Sr^{2+} are all LILEs and have accordingly high solubilities. U^{6+} also is very soluble compared with U^{4+}. Many other higher valency ions, such as Y^{3+}, REE^{3+}, Th^{4+}, Zr^{4+}, Hf^{4+}, Ti^{4+}, Nb^{5+} and Ta^{5+}, all have smaller ionic radii and are insoluble in aqueous solutions.

- Do these differences in trace-element behaviour correspond to selective enrichments in the IAB relative to a MORB?

- In part, yes. Elements such as Cs, Rb, Ba, Sr and U are highly enriched relative to their concentrations in MORBs, and this may be due to the transfer of these elements from the slab in an aqueous fluid.

The observations that some elements are added to the mantle wedge by an aqueous fluid is not surprising, because IABs have elevated H_2O contents and it has already been demonstrated that dehydration reactions in the subducting slab release H_2O. However, this does not explain whether the fluid mobile elements come from the sediment or the oceanic crust, or why IABs are enriched in elements that are not fluid mobile, such as Th. To help understand this dilemma, the composition of an average subducted sediment (called **global subducted sediment** or GLOSS) and the composition of an altered MORB are plotted on a primitive mantle-normalised diagram (Figure 5.24).

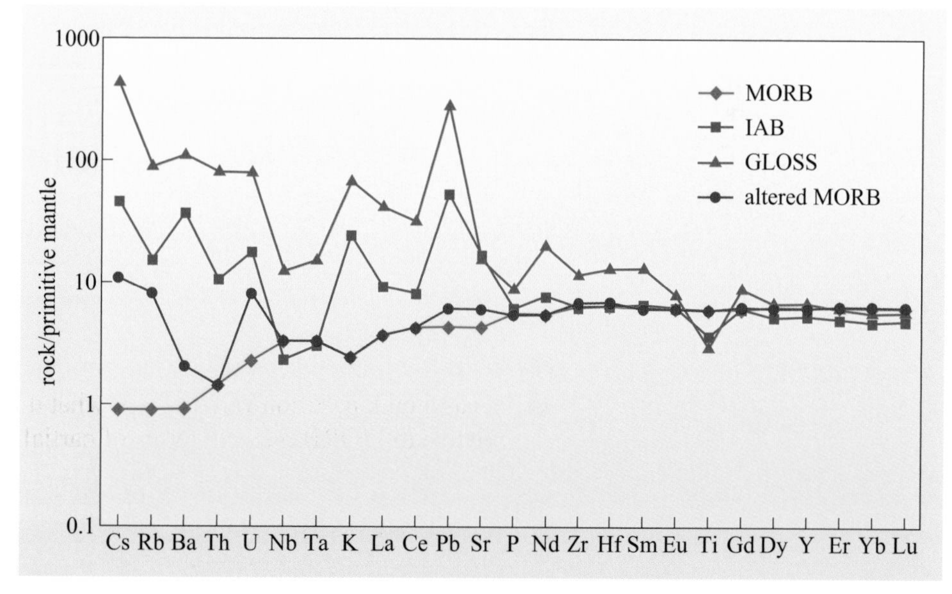

Figure 5.24 Primitive mantle-normalised multi-element patterns of GLOSS and an altered and fresh MORB.

- What compositional similarities exist between the GLOSS sediment, altered oceanic basalt, and IAB relative to fresh MORB?
- There are several clear similarities between the three patterns. In detail, GLOSS is enriched in all of the LILEs, as well as La, Ce, Pb and Th. It also has a marked depletion in Nb and Ta. By contrast, only the elements Cs, Rb, K and U are consistently enriched in the altered oceanic crust, whereas Ba, Pb and Sr are only slightly enriched relative to the fresh MORB.

- What process could have enriched IAB in the element Th which is insoluble in fluids?
- Th could be added in a melt derived from subducted sediment.

In principle, it is possible that sediment is physically removed from the slab and mixed into the mantle wedge. In practice, there is no real evidence for this happening, as the slab seems to remain relatively intact. It is more likely that the sediment partially melts and this melt, which is highly concentrated in these elements, is added to the mantle wedge. Simple mass balance arguments mean that only a small fraction of sediment melt is needed to enrich the mantle (about 0.2% sediment melt to 99.8% mantle). This explains why IABs can still have very low Nb and Ta concentrations, while being enriched in LILE.

The above discussion shows that there are some qualitative similarities between the material being subducted and the trace elements enriched in IAB lavas and in the case of elements that are not mobile in hydrous fluids there is a strong case for melted sediments. However, for the fluid-mobile LILEs, such as Rb and U that are enriched in both altered basalt and sediment, this proposal does not resolve whether these elements are added from either or both of the altered crust or sediments. This issue is the focus of the final section in Chapter 5.

5.9 Subduction zones and recycling

The trace-element patterns of subduction-related magmas are the product of (i) the different components that contribute to the magma source region, (ii) the processes that transport the components into the magma source and (iii) melt generation.

- What four source materials may contribute to the generaton of IAB?
- The mantle wedge; subducted sediments; subducted, altered, basaltic oceanic crust; and subducted, altered, depleted mantle peridotite.

- What processes can modify the source components?
- Melting and dehydration of both the subducted sediment and the basaltic slab could occur. Melting of the mantle wedge is probably the dominant process, along with dehydration of the altered subducted mantle. These processes are not mutually exclusive, and whether any, some or all happen depends on the thermal structure of the subduction zone.

As you have already discovered, the behaviour of an element during subduction processing depends on its geochemical properties, and in particular whether it is mobile in fluids or not and in how it partitions into a melt. When considering recycling it is also critical to assess how much of an element enters a subduction zone (its input flux) and how much leaves during the production of subduction zone magmas. Of the inputs to a subduction zone, altered oceanic crust probably exhibits the least compositional variation. By contrast, sediment input is highly variable both in amount and in terms of its composition, so it is important to assess the amount of sediment input into different subduction zones when considering the evolution and formation of IAB.

5.9.1 Sediment input

Sediments are a very important source for elements that can be subducted and these elements often originate in the continental crust. But can we be sure that sediments actually do get into the mantle and are incorporated into subduction-related magmatism? One of the more dramatic and least ambiguous results in this context has been the observation that the short-lived radioactive isotope ^{10}Be is present in amounts significantly above background levels in young lavas from island arcs (Box 5.2). The only way ^{10}Be can become incorporated in these lavas is if it has been transported into the mantle via subduction, released from the slab into the mantle and then incorporated in the parental arc magma prior to eruption. Moreover, given that ^{10}Be has a half-life of 1.5 Ma, the complete cycle must be completed within about 7.5 Ma (Box 5.2).

Box 5.2 ^{10}Be: the smoking gun for sediment subduction

^{10}Be is an isotope of the element beryllium, which is a Group 2 metal. ^{10}Be is produced continuously in the upper atmosphere by a process called **spallation**. This involves the bombardment of the nuclei of O and N atoms by cosmic rays. Hence, ^{10}Be is known as a **cosmogenic isotope**. ^{10}Be is radioactive and decays to ^{10}B with a half-life of 1.5 Ma. Be is carried by rainfall from the atmosphere on to the Earth's surface and into the oceans, where it becomes incorporated into pelagic sediments on the ocean floor. If these sediments are subducted, then ^{10}Be produced in the upper atmosphere can be transported into the upper mantle. If ^{10}Be is then transferred from pelagic sediments into the source region of subduction-zone magmas and is detected in erupted lavas, it provides an unambiguous tracer of element recycling in subduction zones.

As you might imagine, the amount of ^{10}Be produced in the atmosphere is not great, and with a half-life of 1.5 Ma the chances of finding measurable amounts of ^{10}Be in arc magmas might appear remote. The abundance of ^{10}Be in pelagic sediment is measured in units of atoms per gram, and there are about 5×10^9 atoms per gram of ^{10}Be in an average pelagic sediment. This is a tiny amount, but given that modern accelerator mass spectrometers can measure as low as ~10^6 atoms per gram of ^{10}Be, this is still about 1000 times more than the detection limit. ^{10}Be starts to decay effectively as soon as sediment is subducted, and after 10 half-lives (15 Ma) only $\frac{1}{1024}$ of the original parent isotope will be left, making its concentration too low to measure precisely and be of any scientific use.

Question 5.3

If a plate is subducted at 60 mm y^{-1} and carries down pelagic sediment with recently acquired ^{10}Be, how long would it take to reach a depth of 100 km assuming the subduction zone dips at 45°? How long would it take to reach 100 km if the plate was subducted at only 10 mm y^{-1}? What consequences does the difference in subduction rate have on whether a ^{10}Be signature can be detected in the arc lavas?

Figure 5.25 is a plot of ^{10}Be in MORBs, ocean island basalts (OIBs) and selected arcs. Both MORBs and OIBs have $\sim 10^6$ atoms per gram of ^{10}Be, which is indistinguishable from laboratory blanks (controls) that are nominally ^{10}Be free. By contrast, arc lavas can have up to $(15–20) \times 10^6$ atoms per gram of ^{10}Be, the Central American arc having the highest recorded ^{10}Be contents. The subduction rate along the Central American Trench is 80 mm y^{-1}, and so the most likely way in which the ^{10}Be was incorporated into these lavas is by recycling of ^{10}Be from subducted sediments.

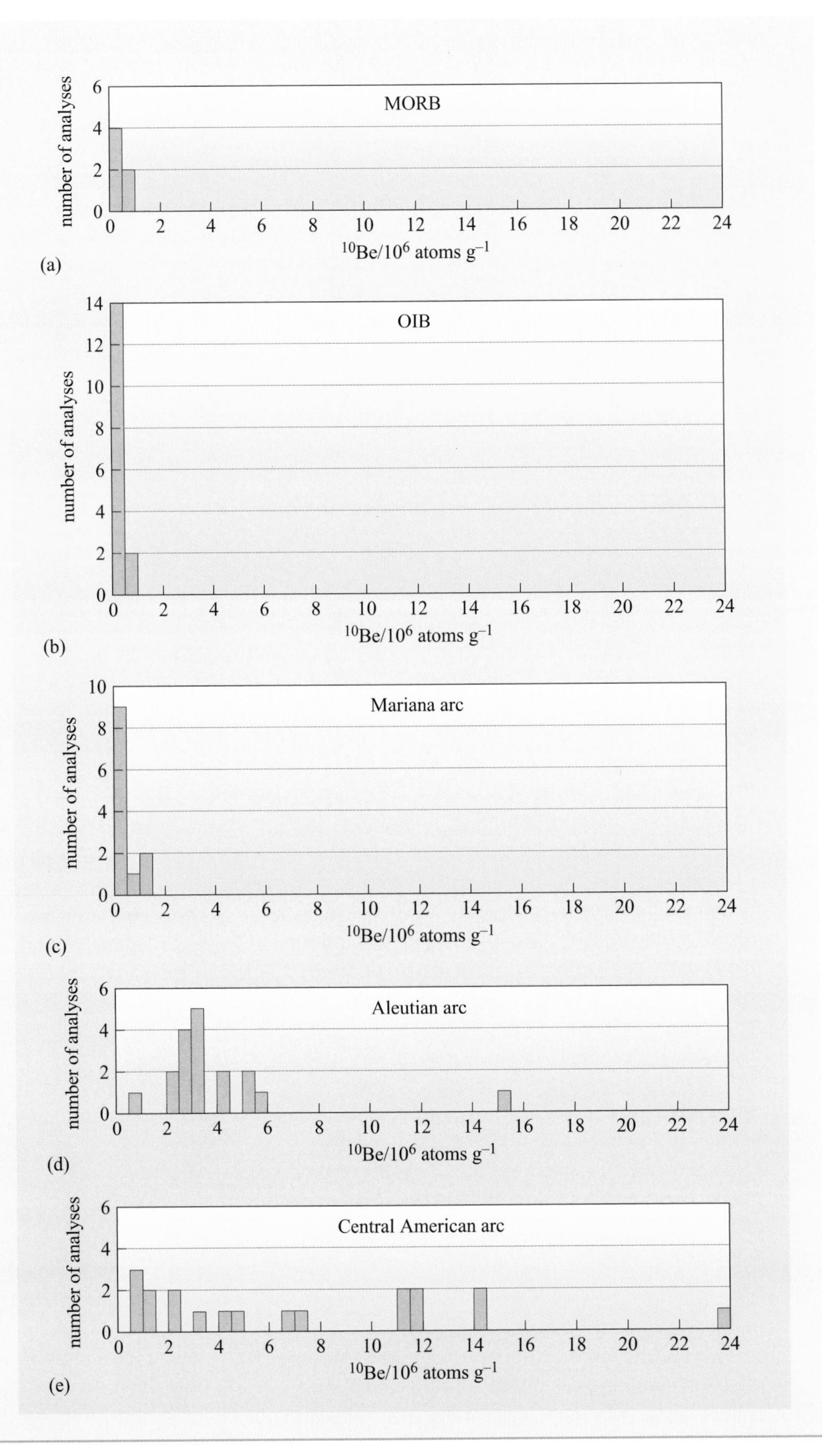

Figure 5.25 Histograms of ^{10}Be contents from: (a) MORB; (b) OIB; (c) Mariana arc; (d) Aleutian arc; and (e) Central American arc.

Despite the undoubted success of ^{10}Be for detecting some subducting sediments, not all sediments have the same composition, a similar number of arcs do not have a ^{10}Be signature, and in order to understand the link between ocean-floor sediments and subduction it is critical to understand the composition of oceanic sediments in more detail. Figure 5.26 is a map of the distribution of oceanic sediment types on the ocean floor. The composition of the sediment varies depending on differences in ocean productivity, which primarily relate to latitude, and to ocean depth, which controls carbonate stability. The proximity to oceanic ridges and continents also influences the distribution of metalliferous sediments (at ridges) and **terrigenous input** (from the continents). Close to the trench some volcanic-derived material (**volcaniclastic sediment**) from the volcanic arc may also be subducted. Therefore, as a plate moves away from a spreading ridge, the sediment on the oceanic crust will get thicker and its composition will vary depending on the position of the plate on the Earth's surface. Figure 5.26 is a useful guide to the types of sediment being deposited at a given point on the Earth's surface, but a more representative assessment of the bulk composition of sediments being subducted can be gained by analysing cores through the sediment section close to the trench. Figure 5.27 illustrates simplified core sections from eight subduction zones where deep-sea drilling has sampled the sediments in enough detail to provide a reasonable estimate of the average sediment type being subducted. There are significant variations between trenches. For example, sediment entering the Guatemala Trench is dominated by carbonate ooze, whereas the sediment being subducted beneath Vanuatu is mainly volcaniclastic.

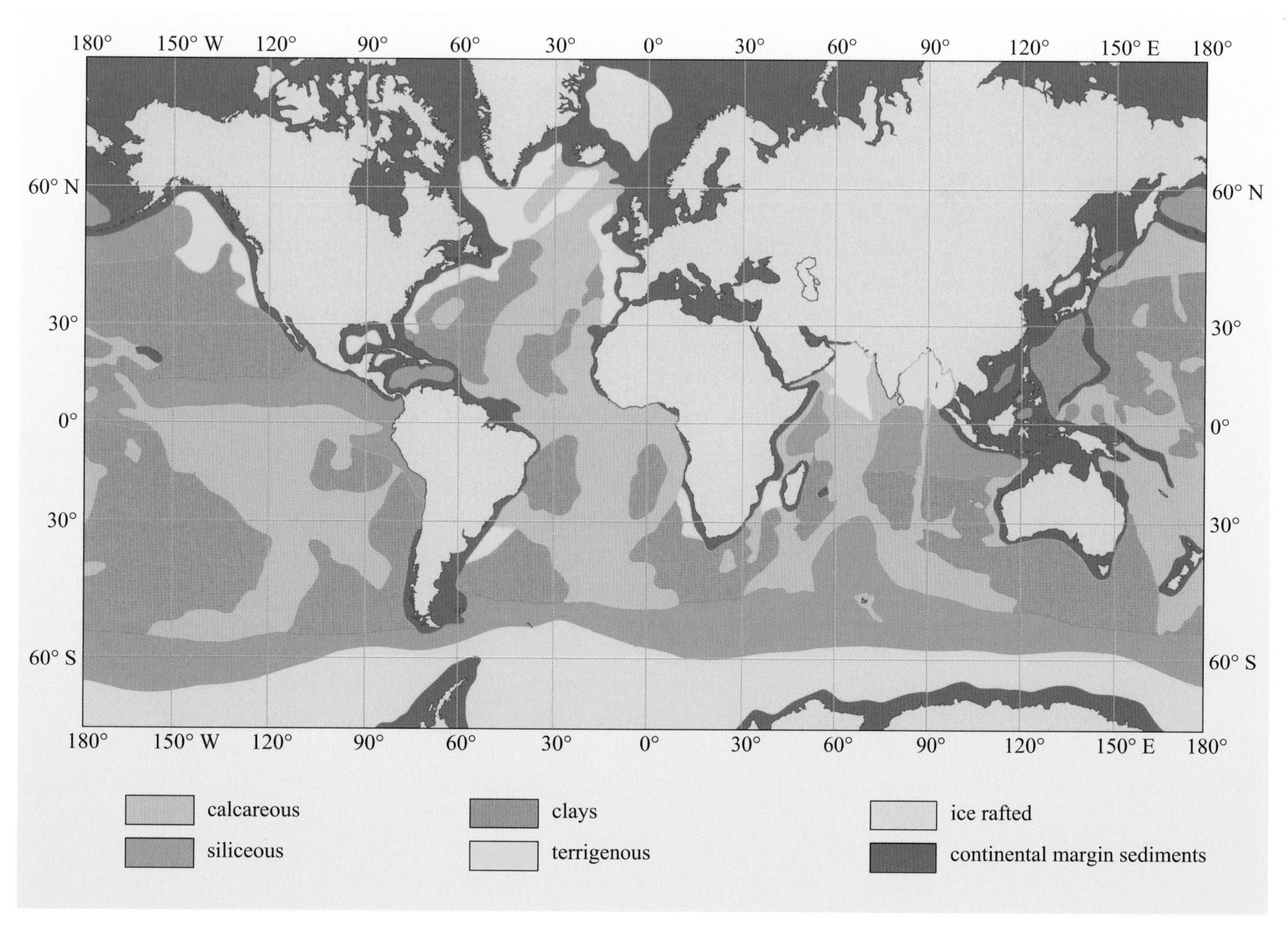

Figure 5.26 Map of the global distribution of sediment types currently being deposited on the ocean floor.

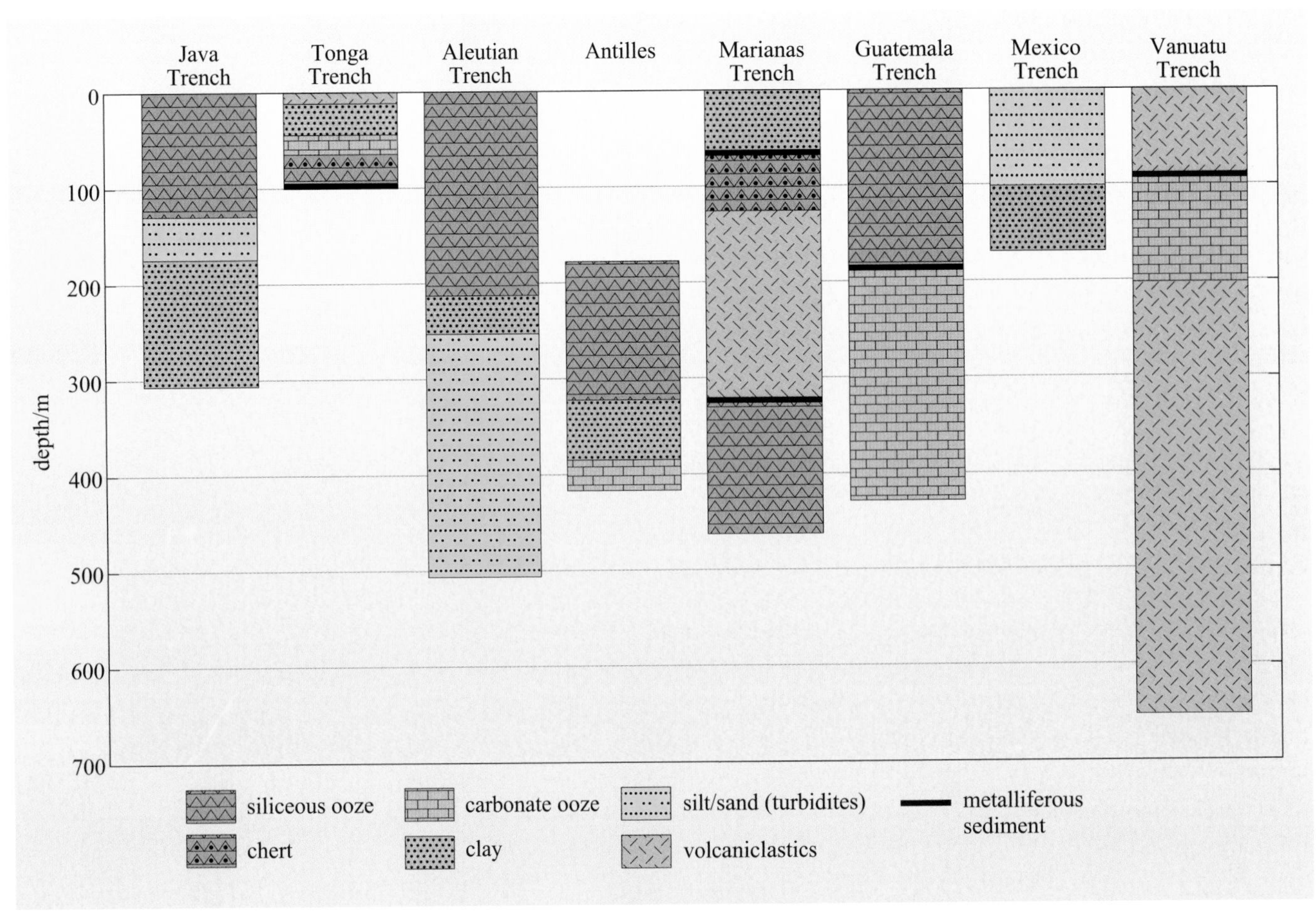

Figure 5.27 Average sedimentary columns being subducted at eight trenches. (Plank and Langmuir, 1993)

Two US geochemists, Terry Plank and Charlie Langmuir, analysed the sediment sequences and calculated an input for each element by taking the average chemical composition and multiplying it by the convergence rate and the mass of sediment. This yields an input for any given element in terms of grams per year per centimetre of arc length, which can then be compared with the composition of the lavas erupting in the adjacent arc.

Estimating the amount of a given element produced by arc magmatism for each subduction zone is rather tricky, not least because there are many submarine volcanoes that are not accessible and only the magma that is erupted can be easily sampled. So, rather than try to calculate the total output of an element it is easier to use a **proxy**. A proxy, in this case, is a geochemical indicator that varies in direct response to variations in the input. Having already established that some elements, such as Sr, Ba, Th, and U, are enriched in arc magmas relative to MORBs, the challenge is to quantify this enrichment and see whether the amount of enrichment in a given arc corresponds to the input of that element into the trench. However, a simple measure of the element concentration in a lava from a subduction zone is not directly related to its input.

■ What general processes affect the concentration of elements such as Ba in lavas?

■ Fractional crystallisation and partial melting.

Fractional crystallisation tends to increase the concentration of incompatible elements in magmatic rocks, while increasing degrees of partial melting decrease their concentration. Ideally, the effects of fractional crystallisation should be minimised by restricting attention to primary basaltic magmas but, as discussed earlier, these are generally rare in arc settings. So, to compare arcs on an equal basis, the effects of both fractional crystallisation and partial melting need to be evened out. Plank and Langmuir achieved this by taking data from a number of volcanoes from each arc, and for each volcano they plotted the variations of selected trace elements against MgO as an index of fractionation. The data define a linear or curved trend and the value of Ba at 6% MgO was used as a proxy for the composition of the primary melt. So, for Ba the corrected value becomes $Ba_{6.0}$. Plank and Langmuir also found that the Na_2O content at 6 wt% MgO provides a useful indicator of the degree of partial melting because Na is an incompatible element. High $Na_{6.0}$ contents indicate low degrees of partial melting, whereas low $Na_{6.0}$ contents indicate high degrees of partial melting. Therefore, dividing the trace-element value (e.g. $Ba_{6.0}$) by the sodium value at 6 wt% MgO (i.e. $Ba_{6.0}/Na_{6.0}$) gives a ratio that is a measure of the enrichment of a given element in the mantle source of the subduction zone melt. Figure 5.28 is a plot of input flux of Ba and Th versus their respected output flux ($Ba_{6.0}/Na_{6.0}$ and $Th_{6.0}/Na_{6.0}$) for seven different arcs. As you can see, there is a good correlation between the input sediment flux and the geochemical proxy for the output flux.

■ What does Figure 5.28 tell you about the relationship between the sediment subducted and the composition of arc lavas?

■ There is a direct relationship between the composition of the sediment subducted at a trench and the composition of the arc lavas erupted in the arc adjacent to the same trench. Therefore at least some of the sediment subducted must be recycled back into the source region of island-arc magmas.

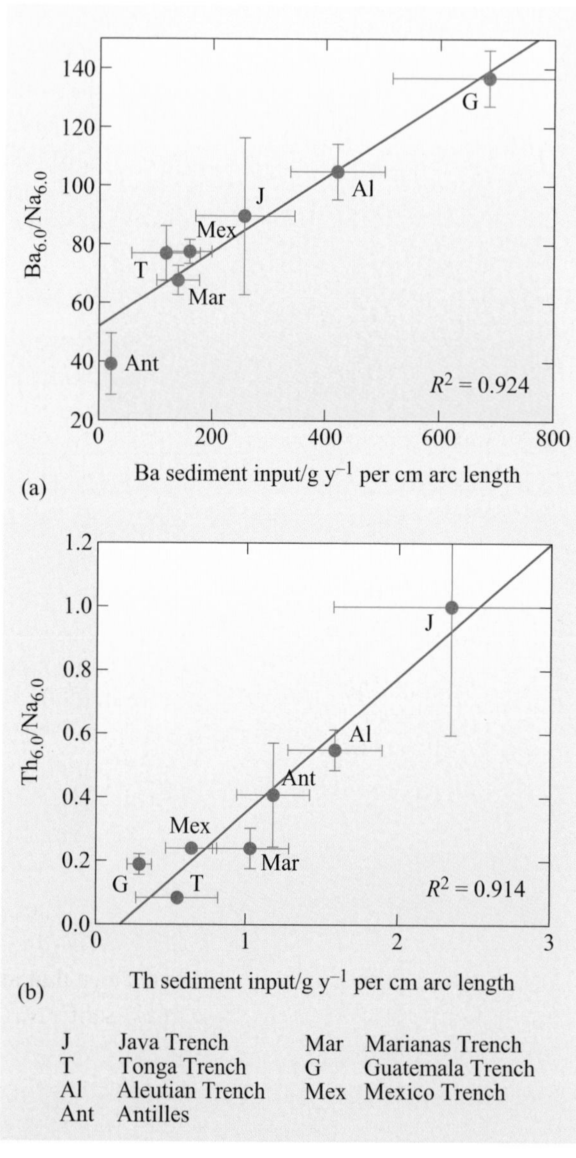

Figure 5.28 Correlations between trace-element sediment inputs (in grams per year per centimetre arc length) and trace-element enrichment in arc basalts for: (a) Ba, and (b) Th. R^2 is the correlation coefficient between the two variables where a value > 0.9 indicates a significant correlation. (Plank and Langmuir, 1993)

Other elements, such as Sr, Rb, K and U, also reveal positive correlations between sediment input flux and (element)/$Na_{6.0}$ ratios. This partly answers a question raised earlier, which was whether these fluid mobile elements come from sediment or altered oceanic crust. The correlations suggest that they mainly come from sediment although they do not rule out an input from dehydration of the basaltic slab.

Central American arc lavas have some of the most unambiguous evidence for sediment subduction and incorporation into the source region of arc magmas. They have high ^{10}Be signals and, because the sediment being subducted into the Central American (i.e. Guatemala) trench is very distinct in composition, the trace-element composition of the sediments is recorded in the lava composition. For example, the sediment has a high Ba/Th ratio (Figure 5.28), and this high Ba/Th signature is found in the erupted lavas. Lavas from Nicaragua provide one last illustration of the link between sediment composition and volcanism in subduction zones. The uplift of the Panama peninsula cut off the flow of tropical waters from the Atlantic to Pacific Oceans 10 Ma ago. This dramatically changed the chemistry of the seawater in the Panama Basin, while the sediments being deposited in the basin changed from carbonate to siliceous ooze. More importantly, there was a factor of four increase in the U content of the sediment 10 Ma ago while the Th and Ba contents of the sediments do not change by more than 10%. Remarkably, this change in U/Th ratio is recorded in the composition of the lavas erupted in the Nicaraguan arc, in which lavas erupted after 7.5 Ma have elevated U/Th ratios compared with those erupted previously.

Question 5.4

What do these elevated U/Th ratios suggest about the Central American subduction zone? Is it consistent with other information about the Central American arc?

Understanding the details of element recycling through subduction zones is complicated, but by combining trace element and isotopic data there is strong evidence that sediments provide a significant flux of highly incompatible elements to most arc lavas. The oceanic crust may also contribute to the U and Sr budget (and H_2O is recycled from all parts of the subducted slab). Calculating absolute fluxes of elements is much more difficult, but, in the example from Central America given above, at least 75% of the sediment at the trench has to be subducted to balance the Th budget of the Nicaraguan lavas. Similarly, it has been calculated that all of the Be that is subducted is returned back to the crust by magmatism, illustrating that some element recycling is highly efficient within the upper 200 km of the mantle.

Summary of Chapter 5

Subduction zones are the major sites of recycling of surface material back into the mantle. The most dramatic effects of subduction are two of the Earth's major natural phenomena, namely earthquakes (particularly deep earthquakes) and explosive volcanism. H_2O is the key component in subduction zones because it influences chemical reactions in the subducting slab, the transport of elements from the slab, the viscosity of the overlying mantle wedge and the chemical evolution of arc magmas. This chapter has considered the recycling of material through the upper 200 km of the mantle; but, clearly, subducting slabs reach and penetrate the 670 km discontinuity, and many scientists think the slabs sink to the core–mantle boundary. Although some elements (Be, Th) are efficiently transported from the slab back into the mantle wedge, the extraction process, be it melting or dehydration, is never 100% efficient. Consequently, many elements are returned to the deep mantle, including some subducted H_2O. Chapter 7 explores what happens to the subducted slabs over geological time and how this influences the large-scale chemical and mechanical evolution of the mantle.

Learning outcomes for Chapter 5

You should now be able to demonstrate a knowledge and understanding of:

5.1 The physical, internal and thermal structure of subduction zones.

5.2 How changes in the thermal structure of subduction zones influence location and physical and geochemical processes involved in the generation of melts.

5.3 How fractional crystallisation processes can account for the range in major element compositions, and the mineralogical and geochemical evolution of arc magmas.

5.4 How distribution (partition) coefficients of trace elements and mineral assemblages can be used to investigate the geochemical evolution of arc magmas.

5.5 The recycling of sediments and atmospheric isotopes through subduction zones and how they can provide an insight into the timescale of subduction as well as where and how fluids and melts form within this tectonic setting.

Processes during continental collision

Chapter 6

We live on an unusual planet. Underlying the ocean basins, which cover two-thirds of the Earth's surface area, is a dense crust, about 7 km thick, forming the outer layer of the lithosphere. But what makes the Earth's structure extraordinary is the remaining third of the Earth's surface area, where the outer layer is a silica-rich, low-density crust, with an average thickness of about 40 km, which supports the continents on which we live. Such continental crust is almost certainly unique within our Solar System because, as you know from previous chapters, it requires both the subduction of tectonic plates and the presence of water locked up within mineral structures of subducted surface rocks for continental crust to form.

The rocks that make up the continental crust have, on average, a density of 2800 kg m^{-3}, which compares with a density of 3000 kg m^{-3} for oceanic crust. This is because much of the continental crust is made up of granitic rocks together with sediments such as sandstones, mudstones and conglomerates, all of which have a low density, largely because of the abundance of quartz and feldspar. Indeed, the average composition of the continental crust is similar to that of an intermediate igneous rock, such as andesite, which is significantly less dense than that of basalt or gabbro. As a consequence, continental lithosphere is too buoyant to descend far at a subduction zone. Hence, continents cannot be destroyed, and since their formation is irreversible they must grow through geological time. This is why the oldest rocks known from the oceanic floor are a mere 180 Ma old, whereas rocks almost 4000 Ma old have been recovered from regions of stable continental crust in Canada and Greenland.

The constant movement of the Earth's tectonic plates causes the expanding continents to be shunted around like froth on the surface of a pond. As they grow through time, it is inevitable that sooner or later there will be collisions between two continental masses. Descent of the ocean crust into subduction zones at the edge of an ocean basin (Figure 6.1a) draws the continents either side of the ocean closer together (Figure 6.1b), ultimately closing the ocean basin completely and leading to continental collision (Figure 6.1c and d).

The consequences of continental collision are far reaching. Neither continent can be subducted back into the mantle due to the buoyancy of continental crust, so the forces that drive the plate movement prior to collision are brought to bear directly on the continental lithosphere itself. At this stage, further convergence of the plates must be taken up by deforming one or both of the plates of continental lithosphere. Under such strong compressive forces, the lithosphere contracts through folding and faulting of the rocks that make up the continental crust. In the upper part of the crust, the sedimentary sequences along the original continental margins are thickened, as sheets of rock are thrust under one another along low-angle faults (Figure 6.1d). The lower crust, being hotter, will deform in a more ductile fashion. Here, the geothermal gradient will steepen due to the effects of thick sequences of continental rocks with high concentrations of the heat-producing elements. Lower crustal rocks will also be subject to high pressures. These changes in pressure and temperature will result in mineral recrystallisation (metamorphism) beneath collision zones. If the temperature rises above the solidi for common crustal rocks, then the crust will start to melt, resulting in a drop in mechanical strength of the lower crust and allowing further ductile deformation and thickening.

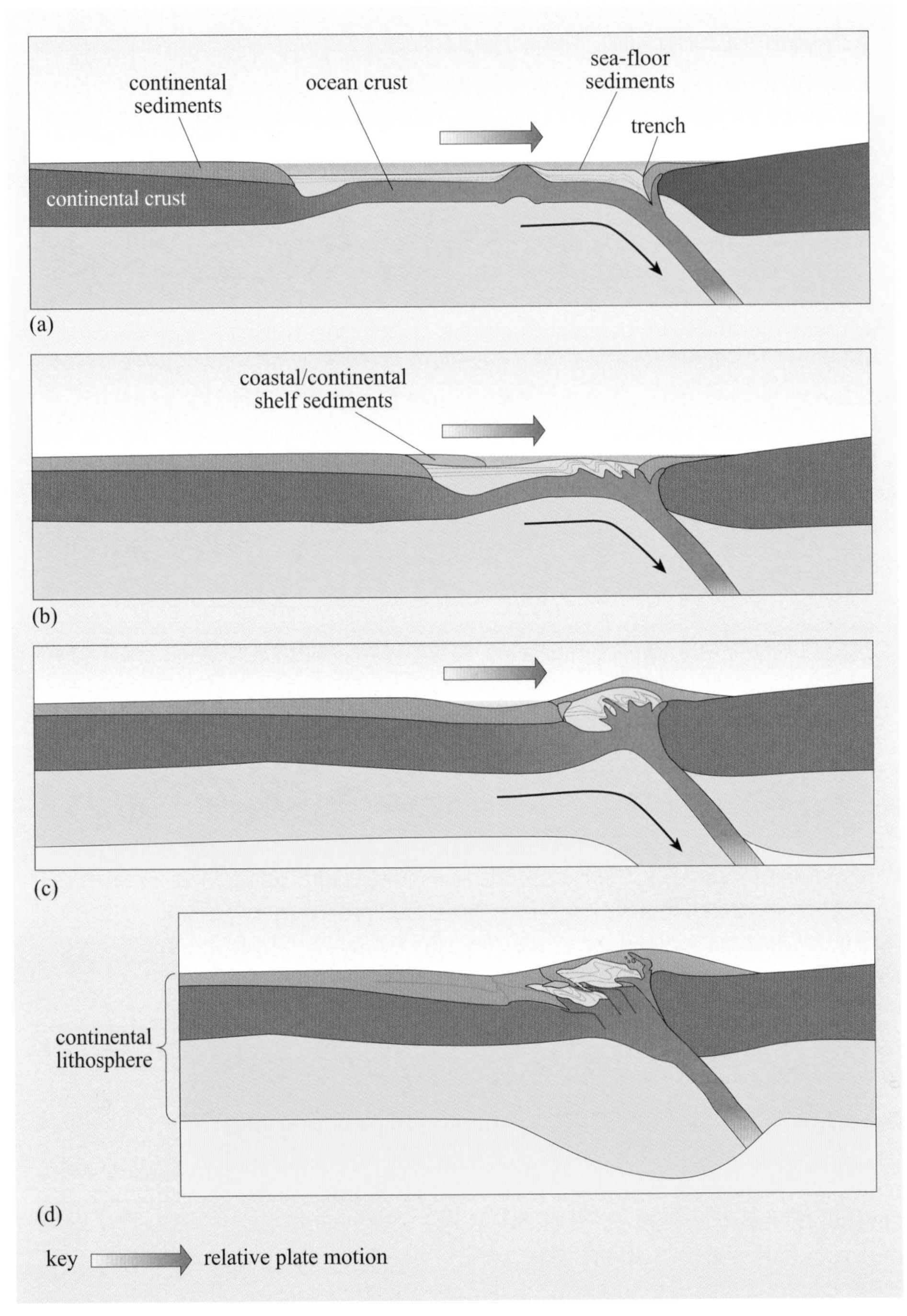

Figure 6.1 Schematic cross-sections showing the formation of a mountain range: (a) subduction at a destructive boundary (compare with Figure 4.3) causes (b) contraction of the ocean basin, which (c) leads to collision, and (d) thickening of the continental lithosphere.

- Bearing in mind what you read in Chapter 1 about the tendency for the lithosphere to 'float' in isostatic equilibrium, suggest what the consequences of thickening the crust might be.

- A thicker crust means a thicker lithosphere with a lower average density. Buoyancy forces in the mantle will cause such a lithosphere to rise, leading to land surface at high elevations.

The strong correlation between high surface elevation and a large crustal thickness (e.g. as seen in the case of the Andes or the Himalaya) is a clear indication of the effect of increased buoyancy forces on thickened crust.

This chapter will examine the tectonic processes that result from continental collision. In particular, it will look at the processes of metamorphism and partial melting that occur deep within the continental crust during continental collision as exemplified by the Himalayan orogeny.

6.1 Heating of the continental crust

All heat generated within the Earth results from radioactive decay. The principal isotopic systems responsible for internal heating stem from the decay of isotopes of three elements, K, U and Th, as was detailed in Chapter 2. All three of these elements are generally incompatible with respect to silicate mineral assemblages and so are concentrated in continental crust, which is initially made up of a high proportion of granitic rocks. Consequently, thick continental rocks sustain steep geothermal gradients, leading to metamorphism.

Metamorphism is a term used to describe the changes that affect existing rocks when they are subject to a change in pressure and/or temperature. It usually refers to chemical and physical reactions that take place in the *solid state*, although many metamorphic reactions are greatly assisted by the presence of any fluid that may be present along the grain boundaries or released by the reaction itself. Metamorphism covers a great range of processes, from subtle changes in the structure of clay minerals that occur in sediments after their burial, to the mineral reactions that occur deep in the crust at high pressures and temperatures. The field of metamorphism extends to the partial melting of rocks, so that, as you will see, there is overlap between the physical conditions under which igneous and metamorphic processes occur.

The simplest form of metamorphic reaction involves the change of a single mineral from one structure to another. If two or more minerals have the same composition, but contrasting structures, they are known as polymorphs. Carbon has two polymorphs, diamond and graphite, and their stability fields are plotted in Figure 6.2. Both polymorphs may coexist along the phase boundary that separates their stability fields.

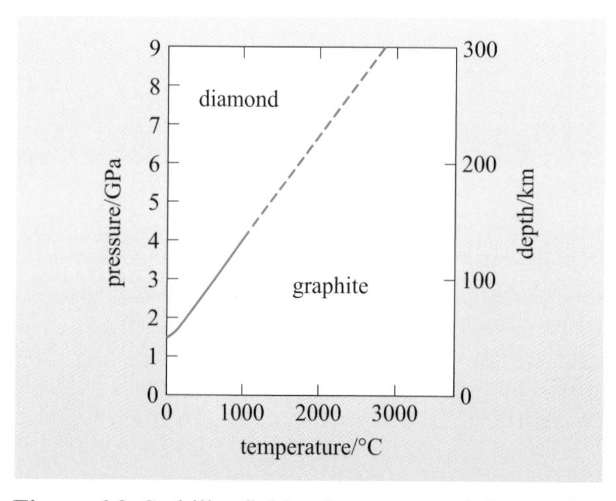

Figure 6.2 Stability fields of graphite and diamond in pressure–temperature space.

Increasing the pressure of a rock will lead to reactions by which minerals of low density will be replaced by minerals of comparatively high density. Diamond is denser than graphite

(a)

(b)

(c)

Figure 6.3 (a) Blue blades of kyanite growing within a quartz lens in mica schist (Glen Esk, Scotland). (b) Prismatic white andalusite growing within slate in the aureole of the Skiddaw Granite (Cumbria, England). (c) Lenses of fibrous white sillimanite within biotite gneiss (Langtang Valley, Nepal). (Nigel Harris/OU)

(3500 kg m^{-3} compared with 2000 kg m^{-3}), and so high pressures favour the graphite → diamond reaction. In other words, the transition from graphite to diamond results in a decrease in volume.

The graphite → diamond reaction is an example of a **geobarometer**, because the occurrence of either phase provides information on the pressure at which a rock has formed. For example, if graphite-bearing rocks are formed at a temperature of 1000 °C, then the maximum pressure the rock formed at was about 4 GPa, equivalent to a depth of about 120 km (Figure 6.2). It is not a very useful barometer though, because graphite is stable for all crustal pressures. Much more useful for petrologists who study crustal rocks are minerals that undergo metamorphic reactions under conditions commonly found within the crust. A particularly useful example is provided by three minerals that occur as minor components in many rocks that have aluminium-rich compositions: the **aluminosilicates**.

All three minerals are polymorphs with the same chemical formula, Al_2SiO_5, but their internal structures and external appearances are markedly different. The three polymorphs are called **kyanite**, **andalusite** and **sillimanite** (Figure 6.3). Each aluminosilicate is typical of a different range of pressure–temperature conditions, and their stability fields are shown in Figure 6.4. Two polymorphs can coexist along the phase boundaries that limit the stability field of each polymorph, and all three polymorphs can coexist at the point where the three phase boundaries converge. Such diagrams are the results of experiments performed in a laboratory, and the precise position of each boundary is subject to experimental uncertainty. This is particularly true for polymorphic reactions that have sluggish reaction rates.

■ From Figure 6.4, estimate the relative densities of the three aluminosilicate minerals.

■ For each phase boundary, the polymorph on the high-pressure side has the highest density. Thus, the order of decreasing density is kyanite, sillimanite, andalusite. (In fact, the densities are kyanite: 3600 kg m^{-3}, sillimanite: 3250 kg m^{-3} and andalusite: 3150 kg m^{-3}.)

The stability fields of the aluminosilicates (Figure 6.4) can be used to provide information about the physical conditions under which metamorphic rocks formed. For example, it is known that the sillimanite-rich rock shown in Figure 6.3c could not have formed at temperatures less than about 500 °C (the minimum temperature for the sillimanite stability field). It is also known that the andalusite in Figure 6.3b could not

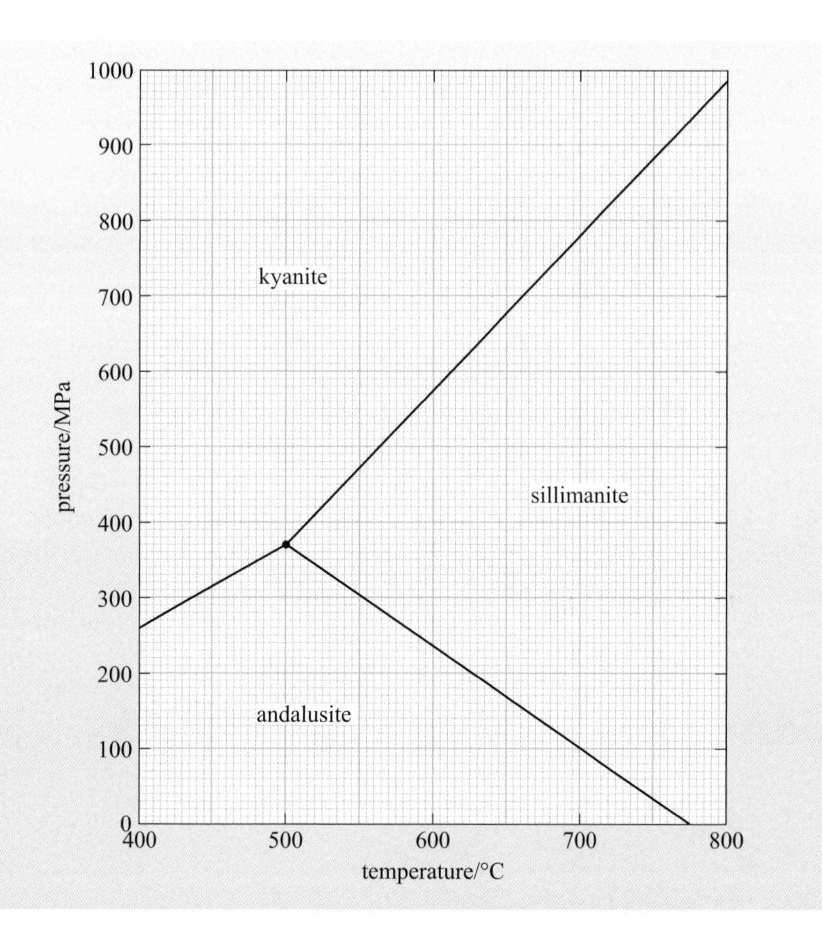

Figure 6.4 The stability fields of the three aluminosilicate minerals in pressure–temperature space. Note the pressure units are MPa. 1 GPa = 1000 MPa.

have formed at pressures greater than about 380 MPa (Figure 6.4). Since these andalusites are from baked sediments at the contact of a large granite in Cumbria, the granite must have been intruded at pressures of less than 380 MPa, equivalent to depths less than about 12 km. Thus, the aluminosilicates allow the pressure–temperature space of metamorphic rocks to be divided up into high-pressure (kyanite), low-pressure (andalusite) and high-temperature (sillimanite) regimes. Other minerals can be used in similar ways, and the conditions under which metamorphic rocks formed can sometimes be pinpointed quite accurately.

So far, this chapter has concentrated on the consequences of increasing pressure on the stability of single minerals and their polymorphs. In most natural rocks, several minerals take part in a metamorphic reaction in response to changing pressure *and* temperature. For example, if a limestone containing mostly calcite ($CaCO_3$) with minor quartz (SiO_2) is heated, a metamorphic reaction occurs:

$$\underset{\text{calcite}}{CaCO_3} + \underset{\text{quartz}}{SiO_2} = \underset{\text{wollastonite}}{CaSiO_3} + \underset{\text{carbon dioxide}}{CO_2} \qquad (6.1)$$

In the metamorphosed limestone (known as a marble), small needles of the mineral wollastonite ($CaSiO_3$) are found along the grain boundaries of calcite and quartz (Figure 6.5).

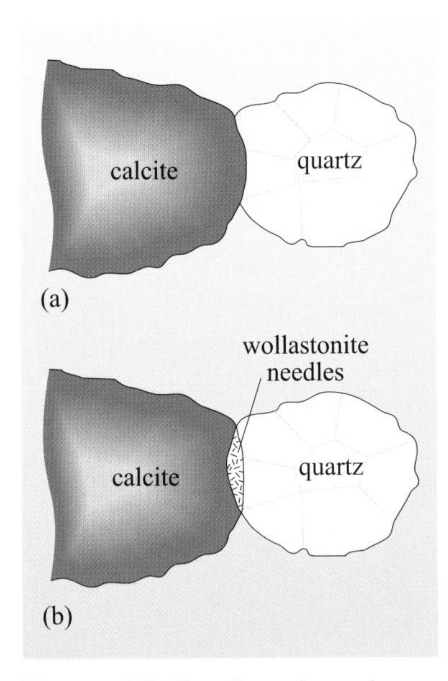

Figure 6.5 Sketch to show the formation of wollastonite ($CaSiO_3$) needles during metamorphism of limestone: (a) shows the minerals present before and (b) shows the mineral present after metamorphism.

The stability fields for calcite + quartz and for wollastonite + carbon dioxide (CO_2) are shown in Figure 6.6. Note that the wollastonite + CO_2 assemblage is favoured by high temperatures.

Just as high pressures stabilise minerals with high densities, high temperatures stabilise minerals and gases with a high degree of disorder in their molecular structure. For polymorphs such as the aluminosilicates, the mineral with the more disordered state will always lie on the high-temperature side of the phase boundary. Thus, sillimanite has the most disordered molecular structure of the three Al_2SiO_5 polymorphs and kyanite has the lowest (Figure 6.4). Since the structures of gases are considerably more disordered than those of solids, any mineral reaction that releases a gas, such as given by Equation 6.1, will be triggered by an increase in temperature. In other words, the phase boundary for the reaction will be steep (at least at higher pressures), as you can see from Figure 6.6. The steep phase boundaries that describe reactions that produce CO_2 (**decarbonation reactions**) or H_2O (dehydration reactions) provide excellent **geothermometers**.

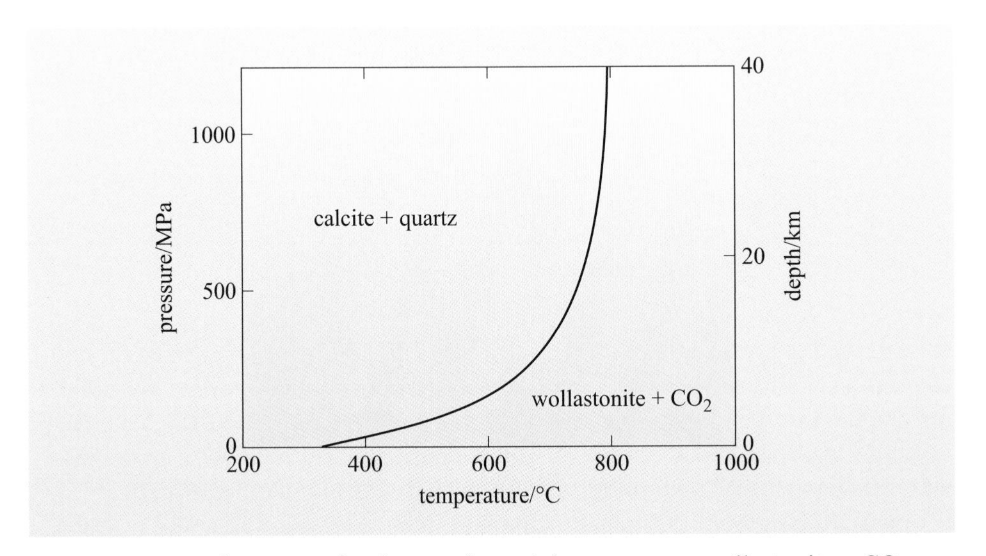

Figure 6.6 Boundary curve for the reaction calcite + quartz = wollastonite + CO_2.

■ A limestone contains wollastonite, calcite and quartz. The adjacent rock is an aluminous sediment that contains kyanite. By considering both Figures 6.4 and 6.6, estimate the minimum pressure at which these rocks have formed.

■ The minerals wollastonite, calcite and quartz must lie on the boundary curve of the decarbonation reaction. This crosses the boundary between kyanite and sillimanite at a pressure of about 1000 MPa (Figure 6.4), which represents a minimum pressure for the assemblage.

Decarbonation reactions are particularly important for the Earth system as they ultimately release CO_2 into the atmosphere. Conditions for such reactions are primarily found within collision zones and, since CO_2 is a greenhouse gas, this is one example of continental collision contributing to climate change.

Different pressures and temperatures can change the minerals that are stable in a rock. It is also true that *different rock compositions will allow different minerals to form under the same pressure and temperature conditions*. Rocks that contain different aluminosilicates, for example, must have been rich in aluminium to form any aluminosilicates at all. Some compositions, like sandstone, made up of just quartz, are quite sterile in terms of mineral reactions because quartz is stable for all normal conditions encountered in the crust. In contrast, aluminium-rich sediments, known as mudstones, provide the raw materials for many mineral reactions. Such aluminous compositions are termed pelitic and their metamorphosed equivalents are referred to as **metapelites**. Mudstones are made up of clay minerals, which are prone to multiple changes during metamorphism on account of their chemical complexity. The following text explores the metamorphism of a mudstone under increasing pressure and temperature.

First, you need to consider the textural changes that are summarised in Table 6.1. The fine-grained clay minerals that make up a mudstone are deposited flat on the floor of a stream or a lake. On compaction, the flakes are rotated to align at right angles to the compacting load, giving the rock a layered appearance. When this compaction is accompanied by heating, the clay minerals begin to recrystallise, and the process is promoted by hot, water-rich fluids being forced through the sediments caused by the compacting load of overlying sediments. In this way, new micaceous (mica-rich) minerals form. The micas will be either biotite, if iron rich, or muscovite, if aluminium rich. All micas are sheet-like or platy in shape and grow perpendicular to the direction of compression acting on the rock in which they grow. This stress is usually vertical due to the compacting load, but it may vary due to non-vertical stress induced by plate movements.

Slates are the result of a small increase in pressure and temperature acting on a mudstone. The minerals are still very fine grained but they are largely micaceous: the preferred orientation of the micas within the rock gives it a well-defined cleavage (exploited in the manufacture of roofing slates, for example). Further metamorphism of mudstones results in an increased grain size. Coarsely crystalline rocks with an aligned micaceous texture are known as **schists**. But micas are not the only minerals that form during metamorphism.

Table 6.1 Classification of metamorphosed sediments (metapelites).

Rock type	Appearance	Grain size
slate	very closely spaced, almost perfectly flat planes	fine
schist	moderately spaced, sub-parallel planes, characteristically with abundant mica	medium
gneiss	widely spaced layers with alternations between mica, amphibole or pyroxene-rich layers and quartz, and plagioclase-rich layers	coarse
migmatite	separation of granitic lenses within darker layers (usually of biotite) or amphibole	coarse

One of the aluminosilicate minerals can also result from a reaction involving micas. For example, a common dehydration reaction in schists is:

$$\underset{\text{muscovite}}{KAl_3Si_3O_{10}(OH)_2} + \underset{\text{quartz}}{SiO_2} = \underset{\text{alkali feldspar}}{KAlSi_3O_8} + \underset{\text{alumino-silicate}}{Al_2SiO_5} + \underset{\text{water}}{H_2O} \quad (6.2)$$

Figure 6.7 Biotite gneiss from deformation of granite (from the Himalaya).

Higher temperatures and pressures will result in a segregation between quartz + feldspar and ferromagnesian minerals, such as biotite or amphibole. Since the former are light in colour and the latter are dark, this gives the rock a banded appearance. Such rock is known as a **gneiss** (which is pronounced 'nice'). Not only mudstones, but also other rock types such as granites and gabbros can form gneisses (Figure 6.7) when metamorphosed to sufficiently high temperatures and pressures. The dark layers may include micas, amphiboles or pyroxenes, depending on the degree of metamorphism and on the bulk composition of the rock.

It is under conditions close to the breakdown of muscovite that a gneiss may begin to melt. This is illustrated in Figure 6.8, where both the granite melting curve and the reaction boundary defined by Equation 6.2 are plotted. The granite melting curve shown is for melting in the presence of quartz, plagioclase, alkali feldspar and water. Note that when this reaction boundary crosses the 'wet' melting curve, muscovite and quartz combine with plagioclase (albite) to form a melt, facilitated by the water released from dehydrating the mica.

■ If you follow the effects of increasing temperature on a typical metapelite at pressures of 500 MPa, what will happen when point X (Figure 6.8) is reached at about 650 °C?

■ Point X lies on the water-present granite solidus, but no reaction will occur in the absence of H_2O. In order to form a granite magma at this point, the three minerals quartz, plagioclase, and alkali feldspar together with water must be present in the rock. Metamorphic rocks at temperatures of several hundred degrees Celsius rarely contain a free water phase and the melting of granite in the absence of water occurs at much higher temperatures. Also absent from many metamorphosed sediments is the mineral alkali feldspar.

■ What will happen when point Y on Figure 6.8 is reached?

■ At point Y, the muscovite dehydration reaction is crossed. This will generate some melt, an aluminosilicate and alkali feldspar. Since quartz and plagioclase are common minerals in sediments, all four phases (plagioclase, alkali feldspar, quartz and H_2O) required for forming a granite minimum-melt composition are now present.

■ Equation 6.2 also has Al_2SiO_5 as a product. Which aluminosilicate will be stable?

■ From Figure 6.8, Y lies within the sillimanite field, and so sillimanite will form.

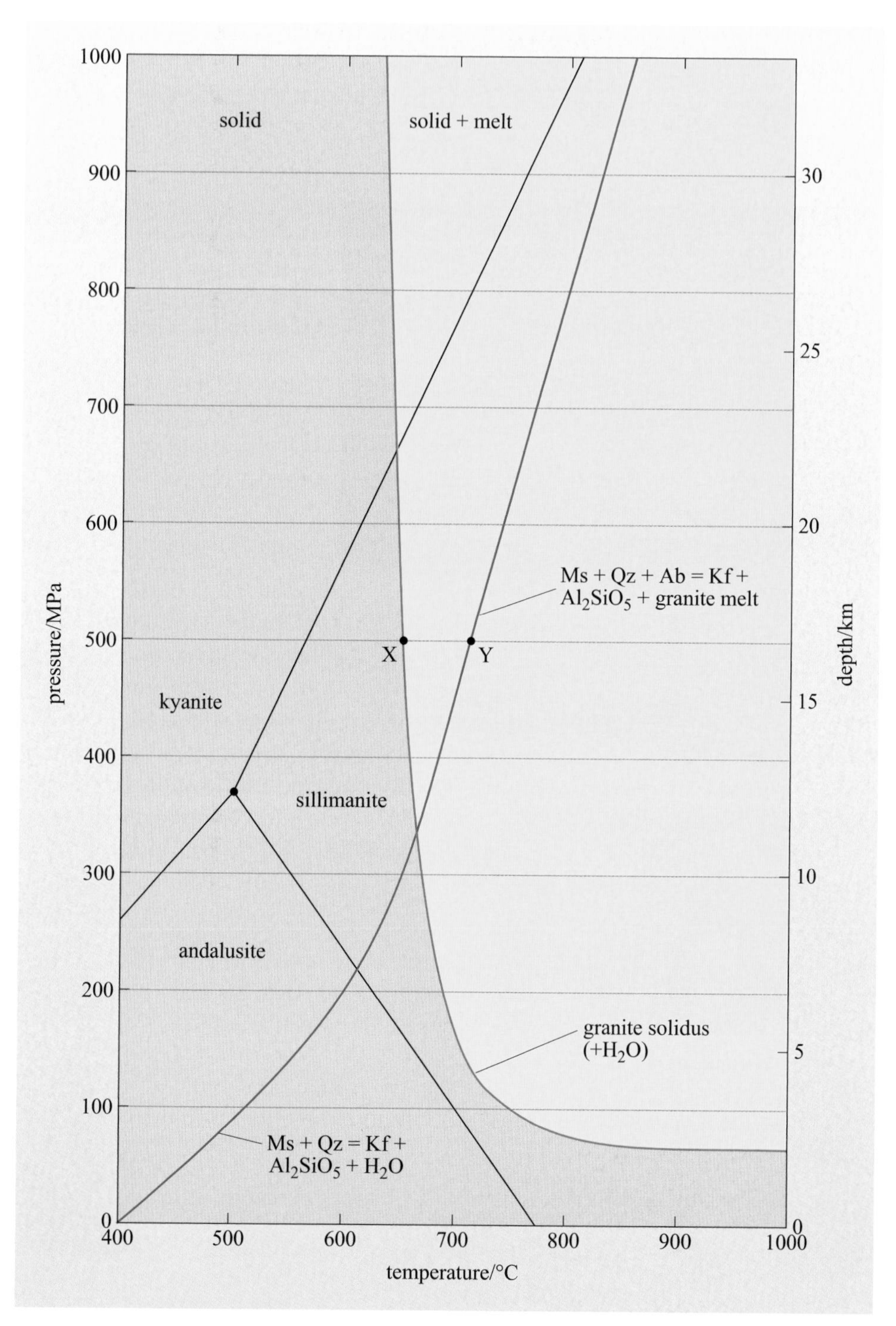

Figure 6.8 Solidus for melting of granite (in the presence of water) and the muscovite and quartz dehydration reaction (Equation 6.2). Also shown are the stability fields for the aluminosilicates. Ms = muscovite; Qz = quartz; Kf = alkali feldspar; Ab = albite.

If only a small proportion of melt is generated, then the granite magma will remain in the rock but be segregated from the micaceous layers. When these melts crystallise, they will form lenses of quartz, plagioclase and alkali feldspar. Typically, such lenses are a few centimetres thick. These mixed rocks are called **migmatites** (Figure 6.9).

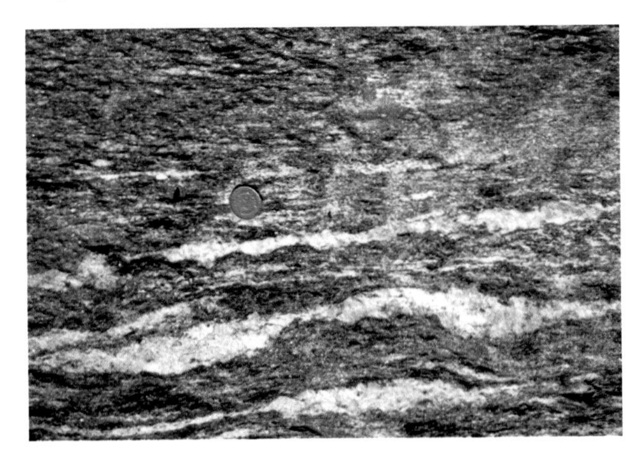

Figure 6.9 An example of migmatite. This is a rock that includes light-coloured lenses of granitic composition within a dark, biotite-rich gneiss (from Langtang Valley, Nepal).

More extensive melting will allow the melts to coalesce and migrate upwards as a granite pluton. With high enough temperatures, the hydrous phases in all crustal compositions will be dehydrated. Initially (above ~700 °C at normal crustal depths), muscovite will melt, leaving an aluminosilicate-rich residue. Biotite will melt at higher temperatures (above ~800 °C) to leave a garnet-rich or pyroxene-rich residue. If igneous rocks of basic and/or intermediate composition are metamorphosed, amphibole will break down to leave a pyroxene-rich residue. The residual dehydrated rocks are known as granulites, and they characterise much of the lower crust, particularly in collision zones, where conditions are too hot for micas or amphiboles to be stable.

Before leaving this introduction to metamorphism, it should be emphasised that metamorphism can occur in any tectonic setting and in both the mantle and the oceanic crust, as well as in the continental crust, in response to changes in temperature or pressure.

■ Where have you already encountered metamorphism and metamorphic reactions?

■ During the subduction of oceanic crust. Changes in mantle mineralogy in response to increasing pressure are also metamorphic reactions.

Andalusite, the low-pressure aluminosilicate, is commonly formed where aluminous sediments are intruded by magma at shallow depths irrespective of the tectonic setting in which the magma forms (e.g. the andalusite shown in Figure 6.3b is from a metamorphosed sediment caused by the intrusion of a granite formed at an active continental margin). Indeed, the continental crust can thicken beneath active continental margins, such as the central Andes, causing metamorphism and deformation very similar to that found in collision zones. But it is in regions of thickened continental lithosphere that crustal rocks are subjected to unusually high temperatures and pressures over a wide area and where evidence for crustal melting is most likely to be found.

6.2 Formation of granites from melting of the continental crust

Ultimately, the origin of all rocks can be traced back to the mantle, although in some cases many cycles of remelting, crystallisation, erosion and deposition may be involved. The great diversity of igneous rocks results from two processes: partial melting and fractional crystallisation. Granite magmas can form from either of these processes.

Before proceeding it is important to clarify the use of the term **granite**. In popular usage it covers a wide range of intrusive rock compositions; indeed, in the building trade, polished gabbro is sometimes referred to as 'black granite'. However, geologists use 'granite' as a more restrictive term for a composition of an intrusive rock that:

- is made up of at least 75% quartz and feldspars
- contains quartz, plagioclase and alkali feldspar in roughly equal proportions (although in some unusual granites alkali feldspar can be significantly more abundant than plagioclase).

For similar rocks, but with a progressively higher plagioclase/alkali feldspar ratio, the terms granodiorite and **tonalite** (or **quartz diorite**) can be used; and if less quartz and more dark minerals, like amphibole and mica, are present, then the term **diorite** is appropriate. We are not concerned here with the precise definition of each rock type, but collectively these three plutonic compositions are intermediate between the compositions of gabbro and granite. Finally, the term 'granitic rock' is a term loosely applied to any light-coloured, plutonic rock containing quartz as an essential component, together with feldspar. This would include not only granites but also most rocks with intermediate compositions.

At destructive boundaries granite represents the end-product of fractional crystallisation of basaltic magma, although rocks of intermediate compositions are much more common than granites in the batholiths of continental arcs. In the Andean batholith, for example, granite constitutes perhaps 5% of the rocks present. In collision zones, granite magma is the first product of partial melting of crustal rocks. The property of granite that allows it to form, either from extreme fractional crystallisation or from the initial stages of partial melting, is that granite in its magmatic state is the lowest temperature silicate melt that can exist. To understand why this is so, the following section explores the melting relationships of the minerals that constitute granite.

6.2.1 The quartz–albite–orthoclase ternary system

Since granites are made up of over 75% of just three minerals, quartz, albite-rich plagioclase and orthoclase, it is possible to make quite accurate comparisons between the compositions of liquids within the quartz–albite–orthoclase (Qz–Ab–Or) system and natural granites. This three-phase system can be investigated by considering the three binary systems that link the three minerals. Mixtures in the binary Qz–Ab and Qz–Or systems exhibit simple eutectic behaviour similar to that shown in Figure 4.22 for fosterite (Fo) and diopside (Di). The relationships between albite and orthoclase, however, are more complex. The system of these two feldspars shows both partial solid solution and a temperature minimum, the characteristics of which are explained below. Albite and orthoclase are the sodium and potassium endmembers respectively of the feldspar system.

Figure 6.10 is a phase diagram of the alkali feldspar system under fixed pressure conditions (100 MPa). For the pressure at which the experiment was undertaken, there is continuous solid solution between albite and orthoclase. The liquidus surface has a minimum value (M′) with a composition of about Or_{30} (70% albite, 30% orthoclase).

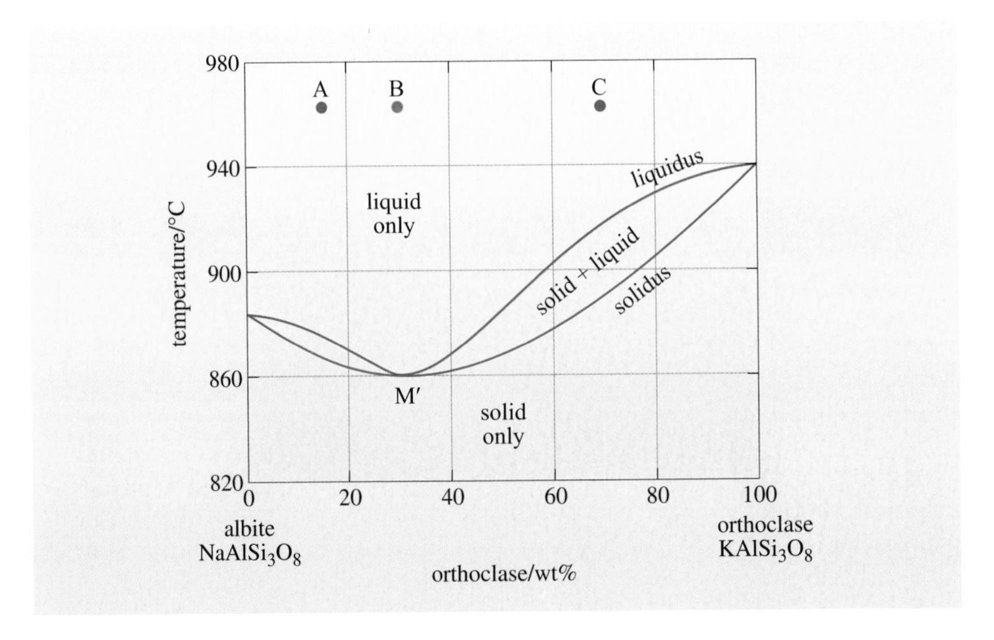

Figure 6.10 The alkali feldspar binary system showing a temperature minimum (M′), determined at a pressure of 100 MPa in the presence of H_2O.

- What is the composition of the first crystals to form on cooling melts of compositions A, B and C in Figure 6.10?

- Melt A has a composition Or_{15}. When this cools to intersect the liquidus, the solid phase that forms is on the solidus at the same temperature and must have the composition Or_7. Melt B has the composition Or_{30}, and when this cools to intersect the liquidus, because the liquidus and solidus coincide, the crystals will have the same composition as the liquid Or_{30}. Melt C has a composition of Or_{70}. The crystals formed when this melt is cooled to the liquidus have a composition of Or_{88}.

- Assuming that bulk equilibrium is maintained between the crystals and the liquid throughout crystallisation, what is the composition of the last crystals to form just prior to complete solidification of samples A, B and C in Figure 6.10?

- If equilibrium between the crystals and the liquid is maintained throughout crystallisation, then the composition of the last crystals to form will be the same as that of the total sample, i.e. Or_{15} for A, Or_{30} for B, and Or_{70} for C.

Contrast your answer with what happens if fractional crystallisation takes place in this system. As more and more crystals are removed, *all* liquids, *regardless of their starting composition*, move towards the minimum point M′. This is arguably the most significant feature of the alkali feldspar system.

The minimum point M′ depicts the composition of the lowest temperature liquid that can exist in the Ab–Or system at this particular pressure. That is a feature it shares with a eutectic point in a binary eutectic system – but can you see what makes a temperature minimum *different* from a eutectic point?

- How many minerals are in equilibrium with the liquid at the eutectic point in a binary eutectic system such as Fo–Di (Figure 4.20)?

- The answer is two. The eutectic point is the only place in a binary eutectic system where *both* minerals (in this case Di and Fo) can coexist with the liquid.

- What is the composition of the mineral(s) in equilibrium with the liquid at the minimum point M′ in Figure 6.10.

- Crystals in equilibrium with a liquid plot on the solidus are at the same temperature as that of the liquid. At the minimum point M′ in Figure 6.10, therefore, they have the *same composition as the liquid.*

Therefore, during equilibrium crystallisation, all crystals at a temperature minimum have the same composition, and thus, unlike the situation at the eutectic point, only *one* mineral can be in equilibrium with the liquid at a binary temperature minimum. All crystals at M′ have composition Or_{30}. So, at a temperature minimum in a binary system, one mineral coexists with the liquid and has the same composition as that of the liquid. In contrast, at a eutectic point in a binary system, two minerals (with distinct compositions) coexist with the liquid.

An additional complication of the alkali feldspar system (not shown in Figure 6.10) is that, for temperatures below about 700 °C, alkali feldspar crystals near the middle of the Ab–Or range may separate into two distinct compositions, i.e. one rich in orthoclase and the other rich in albite. This partial solid solution is due primarily to the disparity in ionic radius between K^+ and Na^+. The radii differ by about 35%, and whereas at high temperatures K^+ and Na^+ will substitute for one another and form a complete solid-solution series, if these crystals cool slowly, then element diffusion will occur between two distinct crystal structures, one of which accommodates Na^+ and the other accommodates K^+. This process is called **exsolution** and is the result of one solid crystal separating into two minerals of distinct compositions. In the alkali feldspars, it results in complex intergrowths between albite-rich and orthoclase-rich crystals (Figure 6.11).

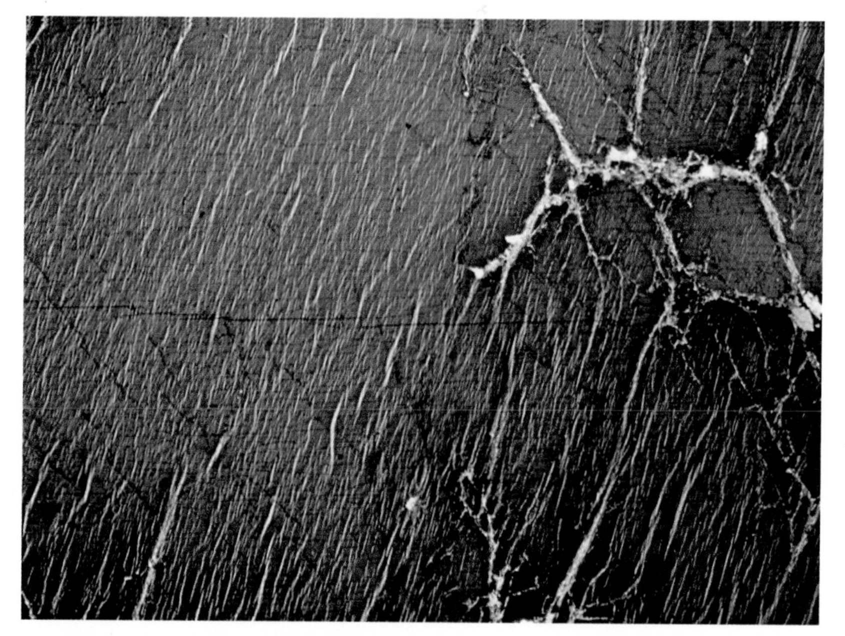

Figure 6.11 Photomicrograph of an alkali feldspar exsolution (magnification ×40). Dark areas are orthoclase, light areas are albite. Each phase is homogeneous. (Mackenzie and Guilford, 1980)

The three binary systems (Qz–Ab, Qz–Or and Or–Ab) can now be put together into a three-component Qz–Ab–Or system. The shape of the liquidus surface in a three-dimensional diagram of composition against temperature is shown in Figure 6.12. However, it is easier to project this surface onto a triangular diagram, as in Figure 6.13, which is the phase diagram (or temperature contour map) of the Qz–Ab–Or system. It consists of two portions, the field in which quartz crystallises first and the field in which alkali feldspar crystallises first.

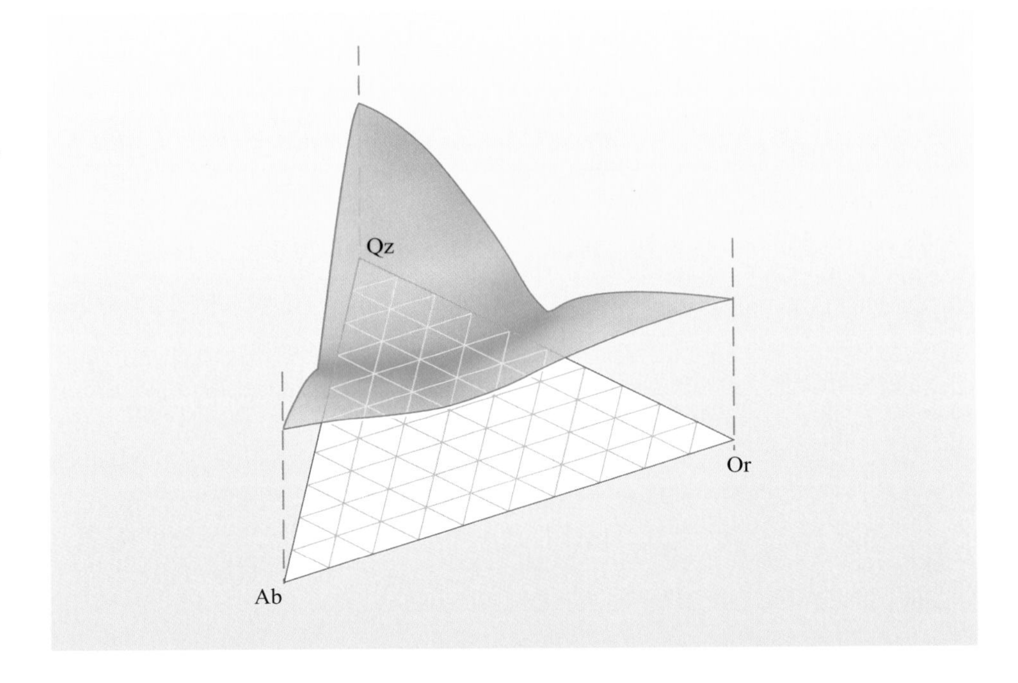

Figure 6.12 Three-dimensional diagram of the liquidus surface of the system Qz–Ab–Or determined at a pressure of 100 MPa in the presence of H_2O.

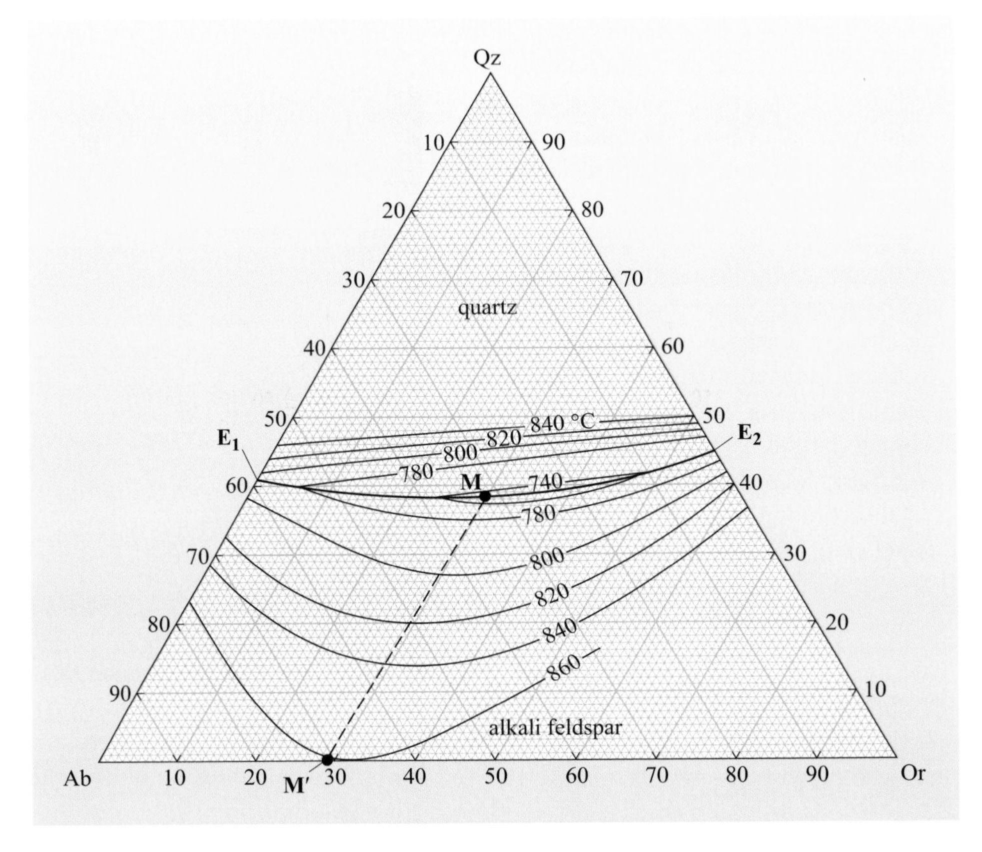

Figure 6.13 Diagram of the Qz–Ab–Or system determined at a pressure of 100 MPa in the presence of H_2O. Temperature contours on the liquidus surface are shown, except for quartz-rich and orthoclase-rich compositions where they are omitted for clarity. M = minimum on curve E_1–E_2; M′ = minimum between albite and orthoclase.

The Qz–Ab and Qz–Or binary systems have *eutectic points* at E_1 and E_2 respectively; these are joined by a cotectic curve E_1–E_2 in the ternary system. This curve slopes 'downhill' towards M from *both* the eutectic points E_1 and E_2; M is the lowest point on the curve and, therefore, the lowest temperature on the entire liquidus surface. This point is known as the **granite minimum**.

The implications of the granite minimum are highly significant. For a melt made up of any combination of quartz, albite or orthoclase, fractional crystallisation will move the composition of the residual liquid firstly towards the cotectic curve E_1–E_2, and then towards the minimum, M. In addition, for a solid made up of any combination of quartz, albite and orthoclase, the composition of the first liquid to form will be that of M.

It follows, therefore, that if naturally occurring granites originated either by fractional crystallisation or by partial melting, their compositions should lie near to that of the granite minimum. To assess this hypothesis, the compositions of a large number of natural granites have been plotted on the phase diagram. The results are shown in Figure 6.14. There is a striking concentration of points within a small area close to the granite minimum, confirming these granites are either the final products of extreme fractional crystallisation or the early magmas formed during partial melting.

There is, however, quite a broad scatter of compositions from the natural rock compositions plotted in Figure 6.14. In addition, there is a displacement between the granite minimum (M) shown in Figure 6.13 and the modal composition of 500 granites in Figure 6.14. This displacement is from $Qz_{38}Ab_{32}Or_{30}$ (M) for the

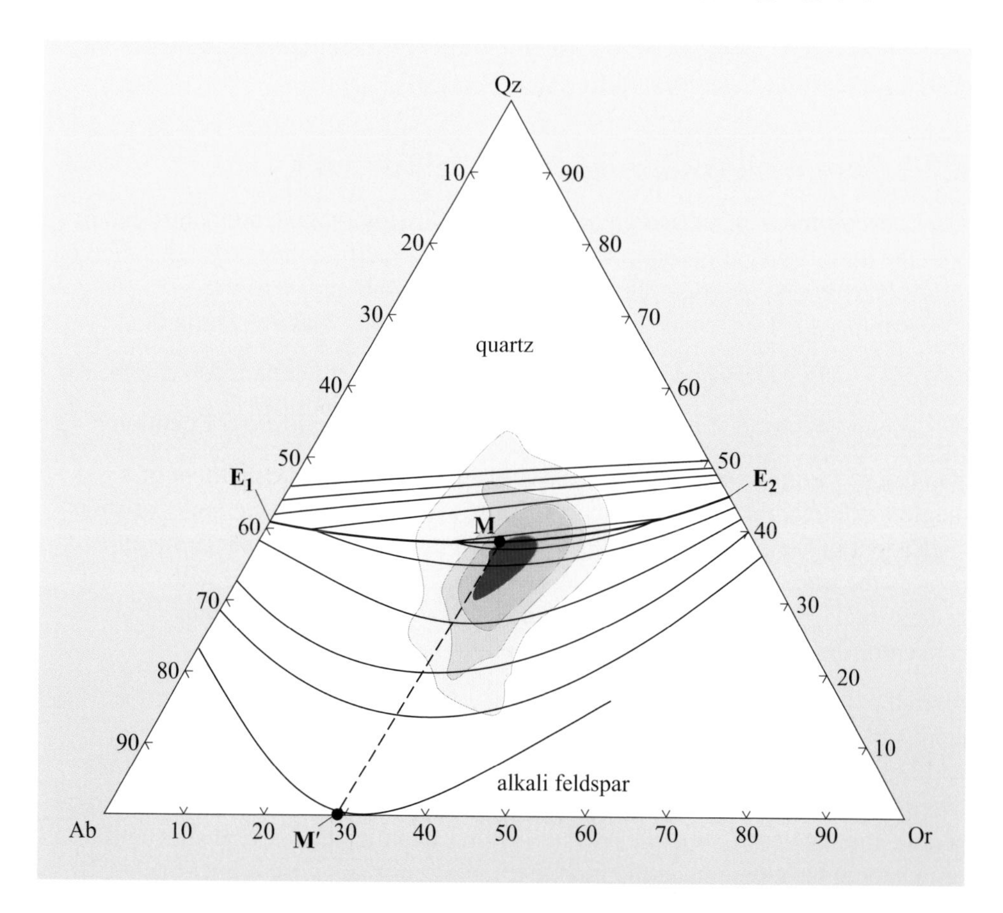

Figure 6.14 Chemical compositions of over 500 granitic rocks in terms of the proportions of quartz, albite and orthoclase. The results have been contoured on the basis of frequency and 90% of the analyses fall in the darkest shaded area in the centre. The curve E_1–E_2 is the cotectic curve for the Qz–Ab–Or system.

experimental plot to $Qz_{36}Ab_{31}Or_{33}$ (within the dark area) for natural granites, towards the orthoclase apex. The reason for both the scatter and the displacement is that natural granites result from a wide range of pressures and H_2O conditions, and reflect small chemical variations in elements not represented by the ternary plot (such as Ca). All these affect the precise location of the granite minimum. The most important variable in this case is probably the amount of H_2O present; the experimental results that are plotted in Figure 6.13 are for a constant H_2O content whereas natural granites are generated over a range of H_2O contents that affect the exact position of M. A second, perhaps equally important, reason for the differences between natural granite compositions and compositions inferred from the topology of the Qz–Ab–Or system is that granites may not always represent the precise compositions of crystallised melts. As granite magma solidifies in the magma chamber, it may include **xenocrysts** of minerals derived from its source region or entrained during ascent or crystallisation. If these are the same types of mineral that are crystallising, e.g. quartz, or feldspar or mica, then it may not be possible to deduce the composition of the pure melt from the composition of the granite.

Whether granites have been derived from fractional crystallisation or by partial melting, their major element composition is much the same. For example, the silica content of melts close to the minimum-melt composition is 70–75%. However, the tectonic implications of the two processes are quite different. In the first case, the granites represent additions to the crust that have formed in a long-lived magma chamber, probably at a destructive boundary. In the second case, the granites are examples of crustal reworking, probably in a collision zone where the crust has thickened and heated sufficiently to melt. In the next section you will look at trace elements as a possible way of distinguishing between granites that result from these different processes.

6.2.2 Trace elements and crustal melting

The concentrations of trace elements in a crystallising magma are controlled by both the initial concentration in the magma and by the fractional crystallisation of minerals from that magma. The relationship between the ratio of the concentration of a trace element in the liquid relative to that present in the source, $\frac{C_L}{C_0}$, the proportion of melt remaining F, and the bulk partition coefficient D is shown graphically in Chapter 5, Figure 5.22 (calculated from Equation 5.7).

You can see from Figure 5.22 that the concentration of a trace element in a magma will only be unaffected by fractional crystallisation if the bulk partition coefficient of the crystallising minerals is $D = 1$. For incompatible elements ($D < 1$) the trace-element concentration in the liquid will increase with decreasing F, and for compatible elements ($D > 1$) the trace-element concentration in the liquid will decrease with decreasing F.

A similar series of curves can be drawn that describe the behaviour of trace elements with varying bulk partition coefficients under conditions of partial melting (Figure 5.21), based on the partial melting Equation 5.6. Note that for these curves, F represents the proportion of melt relative to the original rock. You should note that partial melting will also result in an increase in the concentration of incompatible elements in the melt, particularly for small melt fractions ($F < 0.2$). The melt will be depleted in compatible elements, and for large bulk partition coefficients ($D > 4$) the melt will contain very low concentrations even for quite high melt fractions. In general, it can be concluded that dramatic increases in the concentrations of incompatible elements can be obtained either by extreme fractional crystallisation or by very small melt fractions during partial melting.

The remainder of this section shows how both sets of curves can be used to see whether trace-element concentrations found in andesites and granites from the same igneous complex (Table 6.2) can distinguish whether granite magmas result from fractional crystallisation of an andesitic magma or partial melting of the continental crust.

The first point to note about Table 6.2 is that Rb, with a low bulk partition coefficient, is an incompatible element, whereas Sr is a compatible element. In other words, during melting or crystallisation in the crust, Rb will concentrate in the magma whereas Sr will be more strongly partitioned into minerals, such as feldspar, that coexist with the liquid.

Table 6.2 Rb and Sr concentrations for typical andesite and granite compositions from the Andes, with bulk partition coefficients *D*.

	Concentration/ppm		*D*
	Andesite	**Granite**	
Rb	70	140	0.1
Sr	650	100	3.5

■ Using Figure 5.22 and the data in Table 6.2, determine the proportion of melt left if a liquid of andesitic composition undergoes fractional crystallisation to form a liquid of granitic composition based on (i) Rb and (ii) Sr.

□ C_0 and C_L represent trace-element concentrations in the parental andesite and evolved granite respectively.

(i) For Rb: $\frac{C_L}{C_0} = \frac{140}{70} = 2.$

From the curve for $D = 0.1$ in Figure 5.22, $F = 0.45$ for $\frac{C_L}{C_0} = 2$. In other words, 45% of the magma must be left to form a granitic liquid.

(ii) For Sr: $\frac{C_L}{C_0} = \frac{100}{650} = 0.15.$

It is not easy to obtain a precise value of *F* graphically for $D = 3.5$, but for $\frac{C_L}{C_0} = 0.15$, $F \approx 0.5$ from the curve for $D = 4$, so *F* will be <0.5 for $D = 3.5$

The results could be determined accurately from Equation 5.7, which yields values of $F = 0.46$ and $F = 0.47$ respectively.

Depending on the trace element examined, somewhere between 54% and 53% of the original andesitic magma is required to be removed as cumulates by fractional crystallisation to generate the observed Rb and Sr concentrations in the granite. One reason why the value for *F* differs according to the trace element used is that the assumed value of *D*, the bulk partition coefficient, may be inaccurate, and indeed may vary during fractional crystallisation, partly because the mineralogy of the crystallising phases changes and partly because the mineral/liquid partition coefficients change with the composition of the liquid.

Question 6.1

Using Figure 5.21 and the data in Table 6.2, what proportion of the andesite would be required to melt to generate the granite based on (i) Rb and (ii) Sr trace-element concentrations?

The answer to Question 6.1(ii) suggests that partial melting of the andesite could not generate the granite, and the conclusion is that either the values of *D* are incorrect or that a granite with the trace elements indicated in Table 6.2 could not have formed by partial melting of the associated andesite.

In order to look at the formation of granites by crustal melting more closely, it is necessary to consider what the likely crustal sources might be. The crust is made up of a wide range of compositions, many of which will not melt at all at the temperatures that the continental geotherm can reach in the crust. For example, the temperature at the base of thickened crust is unlikely to exceed about 1000 °C (unless in the immediate proximity of hot mantle melts) and such temperatures will not melt sandstones or indeed a wide range of quartz-rich sediments. You might think that pre-existing granites would be the first lithologies to melt, but this is not the case. This is because, although the three minerals albite, orthoclase and quartz are present, the melting relationships shown in Figures 6.12 and 6.13 have all been determined under H_2O-rich conditions. Dry rocks in the lower crust require much hotter conditions to melt. In fact, the most fertile of common crustal rocks are metamorphosed mudstones (often called metapelites). Prior to melting, these contain quartz, plagioclase and muscovite. The breakdown of muscovite (Equation 6.2) releases both alkali feldspar and H_2O which, together with quartz and plagioclase (albite), forms a granite melt (Figure 6.8). The melting reaction can be written as:

$$\text{muscovite} + \text{quartz} + \text{albite} = \text{granite melt} + \text{sillimanite} + \text{alkali feldspar} \quad (6.3)$$

At pressures appropriate to the mid crust (400–700 MPa) crustal melting can occur at temperatures as low as 680–760 °C provided muscovite is present in the source rock (Figure 6.8).

The next example considers a granite intruding high-grade metamorphic rocks where field relations suggest partial melting from the metapelitic migmatite that it intrudes. The Rb and Sr concentrations for both metapelite and granite are given in Table 6.3.

Table 6.3 Rb and Sr concentrations for metapelitic migmatite and intruding granite.

	Concentration/ppm		D_{bulk}
	Metapelite	**Granite**	
Rb	200	360	0.5
Sr	120	65	2.0

Question 6.2

What is the melt fraction F required to generate a granite of this composition from the metapelitic source?

The trace-element data, therefore, suggest that between 11% and 15% melting of the metapelitic source will yield the appropriate granite composition.

If you try to generate the granite in Table 6.3 by fractional crystallisation of the andesite in Table 6.2, then the required melt fraction (from manipulation of Equation 5.7) is 0.16 for Rb and 0.40 for Sr. Such a large difference suggests that fractional crystallisation from an andesite of this composition is not the correct process by which this granite formed.

It is now possible to draw a general conclusion from considering simple quantitative modelling of the Rb and Sr contents of granites and their possible source rocks. *Either* partial melting *or* fractional crystallisation can explain the high concentrations of incompatible elements like Rb, or low concentrations of compatible elements like Sr. From the two examples considered, granites resulting from fractional crystallisation (Table 6.2) have lower Rb/Sr ratios than those resulting from partial melting (Table 6.3). Although this is often true, it cannot be used on its own to distinguish between the two mechanisms of granite formation, because a *higher* melt fraction would reduce the Rb/Sr ratio of a partial melt. Equally, a *lower* melt fraction would increase the Rb/Sr ratio of a melt residual from fractional crystallisation. However, by applying more than one trace-element ratio it is often possible to exclude one or other of the processes by the inconsistencies between the results from the incorrect model. In practice, it requires a combination of field observations with a range of trace-element analyses to determine the origin of a granite. The most useful tool of all is probably the use of isotopic ratios. Neither fractional crystallisation nor partial melting should change the isotopic ratio, so the granite should preserve the isotopic ratio of its source, whether an andesitic magma or a metamorphosed sediment. This approach will be used later in the chapter.

6.3 The India–Asia collision

The important features of continental collision on the scale of plate tectonics are best illustrated by the boundary between the Indo-Australian and Eurasian plates. This boundary is complex and its character changes along its length, as is exemplified by the northwestern margin of the Indo-Australian plate illustrated in Figure 6.15. Also shown are the epicentres of the 14 largest earthquakes recorded in the region since 1900; all but one are located on, or near, the plate boundary. The largest of these in the southeast corner of the figure was responsible for the tsunami of 26 December 2004. This marks the point where the boundary swings northwards from the destructive boundary of the Sunda Arc (Indonesia) to the strike–slip system of Burma. The region of intense seismicity in northeast India marks the point where the boundary twists sharply to form the eastern boundary of the Himalayan arc. The 2000 km stretch of the plate boundary between the mountain of Namcha Barwa, South Tibet and Nanga Parbat, North Pakistan separates the Himalayan mountains to the south from the Tibetan Plateau to the north, and defines the world's most active continental collision zone.

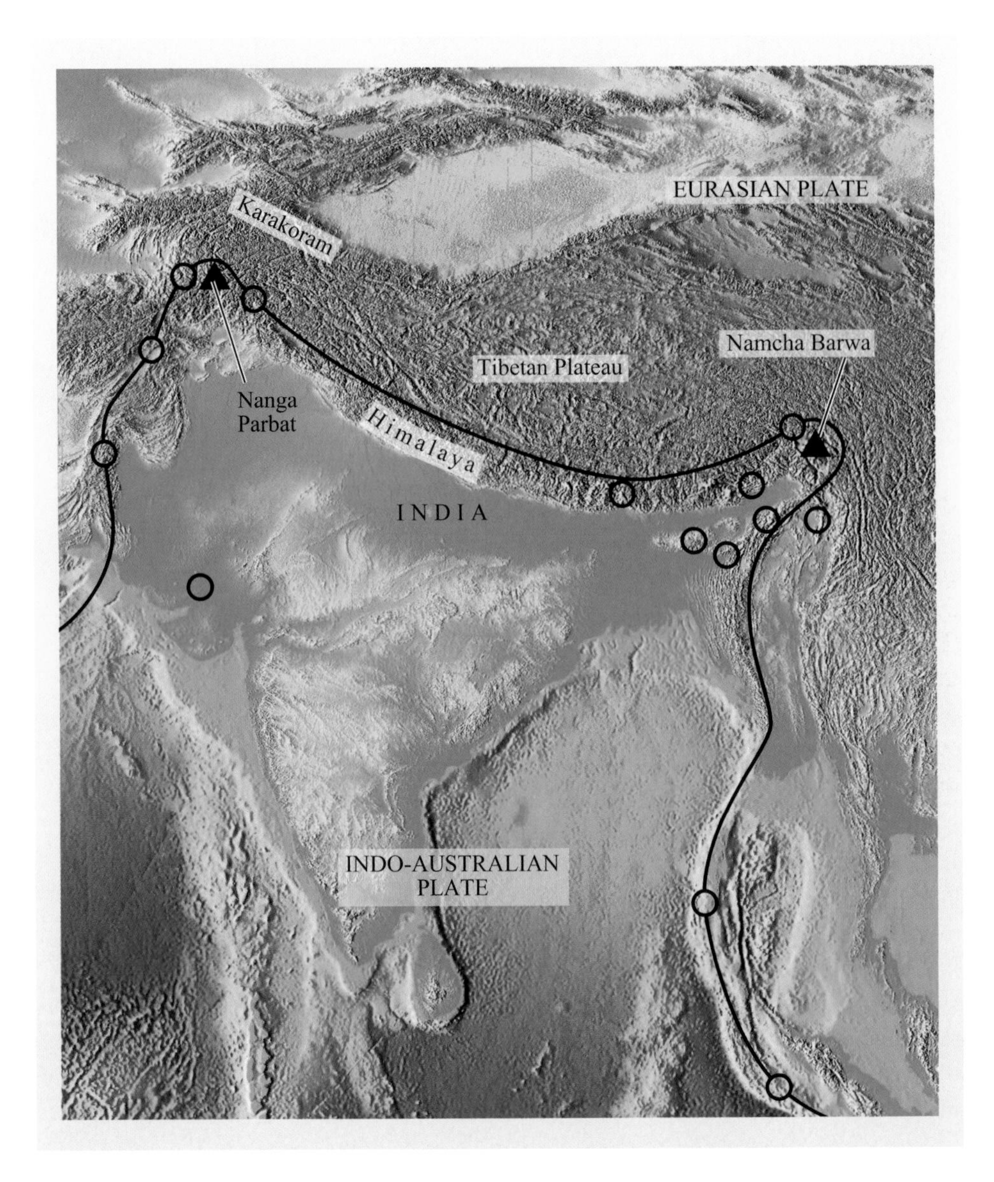

Figure 6.15 Computer-generated map showing the topography of India and Tibet. Grey land areas indicate altitudes >5000 m, red 4000–5000 m (Tibetan Plateau), yellow 500–1000 m, and green <500 m (coastal regions). The solid line indicates the boundary (suture) between the Indian Plate and the Eurasian Plate. The Himalaya stretch from Nanga Parbat to Namcha Barwa. Open circles represent the epicentres of the 14 most powerful earthquakes recorded in the region since 1900.

Tibet is a plateau of 2×10^6 km^2, about the size of western Europe. It is on average 5 km above sea level and includes over 80% of the world's land surface higher than 4 km. The Himalayan and Karakoram Mountains, which define its southern and western margins, include the only peaks on Earth reaching more than 8 km above sea level. Why is this part of the southern Asian continent so anomalous?

India lay well south of the Equator 100 Ma ago. The northern part of the Indian Plate was oceanic crust lying beneath the Tethys Ocean. Subduction of this oceanic lithosphere resulted from the northerly migration of the Indian continental land mass on a collision course with Eurasia (Figures 6.16 and 6.17a). As a result of the subduction, magmas were generated beneath the southern margin of Eurasia, intruding and thickening the crust in the same way that the Andes are being thickened today.

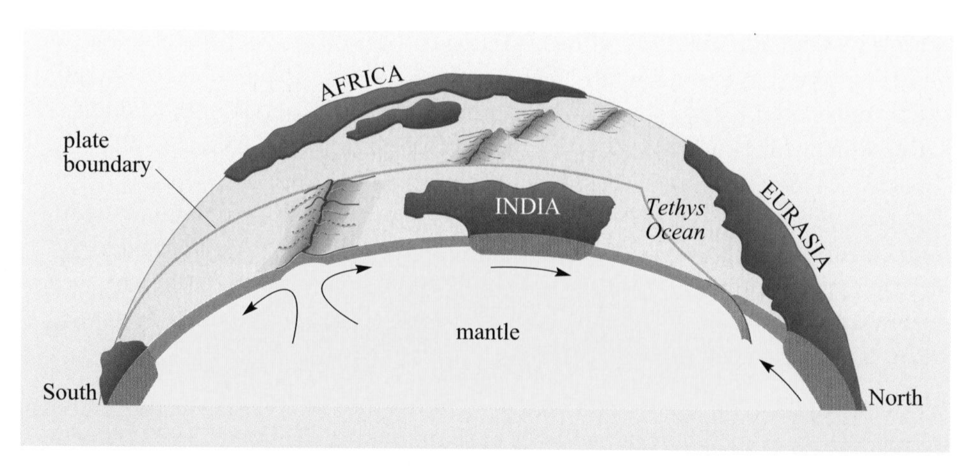

Figure 6.16 Diagram showing the northward movement of India between 100 million and 50 million years ago, causing shrinkage of the Tethys Ocean as India and Eurasia converged. (Adapted from Dietz and Holden, 1970)

By about 50 Ma, i.e. by the mid-Eocene, that part of the Tethys Ocean separating the Indian and Eurasian Plates had completely closed. The two continents then collided, with the result that the continental crust on both sides of the collision zone was thickened and the surface elevation rapidly increased (Figure 6.17b). Today, the join between the Indian and Eurasian Plates can be traced along the northern edge of the Himalayan chain in southern Tibet by a line of outcrops comprising rocks (mainly gabbros and basalts) characteristic of oceanic crust, i.e. ophiolites. These rocks were once part of the oceanic lithosphere underlying the great Tethys Ocean.

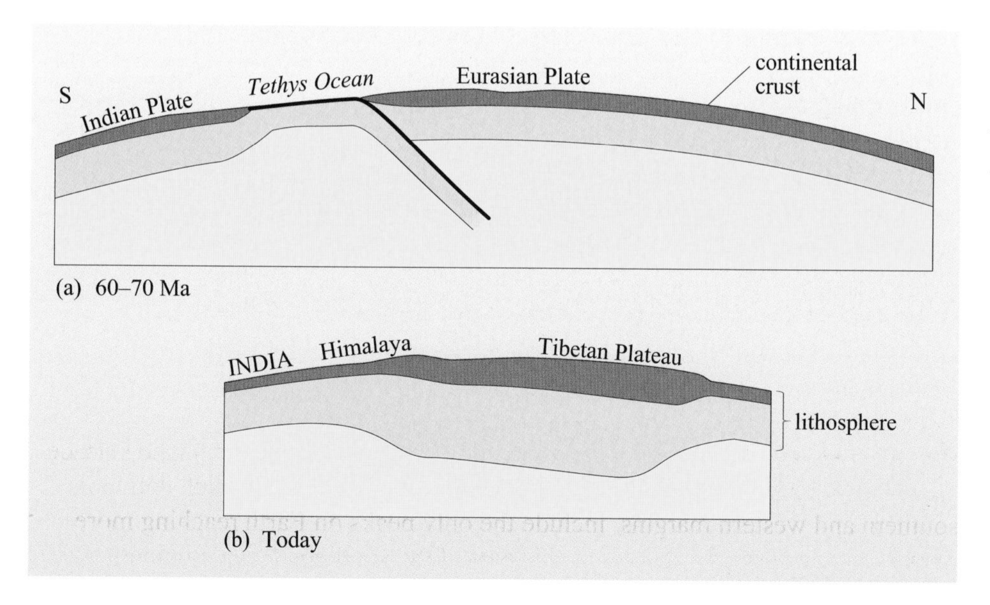

Figure 6.17 The India–Asia collision in cross-section (the scale is realistic, i.e. there is no vertical exaggeration): (a) at about 60–70 Ma; (b) today.

The Himalaya are by no means the only mountains on Earth formed by continental collision. The Alps are the result of a collision between Africa and Eurasia that began 120 Ma ago. During earlier geological times, collisions initiated the building of many of the world's lesser mountain belts, including the Urals and the Scottish Highlands. Why then did the Himalaya, and their hinterland, the Tibetan Plateau, become so high?

The uplift of the Himalaya and of the Tibetan Plateau is a response to considerable crustal thickening. Normal continental crust is 35–40 km thick, whereas the crust beneath the Himalaya and Tibet is 70–100 km thick. Some geologists maintain that the continental edge of Eurasia was unusually thick *before* collision, due to magmatic thickening above the subduction zone. In any event, substantial further thickening occurred after collision, by both folding and faulting of the rocks caught near the leading edges of the colliding plates (Figure 6.1). Such dramatic crustal thickening resulted from two unusual characteristics of the collision. First, it was 'head-on' rather than oblique, with a high closure velocity of ~200 mm y^{-1}. Second, the convergence persisted long after the intervening ocean had closed; India has migrated nearly 2000 km northwards since then, and a good deal of the convergence has had to be taken up by squashing and uplifting the rocks of the Himalaya.

The simple model for uplift outlined above works well for rather narrow mountain ranges like the Himalaya, but has Tibet been uplifted in the same way? It is not immediately clear why an area as broad as the Tibetan Plateau should be uplifted so high following collision. However, in recent years, geophysicists have suggested an additional mechanism for rapid uplift.

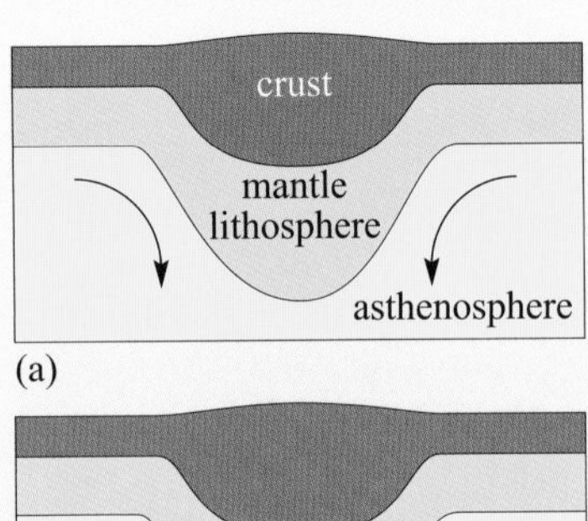

(a)

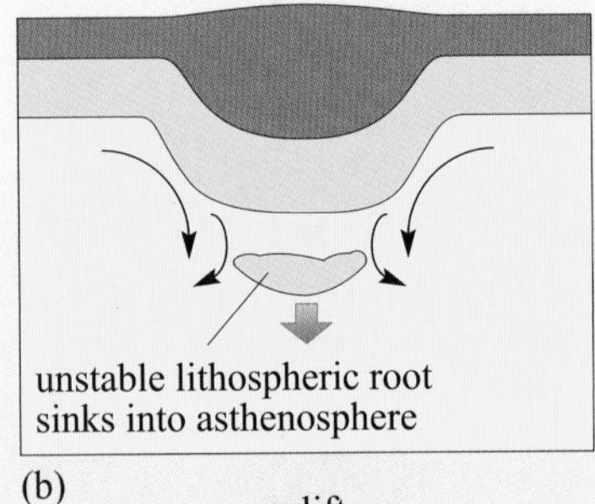

(b)

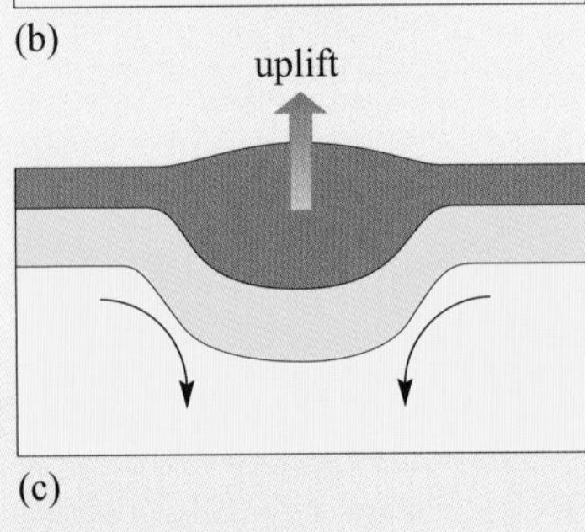

(c)

Figure 6.18 Schematic diagram to illustrate how convective thinning of the lithosphere may have contributed to the uplift of the Tibetan Plateau. (a) The lithosphere thickened as a result of collision. (b) Melting and erosion of the dense lithospheric root by convection currents in the asthenosphere. (c) Uplift of the plateau in response to the removal of the base of the lithosphere.

Although the distribution of major earthquakes is largely restricted to plate boundaries (Figure 6.15), earthquakes of moderate intensity are distributed across the Tibetan Plateau, as indeed is the evidence for deformation resulting from continental collision. This suggests that during continental collision the lithosphere does not behave in a brittle manner, i.e. as a rigid plate deforming only along its edges. Instead, it undergoes internal deformation by ductile thickening because of the higher temperatures generated within the thickened crust by the high concentrations of heat-producing elements. Under compression, the continental lithosphere behaves as a **thin viscous sheet**. Thus, following collision between two continental plates, the entire lithosphere will be thickened creating a lithospheric 'root' that protrudes down into the asthenosphere (Figure 6.18a). Since the early 1980s, geophysicists have been using computer models to study the thermal properties of the lithospheric root and the surrounding asthenosphere. These studies suggest that, following lithospheric thickening, the deep, cold root will eventually become unstable due to convection in the asthenosphere and be recycled back into the convecting mantle. This process, illustrated in Figure 6.18, is known as **convective thinning** of the lithosphere.

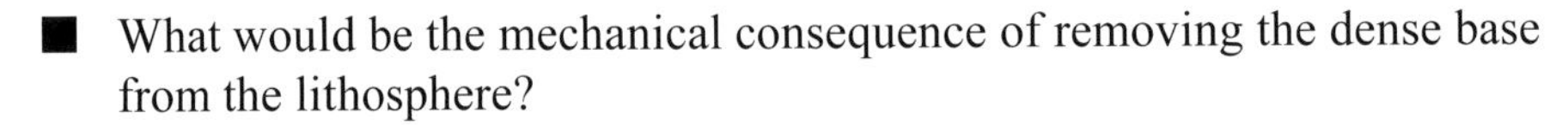

■ What would be the mechanical consequence of removing the dense base from the lithosphere?

■ Isostasy requires that the remaining lithosphere will bob upwards to an elevated position.

Convective thinning, therefore, provides a mechanism for the rapid elevation of the Tibetan Plateau (Figure 6.18c).

So, to summarise: the high-impact velocity between India and Eurasia, followed by continuing compression over tens of millions of years, led to an unusual degree of lithospheric thickening beneath Tibet, and the convective thinning that resulted allowed the lithosphere to rebound. This event may be the principal reason why Tibet is so much higher than other plateaux on the Earth today.

6.4 Metamorphism and melting in the Himalaya

The Himalayan orogen provides evidence of two contrasting examples of magmatism that can be found in collision zones. The first of these is the Trans-Himalayan batholith, which is emplaced into the southern margin of the Asian plate, just north of the suture that marks the boundary between pre-collision Asia and India (Figure 6.19). This forms a vast igneous arc, over 3000 km long and 50 km wide (only a segment of it is shown in Figure 6.19), made up of plutons of granites, granodiorites, diorites and a small proportion of gabbros (Figure 6.20). Rb–Sr isochron ages from the plutons cover a range from about 100 Ma to 40 Ma. These magmas intrude volcanic rocks of similar age and also sediments that were deposited along a continental margin between 100 Ma and 200 Ma ago.

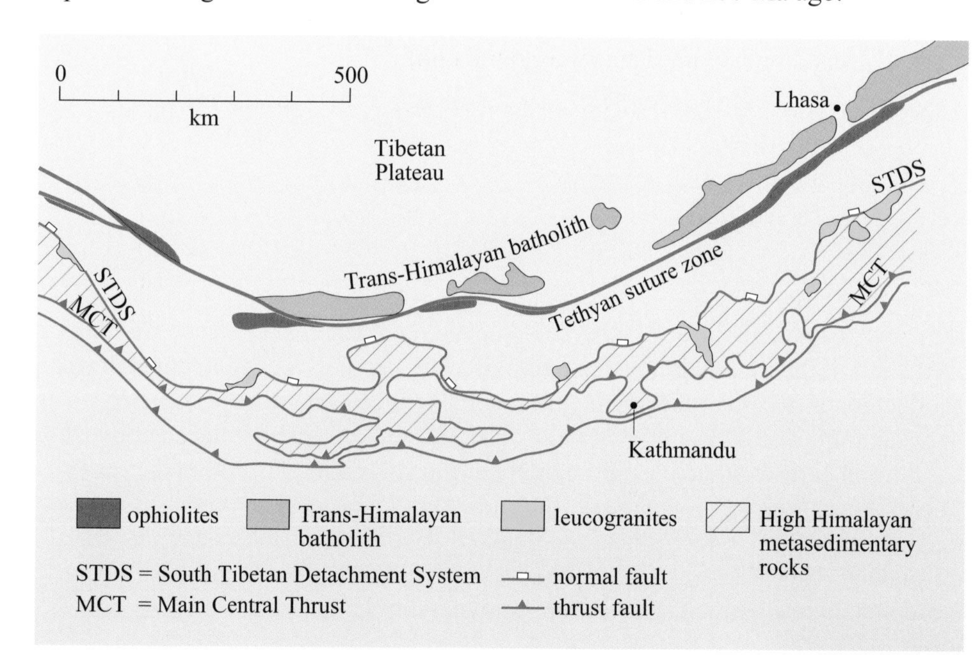

Figure 6.19 Geological sketch map of the central and eastern Himalaya and southern Tibet showing the distribution of plutonic rocks. Hatching indicates the High Himalayan metasediments bound by the South Tibetan Detachment System (STDS) and the Main Central Thrust (MCT). The thick, red line marks the suture zone.

The initial Sr-isotope ratio of a sample $(^{87}Sr/^{86}Sr)_i$ is the isotope ratio at the time of its formation. It reflects the Rb/Sr ratio of the source rock from which it was derived. The initial Sr-isotope ratio of plutons from the Trans-Himalayan batholith, therefore, can shed light on their origin (Figure 6.21).

- Study Figure 6.21 (overleaf); were the granites in the Trans-Himalayan batholith derived ultimately from the mantle or from the continental crust?

- The initial $^{87}Sr/^{86}Sr$ ratios of the Trans-Himalayan granites are closer to the mantle evolution line than that of the sediments they intrude, so they were probably derived from the mantle.

In detail, the early granites formed at about 100 Ma have a similar $^{87}Sr/^{86}Sr$ ratio to the upper mantle and so are likely to result from fractional crystallisation of mantle-derived melts.

Figure 6.20 Inclusion of diorite within Eocene granodiorite from near Lhasa. Hammer handle is ~3 cm wide. (Nigel Harris/OU)

However, the initial ratio of the younger samples shows a range of values; some are similar to the upper mantle, but others are distinctly elevated towards the values of the sediments they intrude. This suggests some contamination by crustal material, probably by partial melting and assimilation of crustal rocks during magma ascent.

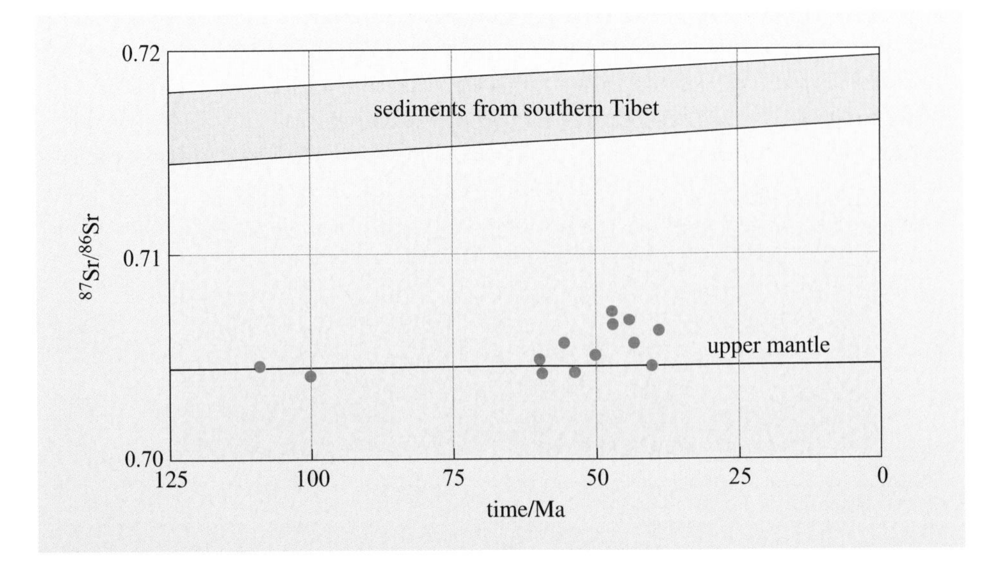

Figure 6.21 Sr-isotope evolution diagram of plutons from the Trans-Himalayan batholith (red circles) and of the sediments that they intrude.

It is now possible to draw together tectonic and geochemical evidence for the formation of the Trans-Himalayan batholith. The magmas were emplaced during subduction of oceanic lithosphere, but magmatism ceased after the collision between India and Asia 40–50 Ma ago. Therefore, they were emplaced above a subduction zone at an active continental margin (Figure 6.17a). From the range of rock types (granites to gabbros), the suite of magmas appears to be similar to magmas generated in the Andes, and detailed geochemical studies support such an analogy. Therefore, the Trans-Himalayan batholith is an example of pre-collision magmatism at an active continental margin.

■ Why do you think mantle-derived granites younger than 40 Ma are largely absent from the Himalaya?

■ Once collision occurred, subduction of oceanic lithosphere stopped, cutting off the supply of fluids into the mantle wedge. As a result, further melting in the mantle virtually ceased.

The second example of Himalayan magmatism contrasts strongly with the Trans-Himalayan batholith. A series of light-coloured granites (the High Himalayan leucogranites) are emplaced as lenses and sheets into the metamorphosed sediments of the Himalayan mountains south of the suture (Figure 6.19). They are known as **leucogranites** because they are light in colour (Figure 6.22), being composed almost entirely of quartz, plagioclase and alkali feldspar. All the intrusions of this type are granites containing 70–75% silica. Rb–Sr isotope data from several of these granites indicate an age of 20 ± 3 Ma and an initial Sr-isotope ratio varying between 0.738 and 0.785 (Figure 6.23). Therefore, the granites are intruded well after collision and have a much higher initial Sr-isotope ratio than any samples analysed from the Trans-Himalayan batholith (Figure 6.21).

■ Suggest a possible source for the High Himalayan leucogranites (based on the isotope data and rock types described so far) and relate this to what you know of Himalayan tectonics.

■ The high initial Sr-isotope ratio strongly suggests a crustal source. The restricted, silica-rich compositions of the granites (70–75% SiO_2), equivalent to minimum-melt compositions, support partial melting in the quartz–plagioclase–alkali feldspar system. From Section 6.3, you know that crustal melting can occur in thickened crust due to the high content of heat-producing elements in crustal rocks. Since the intrusion age of 20 Ma is at least 20 million years after initial collision, it seems likely that the granites formed after the thickened sedimentary pile heated up due to the high concentrations of heat-producing elements in crustal rocks.

Figure 6.22 Pinnacle of High Himalayan leucogranite forming Shivling (6543 m) in the Garhwal Himalaya, India.

The High Himalayan leucogranites are not intruded into unmetamorphosed sediments like the Trans-Himalayan batholith, but into migmatites formed from the partial melting of much older aluminous (pelitic) sediments (Figure 6.9). The sediments found to the north of the suture in southern Tibet have $(^{87}Sr/^{86}Sr)_i < 0.720$ (Figure 6.21), but aluminous metasediments further south in the Himalaya have extremely high $(^{87}Sr/^{86}Sr)_i$ of 0.735–0.790. Careful examination of the minerals in the dark parts of the migmatites shows that muscovite is breaking down to form sillimanite by reaction with quartz. This reaction provides alkali feldspar and a hydrous fluid which, when combined with quartz and plagioclase from the sediment, will form a granite melt (Equation 6.3). Therefore, the mineral reaction that is observed in the migmatite provides a likely source for the granite.

Why rocks should melt at all in collision zones can be understood from Figure 6.24. The aluminosilicate stability fields, the muscovite + quartz breakdown reaction and the granite melting curve have been transferred from Figure 6.8, and superimposed on the diagram are two geotherms, A and B. Geotherm A is a typical geotherm through continental crust of normal thickness. Geotherm B is an elevated geotherm that is a consequence of internal heating within thickened continental crust.

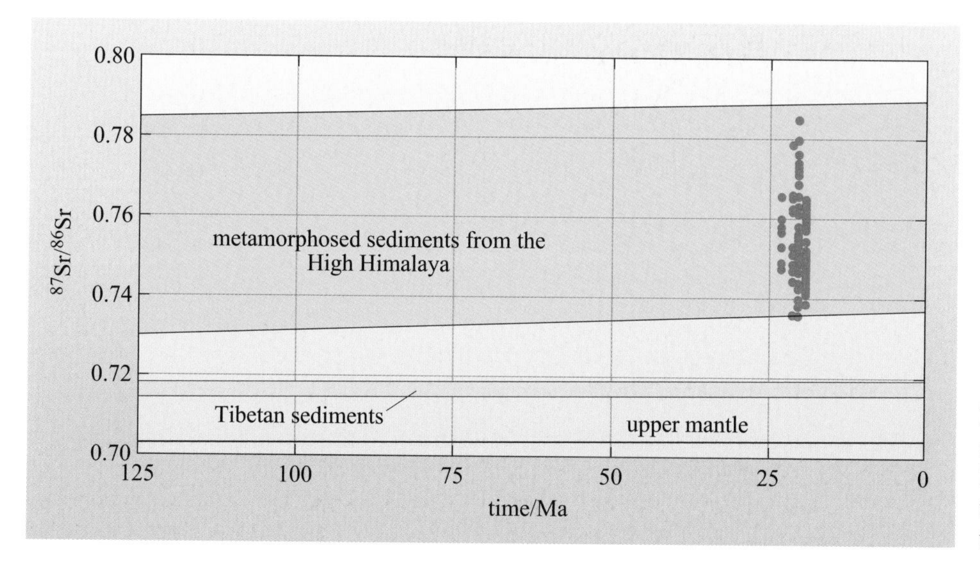

Figure 6.23 Sr-isotope evolution diagram of High Himalayan leucogranites (red circles) and the metamorphosed sediments they intrude.

Figure 6.24 Two continental geotherms A and B plotted on phase boundaries taken from Figure 6.8. Ms: muscovite, Qz: quartz, Kf: alkali feldspar and Ab: albite.

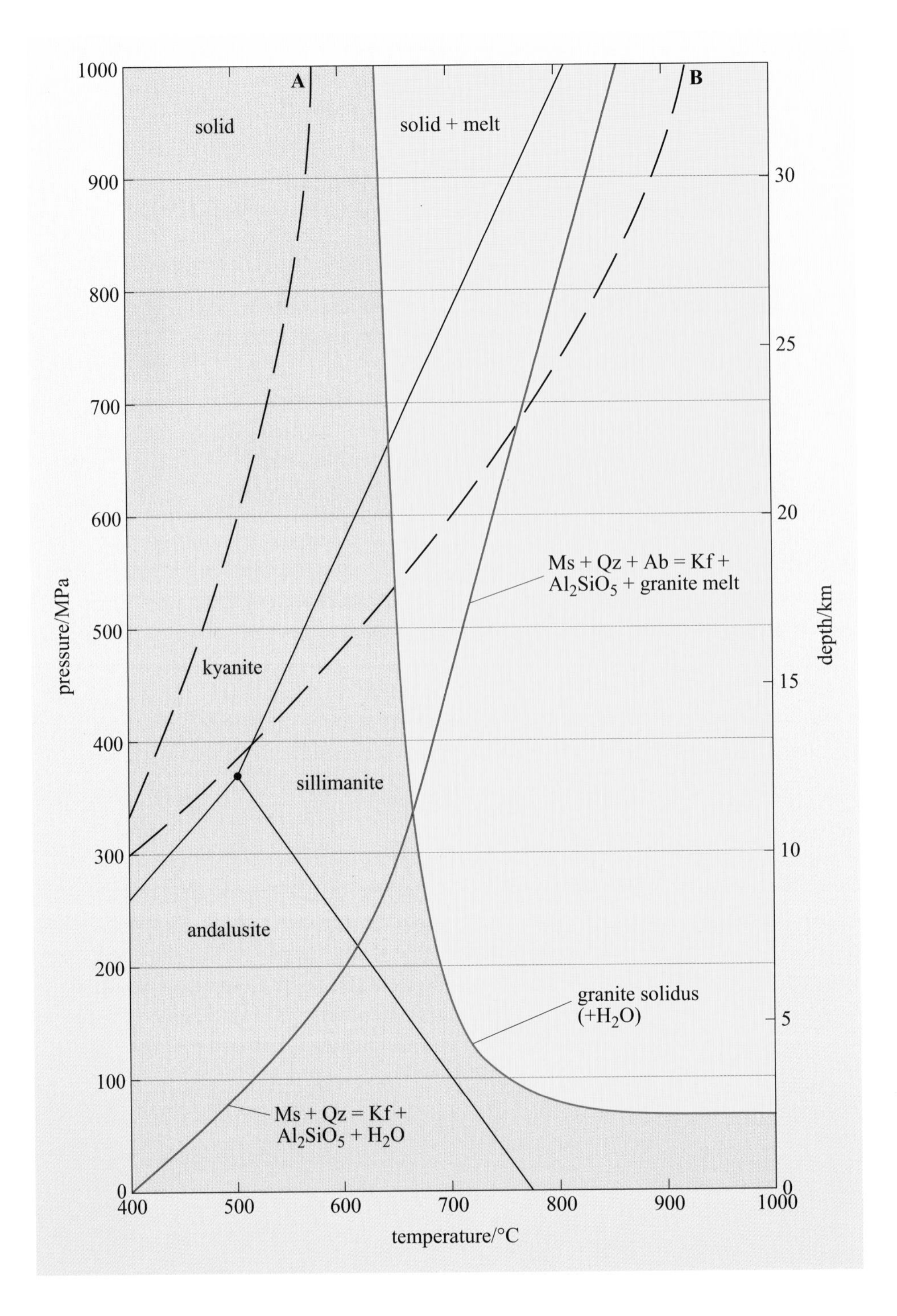

Question 6.3

(a) At which depths would you expect melting to occur in response to (i) geotherm A and (ii) geotherm B in Figure 6.24?

(b) Which polymorph of Al_2SiO_5 will be formed by the melting reaction?

At this stage, some general conclusions can be derived from comparing the two groups of Himalayan granites. At active continental margins, such as the Andes or the Trans-Himalayan batholith, granites form a minor component within a suite of less-evolved plutonic rocks of intermediate compositions, like diorites, tonalites or even granodiorites. They result from fractional crystallisation of mantle-derived basaltic magmas combined with assimilation of crustal rocks. Such suites of granitic rocks, therefore, represent net additions to the mass of continental crust. In contrast, collision-zone granites occur in the absence of less-evolved compositions. They require no contribution from the mantle and provide an example of crustal reworking rather than crustal growth. On a global scale, the latter group are relatively small in volume compared with the great batholiths located along continental margins, but they provide insights into the behaviour of the continental crust when heated above its solidus temperature.

6.5 Modelling the Himalayan orogen

In this chapter so far, the Himalayan orogeny has been treated as the simple consequence of two continental margins colliding and exerting extreme compression on the rocks around the collision zone. However, as the structure of the Himalaya has become better known, it has become clear that the detailed picture is more complex. Not only are there thrust faults thickening up the crustal slices, but also there are normal faults suggesting zones of extension within the overall picture of regional compression.

Schists, gneisses and migmatites of an aluminous composition outcrop across a swathe of the High Himalaya bounded by two faults (Figure 6.19): the South Tibetan Detachment System (STDS) (a normal fault system) and the Main Central Thrust (MCT). Melting of these metasedimentary formations may have been critical in shaping the architecture of the Himalayan orogen. Metamorphic studies of the mineral assemblages from these rocks suggest they have been exhumed from depths of about 20 km, and structural studies suggest they have experienced ductile deformation while being brought to the surface. The mechanism of their exhumation is well established; both the STDS and the MCT that define the boundaries of these rocks have a shallow northward dip (Figure 6.25). Working together, movement on these two faults has resulted in the upward and southward motion of a sheet or a wedge of High Himalayan metasediments (Figure 6.25).

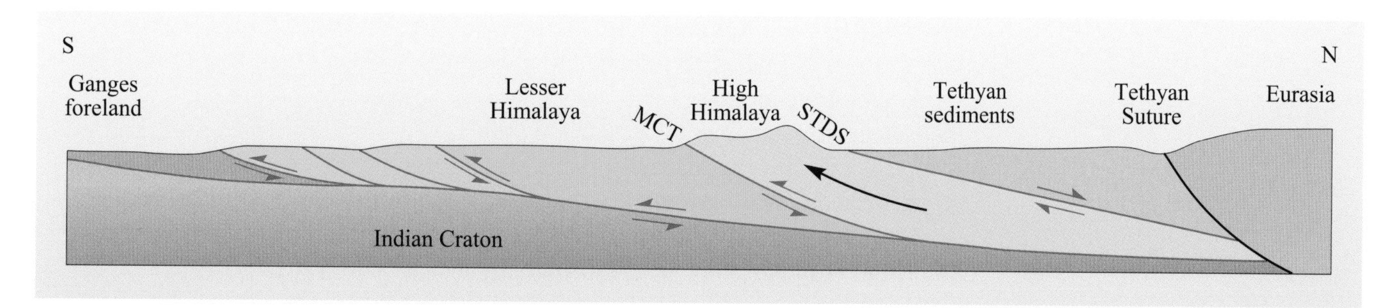

Figure 6.25 Sketch vertical section across the Himalaya. Note that simultaneous movement on the STDS and MCT will exhume the High Himalayan metasedimentary rocks between them (see arrow).

The Himalayan collision is such a clear-cut and recent example of collision tectonics that it has become the focus of intense study by geoscientists. One of their goals is to incorporate the geological characteristics that can be observed at the surface and the deep-level structures imaged by seismic surveys into a thermo-mechanical model of the thickening lithosphere to understand more about the processes that drive lithospheric evolution during continental collision. One such model, the thin viscous sheet model, has been briefly described already. This explains many broad-scale features of the deformation of the Tibetan lithosphere, but it cannot explain why rocks from the High Himalaya are apparently being extruded southwards, as shown in Figure 6.25.

In recent years, the **channel-flow model** has been developed to account for this southern extrusion together with other aspects of the Himalayan orogen. This model takes into account the changing viscosity of crustal rocks with depth. Viscosity is a measure of how freely material is able to flow, and gives an indication of the strength of the material; as rocks are subject to increased temperatures and pressures over time, they may flow like a fluid and so deform in a ductile way, as is implicit in the thin viscous sheet model for deformation of the lithosphere. The viscosity of the lithosphere is determined by the prevailing geothermal gradient and by the nature of the rocks that are present at different depths within it.

A geophysical study of southern Tibet has identified a zone of low seismic velocities at a depth of about 15–20 km within the Tibetan crust.

- Given the steep geothermal gradient that results from radioactive decay within thickened crust, and the ability of pelitic rocks to melt at temperatures as low as 700 °C at crustal depths, can you think of a realistic explanation for this low-velocity layer in the mid-crust?

- One explanation is that it represents a zone of partial melting. Other geophysical properties, such as electrical conductivity, are consistent with this interpretation.

Partial melting initially results in a mix of solid and liquid with the liquid distributed along grain boundaries. Even at low melt fractions ($F < 0.1$) a dramatic decrease in viscosity, and therefore strength, of the rock is observed, once the isolated melt pockets establish connectivity along the grain boundaries. Indeed, the viscosity of partially melted granite varies by about 14 orders of magnitude between its liquidus and solidus temperatures. This process is known as **melt weakening**.

The results of incorporating both the temperature and the viscosity changes into a model of thickened lithosphere are shown in Figure 6.26. By 21 million years after collision, a low-viscosity channel is created by partial melting of the Indian crust at temperatures above 700 °C (Figure 6.26a). By 33 million years after collision the low-viscosity channel is migrating southwards, driven by the differential pressure between the thickened Tibetan crust to the north and the 'normal' Indian crust to the south (Figure 6.26b). At this stage the channel can be likened to toothpaste being squeezed along a tube.

However, the modelling suggested that this channel would not reach the surface unless the topographic rise of the southern Himalaya was being actively eroded.

If the effects of strong erosion are incorporated into the model then the isotherms are folded into an antiformal structure beneath the erosion front (Figure 6.26b) and by 42 million years after collision the low-viscosity channel would break the surface (breaking the cap off the toothpaste tube) (Figure 6.26c). Of course, this model is much simpler than reality. In practice, a partially melted channel would never reach the surface without cooling rapidly, resulting in solidification of the melts followed by brittle deformation as the rocks thrust southwards through the uppermost crust. However, the channel-flow model does explain the southward extrusion of rocks brought up from the mid crust. It also predicts that they will be extruded along the southern topographic front of the Himalaya, where precipitation, from the monsoon, is most intense.

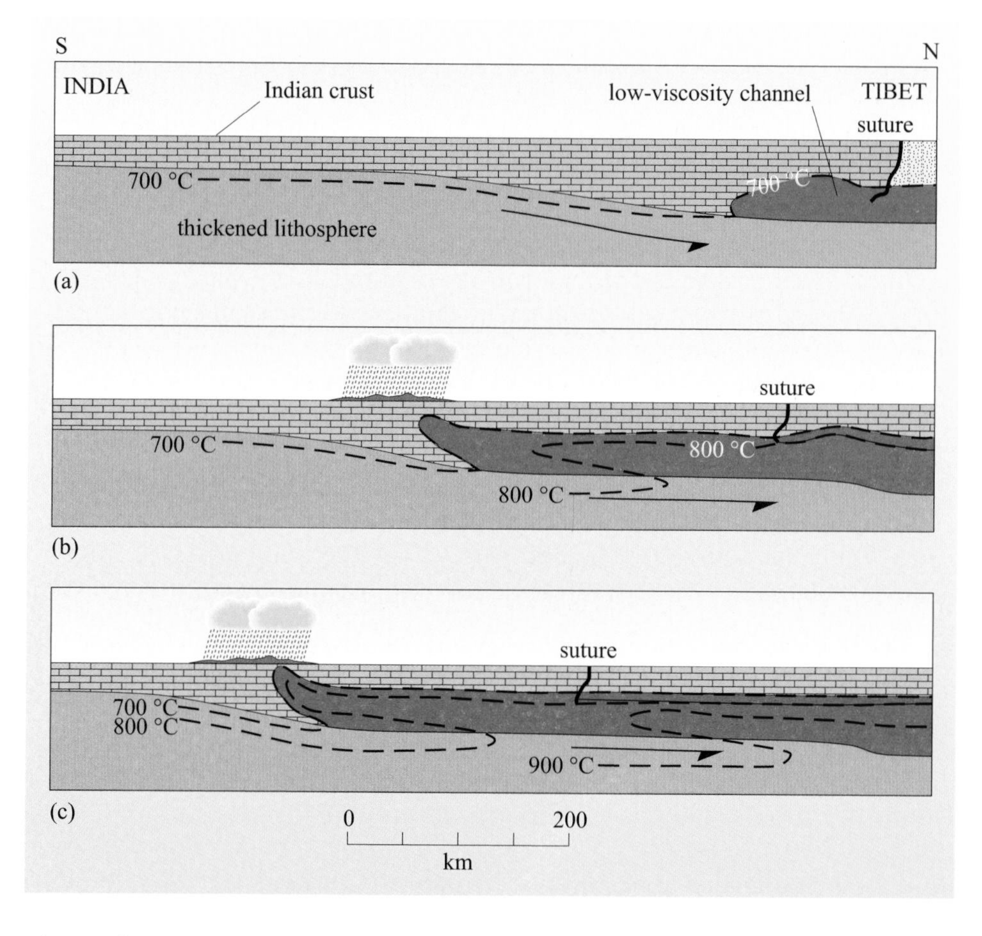

Figure 6.26 Results of modelling the thermal and mechanical behaviour of thickened crust, assuming melting is initiated at 700 °C, and that the southern Himalayan front is undergoing active erosion. A threefold time sequence is illustrated at (a) 21 Ma, (b) 33 Ma and (c) 42 Ma after collision. (Adapted from Jamieson et al., 2004)

The application of the model to the present-day structure of the Himalaya and southern Tibet is shown in Figure 6.27. The low-viscosity channel extruding southwards is represented by the High Himalayan metasediments (in pink), bounded by the STDS above and the MCT below (Figure 6.19). In applying the model to actual geological relationships, the model has yielded some further insights. In order for the channel to be extruded at the surface today, melting must have been initiated by about 20 million years after collision (Figure 6.26a); in the Himalayan context this is equivalent to an age of around 30 Ma. The large High Himalayan leucogranites are about 10 million years younger than this (Figure 6.23) so cannot represent this early melting event. However, dating of zircons from within migmatites in the High Himalaya has revealed that partial melting of the metasediments occurred as early as 30 Ma. Migmatites can result

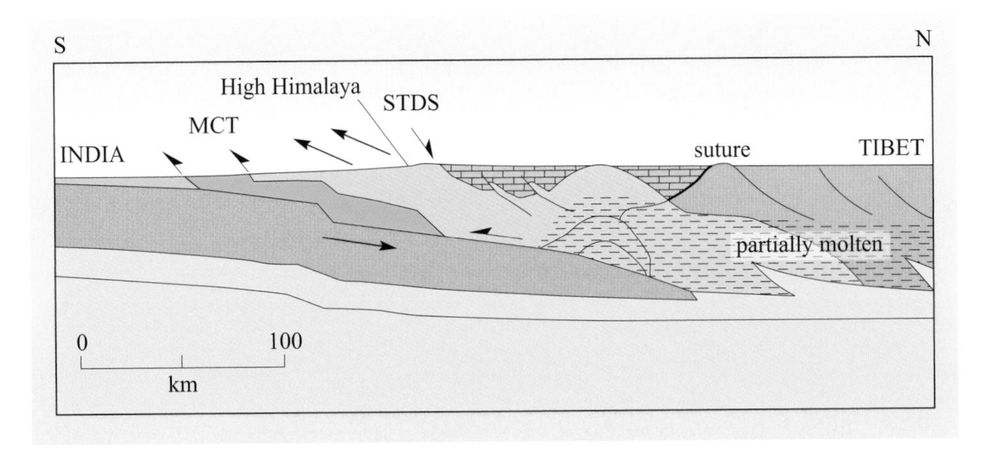

Figure 6.27 Section through the Himalaya and southern Tibet, indicating the southward extrusion of the lower crust from a partial melt zone beneath southern Tibet. (Adapted from Beaumont et al., 2004)

from a very small melt fraction, perhaps as low as 0.05, whereas the larger granites result from higher melt fractions (0.1 to 0.2) being squeezed out of their source rocks into larger magma chambers. This suggests that only small-degree partial melting is needed to initiate the formation of a low-viscosity channel.

Like any scientific model, channel flow offers a simplified view of natural processes. Over the next few years it will be tested and modified by geological and geophysical observations. Other models may eventually supplant it. However, the significance of this particular model for our understanding of continental collision is that it is the first to link deep-seated processes (ductile deformation, metamorphism and magmatism) with surface processes (erosion and precipitation). Until a few years ago, no geoscientist would have considered the possible linkage between atmosphere, hydrosphere and lithosphere, because climate research and the study of tectonics and metamorphism would have taken place in different departments, possibly even at different universities. Now, traditional boundaries between research areas are breaking down and there is an increasing awareness that the Earth behaves as a single system and the components of that system are all interrelated.

Summary of Chapter 6

- Collision between two continental plates results in thickening and metamorphism of the crust.
- Varying pressures and temperatures in the Earth's crust result in reactions between metamorphic minerals.
- The stability field of a mineral, or mineral group, may be represented on a pressure–temperature diagram that then provides information on the pressure or temperature to which a rock containing the mineral(s) has been subjected.
- High pressures favour high-density minerals; therefore, reactions involving large changes in densities provide geobarometers. Geothermometers are provided by reactions that result in dehydration or decarbonation.
- Increasing temperatures and pressures of sediments that contain clay minerals result in the formation of slates, schists, gneisses and migmatites. Gneisses may also form from the metamorphism of igneous bodies.

- The reaction of muscovite and quartz can produce granite melts in sediments of aluminous composition (known as pelites).
- Magmas of granitic composition may be formed either by fractional crystallisation of basic magmas or by partial melting of crustal rocks of pelitic composition.
- The ternary system Qz–Ab–Or is characterised by a liquidus surface that slopes downwards towards a point referred to as the granite minimum. Consequently, fractional crystallisation of all samples in this system drives the composition of the residual liquid towards that of the granite minimum irrespective of their initial composition. Conversely, on heating a rock consisting largely of quartz, plagioclase and alkali feldspar (or a mica that will break down to form an alkali feldspar), the first liquid to form is also close in composition to the granite minimum.
- The trace-element compositions of a granitic rock can be used to distinguish between an origin from fractional crystallisation of an andesitic magma and one from partial melting of a crustal source.
- The collision between India and Eurasia at about 50 Ma was responsible for crustal thickening and uplift of the Tibetan Plateau.
- In the Himalaya, pre-collision magmatism is represented by the Trans-Himalayan batholith that was intruded between 110 Ma and 40 Ma. Its initial Sr-isotope ratio $(^{87}Sr/^{86}Sr)_i$ is indicative of a mantle source with a component of crustal contamination that increases through time. It is typical of magmatism formed at an active continental margin.
- The High Himalayan leucogranites (intruded at about 20 Ma) result from partial melting of mica schists and migmatites in response to thickening and radioactive heating of the continental crust following collision.
- Continuous thickening of the lithosphere beneath Tibet, as proposed by the thin viscous sheet model, may have led to convective thinning followed by rapid uplift.
- The channel-flow model incorporates the effects of reduced viscosity of lower crustal rocks by partial melting, a process called melt weakening. The modelled channel will be extruded laterally by differential pressure exerted by thickened crust and upwards by focused erosion of surface rocks.
- The channel-flow model provides an explanation for the southwards extrusion of High Himalayan metasediments along the MCT and the STDS.

Learning outcomes for Chapter 6

You should now be able to demonstrate a knowledge and understanding of:

6.1 The different mineral assemblages that can form in metamorphosed continental crustal rocks and how they provide information on the pressure–temperature conditions under which the rocks formed.

6.2 How mineral reactions provide information on the conditions of partial melting within the continental crust.

6.3 Why variations in the compositional range of granites occur, by referring to the alkali feldspar binary quartz–albite–orthoclase ternary phase diagrams.

6.4 How the mode of formation (e.g. fractional crystallisation, partial melting) and origin (e.g. source material) of granites can be determined from trace element and isotopic analyses.

6.5 How different approaches to modelling the thermal and mechanical consequences of continental collision provide insight into the timing and interactions between deep-level and surface processes.

The deep mantle and global cycles

Chapter 7

In Chapter 1 you explored the present gross structure of the Earth and in the subsequent chapters you have concentrated on what happens in the upper 100–300 km of the Earth. However, to understand the interactions between the atmosphere, hydrosphere, biosphere, lithosphere and upper mantle it is important to return to a view of the planet as a whole because many of the driving mechanisms that influence these interactions originate deep within the Earth. Perhaps the most obvious process that originates at depth but has a profound influence at the Earth's surface is volcanism. Volcanic eruptions can emit large volumes of volatiles, such as water, carbon dioxide and sulfur dioxide, which can significantly affect the Earth's climate. An obvious question to ask is whether these volatile compounds have always been inside the Earth or whether some proportion of them has been recycled from the Earth's surface back into the mantle by plate tectonics or other processes.

In this chapter you will concentrate on geophysical and geochemical studies that provide information on the nature and composition of the deep mantle. Geophysics provides a means of imaging the inside of the whole of the Earth and provides an insight into the fate of surface materials carried by plates as they are subducted back into the mantle, and their possible return to the Earth's surface. Geophysical studies are limited to providing a snapshot of what the planet's interior is like at the present and perhaps some indication of the last few hundred million years. Geochemistry, by contrast, exploits radiogenic isotopes to provide information about timescales of processes in the mantle from days to billions of years, but it is limited to studying materials that reach the Earth's surface as a result of volcanic or tectonic activity.

The origins of ideas about the deep Earth stem back to the earlier part of the 20th century, when seismic studies allowed the gross structure of the Earth to be defined. At about the same time, Arthur Holmes, who was a great advocate of continental drift, suggested that currents within the mantle might provide a mechanism that could move the continents around the Earth's surface. Unfortunately, his ideas coincided with a rejection of the continental drift hypothesis by the geological community. When plate tectonics became an accepted theory to explain continental drift and sea-floor spreading, researchers started to re-evaluate the idea of solid-state convection within the mantle. Since then, seismic studies have become increasingly sophisticated and with the advent of a global digital seismic network, images of the Earth's interior are improving on a monthly basis. Seismic tomographic images (equivalent to a computer-assisted tomography or CAT scan of a human body) can now resolve slabs subducting deep into the planet and are starting to resolve increasingly narrow zones of hot material upwelling within the mantle. Experimentalists investigate the mineralogy of the deep Earth by taking mantle materials and subjecting them to the pressures and temperatures prevailing throughout the whole 2900 km depth of the mantle. Geochemists study the isotopic and trace element contents of the products of mantle melting and the materials that are recycled into the mantle to build up geochemical models that are the equivalent to the well-known carbon cycle, but on a mantle length-scale and a geological timescale.

During the 1970s and 1980s, the most common model for the deep mantle was the so-called two-layer model in which the boundary between the upper and lower mantle at the 670 km seismic discontinuity represented a physical barrier to material moving both upwards and downwards. It was also thought that the boundary represented a distinct compositional difference between the depleted upper mantle and a lower mantle that had essentially a primitive (chondritic) composition. Mantle convection took place within each of the two layers and hot-spot volcanism originated at the 670 km discontinuity, although later models allowed for material for some hot-spot volcanism to leak periodically from the lower mantle. The two-layer model seemed consistent with most of the geophysical and geochemical data available at the time. However, the two-layer model restricts any recycling of surface materials to a well-mixed upper mantle, and only this reservoir could influence the atmosphere, hydrosphere and biosphere. More recently, there has been a convergence towards a single-layer model in which the whole mantle convects. This allows a much larger reservoir to influence processes at the Earth's surface, and it even includes possible interaction with the Earth's outer core.

7.1 Discontinuities in the deep Earth

The seismic velocity structure of the Earth has been well known since the 1930s and recent work has only served to refine certain parts of the velocity structure. Figure 7.1 is a profile of P- and S-wave velocities in the Earth illustrating the key changes in the seismic velocity structure with depth.

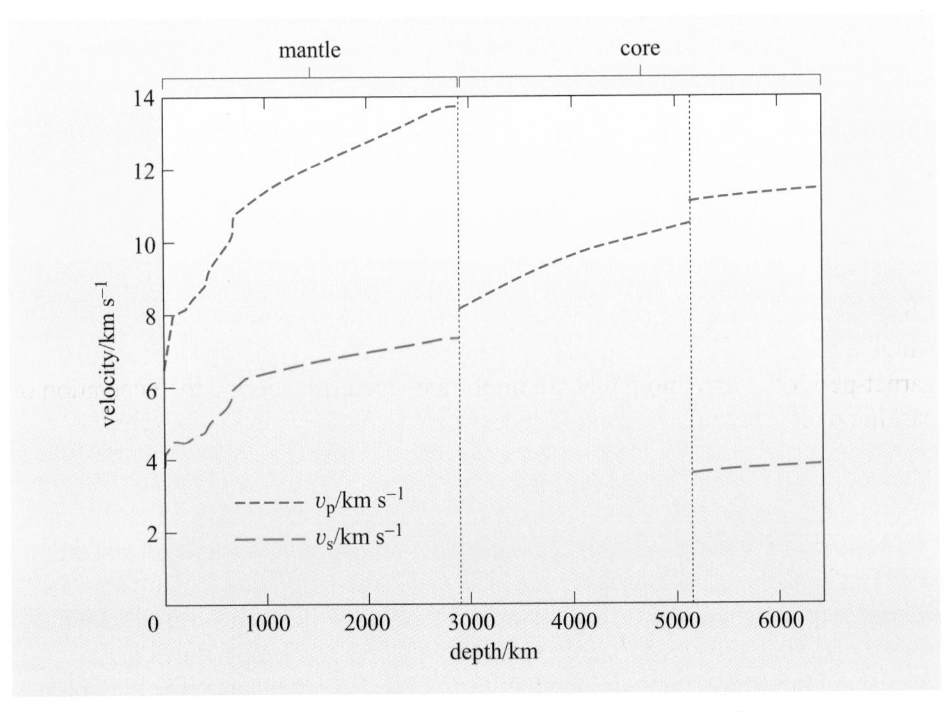

Figure 7.1 P- and S-wave velocity section of the Earth, with a modern Earth reference model. (Fowler, 2005)

- What key velocity changes in the mantle can you pick out from Figure 7.1?
- The major discontinuity occurs at 2900 km with a decrease in P-wave velocity at the core–mantle boundary (CMB) and a corresponding lack of S-waves in the outer core. Within the mantle there is a lower velocity region in the upper mantle at 100–200 km (known rather obviously as the low-velocity zone). There are marked increases in seismic velocity between ~410 km and ~670 km, which is known as the **transition zone**. The lower mantle is rather featureless until just above the CMB, where velocity structure becomes complex, although not apparent in Figure 7.1 because of the scale.

- What are the likely causes of seismic velocity changes in the mantle?
- You should recall from Chapter 1 that seismic velocity is dependent on density and elastic modulus. Also, the presence of any liquid phase will reduce seismic velocities. The discontinuities, therefore, reflect density and possibly phase changes.

The general equation relating seismic velocity, density and elastic modulus is:

$$\text{wave velocity} = \sqrt{\frac{\text{elastic modulus}}{\text{density}}} \quad (7.1)$$

A critical characteristic of most solids is that when the density increases the elastic modulus also increases, but more rapidly. So, if a phase change in the mantle results in a material with a higher density, then the seismic velocity also increases because the elastic modulus increases at a greater rate than the density. At a given temperature, the density of the mantle depends on both its mineralogy and its bulk composition, and it is obviously important to resolve whether simple changes in mineralogy or compositional layering cause the observed stratification.

7.2 The mineralogy of the deep mantle

Chapters 4 and 5 explored phase diagrams for the upper 100 km of the mantle, in which a number of phase changes occur, particularly the spinel-peridotite to garnet-peridotite transition, that are important when considering the generation of basaltic melt. These phase changes occur in a peridotite mantle of constant composition and serve to show that the physical properties of the mantle can change even if the composition remains the same.

To study the next 2800 km of mantle involves some very special experimental equipment in order to replicate the extreme pressures and temperatures expected in the deep interior of the Earth. These conditions can be achieved in the laboratory using **multi-anvil** presses made of very strong materials, such as tungsten carbide and even diamond. To generate pressures equivalent to the depth of the 670 km discontinuity using tungsten carbide anvils required larger and larger presses, the largest of which were the size of a big room. Eventually it was realised that two diamonds cut with perfect facets could be used as an alternative anvil material, and these allow experimentalists to squeeze rock powder to pressures equivalent to those of the outer core, while heating the sample to very high temperatures with a laser. Fortunately, such **diamond anvil** presses fit on a laboratory bench and allow experimental petrologists to characterise the mineralogy of the mantle right down to the CMB; including a phase change at ~2800 km, which was only discovered in 2004. In these experiments, a tiny sample (a few milligrams) of very fine powder of an appropriate mantle composition is enclosed in a metal container and placed between the anvils. Pressure is then applied with a hydraulic press and the capsule is heated using an electric current or, in the case of diamond anvils, a laser. After a period of time at the required temperature and pressure, the experimental charge is cooled rapidly to quench the high-pressure, high-temperature phases allowing the products of the experiment to be characterised.

Figure 7.2a is a summary of the different phase changes that occur in a fertile peridotite (a peridotite that has not melted to produce a basaltic melt and is equivalent to a lherzolite at low pressures) mantle, assuming an adiabatic temperature gradient with a potential temperature of about 1280 °C. The first change sees orthopyroxene dissolve into garnet to produce a new form of garnet called **majorite** at about 5 GPa pressure. This phase change is progressive, and eventually also affects clinopyroxene, such that at pressures of ~13 GPa all pyroxene has been replaced by majorite. Olivine, which makes a significant proportion of a mantle peridotite, undergoes a phase change at ~14 GPa. This involves a reorganisation of the structure of olivine into the so-called β **spinel** structure, producing a mineral called **wadsleyite**, with an associated 10% increase in density. Another phase change occurs at 17–18 GPa, when the β spinel structure changes to a γ **spinel** phase called **ringwoodite** (after the famous experimentalist Professor Ted Ringwood). At ~24 GPa, a series of complicated reactions takes place; the γ spinel phase dissociates to produce **magnesium (Mg)-perovskite**, and **magnesiowüstite (Mg-wüstite)**. The most recent experiments on representative mantle compositions demonstrate that this transformation occurs over a very narrow depth range, and the net effect of all the transformations at this pressure is to produce a 10% increase in the density of the mantle. Majorite breaks down between 20 and 25 GPa, forming **calcium (Ca)-perovskite** and more Mg-perovskite. At higher pressures, peridotite composition consists of 79% Mg-perovskite, 16% magnesiowüstite and 5% Ca-perovskite.

■ How do the pressures of these phase transitions compare with the depth of seismic transitions in Figure 7.1?

■ They correspond remarkably well. The transition of olivine to β and γ spinel structures occurs between ~410 km and 520 km, equivalent to the seismic transition zone, and the appearance of Mg-perovskite occurs at ~670 km, coincident with the boundary between the upper and lower mantle. Finally, the lack of seismic discontinuities in the lower mantle is consistent with a lack of phase changes at pressures greater than 25 GPa.

The mineralogy of the peridotitic lower mantle is essentially constant to great depths, but at 100–300 km above the CMB there are changes in the seismic behaviour, including suggestions of dense layers and considerable seismic anisotropy. This region is known as the D″ (D-double prime) layer, and there has long been speculation over the causes of these seismic anomalies. Recently, experimental studies at extreme pressures (134 GPa), equivalent to 2800 km depth,

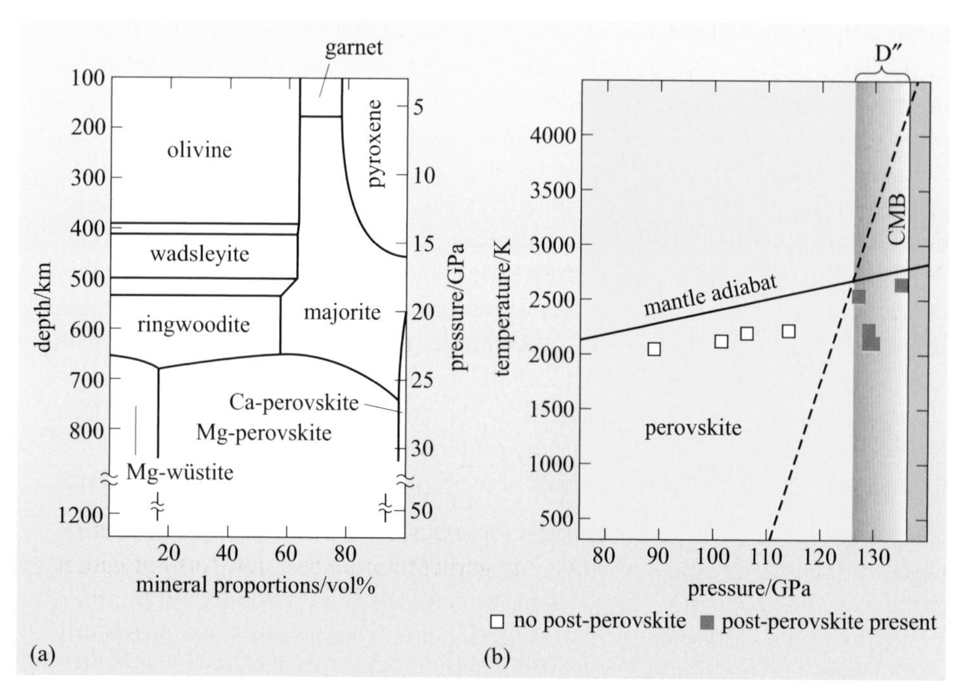

Figure 7.2 (a) Mineral proportions in Earth's mantle for 100–1200 km depth for a fertile peridotite and assuming an adiabatic gradient. The gaps between olivine and wadsleyite, and wadsleyite and ringwoodite respectively are transition regions where both minerals are stable. (b) Plot of experimental data for the perovskite (open squares)–post-perovskite (filled squares) transition in the lower mantle. Shaded area represents the position of the D″ and the sloping dotted line is the best estimate of the position of the perovskite and post-perovskite transition. ((a) Poli and Schmidt, 2002; (b) Helmberger et al., 2005)

have demonstrated that Mg-perovskite transforms to a **post-perovskite phase**, with a 3% increase in density (Figure 7.2b). This phase has a structure that will generate seismic anisotropy, although layers of melt from the outer core and ancient subducted slabs have also been invoked as a way of producing the same seismic effects.

■ Is there a relationship between the depths of the phase changes in Figure 7.2a and the changes in seismic velocities in Figure 7.1?

■ Yes, the 410 km and 670 km seismic discontinuities coincide with the transition of olivine to β spinel and of γ spinel to perovskite.

The key to understanding these phase changes is to remember that they happen without any change in the bulk composition of the peridotite. The phase changes occur over a limited depth range and they correspond to the depths of seismic discontinuities. It is reasonable, therefore, to assume that there is no need for major changes in the chemical composition of mantle at the major transition zones. Although this has been debated for the last 30 years, there is now a consensus that there is no major chemical stratification between the upper and lower mantle.

7.3 Seismic tomography

Seismic tomography is often likened to whole-body CAT scans used in medicine to image the internal structure of the human body. Whereas CAT scans use X-rays, seismic tomography utilises seismic energy from earthquakes. Seismic tomography was first developed in the 1970s, and it has become an increasingly powerful tool to investigate the structure of the Earth as computers have become more and more powerful.

The **preliminary Earth reference model (PREM)** (Section 1.2.4) and the velocity profiles illustrated in Figure 7.1 represent an average velocity structure for the Earth.

Seismic tomography produces a three-dimensional (3D) image of the Earth in terms of deviations in seismic velocity away from this average. When an earthquake occurs (E), seismic waves propagate out from the focus like ripples on a pond. If the position of the earthquake is known accurately, then it is possible to calculate the expected travel time for seismic waves to a seismic station (S) based on one of the Earth reference models (Figure 7.3). If there is a structure in the mantle that either slows down or speeds up the seismic waves, then the actual arrival time will be either later or earlier than expected. By comparing arrival times from different receiving stations, an image of the structure can be progressively built up. From the example in Figure 7.3 it is obvious that the larger the number of seismic stations the better the resolution of the structure that produced the deviations. Because the Earth is a sphere and seismic stations are located all over its surface, tomography images can resolve not only the two-dimensional structure of an anomalous region, as in the schematic section in Figure 7.3, but also its 3D form.

The simplified explanation of seismic tomography given above is a good first approximation to the methodology; but in practice, modern-day tomography uses the arrival times of many other seismic signals in addition to the travel times of P- and S-waves. A modern study can use records from up to 18 000 different earthquakes and hundreds of seismic stations. A best-fit 3D velocity model of the Earth is then calculated using inversion techniques. Calculations that produce global tomographic images involve millions of calculations and can take up to 3 days to run, even on supercomputers!

The resolution of the Earth's interior by seismic tomography is not uniform in either horizontal or vertical planes. This is because seismic stations are more common in the Northern Hemisphere and are concentrated on the continents. Also, fewer ray paths pass through a given volume in the lower mantle than in the upper mantle, so that the resolution of the lower 1000 km of the mantle is more poorly constrained. The increase in ocean-bottom seismographs and specific studies of areas of interest mean that local tomographic images have improved resolutions down to 'cubes' with 100 km sides, which can be imagined as 3D pixels.

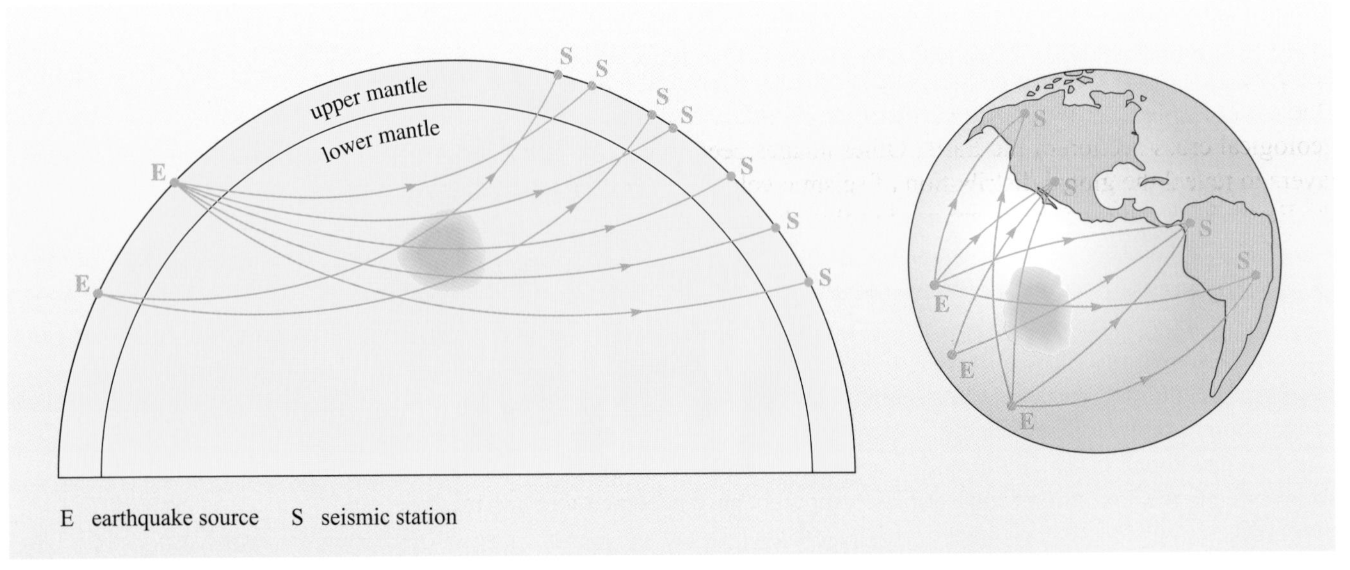

Figure 7.3 The principle of seismic tomography. Information from criss-crossing P-waves (in this case) through the Earth is combined to isolate velocity anomalies such as the shaded area shown here.

Just how good the resolution of the various tomographic images might be and whether they really image structures is a matter of some contention. A formal 'best fit' to millions of calculations is difficult to evaluate, but it is possible to generate 'synthetic' (theoretically generated) seismic data for a given structure, such as a subducting slab. These tests recover the original structure down to a length scale of about 100 km under ideal conditions and suggest that modern tomography is easily capable of resolving features such as subducted slabs and large thermal anomalies. At present discussions are currently focused on the ability of this technique to image narrow (100 km) zones of hot upwelling material in the lower mantle.

Tomography reports the information in terms of percentage difference from the reference models (usually PREM). Seismically slow regions have a negative deviation and seismically fast regions a positive deviation from PREM and are represented on tomographic images as a range of colours. But what do seismically slow and fast regions represent in terms of physical variations in the mantle?

Seismic velocity depends on density, which in turn depends on the temperature and composition of the medium. When a material is heated its density decreases; although density is inversely proportional to the wave velocity (Equation 7.1), the more rapid change in elastic modulus means that the wave velocity also decreases. Likewise, when a material is cooled and its density increases, the wave velocity increases. Therefore, the simplest interpretation of seismic tomographic images is that seismically fast regions represent cool, dense regions and seismically slow regions represent hot, less-dense regions. However, the effects of composition must not be ignored, and the most recent tomographic studies attempt to resolve the effects of temperature and composition on the seismic velocity anomalies. Traditionally, the velocity perturbations are represented using a sliding colour scale, with seismically slow regions being red and seismically fast areas being blue, consistent with their temperature. White regions signify insufficient data to recover useful information. You should also be aware that seismic velocity variations in the upper mantle are generally much greater than those in the lower mantle and often the false colours in tomographic images are 'tuned' to emphasise velocity variations in either one region or the other, but seldom both. Finally, the images are usually sliced in different ways. One method is to slice the image vertically to produce something akin to a geological cross-section of the Earth. Other images peel off constant-depth layers to reveal the global distribution of seismic velocities at different depths in the Earth. Both types of image are used in this chapter.

7.4 Seismic tomographic images

In order to familiarise yourself with seismic tomographic images you will first consider two regions in which you may have some expectations about the likely seismic structure. Figure 7.4 is a seismic tomographic image across the North American continent, where the Farallon Plate has been subducting beneath the continent for a long time.

- What are the characteristics of the subducting slab?
- The slab is obviously seismically fast and, therefore, colder and denser than the surrounding mantle.

- What major feature can you recognise in the lower mantle?
- There is a major zone of seismically fast material that extends from the point where the subducted slab meets the upper–lower mantle boundary. The fast zone extends at a steeper angle than the upper mantle fast zone to depths of greater than 2000 km.

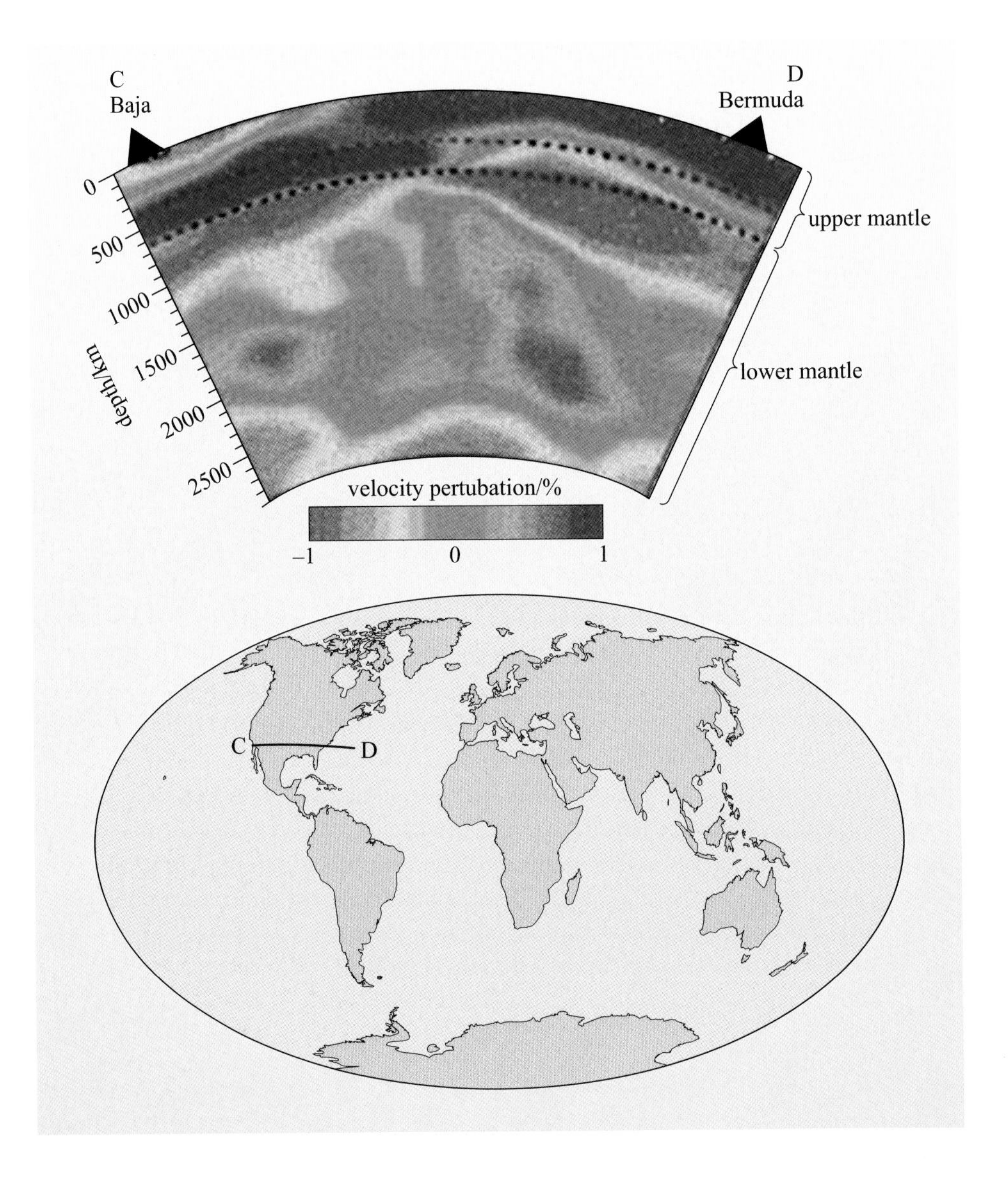

Figure 7.4 Tomographic image along a vertical cross-section across the North American continent. The cross-section runs from Baja to Bermuda and extends to the depth of the CMB. Dotted black lines represent the positions of the 410 km and 670 km discontinuities. Blue (shaded areas to the right of centre in the lower mantle) and red denote the fast- and slow-velocity anomalies respectively. The velocity perturbation scale is shown in per cent. The images are generated from P-waves only. (Lei and Zhao, 2006a)

Figure 7.5 is a N–S seismic tomographic section across the Pacific Ocean from the South Pacific to Hawaii.

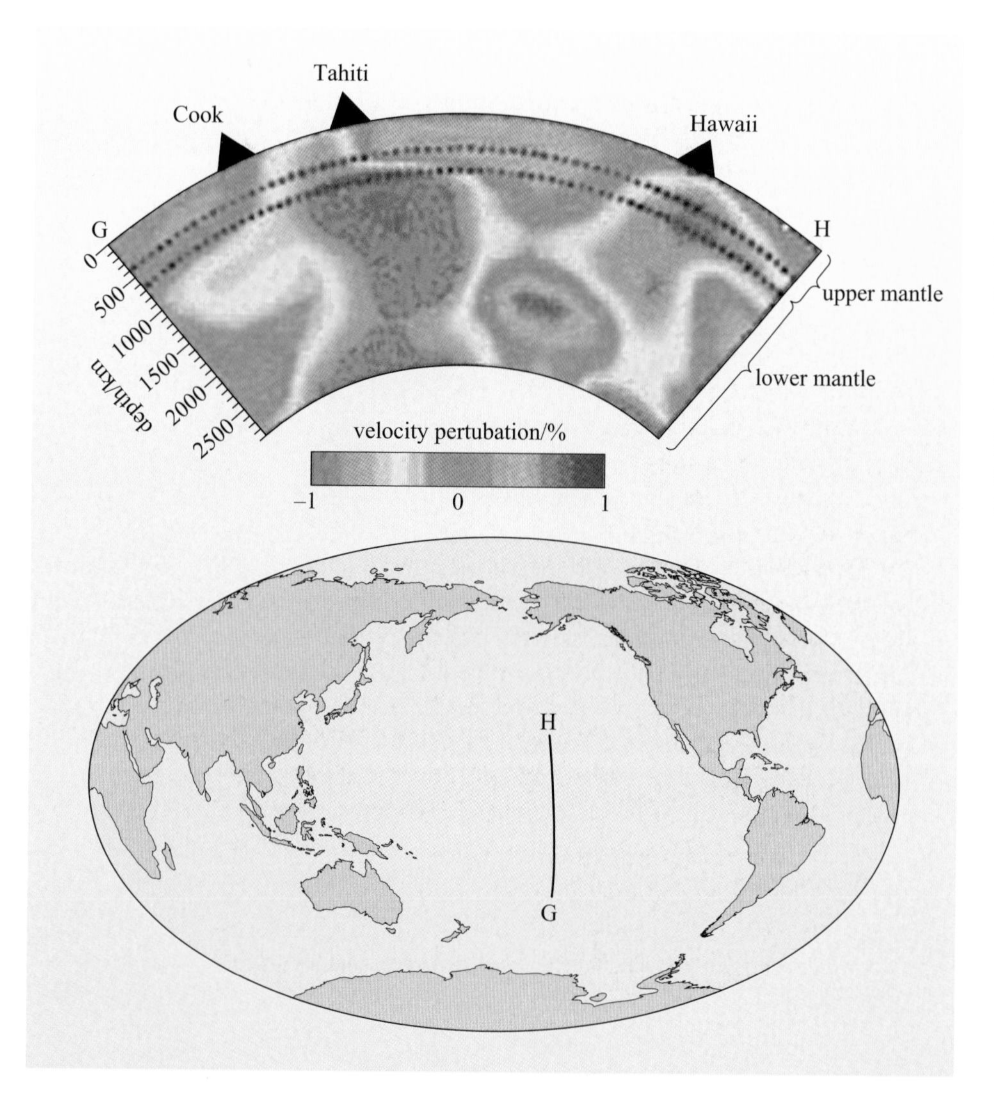

Figure 7.5 Tomographic image along a vertical cross-section across the Pacific Ocean from the South Pacific to Hawaii and extending to the depth of the CMB. Dotted black lines represent the positions of the 410 km and 670 km discontinuities. Red (beneath Tahiti and Hawaii) and blue colours denote the slow and fast velocity anomalies respectively. The velocity perturbation scale is shown in per cent. The images are generated from P-waves only. (Lei and Zhao, 2006a)

- What major features can you recognise in Figure 7.5 in the lower mantle directly below Hawaii and Tahiti?

- There are two major zones of seismically slow material in the lower mantle. One is located beneath the southern Pacific Ocean; the second zone lies beneath Hawaii, but is less laterally extensive and less slow in the middle of the lower mantle.

The large anomaly beneath Tahiti is termed the southwest Pacific super-swell and consists of a broad zone of seismically slow material that extends across the lower mantle. But the question then arises as to what such anomalies might reflect. It is tempting to associate velocity perturbations with just temperature differences, a temptation exacerbated by the use of false colours that have temperature connotations (i.e. blue for cold and red for hot). In the case of seismically fast subduction zones, as in Figure 7.4, the velocity perturbations are consistent with geological and theoretical observations: that slabs should be cold, dense and seismically fast. Tomographic images improve all the time, and the images presented in this chapter use a variety of P-waves, including those reflected off the outer core. By combining these new P-wave phases with the tomographic inversion, the resolution of the lower mantle is significantly improved. Thus, the imaging of slabs in the lower mantle (Figure 7.4) is generally accepted as being a real temperature contrast.

However, there is considerably more debate about whether geophysicists can separate temperature from compositional effects when it comes to seismically slow zones in the lower mantle. The seismically slow zones in Figure 7.5 are commonly interpreted as being associated with increased temperature, and the sizes of the velocity perturbations are consistent with up to a 250 °C increase in temperature, if temperature was the only variable. It has already been noted in Section 7.3 that calculating uncertainties in tomographic images is difficult, but researchers have recently had success in producing models that calculate the probabilities that determine how much temperature and/or composition play a role in the observed seismic anomalies. Such models can successfully separate the effect of temperature from composition and determine their relative importance to buoyancy forces in the mantle. The major conclusion of these studies is that, in the lower 1000 km of the mantle, compositional variations provide the most important component in the buoyancy forces, although higher temperatures are still present.

In the middle to upper portions of the lower mantle, temperature is the most important component in the buoyancy forces. The difference in the importance of temperature and composition to seismic anomalies relates to the effect of pressure on the thermal expansion of mantle materials. A modern interpretation of the southwest Pacific super-swell is that it represents both a compositional (lower density) and a higher temperature anomaly. What causes such large-scale chemical variations is a matter of current discussion; but, as you will explore in Section 7.6.3, the accumulation of subducted slabs may play an important role. Unfortunately, the 'snapshot' aspect of seismic tomography does not indicate whether such large-scale anomalies are stable or long-lived. However, intra-plate volcanism above the current position of the southwest Pacific super-swell has been active for tens of millions of years.

Figure 7.6 shows examples of the most recent (2006) global tomographic images for P-waves, combining a variety of P-wave signals. Each of the 20 images represents a slice of the Earth at a different depth, with the seismic anomalies transposed on to a projection of the Earth's surface. The centre of the Pacific Ocean is in the middle of the image (located somewhere near Hawaii), with continental Europe at the far right and left edges. The positions of the continental margins and the mid-ocean ridges are marked as dark lines. Black triangles mark the positions of known hot spots.

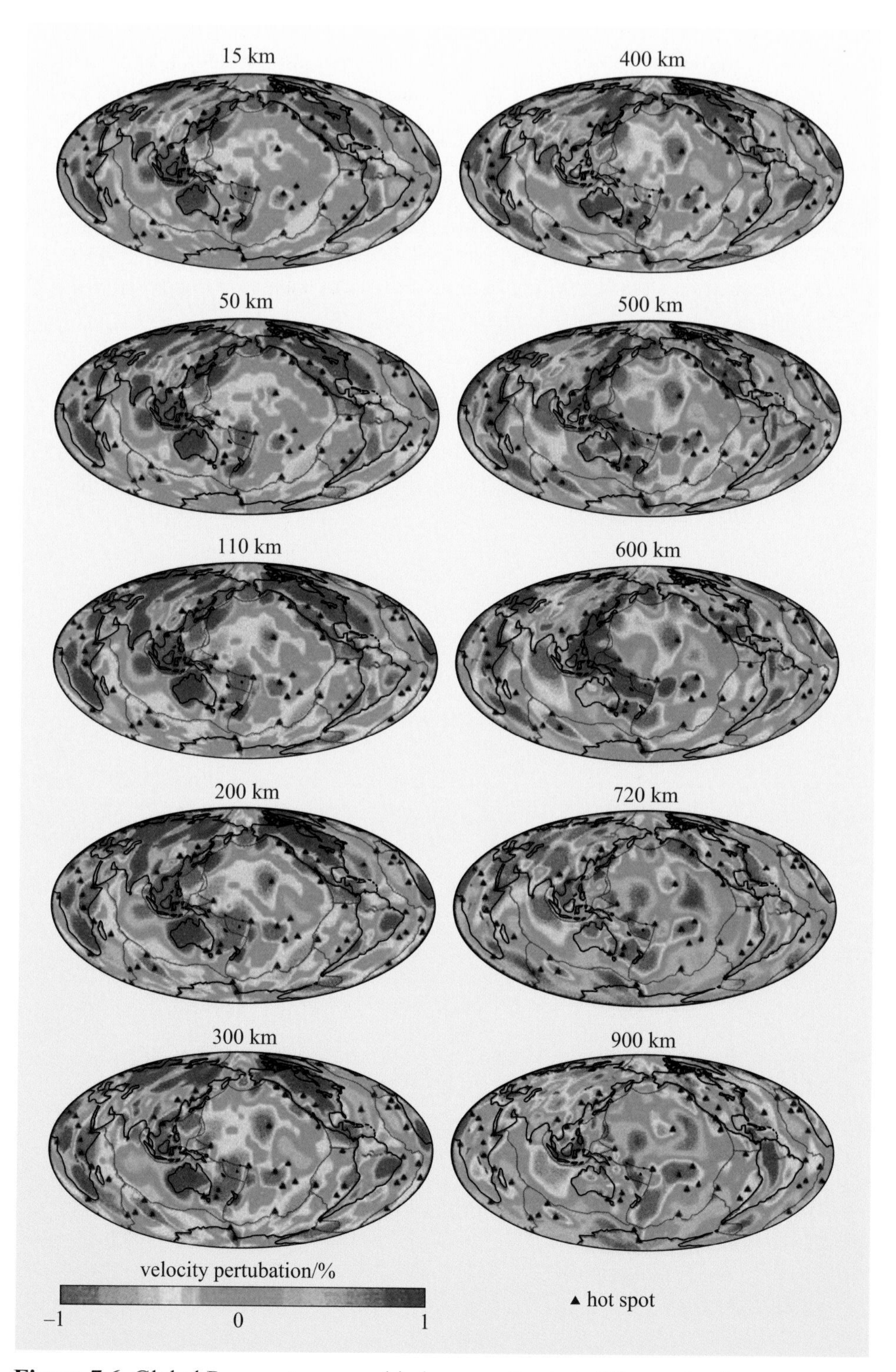

Figure 7.6 Global P-wave tomographic images sliced at different depths throughout the Earth. Red and blue colours denote the slow- and fast-velocity anomalies respectively. Dark lines are continental coastlines and paler lines are tectonic plate boundaries. (Continued overleaf)

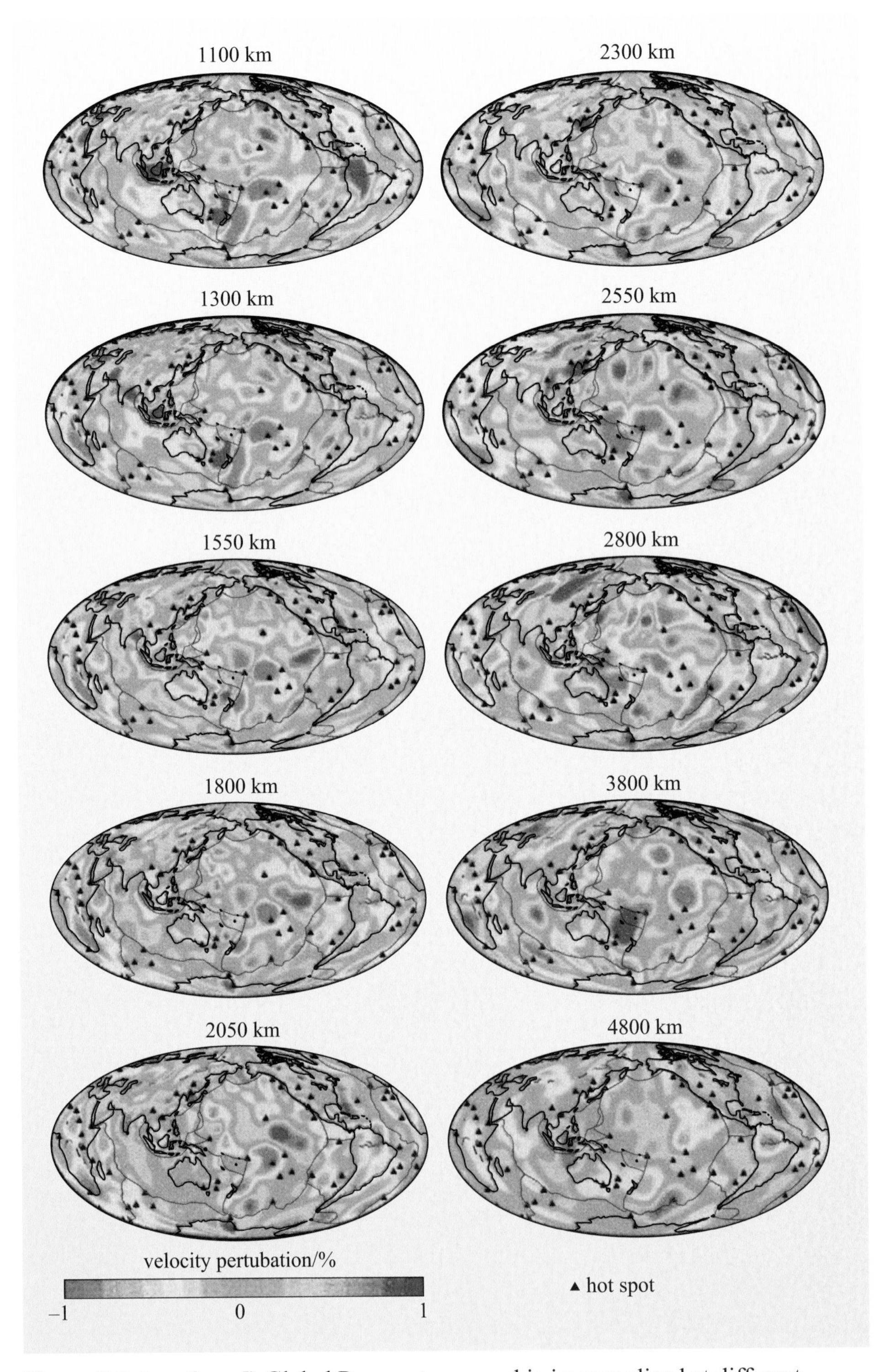

Figure 7.6 (continued) Global P-wave tomographic images sliced at different depths throughout the Earth. Red and blue colours denote the slow- and fast-velocity anomalies respectively. Dark lines are continental coastlines and paler lines are tectonic plate boundaries. There is an enlarged version of part of Figure 7.6 in Appendix D. (Lei and Zhao, 2006a)

Question 7.1

With reference to Figure 7.6:

(a) Describe the distribution of seismic anomalies in the continental lithosphere in the top 110 km of the Earth, across North America, Australia and Africa.

(b) Describe the distribution of seismic anomalies along the SE Pacific Rise, which is a mid-ocean ridge to a depth of 600 km.

(c) Describe the distribution of seismic anomalies along the northern Pacific Rim from North America to Japan and the Marianas in the upper 600 km of the Earth.

(d) Is there a relationship between surface hot spots (e.g. Hawaii) and seismically slow regions in the upper 2800 km of the Earth?

The answers to the above questions should make you realise that there are some connections between seismic tomographic anomalies and surface features that are clear and unambiguous, and others where the connections are less easy to discern. In particular, the upper mantle structure beneath constructive and destructive plate boundaries and the deeper structure of continents stand out clearly, whereas the association between hot spots and the deep mantle is less apparent. Despite these difficulties, seismic tomography provides an alternative view of a dynamic mantle that contrasts strongly with the rather static image presented by the conventional Earth reference models on which they are based.

7.5 Mantle convection

Several of the chapters in this book describe how solid portions of the Earth move and flow like a fluid, so you should be used to the concept of rocks behaving like a very viscous fluid at high temperatures and over geological timescales.

- What drives fluid flow?
- Gravitational potential energy and density contrasts.

Dense materials sink and less-dense materials float; so, if a part of a fluid body becomes either less dense (or more dense) than its surroundings, then it will flow upwards (or downwards) to restore the system to equilibrium. The forces that drive the fluid to flow are known as **buoyancy forces**.

- Name two causes of density changes in the Earth.
- Compositional (chemical) and thermal (temperature) variations.

Changes in the composition of any material invariably change its density, and the same is true of the rocks that make up the Earth's mantle. Similarly, when rocks are heated, they expand and their density reduces.

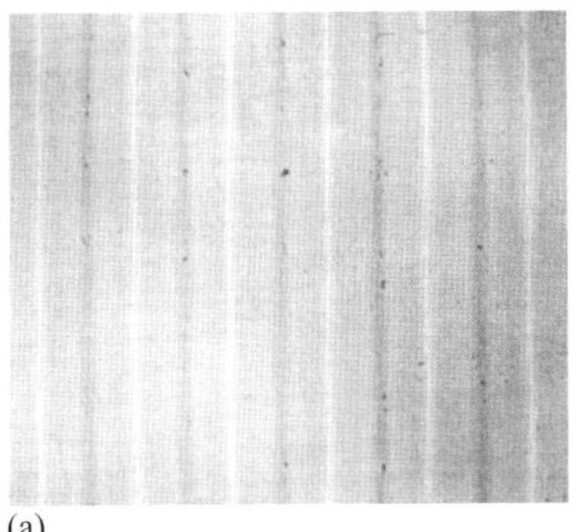
(a)

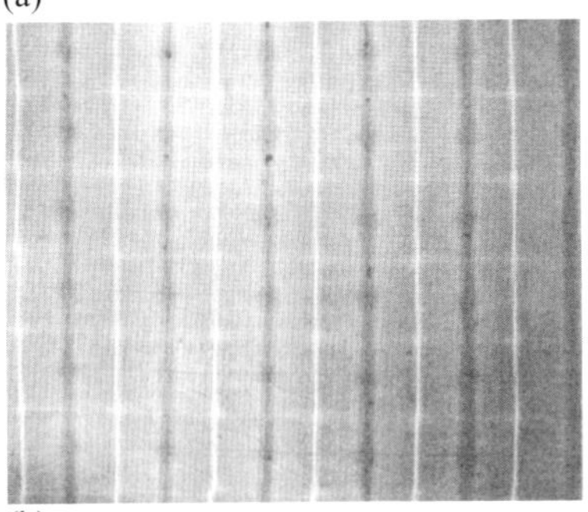
(b)

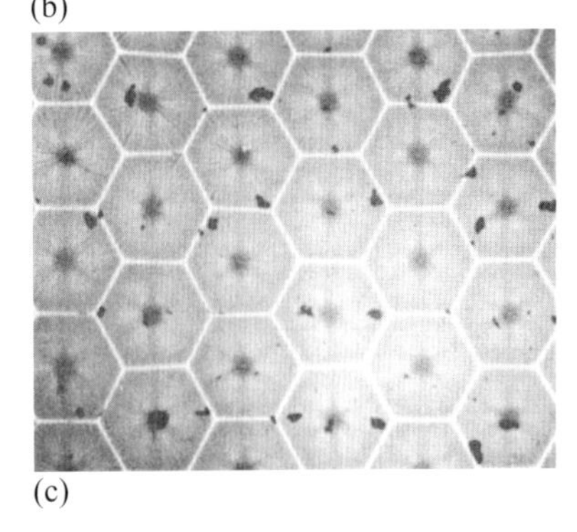
(c)

Figure 7.7 (a–c) Evolution of thermal convection within a fluid. Each image is a view of the surface with the dark regions upwelling and the light regions downwelling. See text for explanation of the surface patterns. (White, 1988)

You will have come across thermal convection in action if you have ever heated a saucepan of soup on a stove. The soup is a fluid that has two thermal boundary layers: a heated layer at the base and a cooling layer at the surface. Movement within the soup is driven by buoyancy forces that result from changes in density caused by heating the lower boundary layer of the soup.

A slightly more sophisticated experiment entails a tank of viscous fluid (golden syrup is an experimentalist's favourite) heated from below. In this set-up it is possible to observe the onset and evolution of thermal convection within the fluid (Figure 7.7). The fluid at the base of the tank is heated and becomes less dense than the overlying fluid, and this density contrast induces convection: sheets of upwelling hot fluid and downwelling cold fluid will form and transport the heat from the base of the pan (lower boundary layer) to the surface of the syrup (upper boundary layer) (Figure 7.7a). With increased heating the fluid forms a second, weaker, perpendicular set of sheets (Figure 7.7b) and ultimately forms coherent hexagonal convective cells with a central rising plume and six sinking sheets (Figure 7.7c). If heating continues, the cells will break and hot material rises at random. This type of convection in a **Newtonian viscous fluid** is known as **Rayleigh–Bénard convection**, and the convection cells have rather box-like geometries such that their **aspect ratio** (the ratio of the width of the cell to the height of the cell) is approximately one. You may be surprised that a laboratory experiment involving a fish-tank full of golden syrup or a numerical experiment on a computer would be useful in understanding convection in the mantle, which takes places over thousands of kilometres and millions of years. The reason for this comes down to some mathematical expressions that describe the flow properties of convecting fluids (see Box 7.1).

7.5.1 Thermal convection

One of the key properties of a fluid is its viscosity. This is just a measure of the ability of molecules in a fluid to move past each other. Water has a low viscosity of 10^{-3} Pa s (the unit is a pascal second) and golden syrup has a viscosity of 10 Pa s, whereas the mantle is highly viscous and has a value of 10^{21} Pa s. For a Newtonian viscous fluid the stress is proportional to the strain rate:

$$\text{strain rate} \propto \text{stress} \tag{7.2}$$

and the constant of proportionality is the **dynamic viscosity**, which gives:

$$\text{strain rate} = \text{dynamic viscosity} \times \text{stress} \tag{7.3}$$

However, the mantle behaves as what is known as a **non-Newtonian fluid**, in which the viscosity varies with the applied strain rate. An everyday example of such a fluid is non-drip paint.

Although the governing equations that describe convection in a heated viscous fluid are well known, they are beyond the scope of this book. However, the flow properties of convective fluids can be described using several **dimensionless numbers** (Box 7.1). Dimensionless numbers are simply mathematical expressions in which the variables (density, times, length, etc.) have been scaled so that the resulting number has no units.

Box 7.1 Thermal convection: the dimensionless numbers

1 The first dimensionless number to consider is called the **Rayleigh number** (*Ra*), which is a measure of the ratio of heat carried by convection relative to heat carried by conduction. (All of the dimensionless numbers described here are named after famous physicists.) The *Ra* value essentially determines whether a fluid will convect or not. It is defined as:

$$Ra = \frac{\alpha g d^3 \Delta T}{\kappa \upsilon} \tag{7.4}$$

where α is the volume coefficient of thermal expansion, g is the acceleration due to gravity, d is the thickness of the layer, ΔT is the temperature difference in excess of the adiabatic gradient across the layer, κ is the thermal diffusivity and υ is the **kinematic viscosity**, which is given by the following equation:

$$\upsilon = \eta / \rho \tag{7.5}$$

in which η is the dynamic viscosity and ρ is the density.

The type of flow at a particular *Ra* is always the same irrespective of the size of the system. Therefore, if a laboratory experiment can be carried out that has the same *Ra* as the mantle, then the results from the experiment can be scaled up (with appropriate time, length and viscosity) to those of the mantle and can provide direct information about mantle convection. Similarly, numerical experiments on computers can be used in the same way. So dimensionless numbers are useful!

2 The governing equations for convection can be solved using different *Ra* and different boundary conditions (usually the temperature or heat flux across the top and bottom boundaries). These calculations are not shown here, but one obvious question needs to be asked – at what *Ra* does a fluid begin to convect? This *Ra* is known as the **critical Rayleigh number** (Ra_c) and is defined as:

$$Ra_c = \frac{\alpha g d^4 (Q + Ad)}{k \kappa \upsilon} \tag{7.6}$$

where Q is the heat flow through the lower boundary, A is the internal heat generation and k is the thermal conductivity. Solutions of the governing equations indicate that although the Ra_c depends on several factors, convection will occur when the Ra_c is about 10^3 within a cell with an aspect ratio of 2–3. Vigorous convection occurs when the *Ra* is 10^5 and more irregular convection occurs at $>10^6$.

3 The final dimensionless number you should consider is the thermal **Péclet number** (Pe_t), which is a guide to the relative importance of heat transport by advection relative to that by conduction. It is defined as:

$$Pe_t = \frac{ul}{\kappa} \tag{7.7}$$

where u is the velocity at which the material is moving, l is the length scale and κ is the thermal diffusion. Equation 7.7 is useful for understanding heat transport where material is being convected upwards, such as under a mid-ocean ridge, where mantle material moves upwards over a discrete distance. Values greater than one indicate that advection rather than conduction is the dominant mechanism of heat transport.

Advection is a special form of convection when a hot region is uplifted by tectonic events (e.g. at a mid-ocean ridge or during isostatic rebound). The heat that is physically uplifted with the rocks is called advected heat.

Dimensionless numbers allow mantle convection to be approached somewhat more quantitatively, e.g. by calculating the value of *Ra* for the mantle.

Question 7.2

Calculate the Ra value of the mantle from the values below for values of $\Delta T = 1$ and 100 K:
volume coefficient of thermal expansion $\alpha = 2 \times 10^{-5}$ K^{-1}
gravitational acceleration $g = 10$ m s^{-2}
depth of the mantle $d = 2900$ km $= 2.9 \times 10^6$ m
thermal diffusivity of the mantle $\kappa = 10^{-6}$ m^2 s^{-1}
dynamic viscosity $\eta = 10^{21}$ Pa s $= 10^{21}$ kg^{-1} s^{-1}
density $\rho = 3300$ kg m^{-3}
ΔT is difficult to determine for the mantle but is unlikely to be <1 K or much more than 10^2 K. Calculate Ra for both cases, giving your answer to two significant figures.

So, does the mantle convect? Calculations for the upper mantle (670 km thick) and the whole mantle (2900 km thick) yield Rayleigh numbers of 10^6 and 6×10^7 respectively, which means that the mantle has an Ra above the Ra_c and will convect. The following question concerns movement of the mantle beneath mid-ocean ridges.

Question 7.3

Mantle decompressing beneath a mid-ocean ridge moves at 0.01 m s^{-1} over 50 km, and the thermal diffusivity κ is 10^{-6} m^2 s^{-1}. What is the thermal Péclet number of this system and how is heat transported?

These simple calculations provide some starting points for understanding heat transport in the deeper mantle; they indicate that the mantle convects vigorously and that convection is the dominant form of heat transport within the mantle.

7.5.2 Convection and plate tectonics

The Earth can be seen as a heat engine trying to lose heat, with this convective heat transport in the mantle driving plate tectonics. An alternative view, as developed in Chapter 3, is that plate forces, such as slab-pull, drive plate tectonics and that mantle upwelling is a passive response to these forces. Clearly, the rigid lithospheric plates are an integral part of the convective system and they dominate the geometry of such a system. Any realistic mantle convection model, therefore, has to incorporate plate-like behaviour.

Experiments with tanks of golden syrup provide important information about some aspects of convection, but are they useful analogies for how the mantle convects? Unfortunately, the answer is no. Convection in these experiments produces cells with aspect ratios of ~1. This is hard to reconcile with plate-scale geometries of 10 000 km (for the Pacific Plate), where the mantle is only 2900 km deep, and more so if the convective system operates only in the upper mantle, which is 670 km thick.

The introduction to this chapter touched on the long-standing debate between whether mantle convection involves two layers of separate convection or a single whole-mantle layer.

- Is there any information from seismic tomography that helps in deciding between the two different models?
- The resolution of slabs subducting through the 670 km boundary reveals a mass flow from the upper to the lower mantle. Assuming the lower mantle is not growing and the upper mantle is not shrinking, there must be a return flow of material from the lower mantle back into the upper mantle. Some tomographic images now resolve seismically slow upwelling regions (e.g. Hawaii, East Africa and Iceland), and it is tempting to interpret these as the upwelling of hot material from the lower mantle.

The images indicate that slabs penetrate the lower mantle. In the upper mantle, subducting slabs are very well defined by seismic tomography, whereas in the lower mantle the signal is rather more dispersed and it can be hard to trace slabs all the way to the CMB. Constant-layer slices at depths of 1700–2000 km, however, do record seismically fast regions whose position is consistent with Cenozoic subduction zones. Many scientists have interpreted the seismically anisotropic behaviour of the D″ layer as being the site of ancient subducted slabs. Therefore, it is clear that slabs do subduct into the lower mantle but that, during the transition through the 670 km boundary and deeper into the deeper mantle, their mechanical behaviour changes.

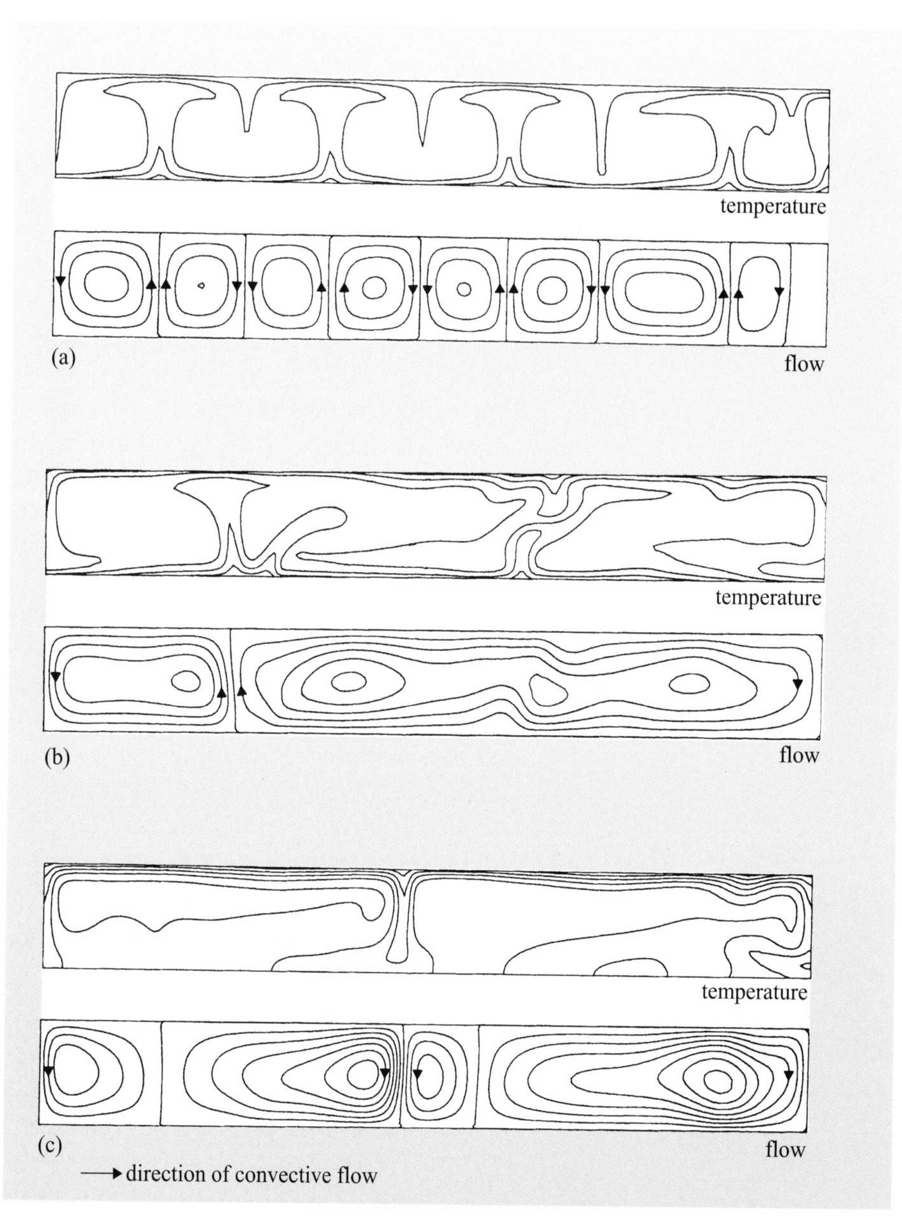

Figure 7.8 (a–c) Temperature distribution and fluid flow for three computer models of convection in the upper mantle. Zones of upward flow relate to upwelling of hot material. The Rayleigh number is 2.4×10^5 and there is no vertical exaggeration. See text for details. (Fowler, 2005)

7.5.3 Modelling mantle convection

Computer models are now the preferred method for understanding mantle convection. To illustrate how sensitive models are to changing parameters, Figure 7.8 illustrates three different models for upper mantle convection in which the Rayleigh number is held constant (2.4×10^5) but the source of heat and/or heat flow is varied. In Figure 7.8a the temperature is held constant at the upper boundary and the heat flow is constant at the lower boundary. This produces convection cells with aspect ratios of ~1 and is analogous to the earlier example of convection in a container of golden syrup. In Figure 7.8b the heat flow is held constant across the upper and lower boundaries, which produces much broader convection cells. The final model (Figure 7.8c) has heat supplied from within the mantle and the heat flow is held constant across the upper surface. This produces broad cells with large temperature variations at the boundaries and rapid convection, characteristic of high Rayleigh-number flow. These models illustrate that it is easy to produce high aspect ratio convection cells that have geometries required by plate tectonics.

Successful models of mantle convection must include a range of factors, including the phase changes that you encountered in Section 7.3. Experimental results and thermodynamic calculations have now defined all the likely major phase transitions that occur in the Earth and how these might affect buoyancy forces and convection. An important property of any phase transition is the so-called **Clapeyron slope**, named after the Clausius–Clapeyron equation in thermodynamics (see Box 7.2). This defines the slope of the phase boundaries on pressure–temperature diagrams and this has a significant effect on the movement of dense material downwards (slabs) and hot material upwards (plumes). If a phase boundary has a positive slope on a pressure–temperature diagram (e.g. the garnet–spinel transition in mantle peridotites discussed in Chapter 4) then it is said to have a positive Clapeyron slope and the reaction is **exothermic**. By contrast, the reaction is **endothermic** for a phase boundary that has a negative Clapeyron slope. The effect on buoyancy is that phase boundaries with negative Clapeyron slopes inhibit slabs penetrating downwards and hotter material moving upwards, whereas a positive Clapeyron slope promotes slab penetration (see Figure 7.9 and Box 7.2). During subduction, as pressure increases, the mineral phases in the subducting slab change progressively to increasingly dense phases. From Figure 7.2, the most important phase changes are γ-olivine (ringwoodite) $\rightarrow$ Mg-perovskite and the breakdown of majorite $\rightarrow$ Mg + Ca-perovskite. The first of these has a negative Clapeyron slope whereas the second has a positive slope. Thus within the cold slab the breakdown of majorite, which may comprise up to 40% of the slab, will occur at a shallower depth than the surrounding mantle and so help to promote deep subduction. By contrast, the negative slope of the breakdown of ringwoodite occurs at a greater depth in the cold slab and may act to prevent a slab penetrating the lower mantle. The net result of these density interactions is that subducting slabs may be stalled at the 670 km boundary. However, the majorite breakdown reaction drives slab penetration if the slab reaches ~800 km depth. The speed of subduction may influence the rapidity of slab penetration with fast-moving old slabs only being stalled at the 670 km boundary for a short period of time (a few million years), but there is no evidence for long-term storage of slabs at the 670 km boundary.

Box 7.2 The Clapeyron law and phase changes

The temperature dependence of a phase transformation is described by the Clapeyron law and is formulated as:

$$\frac{dP}{dT} = \frac{\Delta H}{T\Delta V} \quad (7.8)$$

where dP/dT is simply the change in pressure P as a function of the change in temperature T of a phase boundary (essentially the gradient of the phase boundary on a pressure–temperature diagram), ΔH is the **enthalpy**, ΔV is the volume change of the phase transformation and T is the equilibrium temperature. When a phase transformation has a positive slope ($dP/dT > 0$) (Figure 7.9a) the reaction is exothermic ($\Delta H < 0$), whereas the reaction is endothermic ($\Delta H > 0$) for phase transformations with a negative slope (($dP/dT < 0$). The term ΔV directly relates to the density change that occurs during a phase transformation. The Clapeyron slope of a phase transformation relates directly to the buoyancy forces in a descending cold slab or in ascending hot mantle because phase transformations occur at different pressures to the surrounding mantle. The top panel (Figure 7.9a–c) illustrates the effect of

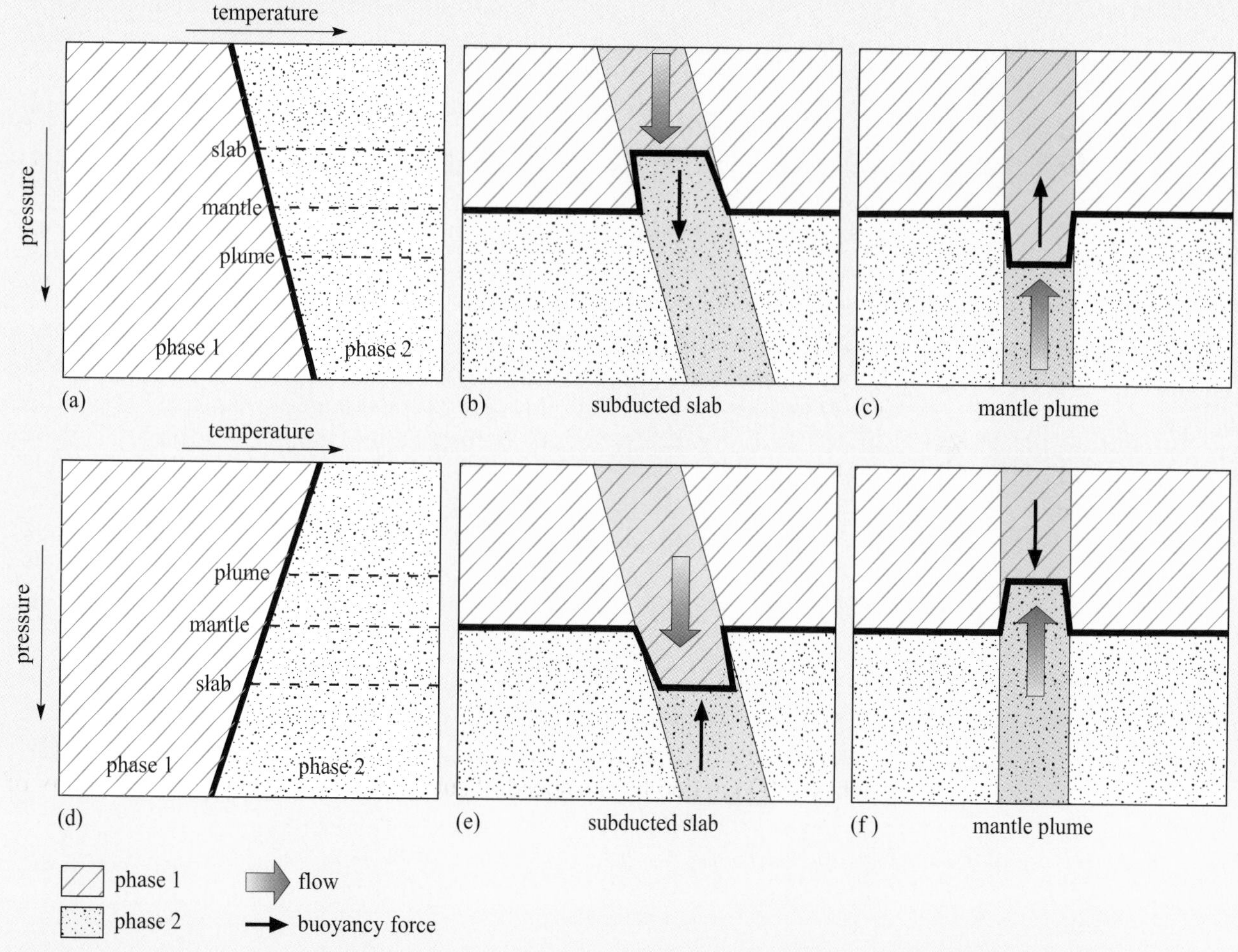

Figure 7.9 Effect of Clapeyron slope on the buoyancy forces. (a) Simplified phase diagram for a phase transformation with a positive Clapeyron slope. (b) Buoyancy forces and the depth of the phase transformation with a positive Clapeyron slope acting on a subducting plate. (c) Buoyancy forces and the depth of the phase transformation with a positive Clapeyron slope acting on a mantle plume. (d) Simplified phase diagram for a phase transformation with a negative Clapeyron slope. (e) Buoyancy forces and the depth of the phase transformation with a negative Clapeyron slope acting on a subducting plate. (f) Buoyancy forces and the depth of the phase transformation with a negative Clapeyron slope acting on a mantle plume.

a phase transformation (phases 1 and 2, assuming the density of phase 2 is greater than phase 1) with a positive slope on a subducted slab and mantle plume. In the cold slab (Figure 7.9b) the phase transformation occurs at a lower pressure than in the surrounding mantle so that the slab has a greater density. This is an additional buoyancy force acting downwards, which will drive the descent of the slab through the mantle phase boundary. In the mantle plume (Figure 7.9c), the phase transformation occurs at a higher pressure than in the surrounding mantle so that the plume has a lower density. This is an additional buoyancy force acting upwards, which is will re-enforce the ascent of the plume through the mantle phase boundary.

The lower panel (Figure 7.9d–f) illustrates the effect of a phase transformation with a negative slope on a subducted slab and mantle plume. In the cold slab (Figure 7.9e), the phase transformation occurs at a higher pressure than in the surrounding mantle so that the slab has a lower density. This buoyancy force acts upwards and operates against the descent of the slab through the phase boundary. For the mantle plume (Figure 7.9f), the phase transformation occurs at a lower pressure than in the surrounding mantle so that the plume has a higher density. This buoyancy force acts downwards and will resist the ascent of the plume through the phase boundary. To resolve whether such phase transformations have a significant effect on convection, modellers have to balance the various forces acting on a slab. Integrating such effects into a convection model is complicated, but modellers now agree that transformations with negative Clapeyron slopes only stall convection for geologically short periods of time.

The other factor that has been added to most models is a depth-dependent viscosity based on theoretical considerations, as shown in Figure 7.10. The viscosity of the lithosphere is high, as it is mechanically strong. Beneath the lithosphere, the viscosity decreases rapidly to a minimum, and then gradually increases with depth to the 670 km discontinuity. Between these two depths the mantle viscosity is lowest, and this corresponds to the asthenosphere. At 670 km there is a viscosity increase of 20–30 times, after which it increases steadily down to the CMB.

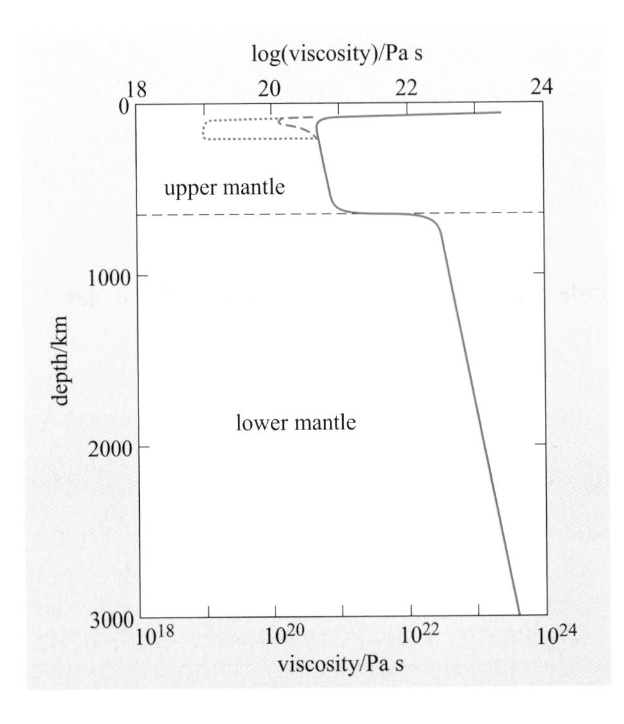

Figure 7.10 A plausible viscosity profile through the mantle. The solid line illustrates the direct, pressure-induced viscosity increase in depth below the lithosphere, and increases due to phase transitions in the transition zone. The dotted lines illustrate the uncertainty in the low-velocity zone in the upper mantle. (Davies and Richards, 1992)

■ How might the viscosity change at the 670 km boundary affect slab subduction?

■ The increase in viscosity will make the transit of slabs into the lower mantle much harder, as it becomes considerably stiffer.

This effect may explain why the tomographic images have the slab more dispersed as it moves below the 670 km boundary, and some images of the Mariana Plate indicate a buckling of the slab at a depth of 670–800 km (Figure 7.11).

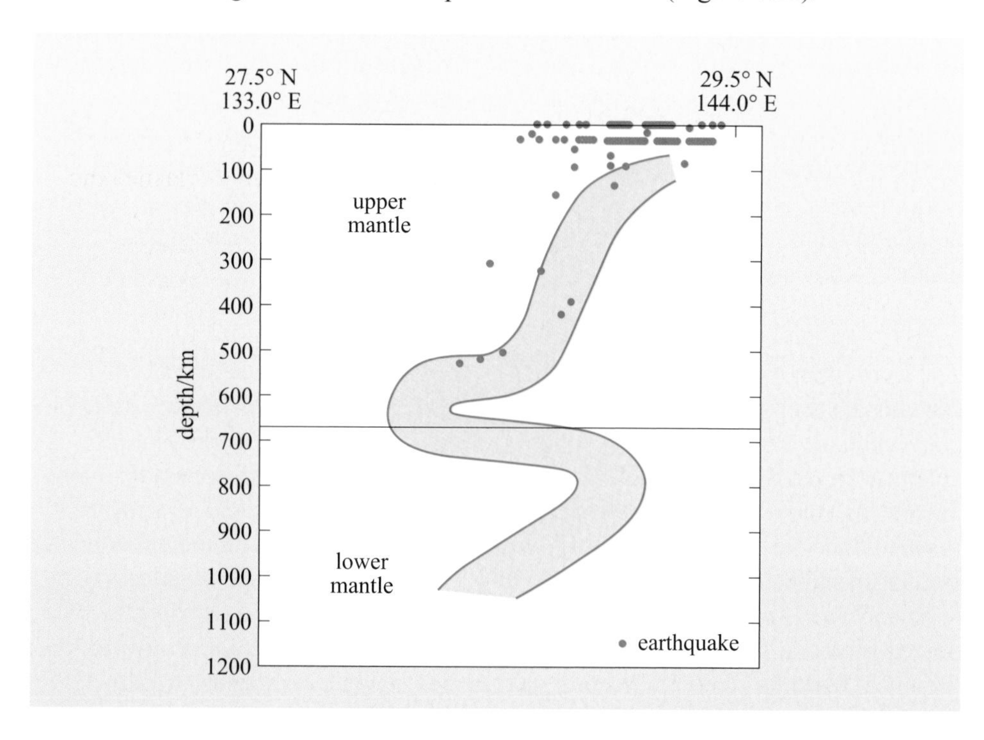

Figure 7.11 Interpretation of the Marianas subduction zone, illustrating the possibility that the slab has buckled upon entering the higher viscosity lower mantle. Shading shows regions of high seismic velocity. (Adapted from Davies and Richards, 1992)

Changes in viscosity and the thermodynamics of mineral reactions are just two of the properties of the mantle that have to be incorporated into realistic models of mantle convection. Other factors include non-Newtonian viscosities, full 3D models and incorporating realistic plate-like features. Linking all these parameters together is mathematically and computationally challenging, but critical results have emerged that impact on our understanding of how the Earth works. These are summarised below.

1 The dominant mode of mantle convection is plate-scale flow with the plates controlling the overall geometry of convection. The plates, therefore, are an integral part of convective flow.

2 Approximately 85% of the Earth's heat loss is through the oceans, and the overall simple topography of the sea floor is consistent with plate-scale flow rather than with many small convection cells in the upper mantle.

3 The 670 km boundary is not a barrier to the flow of slabs into the lower mantle and the upwelling of lower mantle material into the upper mantle. If it were, then the base of the upper mantle would become a thermal boundary layer that would produce a dynamic topography inconsistent with the observed sea-floor topography.

4 An increase in viscosity at the 670 km boundary has several effects. These include: increasing the mixing or stirring time of the lower mantle compared with the upper mantle; producing a time lag between surface plate configurations and deep mantle structure (see tomographic images); complicating the fate of subducted slabs at the bottom of the mantle.

5 Hot upwellings from the CMB can account for 12% of the Earth's heat loss, which is consistent with the heat loss from the core. Such upwellings, known as mantle plumes, are associated with intra-plate volcanism.

Current models suggest that whole mantle convection is the dominant form of mantle flow at present and that it explains many large-scale features of the Earth's interior. There are several points in the summary that need some further clarification. For example, is there any evidence for hot material moving from the CMB to the Earth's surface? What does the stirring time of the mantle mean and why would anyone want to know? What is the ultimate fate of subducted slabs? Various aspects of these questions will be discussed in the following sections, but it is useful to make some general comments first.

Mantle plumes

The word 'plume' is used in several different ways in the geological literature. It can simply mean an upwelling body of fluid, but because early models of mantle convection were based on simple golden syrup experiments that invoked lots of convection cells in the upper mantle, it is often used to explain all manner of slightly anomalous volcanism that occurs within a plate. Most workers would now regard mantle plumes as expressions of deep-seated mantle upwelling that produce extensive volcanism and have obvious geophysical (gravitational, thermal and seismic anomalies) and topographic signatures. The number of plumes that match these criteria is around 15 and includes Hawaii, Iceland and East Africa. Large igneous provinces from the geological record, such as the Ontong Java Plateau, are also regarded as volcanism associated with a plume. For these types of plume there are well-constrained dynamic models that draw material from the CMB and move upwards entraining some lower mantle material. This source material, which is hotter than ambient mantle, ultimately melts in the upper mantle at depths of ~100 km.

Stirring times

Stirring times are simply how long it takes to destroy chemical heterogeneities, such as subducted slabs in the convecting mantle. To continue with the cooking analogies for mantle dynamics, if you add some strawberry jam to a bowl of thick semolina and start to stir, the effect will be to streak out the jam. However, after some time you will distribute the jam throughout the semolina which will eventually turn pink. By contrast, if you add some red food dye to a bowl of water and stirr the two will mix rapidly. The two systems have different mixing times because their viscosities are different. We know from studies of mid-ocean ridge basalts (MORBs) that their upper mantle source region is compositionally homogeneous and has a relatively uniform temperature (Chapter 4), from which you can infer that it is well mixed. Hence, the mixing time of the upper mantle must be substantially less than the age of the Earth. The mixing time of the lower mantle is thought to be longer because of its higher viscosity; therefore, mantle heterogeneities may be preserved for much longer. The actual stirring time for the lower mantle is unknown, but all models produce reasonably well-mixed lower mantles within about 1–2 billion years and upper mantles that are well mixed in much shorter timescales. An example of a computer simulation of mantle stirring is shown in Figure 7.12.

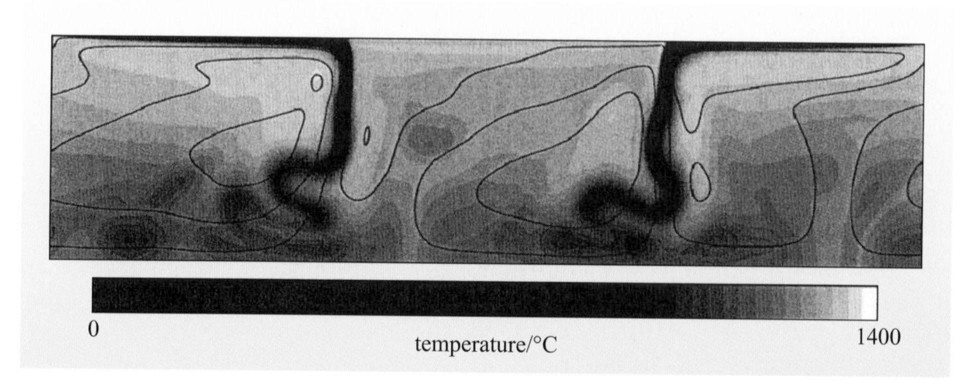

Figure 7.12 Mixing of a passive tracer in a modern, convecting mantle. The tracers are subducted slabs and the mixing is simply shown by temperature variations from 0 °C (black for the slab at the surface) to homogenised higher temperatures (lighter colours). The thin black lines represent streamlines that show the instantaneous direction of fluid flow. (Davies, 2002)

The timescales of stirring have implications for geochemists. The longer a particular compositional heterogeneity is preserved, the more likely it is to generate variations in radiogenic isotope ratios. The observation that some **ocean island basalts (OIBs)** have isotope ratios that are different from those in MORBs is strong evidence that some regions of the Earth's interior can remain separate from the convecting upper mantle for periods of 1–2 Ga. This geochemical constraint requires some mechanism for preserving portions of the mantle with distinct compositions; therefore, the ultimate fate of subducted slabs is crucial in this debate. The transit of most slabs to 1500–1700 km depth in the Earth can be traced readily by seismic tomography, and there is no dynamic reason why they should not traverse the last 1000 km to the CMB. Impressed by this observation, many scientists have suggested that some portions of the subducted slabs reach the CMB and founder in the D″ layer. This slightly ad hoc arrangement allows material to be stored in the D″ for billions of years, where it slowly warms up before being eventually convected back into the mantle.

7.6 Intra-plate volcanism: the Hawaiian connection

7.6.1 Sources and causes of intra-plate volcanism

Volcanism on most of the planet is associated with plate boundaries such as mid-ocean ridges and subduction zones. However, there is an important class of volcanism that is located away from plate boundaries that is grouped under the general term of intra-plate volcanism. Such volcanism can be found on the continents (e.g. East African Rift and Yellowstone, USA) and in the oceans. There are many examples of oceanic intra-plate volcanism, but the best known and most closely studied is located in the Hawaiian Islands.

The Hawaiian Islands are situated in the centre of the Pacific Ocean on oceanic crust that is approximately 80–100 million years old. In Chapter 3, the age progression along the Hawaiian Chain was used to derive the true motion of the Pacific Plate.

- What does this use of hot-spot volcanism tell you about the location of the source of the heat that produces Hawaiian magmatism?
- It must be located beneath the plate.

You should also recall from Chapter 3 that the oceanic lithosphere cools, subsides and thickens as it ages. As a result, the lithosphere beneath Hawaii is about 60–70 km thick.

- If the lithosphere is 60–70 km thick, what does this suggest about the temperature of the asthenosphere beneath Hawaii?

- For partial melting to occur beneath the lithosphere at 60–70 km, the asthenosphere must have a temperature greater than that of the ambient mantle (see Section 4.5).

The lithosphere behaves as a cool lid at a temperature below the mantle solidus, and so for melting to occur beneath this lid the mantle temperature must exceed the solidus at higher pressures than beneath mid-ocean ridges. With a normal mantle potential temperature the mantle starts to melt at about 50 km depth, and so mantle with a normal potential temperature should not melt beneath the 60–70 km thick lithosphere beneath Hawaii. From Chapter 4, you should recall that the control on the maximum depth of melting in a simple decompression melting system is the potential temperature of the upwelling mantle: the higher the potential temperature, the deeper the onset of melting. Thus, for melting to occur beneath Hawaii the mantle must have a higher potential temperature than the mantle beneath ocean ridges. For Hawaii, it has been estimated that the potential temperature is about 1480 °C, which is about 200 °C higher than the ambient mantle temperature. This inferred higher temperature beneath Hawaii and other intra-plate volcanoes is the reason why they are termed hot-spot volcanoes. However, the cause of these higher than average mantle temperatures away from plate margins has been, and continues to be, the source of much controversy.

The realisation that the heat source for hot-spot volcanism lies beneath the plates led to the development of the ideas behind mantle plumes, i.e. focused columns or jets of hot material rising from deep within the mantle. The active volcanic islands are located on a broad topographic high known as the Hawaiian swell, which is associated with a similarly broad negative free-air gravity anomaly. This association has been interpreted as resulting from uplift caused by upwelling in the mantle beneath the islands. The islands themselves are characterised by a large, narrow positive free-air anomaly that results from the excess mass focused within the volcanic islands on top of and flexing the underlying lithosphere.

The plume model provides an explanation for some of the features of ocean island volcanism, but what is the evidence that Hawaiian basalts do originate from hotter than usual mantle? The answer lies in the compositions of the basalts erupted. Decompression melting beneath Hawaii requires that melting starts at higher pressures and temperatures than beneath mid-ocean ridges and produces hotter melts (Figure 4.24d).

- Recall from Chapter 4 which aspect of basalt composition is related to its temperature.

- The MgO content of basaltic magmas increases with temperature.

This observation from experimental studies offers a test of the model that Hawaiian basalts are indeed derived from mantle with higher than normal potential temperatures. If melting beneath Hawaii occurred at higher temperatures, then the primary melts should have higher MgO contents. Submarine glasses from the Kilauea volcano have MgO contents up to 18 wt%, which is higher than the most MgO-rich MORB glasses analysed (10–11 wt%). This information on its own would suggest that the mantle beneath Hawaii is hotter than that below a mid-ocean ridge. However, most MORBs are fractionated, so the comparison is not strictly valid. To assess whether Hawaiian primary melts are generated at higher temperatures than MORB primary melts, the primitive melt compositions have to be calculated, and exactly how this is done is a matter of debate and beyond the scope of this chapter. In summary, most scientists now agree that primitive MORBs have MgO contents seldom greater than 13 wt%, whereas Hawaiian lavas with similarly primitive Mg# have MgO contents of up to 18 wt%, implying significantly higher melting temperatures in the magma source region beneath Hawaii.

A second implication of the plume model for the origin of the Hawaiian basalts is that they originate from greater depths than most MORBs. Again, major element compositions can provide some information in this regard, but it is less easy to relate such compositional variations to pressure than it is to temperature. However, clear pressure indicators can be recognised in the behaviour of certain trace elements in the Hawaiian basalts and their partitioning between the melt and residual mineral phases.

- What is the major difference between the mineralogy of mantle at shallow depths beneath a mid-ocean ridge and that beneath lithosphere up to 75 km thick?

- The shallow mantle has a spinel lherzolite assemblage, whereas at higher pressures the aluminous mineral is garnet (Figure 4.15).

Changes in the nature of the residual minerals during melting have a profound effect on the trace element contents of the melt. Figure 7.13 is a primitive mantle normalised trace element diagram for a primitive Hawaiian lava (from Kilauea) compared with a MORB.

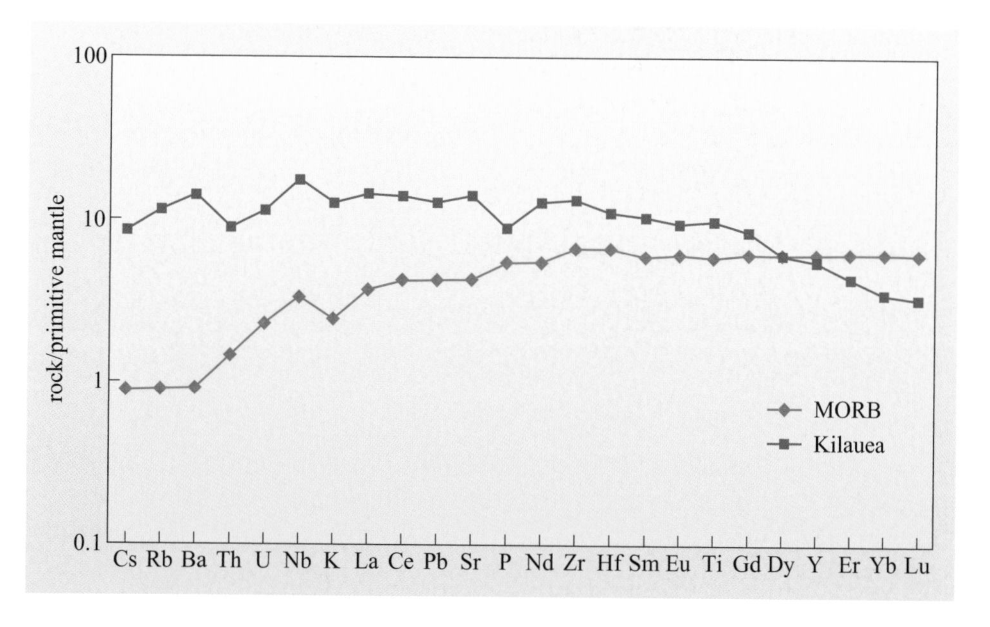

Figure 7.13 Primitive mantle normalised trace element diagram for an average MORB and a primitive lava from Kilauea, Hawaii. See Appendix A for names of elements.

■ How does the Hawaiian trace element pattern differ from that of a MORB?

■ The principal difference from the MORB pattern is that Hawaiian lavas have lower heavy rare earth element (HREE) (Dy–Lu) concentrations and much higher highly incompatible element concentrations (see Section 5.8 for partition coefficients and Appendix A for element names).

The low HREE contents could be explained by higher degrees of partial melting than a MORB, but the higher incompatible element concentrations are more consistent with small degrees of partial melting. So which of these is correct?

If the Hawaiian basalts are generated at depth beneath the lithosphere, then garnet should be an important mineral phase in the source region. Table 5.2 lists the distribution coefficients for a range of elements between garnet and melt.

■ Which element is compatible in garnet?

■ Yttrium (Y).

Yttrium is geochemically similar to the HREEs, and garnet has HREE distribution coefficients that have values between 4 and 7, so the bulk distribution coefficients for the HREEs in a garnet–peridotite are much higher than those for the light rare earth elements (LREEs), so that, at a given melt fraction, the LREEs are more enriched in the melt than the HREEs. By contrast, the bulk partition coefficients for the LREEs for both spinel and garnet lherzolite assemblages are similarly low – the LREEs are highly incompatible during melting at all pressures. The combination of high LREE and low HREE contents relative to MORBs, therefore, implies smaller degrees of melting in the presence of residual garnet. In detail, the trace element patterns are consistent with less than 5% partial melting of the mantle at a depth of 75–120 km beneath Hawaii, which is consistent with melting of hotter mantle beneath the oceanic lithosphere.

Although the amount of melting required to generate the Hawaiian basalts (~5%) is smaller than that required to generate MORBs (15–20%), this does not mean that the flux of magma to the surface is similarly small. At mid-ocean ridges, the relative constancy of the crustal thickness (~7 km) means that the volume of melt extracted from the mantle per unit of time is simply related to the spreading rate. In Hawaii, where there is no sea-floor spreading, the volume of melt generated is related to the amount of mantle that upwells through the melting zone beneath the lithosphere. Observations of eruption rates at Hawaii show that the magma production rate in the source region is about 0.2 $km^3\ y^{-1}$, which compares with a total magma production rate of the entire ocean ridge system of about 20 $km^3\ y^{-1}$. In other words, the single edifice at Hawaii erupts the equivalent of 1% of the volume erupted by the 20 000 km length of the global ocean-ridge system. Generating relatively large volumes of magma in such a focused system implies that upwelling in the mantle plume beneath Hawaii is strong.

Before accepting the hot mantle explanation for Hawaiian magmatism as the only explanation, it is instructive to consider other possibilities. For example, if the mantle solidus is reduced, then it might become possible for melts to be generated by decompression at greater depths without having to invoke higher temperatures.

- From what you have learned in previous chapters, how might the mantle solidus temperature be lowered?
- By lowering the Mg/Fe ratio (Chapter 4) and by adding H_2O (Chapter 5).

Peridotite minerals with reduced Mg# melt at lower temperatures than those with more magnesian minerals, and so a mantle composition that is richer in iron will melt more easily than one with a typical peridotite Mg# of 88–90. Similarly, if the mantle source region is H_2O rich, then it will melt at a lower temperature, as occurs beneath island arcs. In both cases, such source variations will be reflected in the compositions of the primary melts generated, which you would expect to be more water rich and more iron rich than MORBs. For example, primitive lavas from Hawaii have between 0.3 wt% and 0.5 wt% water, which is slightly higher than most MORBs, but much less than in island-arc basalts.

- Give a simple explanation for the higher water content of the Hawaiian melts.
- Because water is an incompatible element, it will be more concentrated in small melt fractions; and, as Hawaiian primary magmas are the result of smaller degrees of melting than MORBs, this might be a reflection of that aspect of melt generation.

Establishing the iron content of the Hawaiian source region is harder to pin down because the iron content of primary melts increases with pressure, and disentangling the effects of pressure from source composition is not straightforward. Suffice it to say that, while the high temperature mantle plume model is widely accepted, it is not universally so. However, any alternative must also explain the geophysical and topographical effects and the magma production rates. At the time of writing, of all the alternatives only the hot mantle plume model comes anywhere close to explaining the majority of the observations.

7.6.2 The source of the Hawaiian mantle plume

As you have read previously, hot upwellings are common features in laboratory simulations of convection. These experiments demonstrate that convective upwelling originates at a thermal or density boundary. In a saucepan of soup, for example, or in a laboratory container of golden syrup, upwelling plume-like structures originate from the bottom of the saucepan.

- From which boundary layers in the Earth could mantle plumes originate?
- Several boundaries are known about from tomographic and seismic profiles of the Earth. The most important are the 670 km discontinuity and the CMB.

The geophysical evidence for the deeper origins of the Hawaiian plume remains unclear. Hawaii has been the location of extensive seismic tomographic studies primarily because it is the place where the plume hypothesis was first put forward. Seismically slow regions are readily identified beneath Hawaii in the upper mantle, demonstrating that the plume has a physical presence at least down to the 670 km discontinuity. Moreover, several studies have located significant seismic anomalies in the D″ layer just above the CMB. But the key question is whether there is a link between the D″ layer and the upper mantle. The most recent tomographic images based on a variety of different waveforms seem to suggest this may be the case

(Figures 7.14 and 7.5), but the connection between the Hawaiian plume (as imaged in the upper mantle) and features in the deep mantle (especially the D″ layer) remain speculative. However, given that the Hawaiian plume is the largest that is currently active on Earth, if any plume were to originate from such great depth it would be that beneath Hawaii.

Petrological and geochemical information about where melting takes place in the mantle is an important constraint on the conditions in the mantle beneath Hawaii; but if Hawaiian lavas originate from some sort of plume or hot spot that originates deep within the Earth, then there should be some differences in the composition of the source material that are reflected in the compositions of the lavas. The trace element patterns of the Hawaiian lavas in Figure 7.13 primarily provide information about partial melting, although there are various trace element

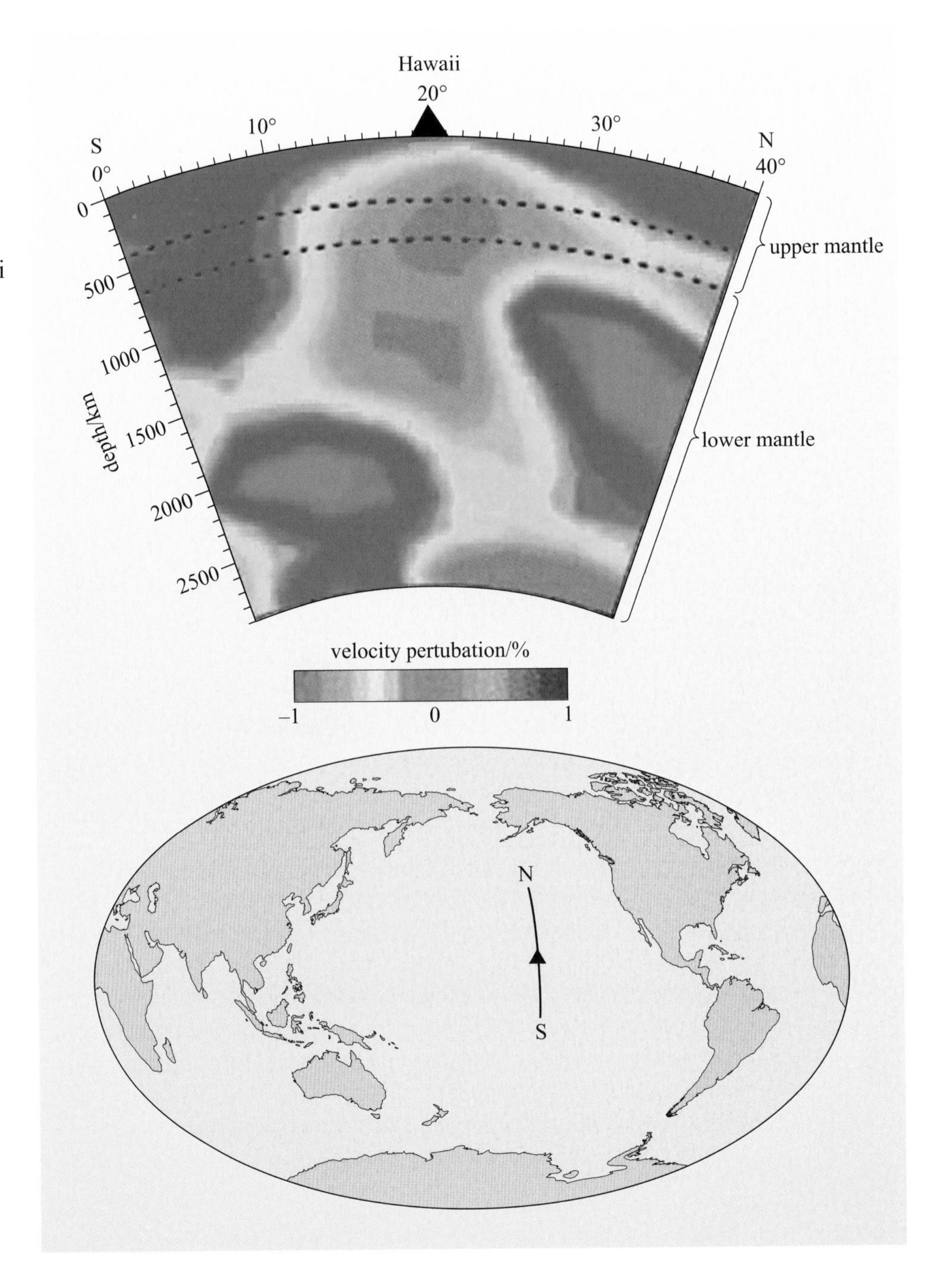

Figure 7.14 P-wave tomographic image along a north–south vertical cross-section passing through the Hawaiian hot spot and extending to the depth of the CMB. Dotted black lines represent the positions of the 410 km and 670 km discontinuities. Red (below Hawaii to bottom right-hand corner) and blue (oval 'blobs' at 40° N and 10° S in the lower mantle) colours denote the slow- and fast-velocity anomalies respectively. (Lei and Zhao, 2006b)

ratios that are used to try to understand whether the mantle source is different from MORB. For example, the ratios of some highly incompatible elements, such as barium (Ba)/niobium (Nb), are different in Hawaiian lavas compared with MORBs, and as these two elements are not fractionated during melting this probably reflects a real compositional difference between the two source regions. However, with trace element ratios there is always the possibility that they have been affected in the tortuous passage from the mantle source to the surface.

By contrast, isotope ratios are less likely to be fractionated from each other during these processes. Consequently, isotope ratios, and especially correlations between isotope ratios and other compositional parameters, have been used to speculate on the origins of the material incorporated in mantle plumes. Radiogenic isotopes are especially useful in this respect because they reflect long-term variations in the parent–daughter elemental ratios in the mantle source, as well as placing constraints on the time required for a particular isotopic signature to develop.

Hawaiian lavas display a much wider isotope composition than MORBs from all three oceans, suggesting they have a more complicated source region than the well-mixed MORB source. Hawaiian lavas also exhibit good correlations between major elements and radiogenic isotopes. For example, FeO and SiO_2 contents (corrected for olivine fractionation back to an Mg# in equilibrium with mantle olivine) correlate with neodymium (Nd) and osmium (Os) isotope ratios (Figure 7.15). These data confirm the different source characteristics of the Hawaiian basalts, but without further information it is difficult to argue for one particular origin for the Hawaiian plume over another. One popular idea is that the compositional variations in Hawaiian basalts reflect the recycling of ancient subducted oceanic lithosphere, the arguments for which are explored in the following section.

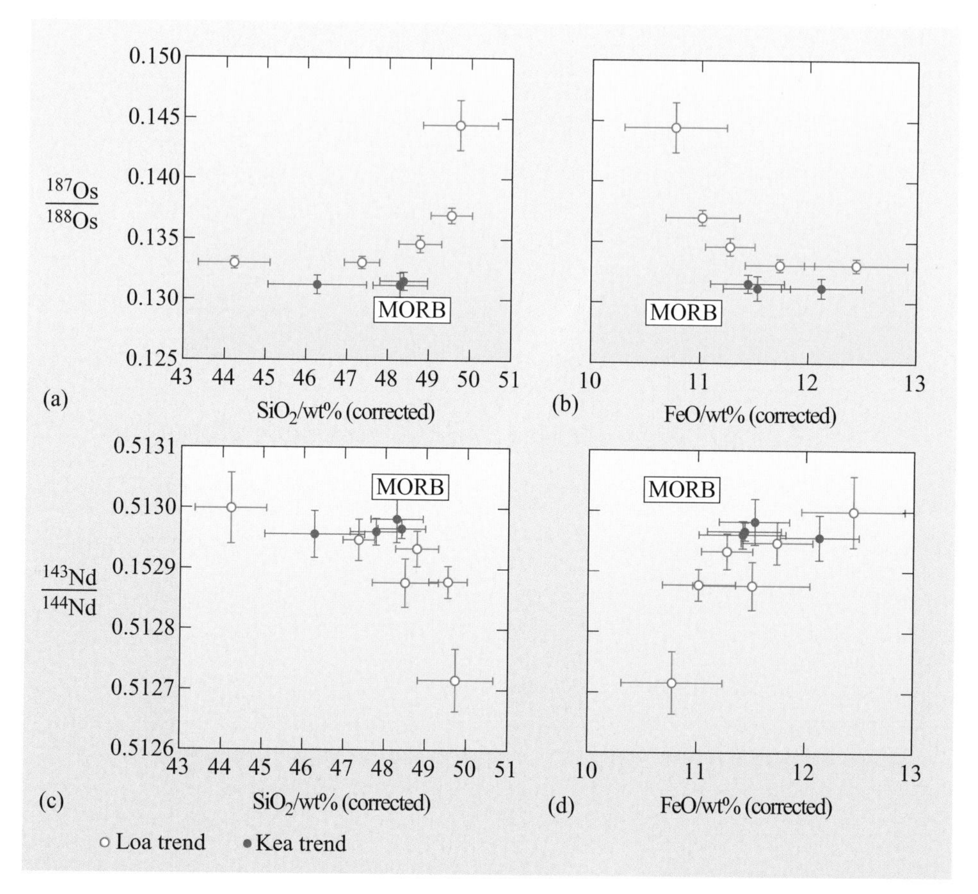

Figure 7.15 Plot of $^{187}Os/^{188}Os$ versus (a) SiO_2/wt% (corrected) and (b) FeO/wt% (corrected) and $^{143}Nd/^{144}Nd$ versus (c) SiO_2/wt% (corrected) and (d) FeO/wt% (corrected). Increasing $^{187}Os/^{188}Os$ and SiO_2, and FeO and decreasing $^{143}Nd/^{144}Nd$ are interpreted as an increased recycled oceanic crust component. (Hauri, 1996)

7.6.3 Recycling of oceanic crust

In Chapter 5 it was demonstrated that a variety of elements are recycled through modern-day subduction zones. Elements such as thorium (Th) and beryllium (Be) may be very efficiently recycled from the slab sediment back into the arc crust but other elements may be transferred to the deeper mantle. The composition of the subducted oceanic lithosphere has the same bulk composition as the upper mantle except that it is divided into a basaltic portion and a residual mantle. The basaltic part of the slab has higher SiO_2 and lower MgO and has a lower Mg# than the residual mantle. It is also enriched in incompatible trace elements compared with the residual mantle. In terms of any additional return of elements to the mantle, the key aspect is whether hydrothermal alteration of the oceanic crust and subsequent processing during subduction adds or modifies elements or element ratios.

- Referring back to Chapter 5, which elements are added to the altered oceanic crust during hydrothermal alteration at the ridge crest?

- The elements uranium (U), caesium (Cs), rubidium (Rb) and potassium (K) are added in significant quantities, and barium (Ba), lead (Pb) and strontium (Sr) to a lesser degree.

In Chapter 5 it was demonstrated that, although these elements may be recycled from the basaltic slab by dehydration reaction back into the mantle wedge, there was strong evidence that subducted sediment was the major source of U, Ba and Rb in arc lavas. This means that the subducted slab may well carry its high U/Pb ratio through the subduction zone and back into the mantle. The generation of high U/Pb ratios in subducted oceanic crust has important implications for Pb isotopes (Box 7.3)

Box 7.3 Lead isotopes

You have already met Pb isotopes in Chapter 2, because they produce the most accurate dates for the age of the Earth. The Pb isotope system is complicated because there are three decay schemes that ultimately produce Pb isotopes: two based on the decay of U isotopes and one based on the decay of a Th isotope. The three decay systems are as follows:

- ^{238}U decays to ^{206}Pb by a combination of alpha (α) and beta (β) decay with a half-life of 4.468 Ga.
- ^{235}U decays to ^{207}Pb by a combination of α and β decay with a half-life of 0.7038 Ga.
- ^{232}Th decays to ^{208}Pb by a combination of α and β decay with a half-life of 14.010 Ga.

^{204}Pb is a stable isotope of Pb, so all Pb isotope compositions are reported as ratios with respect to ^{204}Pb, i.e. ^{206}Pb/^{204}Pb, ^{207}Pb/^{204}Pb and ^{208}Pb/^{204}Pb.

Geochemists like Pb isotopes because they evolve by the decay of two different parent elements that have contrasting geochemical behaviour. Also, the three decay schemes have very different half-lives. This allows processes to be identified that happen over different timescales. ^{238}U decays with a half-life similar to the age of the Earth, whereas ^{235}U has already decayed over six half-lives, so there is now only about 1/64 of the original ^{235}U left in the Earth. Therefore, any process that fractionates U from Pb early in Earth history will potentially produce different $^{206}Pb/^{204}Pb$ ratios relative to $^{207}Pb/^{204}Pb$ compared with a process that happened later in Earth history because of the more rapid decay of ^{235}U relative to ^{238}U.

You are not expected to become an expert in the interpretation of Pb isotopes, but consider the following question:

- What would happen over time if the slab has a higher U/Pb ratio than the surrounding mantle?

- Over time the slab will evolve to have high $^{206}Pb/^{204}Pb$ and $^{207}Pb/^{204}Pb$ ratios relative to the mantle.

The fractionation of U from Pb in the oceanic crust is due in part to alteration that occurs during the hydrothermal interaction close to mid-ocean ridges, and so the presence of high $^{206}Pb/^{204}Pb$ and $^{207}Pb/^{204}Pb$ ratios in OIBs is frequently interpreted to reflect the presence of recycled altered oceanic crust in the mantle source region.

Hydrothermal alteration is only one of the processes that control the parent–daughter ratios of subducted oceanic crust. Another occurs during mantle melting in the generation of MORB magma. This fractionates most radioactive parent–daughter couples by only modest amounts but one is affected more strongly than the rest. During partial melting of the mantle, the elements rhenium (Re) and osmium (Os), which make up a long-lived isotope system (Table 2.1), are strongly fractionated. Re is incompatible and is concentrated in the oceanic crust, whereas Os is compatible and is concentrated in the residual mantle. The ^{187}Re isotope decays to the ^{187}Os isotope by β-decay with a half-life of ~42 billion years, so variations in the ratio of ^{187}Os to another non-radiogenic isotope of Os, usually ^{188}Os, reflect both the time since Re/Os fractionation and the Re/Os ratio. What makes Os isotopes particularly useful is the extreme fractionation of the Re/Os parent/daughter ratio during partial melting. Consequently, MORBs have Re/Os ratios of several hundred, whereas the mantle has an Re/Os ratio of 0.08. Even over tens of millions of years the subducted slab can double its $^{187}Os/^{188}Os$ ratio, and over a billion years the ratio may increase by a factor of 40. These large shifts in isotope composition are very unusual in most materials that might be recycled into the mantle.

The obvious question to ask is whether there are any lavas that have high $^{206}Pb/^{204}Pb$, $^{207}Pb/^{204}Pb$ and $^{187}Os/^{188}Os$ ratios. The answer is yes (Figure 7.16 overleaf). They are found on the small ocean islands of Tubuai, Mangaia and Rurutu in the southwest Pacific, associated with the Cook–Austral islands hot spot, and on the island of St Helena in the South Atlantic. The isotope ratios of the lavas from these islands are such that they require up to 2 Ga to evolve from values typical of the source region of MORB to their present-day values and they provide convincing evidence that the mantle is capable of preserving compositional heterogeneities for this period of time.

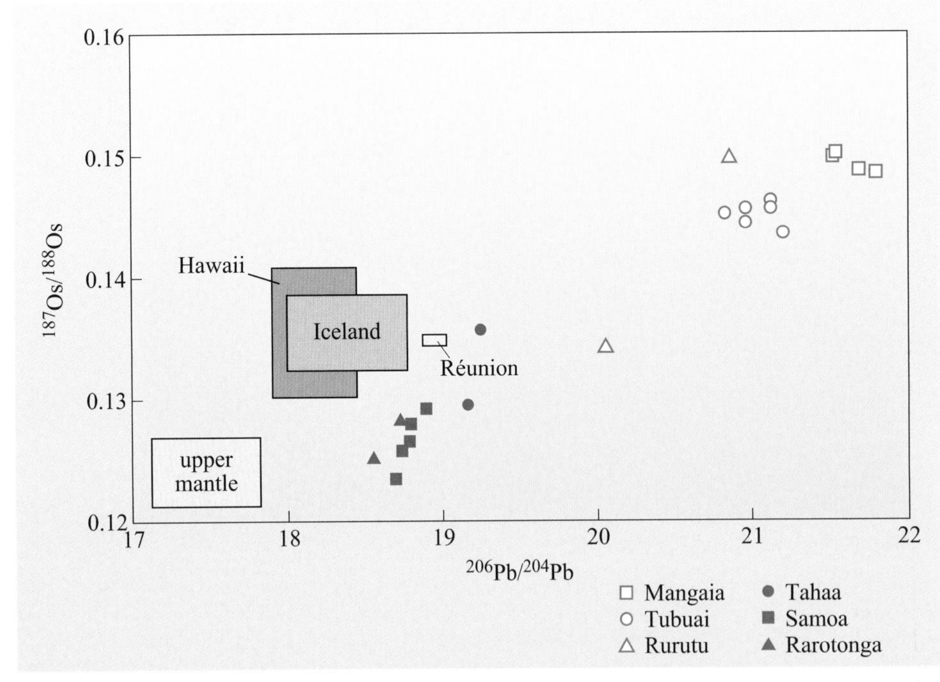

Figure 7.16 Plot of $^{187}Os/^{188}Os$ versus $^{206}Pb/^{204}Pb$ for OIBs. MORB is represented by the upper mantle. The three boxes include data from Hawaii, Iceland and Réunion. (Adapted from Hauri and Hart, 1993)

Compared with basalts from the Hawaiian Islands, the lavas from Tubuai and Mangaia have a much stronger and more uniform isotope signature. In Hawaii, the large range in isotopic data and the correlation of various isotope ratios with major elements is thought to represent mixing of melts from recycled slab material and mantle peridotite. Increasing $^{187}Os/^{188}Os$ and SiO_2, and decreasing $^{143}Nd/^{144}Nd$ are all consistent with the compositional properties of (melts derived from) recycled oceanic crust, and the trends illustrated in Figure 7.15 indicate mixing between recycled oceanic crust and surrounding mantle.

It should be noted that, although the association of isotope variations in plume-related basalts with recycled oceanic crust and lithosphere is currently very popular, it is not a unique interpretation of the data. For example U/Pb and Re/Os fractionation can be effected by processes other than those occurring at mid-ocean ridges. Any melt generated within the mantle can develop high U/Pb and Re/Os ratios; the separation of sulfide minerals and liquids is known to fractionate the Re/Os ratio and will also fractionate lithophile U from chalcophile Pb, albeit to an unknown extent. But regardless of the process of element fractionation, the degree of isotope variation and the lengths of the parent isotope half-lives require that certain parts of the mantle remain isolated from the well-mixed asthenosphere for periods of up to 2 Ga.

The problem still remains as to where material might be stored for such periods. The great temptation is to link the images from seismic tomography to the conclusions from isotope geochemistry. Perhaps subducted slabs that penetrate the 670 km discontinuity accumulate at the base of the mantle where they constitute the D″ layer. After a period of time, determined by its density and thermal properties, material in the D″ layer becomes buoyant and re-enters the mantle convection cycle. Testing such ideas using models of mantle circulation, isotope geochemistry and mantle tomography is currently at the leading edge of scientific investigation and, even though Hawaii is perhaps the most studied group of islands on the planet, there is still as yet no consensus on the origins of the Earth's most active mantle plume.

7.7 The water cycle in the solid Earth

So far in this chapter, the emphasis has been on the way in which heat and gravity drive the inexorable evolution of the Earth, concentrating low-density, incompatible and volatile components in its outer layers. The dominant expression of that process today is plate tectonics, which cycles material from the Earth's interior to the surface and then back into the interior. Images from seismic tomography graphically illustrate the large scale of the plate tectonic cycle and the way in which oceanic lithosphere is returned to the deep mantle, possibly to the CMB. The implication of such a dynamic view of the Earth is that even the deeper regions of the Earth's interior communicate with the outer layers. This then raises the question of whether conditions at the Earth's surface might influence the way the deeper Earth works. Is it possible for a thin veneer of water only a few kilometres deep to have an effect on the vastly more massive and thermally potent mantle? More intriguingly, is it possible that the presence of an oxidising atmosphere, which is a product of the biosphere, affects the workings of the mantle? The remaining sections of this chapter focus on the role of water within the Earth and reach what you may consider to be some startling conclusions.

In order to understand the behaviour of H_2O in the solid Earth today, and to make inferences of its effects in the past, it is necessary to investigate its behaviour in a number of different processes within the plate tectonic cycle. These processes are summarised in Figure 7.17.

The H_2O cycle can be divided into three key stages: melting, subduction and hydrothermal processes.

Melting

You should recall from Chapter 5 that the presence of H_2O reduces the melting point of peridotite by some 100–200 °C.

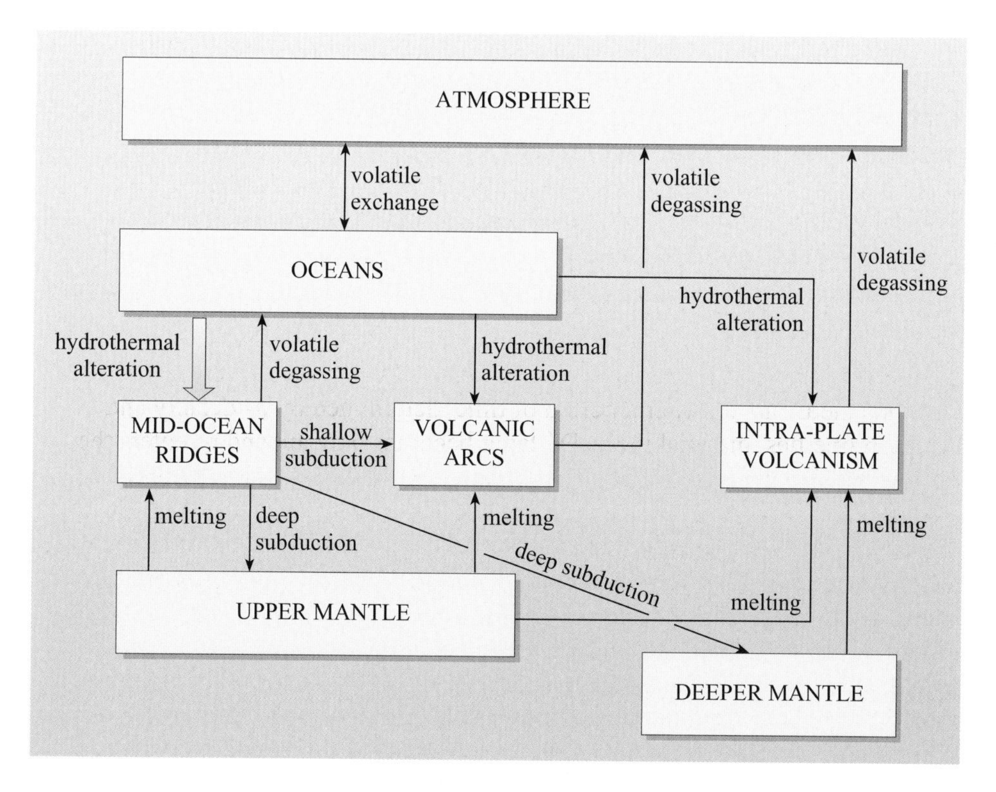

Figure 7.17 Schematic illustration of the H_2O cycle in the Earth. Note the size of the arrow between the oceans and mid-ocean ridges reflects the importance of this process of returning H_2O to the mantle.

- Given this observation, do you think H_2O is compatible or incompatible during melting?
- Incompatible. H_2O reduces the melting point of the mantle because it is highly soluble in silicate melts at mantle pressures.

- What will the effect of melting be on the H_2O content of the residual mantle?
- Melting will effectively strip the H_2O from the mantle; the residue will be virtually anhydrous.

As with any incompatible element, during melting the H_2O in the mantle enters the melt phase and is removed as magma is extracted. Analyses of MORB and OIB suggest that it is about as incompatible as cerium (Ce).

Question 7.4

Using Equation 5.6, calculate the relative amount of H_2O left in the mantle residue after 1, 5 and 10% melting. Assume there is equilibrium melting, a bulk partition coefficient of 0.005 and an initial H_2O content of 1 wt%.

These simple calculations reveal that, with a partition coefficient similar to that of Ce (0.005), only a few per cent melting is required to extract >90% of the H_2O into the melt phase.

- What will be the fate of H_2O dissolved in the magma on cooling?
- As the magma rises towards the surface, cools and crystallises, the H_2O is released and is largely exhaled into the ocean (MORB) or the atmosphere (OIB).

Thus, melting effectively strips H_2O from the mantle and transports it to the surface, leaving the residual mantle dry.

Hydrothermal processes

As you saw in Chapter 4, the magmatic heat released as new ocean crust is produced drives hydrothermal systems around the ocean ridge, and water from the oceans penetrates and interacts with the newly generated solid oceanic crust.

- What is the effect of this circulation on the rocks of the ocean crust?
- Individual mineral phases react with the water to form new mineral phases that are stable at lower temperatures and contain a substantial amount of H_2O.

The end product of hydrothermal activity is a metamorphosed, hydrated basalt that contains up to 5 wt% H_2O. In other localities adjacent to ocean fracture zones and slow-spreading ridges where the uppermost mantle is exposed to the oceans, minerals such as talc and serpentine are also produced from interaction with mantle peridotite. Hydrothermal activity transports many other elements, some of which are fixed in the oceanic crust in addition to H_2O, whereas others are leached and released to the oceans. For the sake of this example, you need only consider H_2O; however, when considering other elements, their mobility or otherwise in hydrothermal systems is a critical part of their geochemical cycle.

Thus, hydrothermal circulation replaces some of the water lost to the oceans during the cooling and solidification of basaltic magma.

As an oceanic plate ages and cools, it interacts less and less with the overlying ocean. Consequently, once the ocean crust is altered it remains altered and carries the inventory of H_2O and other elements acquired at the mid-ocean ridge towards the subduction zone and eventually back into the mantle.

Subduction

During subduction, H_2O is progressively released from the various hydrated phases in the altered mafic and ultramafic rocks, as well as any accumulated sediments, as the slab descends into the mantle. Some H_2O release occurs at relatively low pressures and temperatures associated with the breakdown of clay minerals and other phases that are only stable at low temperatures, whereas H_2O can also be transported to greater depths in phases that have higher temperature stability fields.

- Name two H_2O-bearing silicate minerals that are stable at higher temperatures (>400 °C).
- Amphibole and mica are two hydrous minerals that are stable to many hundreds of degrees in the crust and the upper mantle.

You saw in Chapter 5 how subduction-related magmatism is triggered by the release of H_2O in the mantle at 100–150 km depth, and this is thought to be related to the breakdown of hydrous phases in the subducted slab. Clearly, a substantial amount of H_2O trapped in the hydrated oceanic crust penetrates to mantle depths. One of the critical questions in this discussion, however, is whether or not water can be carried beyond the depths of arc magma generation and back into the deep mantle.

During the hydrothermal alteration of mafic and ultramafic rocks at mid-ocean ridges, two of the products are the minerals serpentine and talc, both magnesium silicates with up to 10% H_2O bound in their crystal structures. These two minerals are stable at low pressures and temperatures below about 600 °C. Consequently, during subduction, as the slab gradually heats up, they break down into other mineral phases, releasing their chemically bound H_2O. However, when experimentalists studied these decomposition reactions in some detail they discovered that mineral dehydration is a complex process and that the breakdown products of serpentine and talc may also be hydrous. These mineral phases are known collectively as **dense hydrated magnesium silicate (DHMS) minerals** and are stable at mantle pressures and temperatures of up to 1000 °C.

Subducted slabs follow a variety of temperature–depth paths, depending on the age of the plate when it is subducted (see Figures 5.9 and 5.11).

As DHMS minerals are unstable above 1000 °C, H_2O can be transported to just over 200 km depth by young (hot) slabs, but to depths in excess of 600 km by old (cold) slabs.

- Can H_2O be stored in DHMS minerals in the mantle?
- No. DHMS minerals break down above 1000 °C and, as all the mantle is hotter than 1280 °C at depth, they are not stable mantle mineral phases.

Eventually, all subducted slabs heat up above 1000 °C and any H_2O stored within them is released into the mantle. If the mantle is hotter than the H_2O-saturated solidus then it will melt, but if it is cooler then it is possible that H_2O may even exist as a free vapour phase. Thus, DHMS minerals provide a means of transporting into the mantle, possibly into the lower mantle, but they do not offer a long-term storage reservoir because of their limited temperature stability. Given that this is the case, where is the H_2O in the mantle located?

H_2O in the mantle

In the mantle, water can be stabilised in phases that contain potassium. Notable among these are magnesium-rich forms of mica and amphibole. Both minerals have potassium as an essential element in their crystal structures, and their stability depends on the availability of both potassium *and* H_2O. You should recall that many arc-related magmas are enriched in the large ion lithophile elements (Chapter 5) that are mobile in H_2O and that potassium is one of these elements, so the conditions required for the development and stability of mica and amphibole in the mantle are not unreasonable. Indeed, both amphibole and mica have been recognised in mantle xenoliths from kimberlites and basalts, but they are not common.

Away from subduction zones, in regions of the mantle with more normal (i.e. low) potassium contents, more likely hosts for H_2O are, surprisingly, phases such as olivine, pyroxene and garnet and their high-pressure equivalents, such as β–Mg_2SiO_4. As with silicate melts at high pressure, nominally anhydrous silicate phases can incorporate small but significant amounts of H_2O in their crystal structures. Analysis of experimental charges has shown that clinopyroxene can contain between 200 and 500 ppm H_2O, whereas β–Mg_2SiO_4 can contain up to 4000 ppm (0.4 wt%). Indeed, the evidence for H_2O in the nominally anhydrous mantle is probably most convincing in the H_2O content of MORB itself, bringing this discussion of the H_2O cycle neatly back to where it began.

Earlier, you considered the relative proportion of H_2O extracted from the mantle during small degrees of partial melting, but observations of the H_2O content of mantle-derived magmas allow the determination of a more quantitative estimate of the content of H_2O in the mantle. Measurements of MORBs show that they typically have an H_2O/Ce ratio of ~200, and this ratio appears to be characteristic of MORBs globally. The similarity of this ratio in MORBs worldwide is strong evidence that H_2O and Ce have similarly low partition coefficients during melting, and from this two important aspects of the mantle geochemistry of water can be deduced:

- the H_2O/Ce ratio of the mantle is ~200
- melting effectively strips >95% of Ce and H_2O from the mantle.

Armed with this information, you should now attempt to answer the following question.

Question 7.5

Given that primary MORB has a Ce content of about 7.5 ppm and assuming MORB is derived by 10% melting of the mantle, what is the H_2O content of the mantle?

This number probably represents a minimum concentration, but it gives a broad indication of the amounts of H_2O that are probably present in the nominally anhydrous mantle source region of MORB. The H_2O contents of basalts from Hawaii and other OIBs appear to be significantly higher than those of MORBs.

■ How can this observation be explained from what you know about the differences between the melting processes that generate MORBs and OIBs?

■ Hawaiian basalts and other OIBs are generated from smaller amounts of melting than MORBs. As H_2O behaves like an incompatible element it will be more concentrated in basaltic magmas generated by smaller melt fractions.

H_2O/Ce ratios provide the key to understanding the composition of the source regions of Hawaiian and other OIBs. As the H_2O/Ce ratios of Hawaiian and other OIBs are similar to those of MORBs, this suggests that their source regions contain similar amounts of H_2O.

From analyses of oceanic basalts, estimates of mantle H_2O concentrations from OIBs and MORBs vary between 100 and 500 ppm. Neither number sounds like very much, but remember that the mantle represents 66% of the mass of the Earth; so, if this concentration is representative of the mantle as a whole, then even these small concentrations may amount to a considerable mass and volume of water locked away in the deep Earth.

Question 7.6

Using the masses of the mantle and the oceans given below, calculate how many ocean masses there are in a mantle if it contains (a) 100 ppm H_2O and (b) 500 ppm H_2O. Mass of the mantle is 40×10^{23} kg; mass of the ocean is 1.4×10^{21} kg.

Thus, the nominally anhydrous mantle may contain as much H_2O as there is in the present-day oceans. You may find this result surprising, especially considering conclusions about the early extensive degassing of the Earth's atmosphere that resulted from the Earth's hot and violent early history. The continual evolution of the Earth has also entailed extensive melting of the mantle both at plate margins and associated with mantle plumes throughout Earth history (Box 7.4), which must have led to the progressive dehydration of the residual mantle. Yet the current estimates suggest that much of the upper mantle contains substantial amounts of chemically bound H_2O in nominally hydrous mantle minerals. Some authors have even proposed that the upper mantle is saturated in H_2O – it can hold no more without it forming a separate vapour phase, although such statements are continually subject to challenge as experimental evidence improves. Notwithstanding these arguments, the corollary of these rough and ready calculations is that to sustain the present-day concentration of H_2O in the mantle, then H_2O *must* be recycled from the surface back into the interior via subduction.

H_2O released from the mantle via volcanism can find its way back into the mantle via subduction because it reacts with the constituent mineral phases of basalt and peridotite in a complex way. The H_2O cycle in the solid Earth, therefore, is a true cycle, a reversible process and, as it turns out, a critical aspect of what makes the Earth work in the way it does.

Box 7.4 How much of the mantle has been melted?

Apart from geochemical arguments about the depleted composition of the upper mantle, you can estimate how much mantle material must have been processed throughout geological time based on present-day plate production rates. Most modern-day melting occurs beneath mid-ocean ridges. If we assume that the present-day rate of spreading, about 2.7 km^2 of new crust per year, is representative of Earth history, then this is equivalent to ~20 km^3 of magma generation per year, assuming an average oceanic crustal thickness of ~7 km (Chapter 4). This in turn implies that about 200 km^3 of mantle is melted each year (assuming 10% melting). Over the course of geological time, 4.5×10^9 years, this means that a total of $4.5 \times 10^9 \times 200$ km^3 of mantle have been processed at mid-ocean ridges, equivalent to 9×10^{11} km^3. Remarkably, this is the same as the volume of the mantle, which is 9×10^{11} km^3.

You might, quite rightly, question the assumptions in this calculation. For example, it is most unlikely that plate and magma production rates have always been the same. However, as you have seen in Chapter 4, production rates depend largely on mantle temperature; so, as the mantle temperature was greater in the past (Chapter 2), magma production rates were also higher and the volume of mantle processed per year was probably higher. Furthermore, this calculation ignores the volumes of magma generated in mantle plumes; although, today, this volume is <10% of that generated at mid-ocean ridges, on occasions in the past, during the development of large igneous provinces, it may have equalled that of the mid-ocean ridge system. Given these considerations, the estimate that a volume of mantle equivalent to the whole mantle has been processed by partial melting may be conservative!

This is not to say that the whole mantle has at some time experienced melting, but if one part of the mantle has escaped melting then another part must have melted more than once. Notwithstanding all of these possibilities, the rates of melting today and the incompatible behaviour of H_2O during mantle melting require that, in the absence of recycling, the mantle should contain virtually no H_2O at all.

7.7.1 H_2O in the solid Earth

By now you should appreciate that H_2O is a critical component of the whole Earth; not just the hydrosphere, but also the crust and the mantle. It controls many aspects of melting, the dominant process of Earth differentiation, and is cycled from the interior of the Earth to the outer layers and back into the interior. The amount of H_2O hidden in the Earth's interior may well be equivalent to the water in the oceans, despite the outgassing associated with melting and volcanism at the Earth's surface, and the oceans now appear to have been a feature of the Earth from Hadean times onwards.

So what other effects does H_2O have on the way the Earth works? One of the most intriguing is its effect on the strength of minerals. In the previous section you discovered that nominally anhydrous minerals such as olivine and clinopyroxene are capable of dissolving small but significant quantities of H_2O. Incorporation of H_2O in silicate lattices is accomplished through two substitution mechanisms involving hydrogen ions (H^+). The first of these involves the direct replacement of Mg^{2+} with $2H^+$. This is the most common substitution, because all mantle silicates contain Mg. The second mechanism involves the coupled substitution of H^+ and Al^{3+} replacing Si^{4+} in aluminium-bearing minerals, such as pyroxenes, particularly garnet. The introduction of water into the crystal structure of olivine may also replace oxygen atoms with OH^-. All of these substitutions serve to modify the structure of the minerals and so reduce their strength. Various experiments have investigated the strength of minerals with the amount of dissolved H_2O, and the results of selected experiments are summarised in Figure 7.18.

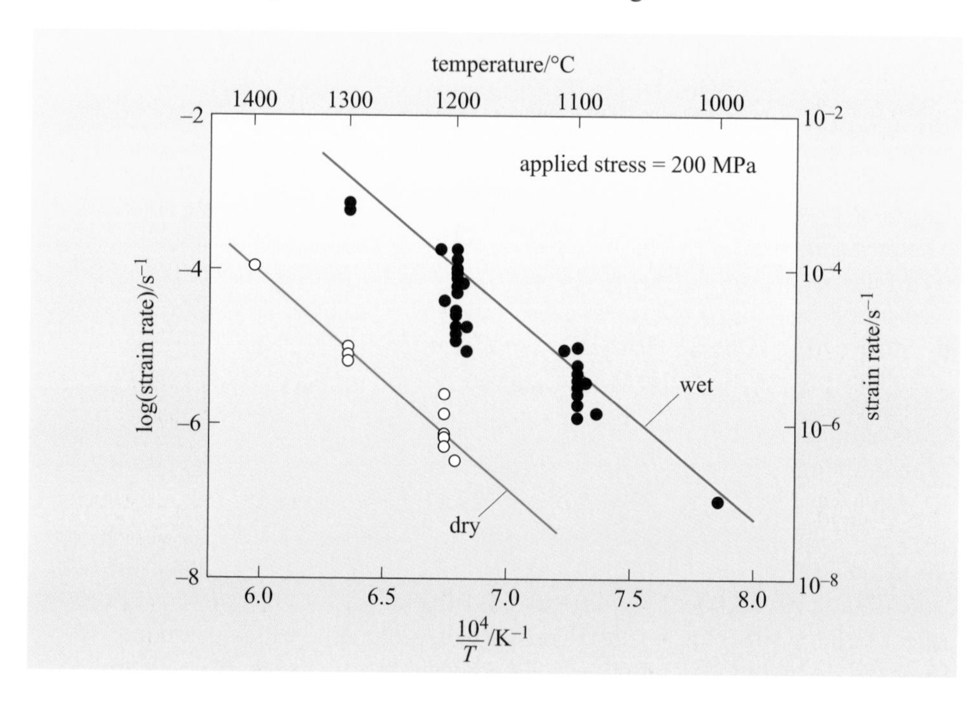

Figure 7.18 A plot of the strain rate of olivine against temperature for dry and H_2O-bearing (wet) systems. The strain rate is a measure of the strength of a mineral under a particular stress. In this case, the applied stress is 200 MPa. The higher the strain rate at a constant stress, the weaker the mineral. (Hirth and Kohlstedt, 1996)

- Using the information in Figure 7.18, by what factor is the strength of olivine reduced by the addition of water at 1200 °C?

- At 1200 °C the dry olivine curve has a value of 10^{-6} s^{-1}, whereas the wet olivine curve has a value of 10^{-4} s^{-1}. Thus, dry olivine is 100 times stronger than wet olivine.

- What temperature change produces the same change in strength in dry olivine?

- The dry olivine curve has a strain rate of 10^{-4} s^{-1} at 1400 °C, and a value of 10^{-6} at 1200 °C, so a similar change in strength is produced by a 200 °C temperature change.

The amount of H_2O required to reduce the strength of olivine significantly is surprisingly small, of the order of 50 ppm.

- How does this value compare with your estimate of the H_2O content of the mantle determined in Question 7.5?

- It is smaller: you calculated previously that the upper mantle has between 100 ppm and 500 ppm water.

Thus, the mantle source region of both MORBs and OIBs has more H_2O than is necessary to reduce the strength of olivine by a factor of ~100. This has clear implications for the way the mantle behaves under stress because olivine comprises 50–60% of the upper mantle. For example, the strength of olivine determines how the mantle responds to isostatic loading by the overlying crust and lithosphere. Clearly, a weaker mantle will respond more rapidly to loading or unloading than a strong mantle. Weakening olivine allows the mantle to undergo solid-state flow more easily, and so aids mantle flow. To put it another way, water reduces the dynamic viscosity of the mantle. The details of calculating the effects on the dynamic viscosity are beyond the scope of this chapter, but the overall effect of more than 50 ppm water in olivine is to reduce the dynamic viscosity of the mantle by two to three orders of magnitude (a factor of 100–1000).

You should recall from earlier in this chapter that the viscosity of the mantle determines how vigorously the mantle convects. The style of mantle convection is determined by the Rayleigh number Ra of the mantle, and mantle viscosity is one of the parameters used to determine Ra.

■ How will an increase in mantle viscosity effect Ra?

■ Dynamic viscosity η is a factor in the denominator of the equation for Ra (see Equation 7.7 in the answer to Question 7.2). Its effect is to dampen motion rather than drive it – so an increase in viscosity will decrease the Rayleigh number.

For the mantle to convect, the value of Ra must be significantly greater than the critical Rayleigh number Ra_c. The current value of Ra for the mantle is of the order of 10^5 or 10^6. This is substantially greater than Ra_c, which is around 10^3. Thus, the situation today allows mantle convection and, hence, plate tectonics. Increasing the viscosity by a factor of 10^2–10^3 will decrease the value of Ra by a similar amount, bringing it much closer to the critical value.

■ What would be the effect of reducing the Rayleigh number of the mantle?

■ A lower Ra means that mantle convection would become more sluggish.

As a final point, consider the following question:

■ Given that the lower parts of oceanic plates are made up of residual harzburgite, how will their strength and viscosity be affected?

■ As harzburgites are the product of significant melt depletion, they should have low H_2O contents and so be much stronger than the underlying asthenosphere.

The loss of H_2O from the mantle during the generation of new oceanic lithosphere increases its strength considerably, thus encouraging the thermal development of plate structure.

It should now be apparent that the H_2O content of the mantle has a controlling effect on both the style and vigour of mantle convection and the development of lithospheric plates, even at the small concentrations present in the mantle today. But the only way in which that H_2O content can be maintained over geological time is as a result of recycling of water from the oceans through hydrothermal modification of the ocean crust and subduction of the resultant hydrated minerals.

7.7.2 Positive and negative feedbacks

On the Earth today, the presence of the oceans allows the mantle to remain hydrated to a sufficient degree to allow it to convect relatively easily and, geologically speaking, rapidly. In addition, the dehydrated residue after melting is significantly stronger than the fertile mantle. As the mantle directly beneath the crust is depleted harzburgite (Chapter 4), this adds to the strength of the oceanic lithosphere, thus aiding the generation and preservation of lithospheric plates. Both effects are essential for maintaining plate tectonics. However, plate tectonics is also the mechanism whereby the mantle is rehydrated via subduction. Thus, it is a system in which the very actions of the system keep that system operating.

■ What would happen to the viscosity of the mantle if the rate of subduction increased?

■ More H_2O would be returned to the mantle; the viscosity would be reduced.

If mantle viscosity is reduced then subduction would be easier, forcing more H_2O back into the mantle, thereby further decreasing mantle viscosity, and so on. Such a phenomenon is known as a **positive feedback** because the operation of a process has an effect that helps that process to operate more efficiently.

So why isn't there an ever-increasing rate of plate motions? The answer lies in the way in which the rate of subduction controls the rate of ocean spreading. You should recall from Chapter 3 that the dominant driving force for plate motion is slab-pull, so if the rate of subduction increases, so does the rate of spreading. An increase in the rate of spreading leads to a concomitant increase in the volume of magma generated at mid-ocean ridges. This, in turn, dehydrates a proportionately larger volume of mantle, increasing the overall viscosity of the mantle and hence slowing convection. Melting, therefore, provides a **negative feedback** on the system, acting to increase mantle viscosity and so slow the system down. Feedbacks, both positive and negative, are important concepts in adopting a systems approach to studying the Earth. In the case of plate tectonics and mantle convection, it is the balance between two processes, i.e. dehydration by melting and rehydration through subduction of 'wet' oceanic plates, that keeps the mantle operating in the way it does today and, as far as we can judge, has done for much of geological history.

7.7.3 The Earth without an ocean

It is instructive now to consider the consequences if the Earth did not have a significant hydrosphere, but had a surface dominated by bare rock and no major oceans. Clearly, the surface of the Earth would be very different. For a start it would be lifeless: the evolution of life as we know it depends on water; and the surface temperature would swing between much greater extremes: the oceans have a controlling and buffering effect on atmospheric temperatures because of their large heat capacity. But in this chapter you are considering the way in which the solid Earth system works, so how would this be affected?

Assume that the initial H_2O content of the mantle and its temperature gradient are both similar to that of today. With this concentration of H_2O the mantle will convect and hot material will rise towards the surface and melt. As with the plate cycle discussed at length above, melting in a ridge or plume environment will extract H_2O from the mantle and transport it towards the surface. As the magma cools and solidifies, the H_2O will escape to the surface because H_2O is incompatible in most silicate minerals. However, on the waterless planet there is now no ocean to trap it, and so it will escape into the atmosphere.

The next stage in the dry Earth's water cycle should be hydrothermal interaction with the oceans; but, because the oceans are missing, there can be no further interaction. Hence, the plate will cool and subside as now, but it will retain its largely anhydrous igneous composition. Already the cycle has been broken and the path followed by H_2O through the Earth has changed.

As the plate ages and moves towards subduction, it does not accumulate a significant sedimentary layer, again because there is no active surface H_2O cycle to drive erosion and deposition as there is on Earth today. On subduction, therefore, the material being taken down into the mantle is dry: there is no water in sediments and no water in the crust – just dry igneous rock of broadly basaltic composition.

■ What is the consequence for destructive plate boundary processes on our dry Earth?

■ There will be no magmatic activity associated with subduction.

Subduction-related magmatism is triggered by dehydration reactions at depth in the mantle. Given that the subducting plate is cold and the mantle is not hot enough to melt without decompression, there will be no melting because the subducting plate is not carrying H_2O into the mantle and will simply act to cool the surrounding mantle wedge.

As subduction proceeds further the plate sinks into the deep mantle and gradually equilibrates with its surroundings.

■ What is the next consequence for our dry Earth?

■ No H_2O is recycled into the mantle because, once again, the plate is dry.

In summary, without an ocean covering the ocean ridge to capture the water released by cooling magma and to rehydrate the plate via hydrothermal and sedimentary processes, there would be no magmatism at destructive plate boundaries and the mantle would become progressively drier. The feedbacks that maintain the current convective movements and plate cycling in the mantle would not be balanced and new continental crust could not be generated because of the lack of subduction-related magmatism. Thus, the hypothetical dry planet will eventually lose any H_2O trapped in the mantle, plate tectonics will cease and there will be no new continental crust formed above subduction zones.

But just how hypothetical is this planet? Can these predictions be tested against another planetary system? Returning to Section 1.1, you should recall that, despite the similarity in radius and mass, observations of Venus have revealed that its surface is too hot for water to exist, so perhaps this is a model for the Earth without water.

- Can you recall from Section 1.1 the differences between the geological structure of the lithosphere of Venus and Earth?
- Venus shows no evidence for plate tectonics and no evidence for a bimodal hypsometric plot.

Although Venus is clearly an active planet, as reflected by the geological youth of many of its surfaces, there is no evidence for active plate tectonics, and the hypsometric plot does not suggest distinctive regions of oceanic and continental crust. Thus, first-order predictions for a dry planet are borne out by primary observation.

Whether Venus has always behaved as it does today or not is a matter of active research and discussion. Remote spectroscopic measurements of the isotopic ratio of traces of water in the atmosphere show it to be much richer in deuterium than water is on Earth, suggesting that Venus may originally have had as much water as the Earth had, but subsequently lost it. Alternatively, the high surface temperature may have allowed its inventory of H_2O to be locked deep in the mantle. Numerous hypotheses abound; but what remains clear is that, without a surface hydrosphere, plate tectonics and the differentiation of the continental crust become much less likely.

Summary of Chapter 7

This chapter gives a taste of how the Earth operates as a system and how processes that operate at great depth can influence, and be influenced by, surface processes. The whole Earth system determines the geochemical and geophysical evolution of the Earth. Some important points are summarised below.

- The physical properties of the minerals are known throughout the whole 2900 km depth of the mantle, based on modern experimental methods. Physical properties, such as density and elastic modulus, control the seismic velocity of the mantle.
- Physical properties of the mantle can change even if the composition remains the same (e.g. spinel–peridotite to garnet–peridotite in the upper mantle) and there is now a consensus that there is no major chemical stratification between the upper and lower mantle.
- Major seismic transitions correspond to phase transitions: olivine to β spinel and β spinel to γ spinel at ~410 km and 520 km respectively; Mg-perovskite and Mg-wüstite occur at ~670 km and, along with Ca-perovskite, make up the mineralogy of the lower mantle. The lack of seismic discontinuities in the lower mantle is consistent with the lack of phase changes at pressures greater than 25 GPa (~670 km). The transition from the lower mantle to the outer core, the D″ layer, is complex and involves a further phase change of Mg-perovskite to a post-perovskite phase at a depth of 2800 km. The core–mantle boundary is marked by a change from solid peridotite to the liquid, metallic outer core.
- The physical properties of the different phases in the mantle feed into seismic tomographic images of the Earth and computer models of mantle convection. Seismically slow regions of the mantle have a negative deviation from PREM

(red areas on the tomographic images in Figure 7.6); seismically fast regions have a positive deviation (blue areas in Figure 7.6). Seismic velocity depends on density (and hence composition and temperature). Thus, in general, hotter materials have lower density and slower wave velocities while cooler materials have higher density and faster wave velocities.

- Seismic tomography resolves subducting slab to depths of 2000 km (e.g. the Farallon Plate below North America), and hotspots (e.g. the Hawaiian plume, with temperatures up to 250 °C above ambient mantle). Seismically slow zones may result from both temperature and compositional variations and their relative importance is disputed. However, mantle plumes probably relate to both temperature anomalies and are compositionally distinct (possibly comprising of ancient subducted slabs that have spent time at the core–mantle boundary).
- Buoyancy forces, which comprise gravitational potential energy and density contrasts, drive fluid flow. Density is controlled by composition and temperature. Convection is described by dimensionless numbers, which is useful because the results of small-scale laboratory and computer experiments can be scaled up and hold good for the entire mantle.
- A key parameter in understanding convection in the mantle is viscosity. Viscosity is a measure of how easily a fluid can flow; the mantle is a non-Newtonian fluid with a viscosity of about 10^{21} Pa s, making it a very sticky fluid.
- Key dimensionless numbers are the Rayleigh number Ra (the ratio of heat carried by convection relative to the heat carried by conduction) and the critical Rayleigh number, Ra_c, which is the value above which convection will occur. Convection will occur when the Ra_c is about 10^3 for a convection cell with an aspect ratio of 2–3; vigorous convection occurs when the Ra is above 10^5. Rayleigh numbers over 10^6 for the mantle suggest that the mantle will convect vigorously.
- Plates are an integral part of the convective system and they dominate the geometry of mantle convection. Computer models of mantle convection support some form of whole mantle convection in which subducted plates cross the upper/lower mantle boundary and may reach the core–mantle boundary, consistent with tomographic images.
- The Clapeyron slope has a significant effect on the movement of dense materials (slabs) downwards and hot material (plumes) upwards. Phase boundaries with a positive Clapeyron slope promote slab penetration, whereas reactions with a negative slope inhibit deep subduction. Slabs may be stalled at the 670 km (depending on the slab velocity) but there is no evidence for long-term storage of slabs at the 670 km boundary, suggesting the reactions with positive Clapeyron slopes ultimately mean that slabs will penetrate the lower mantle.
- The viscosity of the mantle increases at the 670 km boundary, which increases the mixing or stirring time of the lower mantle (1–2 Ga) compared with the upper mantle (<1 Ga). This produces a time lag between surface plate configurations and deep mantle structure (Figure 7.6) and complicates the fate of subducted slabs at the bottom of the mantle.

- OIBs require high mantle temperatures and are potentially produced by mantle plumes. For example, melting beneath Hawaii requires a mantle potential temperature of 1480 °C. This is consistent with the higher MgO contents in primitive Hawaiian basalts (18 wt%) compared with MORB (13 wt%), and trace element patterns that suggest a deep garnet-bearing source. Isotope compositions in some OIBs are consistent with a component of old (1–2 Ga) subducted slab in their mantle source. This suggests that mantle plumes may originate at the core–mantle boundary.
- Subducting slabs play a key role in keeping the planet in a dynamic state because they recycle H_2O back into the mantle.
- H_2O plays a crucial role in the inner workings of the Earth. H_2O in the geosphere is cycled through melting, subduction and hydrothermal processes. It is incorporated into the oceanic plate by hydrothermal alteration at mid-ocean ridges, is partially released during subduction triggering melting and is carried into the deeper mantle in K-rich and nominally anhydrous phases.
- H_2O lowers the solidus of the mantle, behaves as an incompatible element (leaving an anhydrous residue) and lowers the dynamic viscosity of the mantle, thus increasing its Rayleigh number and invigorating convection. The anhydrous mantle residue from melting at spreading centres is mechanically very strong and encourages the thermal development of plate structure. Water in the mantle, therefore, controls the style and vigour of mantle convection and the development of plates.
- Melting is a negative feedback that holds the system in check. Increasing subduction (driven mainly by slab-pull) leads to an increasing spreading rate and an increase in the volume of MORB. This dehydrates a larger volume of mantle, increasing the overall viscosity of the mantle and slowing convection.
- A world without water would be lifeless, with extreme surface temperature variations. H_2O in spreading centre magmas would escape to the atmosphere, due to a lack of oceans, and there would be no hydrothermal circulation to replenish the lost water. Dry oceanic crust would be subducted, but no subduction-related magmatism would be generated. With no water cycled back into the mantle, viscosity would increase and convection would become less vigorous and stop. No more plate tectonics means no new continental crust would be formed above subduction zones. Venus is a possible analogy to this scenario.

Although there are clearly areas where there is much speculation and controversy, an understanding of how the planet operates as a dynamic system will be one of the main areas of research during this century.

Learning outcomes for Chapter 7

You should now be able to demonstrate a knowledge and understanding of:

7.1 The discontinuities (e.g. transition zones) within the Earth and how these are related to sequential changes in the mantle mineralogy.

7.2 How understanding of the structure of the Earth and the location of melting in the mantle has been enhanced through three-dimensional imaging of changing seismic velocities within the Earth by seismic tomography.

7.3 Methods of modelling mantle convection as a means of determining how materials are cycled between different layers within the Earth and over what timescales.

7.4 The potential source and cause of intraplate volcanism, and how this can be identified by changes in the major and trace element geochemistry of resultant magmas, in comparison with normal mantle melts.

7.5 How the recycling of altered oceanic crust into the deep mantle via subduction zones can be recognised by distinctive major and trace element and isotopic signatures.

7.6 The role of water in the solid Earth and its controlling influence on magmatic and tectonic processes (e.g. melting, subduction, hydrothermal alteration, and plate tectonics) as modelled by positive and negative feedback mechanisms.

The continental crust

Chapter 8

At the start of this book the continental crust was recognised as one of the defining features of what makes the Earth unique in the Solar System. So it is fitting that the book should end with a summary of modern ideas of how and when the continental crust formed, and whether or not it is an end-product of planetary differentiation or part of a reversible cycle of material within the Earth.

The continents literally form the bedrock of our existence; they are the source of all our raw materials and provide a platform above the surface of the oceans on which terrestrial life has evolved. Without the continents the Earth would be covered with water because so little of the crust would have sufficient buoyancy to push the solid surface above the globe-encircling ocean. As Bill Bryson puts it in his book *A Short History of Nearly Everything* (2003), without the continents 'There might be life in that lonesome ocean, but there certainly wouldn't be football.' More significantly, however, the continents define the inorganic chemical environment within which all life has evolved, even marine life, because critical aspects of ocean chemistry are controlled by continental river run-off.

Elements that have a trace abundance in the crust, because they are chalcophile or siderophile, are more likely to be toxic than those that are lithophile and hence in greater abundance in crustal rocks. For example, selenium is a volatile element during planetary accretion and has strongly chalcophile properties. As a result there is very little Se in the crust, but what there is tends to be soluble in water and finds its way into living organisms. Organisms have subsequently evolved not only to tolerate that small amount of Se but also to rely on its presence for particular aspects of their metabolism. Again Bill Bryson puts this much more engagingly:

> Selenium is vital to all of us, but take in just a little too much and it may be the last thing you ever do … We have evolved to expect, and in some cases actually need, the tiny amounts of rare elements that accumulate in the flesh or fibre that we eat. But step up the doses, in some cases by only a tiny amount, and we can soon cross a threshold.
>
> (Bryson, 2003)

Consequently, understanding the evolution of the continental crust is to understand a critical influence on our own evolution.

The continents contain the oldest rocks but so far have been regarded simply as passengers that slowly migrate across the Earth's surface in response to internal processes, currently dominated by plate tectonics and mantle convection. What has not yet been discussed is how the continental crust relates to this dynamic interior and whether the processes operating today lead to the evolution of the continental crust. Is the crust simply the end-product of planetary differentiation or is it part of a reversible cycle?

8.1 Reversible cycles and irreversible processes

By now you should appreciate that far from being a solid, static sphere of rock and metal with a thin covering of water and gas, the Earth is a dynamic evolving system that has changed over geological time. After cooling from an initially largely molten state during which period the core, earliest crust, atmosphere and hydrosphere separated from the mantle (Chapter 2), the dominant process in the evolution of the Earth has been plate tectonics (Chapter 3). Material brought to the Earth's surface at mid-ocean ridges carries with it a fraction of the internal heat, which is lost to the ocean by conduction and advection of hydrothermal fluids, especially close to spreading centres (Chapter 4). It is here that the geosphere interacts most directly with the hydrosphere and material exchange takes place. Once the plate cools, interaction becomes more limited and after ~200 Ma, it thickens and subsides into the underlying mantle and is subducted back into the interior (Chapter 5). Sediments are scraped from the plate surface and part of the inventory of volatile and fluid-mobile elements introduced at the mid-ocean ridge are released as the plate warms up during its passage back into the mantle. This in turn induces magmatism in volcanic arcs and the material so generated becomes incorporated into the continental crust during continental collision (Chapter 6). If elements can become fixed in the subducting slab then they can be returned to the deep mantle and evidence of such recycling of surface materials can be found in the magmatic products of mantle plumes (Chapter 7). The most visually compelling evidence for communication between the surface and the deep Earth is apparent in the images generated from seismic tomography that reveal the movement of subducted plates through the upper mantle and into the lower mantle. The overall image is one of a planet in internal turmoil, albeit on a very long timescale, kept mobile by its internal heat but driven by gravity.

The solid Earth system involves a material cycle in which plates generated at mid-ocean ridges are returned to the mantle at subduction zones. Segments of oceanic crust and lithosphere are therefore temporary additions to the outer part of the Earth – oceanic crust is never older than about 180 Ma because after that time it is reincorporated into the mantle, heats up and becomes indistinguishable from the rest of the mantle. The generation of oceanic crust and lithosphere can therefore be regarded as part of a **reversible cycle** in which the mantle cools by gravity-driven convection.

By contrast, segments of the continental crust and lithosphere are much older, and the record in continental rocks extends back to almost 4.0 Ga as individual rocks in Greenland and Northern Canada, while very rare mineral ages extend back to 4.4 Ga.

■ Can you recall why this should be?

■ Continental crust is composed of low-density rocks that cannot be subducted back into the mantle (Chapter 6).

Thus the generation of continental crust may be regarded largely as an **irreversible process** that leads to the permanent differentiation of the planet.

- Can you think of other material or energy cycles within the Earth that are irreversible?
- Irreversible processes include the loss of heat (cooling) and the differentiation of the atmosphere from the solid Earth.

The more you consider the workings of the Earth's interior, the more difficult it becomes to define irreversible processes. As you have discovered from previous chapters, simply because a component of the Earth resides at the surface does not mean that it has completely escaped from the interior. The water in the oceans is recycled into the Earth's interior and exerts a controlling influence on the behaviour of the mantle and mantle melting. Similarly in Chapter 5 you explored some of the evidence from the compositional variations of subduction-related magmas that indicate the incorporation of ocean sediments in their source regions, extending the rock cycle to mantle depths. So even material from the continental crust can be recycled into the mantle, although whether that material descends beyond the depths of magma generation is the subject of debate and uncertainty.

8.2 How continental crust is formed

Despite the significance of the continents for our very existence, their evolution and even their bulk composition remain the subjects of scientific debate. From what you have read earlier in this book, you should now appreciate that the oldest rocks are slightly less than 4.0 Ga old; and there is evidence in zircons and from minuscule variations in the radiogenic isotopes of Nd that there was a form of continental crust during the Hadean, although the volume and extent of such primordial crust remains unknown.

The age of the Earth's earliest crust depends on how crust is defined. Starting with the primitive Earth enveloped by a magma ocean, the first crust would be analogous to a chilled skin on the surface of the magma ocean, possibly with a composition modified by the floatation of minerals less dense than the liquid magma. An example of ancient primary crust is seen on the Moon in the anorthosites of the lunar highlands. These are characterised by some of the oldest ages yet obtained from lunar material and probably formed by floatation of light plagioclase feldspar on the surface of the lunar magma ocean. Such **primary crust** is no longer represented in the geological record on Earth and it is assumed that it was eradicated by the effects of impacts during the waning stages of accretion while vigorous convection in the underlying hot mantle ensured its rapid recycling back into the interior.

Subsequent to the destruction of the Earth's primary crust, mantle convection resulted in mantle melting to form **secondary crust**.

- Is secondary crust being formed today and, if so, what is it?
- Today, partial melting of the mantle generates oceanic crust that is considered to be secondary crust.

The modern-day process of oceanic crustal formation is part of the reversible plate tectonic cycle within the Earth. As stated above there is no oceanic crust surviving beneath the oceans older than 180 Ma, although older fragments do occur as ophiolites in orogenic belts dating back to the late Proterozoic, and may be even older.

The oldest surviving crust is found in the continents and this is known as **tertiary crust**. In considering the evolution of the continental crust it is first important to define where continental crust is created. Today, the destruction of secondary crust by subduction leads to the generation of magmatic rocks in island arc and Andean margins and these are recognised as the current sites of continental crust generation. In the discussion of the evolution of the Himalayan orogen in Chapter 6, a distinction was made between two different types of granite intrusion.

- Can you recall what they are?
- The Trans-Himalayan batholith evolved during a period of Andean volcanism prior to continental collision while the High-Himalayan granites formed from crustal melting during collision.

- Which of these two intrusions represented examples of crustal growth and why?
- The Trans-Himalayan batholith has initial $^{87}Sr/^{86}Sr$ ratios close to that of the upper mantle and so has an origin in the upper mantle. It therefore represents an example of recent crustal growth.

The large volume Trans-Himalayan batholith comprises a range of igneous rocks related to each other by fractional crystallisation and so represents new material added to the continental crust from the mantle. By contrast the isotopic similarity of the High-Himalayan granites with pre-existing crustal rocks, and their origin as partial melts with no associated mafic counterparts, shows that they are a product of the rock cycle within the crust. While they do not represent a new addition to the crust, they reflect an important refining process within the crust whereby silica-rich granitic melts migrate upwards leaving a more mafic residue in the deeper crust. Destructive plate margins are therefore regarded as the most important sites for the growth of new continental crust while collision zones allow the further refinement of the crust into a granitic upper crust and a more mafic mid- and lower crust.

While this appears to be the dominant crust-forming process today, it is not clear whether the same is true for all of geological time. For example, voluminous crustal rocks formed during Proterozoic times do not share the same compositional characteristics as are typical of subduction-related magmas today and there is a growing body of evidence to suggest that at some time in the past crustal growth was dominated by processes occurring within plates rather than at their boundaries, probably controlled by mantle plume activity.

- What is the composition of the magma transferred from the mantle to the crust at both destructive plate boundaries and at locations within plates?
- Basalt.

Notwithstanding the possible different causes of melting, the flux of material from the mantle to the crust throughout geological time is dominated by basalt. The crust itself has formed by a multiplicity of processes related to mantle and crustal melting, fractional crystallisation, and mixing between new material from the mantle and pre-existing continental crust.

8.3 The composition of the continental crust

Estimates of the composition of the continental crust are surprisingly varied. In part this relates to the variety of rock types exposed in the continents, but it is also because the deeper layers of the crust are rarely exposed at the surface. Table 8.1 lists estimated major element compositions of the upper and lower continental crust and an estimate of the composition of the total crust. The composition of the upper crust is the most accessible and the average composition is based on two methods:

- a weighted average of the composition of rocks exposed at the surface
- using the abundances of insoluble elements in fine-grained clastic sediments to infer bulk composition.

The first method involves large-scale sampling of the crust whereas the second exploits the processes of weathering and erosion, transport and sedimentation that serve to average the composition of crustal source regions. The average compositions in Table 8.1 are based on information from both of these methods.

Table 8.1 Average composition of upper, lower and whole crust. (Adapted from Rudnick and Gao, 2005)

	Upper crust/wt%	Lower crust/wt%	Whole crust/wt%
SiO_2	66.6	53.4	60.6
TiO_2	0.64	0.82	0.72
Al_2O_3	15.4	16.9	15.9
FeO	5.04	8.57	6.70
MnO	0.10	0.10	0.10
MgO	2.48	7.24	4.70
CaO	3.59	9.59	6.40
Na_2O	3.27	2.65	3.10
K_2O	2.80	0.61	1.80
P_2O_5	0.15	0.10	0.10
Total	100.07	99.98	100.12

The composition of the lower crust is based on averages of granulites from around the world. You should recall from Chapter 1 that the lower crust is thought to be composed dominantly of relatively anhydrous, high-grade metamorphic rocks known as granulites and these are found both as xenoliths in volcanic pipes and exposed at the surface as a consequence of uplift and erosion. In addition the more uniform physical properties of the lower crust as reflected in seismic velocities can be matched against compositions as these properties vary with both composition and mineralogy.

- What are the major differences between the upper and the lower crust?
- The upper crust is richer in silica, potassium and sodium than the lower crust, whereas the lower crust is richer in iron, magnesium and calcium.

These averages clearly show that the lower crust has a more mafic, i.e. less differentiated, composition than the upper crust. The lower crust has a composition similar to that of a basaltic andesite whereas the upper crust is closer to dacite.

- Thinking back to earlier chapters, can you suggest how this layering might have developed?
- Mafic residues are produced when magmas evolve and differentiate. Also, the extraction of a granitic melt from a crustal precursor will produce a more mafic residue.

Magmatic processes, be they crustal melting or fractional crystallisation, will produce a more evolved upper crust while leaving the lower crust richer in iron, magnesium and calcium.

The final column of Table 8.1 includes an estimate of the bulk major element composition of the whole crust.

- What igneous rock type most closely represents the composition of the whole continental crust?
- The bulk crust has an SiO_2 content of 60.6 wt% and an MgO content of 4.7 wt%. Looking back at Figure 5.13, it most closely resembles the composition of andesite (or diorite).

Somewhat unsurprisingly, the composition of the bulk crust is intermediate between that of the upper and the lower crust and resembles that of an andesite. However, this rather simple observation lies at the root of a fundamental problem in Earth sciences.

- How does the bulk composition of the crust compare with the composition of material supplied from the mantle?
- The bulk crust is andesitic and is therefore more evolved than the basaltic material that is supplied from the mantle.

Despite the association of andesites with the Andes and other regions of modern-day and recent crustal growth, andesites are derived from magmas that have fractionated from a basaltic parent (Chapter 5). The flux of material from crust to mantle at subduction zones, as at mid-ocean ridges and above mantle plumes, is dominantly basaltic. Why, then, does the crust have an overall andesitic composition?

The short answer to this is that no one really knows, but a number of possibilities have been suggested. One is that the estimated bulk composition of the continental crust is incorrect and that rather than being andesitic it is basaltic. If this is the case then the lower crust is a mafic complement to the dacitic upper crust. You should recall from Chapter 5 that to produce dacite from a parental basalt requires a considerable degree of fractional crystallisation. Calculations based on both major and trace elements requires up to 85% fractionation before a dacite composition is produced.

- Assuming the upper crust has a thickness of 12 km, how thick would the total crust be if its overall composition was basaltic?
- If 12 km of upper crust is equivalent to 15% (100% – 85%) of the total mass of the crust, the total crustal thickness should be $\frac{12\ \text{km}}{0.15} = 80\ \text{km}$.

The calculation is the same if the upper crust developed as a consequence of partial melting because a given amount of basalt can only produce this small amount of melt with a dacite composition. Greater amounts of melting of a basaltic precursor will produce less siliceous compositions. The conclusion from chemical mass balance arguments therefore is a 12 km thick upper crust must be balanced by a lower crust almost 70 km thick.

- How does this total crustal thickness compare with the thickness of the continental crust from seismic refraction (Chapter 1)?
- It is larger.

This simple calculation shows that, regardless of the composition of the lower crust, if the flux from mantle to crust is basaltic and the upper crust is dacitic, the total crustal thickness should be 80 km. This is much thicker than the average seismic thickness of the continental crust above the Moho, which is about 40 km.

The fate of the mafic residue from the formation of the upper crust is a matter of current scientific debate but a popular model to explain its absence invokes recycling into the mantle. As continental crust differentiates into a granitic upper crust, the lower crust takes on a more mafic composition.

- What mineralogical transformation occurs in mafic rocks at high pressures?
- The major aluminium-bearing mineral phase changes from low-density plagioclase to high-density garnet, while other chemical components combine to form a dense, sodium-rich clinopyroxene.

You should recall from Chapters 5 and 7 that during subduction, as pressure increases, the ocean crust is transformed into eclogite, which is a metamorphic rock dominated by high-density garnet and clinopyroxene. After repeated melting and dehydration, the altered crustal rocks in the slab become more mafic and, under high-pressure, metamorphic conditions, are transformed into granulites and eclogites with a dense garnet-bearing mineralogy. Similarly, during arc magmatic processes, the fractionation of basaltic magma to an andesitic composition produces large volumes of mafic cumulates dominated by olivine and pyroxenes. Under suitable metamorphic conditions these too can take on a high-density eclogitic mineralogy of garnet and clinopyroxene. The density of eclogite is generally greater than that of peridotite, and so eventually mafic lower crust becomes dense enough to founder into the mantle. This process is frequently referred to as crustal delamination, and is analogous to the convective removal of mantle lithosphere that may occur during continental collision (Chapter 6).

8.4 Crustal evolution through time

The observation that the continental crust contains rocks ranging in age from almost 4 Ga to the present day and that material is still being transferred to the crust from the mantle as a result of melting implies that the volume of crust must have changed through time. But how quickly did the crust grow? Archaean rocks constitute only ~1% of the exposed crust, suggesting that much of the crust has grown since the Archaean.

- Why do you think this conclusion might be incorrect?
- Ancient crust is reworked during continental collision and so old ages can be reset by subsequent metamorphic and igneous events.

As you should appreciate by now, judging rates of major Earth evolution processes from surface observations is fraught with problems and none more so than estimating the rate of crustal growth. Even estimating the rate of crustal growth at the present day is far from simple, largely because much crustal growth happens below the surface with the intrusion of mantle-derived magmas and their evolution towards granitic compositions. The details of how physical growth estimates are made is beyond the scope of this book, but current estimates of crustal growth at destructive plate boundaries lie between 1 $km^3\ y^{-1}$ and 4 $km^3\ y^{-1}$, with a preferred value of 1.6 $km^3\ y^{-1}$.

- The volume of the continental crust is $1 \times 10^{19}\ m^3$. Could this volume have been generated over the age of the Earth at the present rate of crustal growth?
- $1 \times 10^{19}\ m^3 = 1 \times 10^{10}\ km^3$ (remember $10^9\ m^3 = 1\ km^3$), so at 1.6 $km^3\ y^{-1}$, it would take:

$$\frac{1 \times 10^{10}\ km^3}{1.6\ km^3\ y^{-1}} \approx 10^{10}\ y$$

to generate the present volume of continental crust, i.e. twice the age of the Earth. The answer is therefore no, the current rate of crustal growth is insufficient to generate the present-day volume of the continental crust over the span of geological time.

Even if the crustal growth rate today is as high as 4 $km^3\ y^{-1}$, it is still barely sufficient to generate the total volume of the continental crust in the full span of geological time. Moreover the problem is exacerbated by the previous discussion of the compositional mass balance that reveals that mafic crustal material must have been recycled into the mantle to balance the dacitic composition of the upper continental crust.

Perhaps this result is not so surprising if you consider the effect of the Earth's internal heat. Heat production was significantly greater early in Earth history leading to greater amounts of mantle melting. As melting is the dominant process for crust generation, the rate of crustal growth must have been greater in the past than it is today.

Although crustal growth rates both today and through geological time are difficult to measure directly, important insights into growth rates can be derived from the secular evolution of radiogenic isotope ratios in both the mantle and the crust. The details of these models and calculations are complex, but the results can be summarised simply to show the proportion of crust generated through geological time (Figure 8.1).

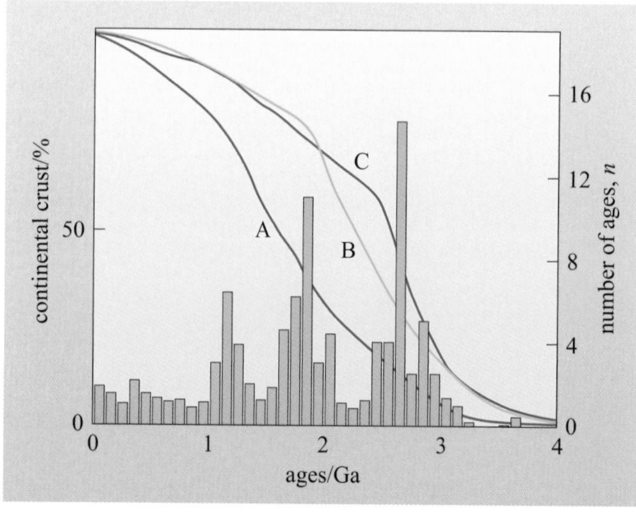

Figure 8.1 The growth of the continental crust through time as estimated from: (A) Nd isotopes in shales; (B) Pb isotope evolution of the mantle; (C) secular changes in the trace element compositions of shales. The histogram shows the ages of individual igneous bodies that contributed to crustal growth.
(Hawkesworth and Kemp, 2006)

- Do the three curves in Figure 8.1 show constant or variable rates of crustal growth throughout geological time?

- Constant growth rates would result in a straight line, but all three models show curves that are steeper after 3 Ga and then become shallower towards the present day.

While the three curves shown in Figure 8.1 differ in detail they all show a steeper gradient between 1 Ga and 3 Ga than is apparent today, implying greater crustal growth during the late Archaean and Proterozoic. The lower growth rates prior to 3 Ga may be counter-intuitive – after all, the Earth was hottest during the Archaean – but may reflect greater rates of recycling of less-evolved crust during this period.

An alternative method of investigating crustal evolution is to assess the age of new additions to the crust, be they volcanic or plutonic. This is less easy than it sounds because it requires the distinction between new and recycled material, although the use of isotope tracers such as initial $^{87}Sr/^{86}Sr$ and other discriminants provides useful indicators. However, a further difficulty arises from the non-random distribution of crust across the surface of the Earth and the variable detail in which different parts of the continents have been explored. First attempts to make such assessments in the 1960s and 1970s revealed distinct peaks in the distribution of ages of new crustal additions as exemplified by large plutons and thick sequences of mantle-derived volcanic rocks. Originally it was thought that these peaks reflected the pattern of preservation of ancient rocks rather than real variations in crust formation ages. However, subsequent additions to this global database, rather than evening out the peaks, have enhanced them, giving the current distribution of crust formation ages shown in Figure 8.1. These ages are based largely on the U–Pb (zircon) technique on individual plutons and volcanic units. Each age in Figure 8.1 represents only the most precise age from a unit or pluton to avoid weighting towards well-studied regions of the crust. As is apparent from this diagram, the distribution of crust formation ages is far from uniform – crust formation ages show peaks during the Archaean (2.7–3.2 Ga), the mid-Proterozoic (1.8–2.2 Ga) and in the late Proterozoic (1.2–1.5 Ga).

The pulsed pattern in these data clearly contrasts with the smooth curve of crustal growth derived from sediments. Superficially these two results appear at odds with each other, but they may be reconciled by considering the formation and development of the sedimentary record. As emphasised above, sediments represent an average of their eroding source regions and so their radiogenic isotopes are the sum of all of the contributing materials. Thus the very process of formation tends to smooth out any spikes that may be present in the region of crust sampled and the 'age' of any one sample may well be a mixture of the ages of the source terrain. By contrast, the crust formation ages are derived from specific rock units and record the timing of a specific igneous event during which new continental crust was generated.

What could the pulses in crust formation represent? One possibility is that they could reflect plate tectonic Wilson cycles, as discussed in Chapter 3.

- Can you see why Wilson cycles may not be the cause of peaks in crustal growth?
- Wilson cycles have dominated the geological history of the past 1000 Ma of Earth history, yet there are no peaks of crust generation during this time period.

In addition, Wilson cycles occur with a periodicity of about 500 Ma and, even though there are no crust formation peaks during the past 1000 Ma, the periodicity before this time is still not 500 Ma.

For evidence of the cause of crustal growth during the peak periods it is necessary to look into the composition of the dominant magmatic rocks in crust formed at those times. For example, large areas of the crust of West Africa were formed during the early and middle Proterozoic, around 2 Ga ago. They developed remote from any pre-existing continental crust and strongly imply the involvement of mantle plumes. Such evidence, although complex, illustrates that crust formation processes have probably not been constant throughout geological time. Since 1000 Ma, crust has been generated from the mantle from the quasi-continuous process of subduction at destructive plate margins. Prior to 1000 Ma, the evidence for plate tectonics as the dominant process for cooling the Earth becomes less clear and the periodic pulses of crust formation may relate to another process, possibly related to large-scale mantle convection and enhanced mantle plume activity.

Summary of Chapter 8

At present our understanding of the formation, evolution and growth of the continental crust remains vague and is defined more by what is not, rather than what is, known. Mantle melting is the dominant process for transporting material from within the planet to its surface. The composition of that material is basaltic, yet the average crust is andesitic and the upper crust is closer to a dacite. Mass balances of current crustal compositions require some crustal material to have been returned to the mantle, but where and how that happens remains controversial. Today destructive plate boundaries are the major sites of crust generation but that may not have been the case for the distant geological past when mantle plume-related activity may have played a significant role.

The continental crust is a product of planetary evolution and has accumulated over the length of geological time. The upper continental crust at least is an end-product of an irreversible process in the evolution of the planet. By contrast the more mafic lower crust is a remnant of the mafic residue that must have been produced during the formation of the upper crust, most of which has been removed and returned to the mantle. So, to answer the question posed at the start of this chapter, the continental crust is both the product of an irreversible process and a component in a long-term cycle of material in the solid Earth. Resolving the details of the development of the continental crust remains one of the important challenges facing modern Earth sciences, not least because thick continental crust is a unique terrestrial feature not seen on the other rocky planets of our Solar System.

Learning outcomes for Chapter 8

You should now be able to demonstrate a knowledge and understanding of:

8.1 The relative timing and processes involved in the formation of primary, secondary and tertiary crust on the Earth.

8.2 How the composition of the upper, lower and whole continental crust can be determined using weighted average compositions and/or relative abundances of insoluble elements in clastic sediments.

8.3 The different types of evidence that indicate the lower continental crust is recycled back into the mantle.

8.4 How models of continuous versus episodic crustal formation differ and can be used to explain variations in the rates of crustal formation throughout the evolution of the Earth.

Answers to questions

Question 2.1

(a) From Equation 2.2:

$$\Delta T = \frac{mv^2}{2MC}$$

where $m = 10^{15}$ kg (given), $v = 10^4$ m s^{-1} (given), $C = 7.5 \times 10^2$ J kg^{-1} K^{-1} (given), and $M = 6.0 \times 10^{24}$ kg (given).

$$\Delta T = \frac{10^{15}\ \text{kg} \times (10^4\ \text{m s}^{-1})^2}{2 \times (6 \times 10^{24}\ \text{kg}) \times (7.5 \times 10^2\ \text{J kg}^{-1}\ \text{K}^{-1})}$$

$$= 1 \times 10^{-5}\ \text{K (to 1 sig. fig.)}$$

(b) The number of planetesimals of mass 10^{15} kg each required to construct the Earth is:

$$\frac{6 \times 10^{24}\ \text{kg}}{10^{15}\ \text{kg}} = 6 \times 10^9$$

If each of these produced the temperature rise in part (a), the total temperature rise in the Earth would be $(6 \times 10^9) \times (1 \times 10^{-5}\ \text{K}) = 60\ 000$ K.

This is a very crude, but very interesting, calculation. The temperature rise given by part (a) was based on the impact of a 10^{15} kg planetesimal on an Earth of current size. In practice, most of the planetesimals going into the construction of the Earth would have impacted a *growing* Earth (i.e. one smaller than the present Earth). As Equation 2.2 shows, if M were smaller, ΔT would be bigger. In other words, the 60 000 K temperature rise obtained above can be regarded as a *minimum* temperature rise – and it is already well over the vaporisation temperature of refractory elements (see Figure 1.16).

Question 2.2

The number of seconds in a year = 3.2×10^7 s y^{-1} (to 2 sig. figs).

So the rate of heating is:

$$\frac{3.0 \times 10^{19}\ \text{J y}^{-1}}{3.2 \times 10^7\ \text{s y}^{-1}} = 9.375 \times 10^{11}\ \text{J s}^{-1} = 9 \times 10^{11}\ \text{W (to 1 sig. fig.)}$$

The rate of tidal heating generated within the Earth is approximately:

$$\frac{9 \times 10^{11}\ \text{W}}{6 \times 10^{24}\ \text{kg}} = 1.5 \times 10^{-13}\ \text{W kg}^{-1}$$

$$= 2 \times 10^{-13}\ \text{W kg}^{-1}\ \text{(to 1 sig. fig.)}$$

The data in Question 2.2 is to one significant figure, therefore your answer should be to one significant figure.

Question 2.3

(a) The age of the Earth is 4.6×10^9 y. To calculate the proportion of the original radiogenic isotope remaining today, you first need to determine how many half-lives have occurred since the Earth was formed:

$$\text{number of half-lives} = \frac{\text{age of Earth}}{\text{isotope half-life}}$$

The proportion remaining after one half-life is $(\frac{1}{2})^1 = 0.5$, after two half-lives it is $(\frac{1}{2})^2 = 0.25$, and after three half-lives it is $(\frac{1}{2})^3 = 0.125$.

For ^{40}K, the number of half-lives =

$$\frac{4.6 \times 10^9 \text{ y}}{1.25 \times 10^9 \text{ y}} = 3.59 \text{ half-lives (to 3 sig. figs)}$$

So the proportion remaining today is $(\frac{1}{2})^{3.59} = 0.083$ (to 2 sig. figs).

Note: on many calculators this function can be performed by the key labelled x^y, which is often operated by the SHIFT key. Your calculator may differ in detail, but for many common calculators this calculation can be performed by keying:

0.5 [SHIFT] [x^y] 3.59 [=]

For ^{232}Th, the number of half-lives

$$\frac{4.6 \times 10^9 \text{ y}}{1.39 \times 10^{10} \text{ y}} = 0.331 \text{ half-lives (to 3 sig. figs)}$$

So the proportion remaining today is $(\frac{1}{2})^{0.331} = 0.79$ (to 2 sig. figs).

The data in Question 2.3a are to two significant figures, therefore your answer should be to two significant figures.

(b) The amount of radiogenic heating is determined by utilising the proportion of the different radiogenic isotopes remaining today (you need to calculate the values for ^{235}U and ^{238}U as above) to calculate the original heat budget.

If only 0.083 of the original ^{40}K remains today and its present rate of heat generation is 2.8×10^{-12} W kg^{-1} (from Table 2.1), its original heat budget is:

$$\frac{2.8 \times 10^{-12} \text{ W kg}^{-1}}{0.083} = 3.37 \times 10^{-11} \text{ W kg}^{-1} \text{ or } 34 \times 10^{-12} \text{ W kg}^{-1} \text{ to 2 sig. figs}$$

Note: the lowest level of accuracy is 2 sig. figs so the answer should be given to 2 sig. figs.

Likewise:

$$\text{initial heating by } ^{232}\text{Th decay} = \frac{1.04 \times 10^{-12} \text{ W kg}^{-1}}{0.79} = 1.3 \times 10^{-12} \text{ W kg}^{-1} \text{ (to 2 sig. figs)}$$

$$\text{initial heating by } ^{235}\text{U decay} = \frac{0.04 \times 10^{-12} \text{ W kg}^{-1}}{0.0113} = 3.6 \times 10^{-12} \text{ W kg}^{-1} \text{ (to 2 sig. figs)}$$

$$\text{initial heating by } ^{238}\text{U decay} = \frac{0.96 \times 10^{-12} \text{ W kg}^{-1}}{0.492} = 2.0 \times 10^{-12} \text{ W kg}^{-1} \text{ (to 2 sig. figs)}$$

Note: the lowest level of accuracy is 2 sig. figs so the answer should be given to 2 sig. figs.

By adding the four answers, the total initial heating was therefore about 4×10^{-11} W kg^{-1} compared with a modern total of about 5×10^{-12} W kg^{-1} (obtained by adding the four present-day heating rates given in Table 2.1), in other words it was at least eight times greater (we have not accounted for other short-lived isotopes, such as ^{26}Al, and the heat from their decay). This initial decay would have originally provided a major source of internal heating even in relatively small planetary bodies, and so would have contributed significantly to early planetary differentiation.

(c) By summing the four values quoted in Table 2.1, the present rate of radiogenic heating is about 5×10^{-12} W kg^{-1}, which is about 30 times greater than that attributable to estimates of tidal heating, and so represents the present major source of the Earth's internal heating.

Question 2.4

From Equation 2.13:

$$k = \frac{a^2\Phi}{24\pi}\ \text{m}^2$$

where $a = 10^{-3}$ (0.001 = 1 mm), $\Phi = 0.1$

$$k = \frac{(0.001\ \text{m})^2 \times 0.1}{24\pi}$$

$$= \frac{1\times10^{-7}\ \text{m}^2}{75.398} = 1.33\times10^{-9}\ \text{m}^2$$

Using this estimate of permeability (k) in Equation 2.12:

$$\upsilon = \frac{k}{\eta}\Delta\rho g$$

where $\eta = 0.005$ Pa s (1 Pa s = 1 kg m^{-1} s^{-1}), $\Delta\rho = 3500$ kg m^{-3}, and $g = 9.8$ m s^{-2}

$$\upsilon = \frac{1.33\times10^{-9}\ \text{m}^2}{0.005\ \text{Pa s}} \times 3500\ \text{kg m}^{-3} \times 9.8\ \text{m s}^{-2}$$

$$= \frac{1.33\times10^{-9}\ \text{m}^2}{0.005\ \text{kg m}^{-1}\ \text{s}^{-1}} \times 3500\ \text{kg m}^{-3} \times 9.8\ \text{m s}^{-2}$$

$$= \frac{2.66\times10^{-7}\ \text{m}^3}{\text{kg s}^{-1}} \times 34\,300\ \text{kg m}^{-2}\ \text{s}^{-2}$$

$$= 9.1\times10^{-3}\ \text{m s}^{-1}$$

There are 3.2×10^7 seconds in a year and 10^3 metres in a kilometre.

$$\text{So the migration velocity} = \frac{(9.1\times10^{-3}\ \text{m s}^{-1})\times(3.2\times10^7\ \text{s y}^{-1})}{1000\ \text{m}}$$

$$= 291.2\ \text{km y}^{-1}$$

$$= 290\ \text{km y}^{-1}\ \text{to 2 sig. figs}$$

The distance between the Earth's surface and the outer liquid core is 2900 km, which suggests that it would take about 10 years for metallic melt to sink to the core. During accretion, the Earth (and core) would have been growing, and gravitational acceleration would also be changing. If gravitational acceleration were smaller then this would reduce the migration velocity of the melt. For example, for a Mars-sized body with $g = 3.7\ \text{m s}^{-2}$, the migration velocity would be about 110 km y^{-1}, but the depth to the core would be less.

Question 2.5

The $\varepsilon^{182}\text{W}$ value of the meteorite is:

$$\begin{aligned}\varepsilon^{182}\text{W} &= \left[\left(\frac{0.864523}{0.864696}\right) - 1\right] \times 10^4 \\ &= (0.999799 - 1) \times 10^4 \\ &= -0.000200 \times 10^4 \\ \varepsilon^{182}\text{W} &= -2.0 \text{ (to 2 sig. figs)}\end{aligned}$$

Question 2.6

Compared with chondritic composition, the Earth's mantle is depleted in Rb, K and Na (Figure 1.22). This cannot be explained in terms of segregation into the core, as was the case for Fe and Ni depletion, since the core does not contain these elements. Instead, these depletions can be best explained using a similar argument to that offered to explain the Moon's depletion in volatiles. During the final stages of the assembly of Earth, the amalgamated planetary embryos were beginning to differentiate into layered bodies with dense cores, and with elements such as Rb, K and Na segregating towards the surface. Continuing impacts would have partially vaporised the surface layers, and some of the more volatile elements, including Rb, K and Na (and probably Zn) may have been lost into space. Later in Earth history, the segregation of material to form oceanic and continental crust would have further augmented removal of these elements from the mantle.

Question 3.1

The coalfields formed from plants that grew in the tropical and subtropical climate belts. They must have drifted to their present positions from these latitudes either south or north of the Equator (i.e. 23° S–23° N).

This represents a latitude drift of between 23° S to 55° N (i.e. 78°) and 23° N to 55° N (i.e. 32°).

If the Earth's radius is 6370 km, its circumference must be $2\pi \times 6370\ \text{km} =$ 40 024 km.

1° of latitude (assuming the Earth is a sphere) is therefore $\dfrac{40\,024\ \text{km}}{360} = 111\ \text{km}$.

Therefore the *minimum* distance that Britain can have drifted since the late Carboniferous is $32 \times 111 = 3552$ km (i.e. about 3500 km at the level of accuracy of this information), which gives a rate of:

$$\frac{3500 \times 10^6 \text{ mm y}^{-1}}{300 \times 10^6} \approx 12 \text{ mm y}^{-1} \text{ or } 10 \text{ mm y}^{-1} \text{ to 1 sig. fig.}$$

The *maximum* distance is at least 8672 km (i.e. about 8700 km), which gives a rate of:

$$\frac{8700 \times 10^6 \text{ mm y}^{-1}}{300 \times 10^6} \approx 29 \text{ mm y}^{-1} \text{ or } 30 \text{ mm y}^{-1} \text{ to 1 sig. fig.}$$

Both these values could be larger if Britain drifted in terms of longitude as well – in other words, if its course was not in a straight line.

Question 3.2

The APW parts of Europe and Siberia are the same as far back as the Triassic, but before this time the Siberian pole was to the west of the European pole. This indicates that the two regions were part of different land masses until the Triassic; at this time they must have collided, and afterwards they continued to move as a single unit.

Question 3.3

For the South Atlantic Ocean:

$$\text{half spreading rate} = \frac{38 \text{ mm y}^{-1}}{2} = 19 \text{ mm y}^{-1} \text{ or } 19 \text{ km Ma}^{-1}$$

The age of ocean crust (t) adjacent to the South American continental shelf is derived from:

$$\text{age} = \frac{\text{distance}}{\text{half spreading rate}} = \frac{3100 \text{ km}}{19 \text{ km Ma}^{-1}} = 163 \text{ Ma (or 160 Ma to 2 sig. figs)}$$

The age of ocean crust adjacent to the southern African continental shelf

$$= \frac{2700 \text{ km}}{19 \text{ km Ma}^{-1}} = 142 \text{ Ma (or 140 Ma to 2 sig. figs).}$$

Question 3.4

(a) (i) Rearranging Equation 3.1:

$$t = \left(\frac{d - 2500}{350}\right)^2 \text{ Ma} = \left(\frac{4700 - 2500}{350}\right)^2 \text{ Ma}$$
$$= 6.285^2 \text{ Ma} = 39.5 \text{ Ma} = 40 \text{ Ma (to 2 sig. figs)}$$

The age of the crust at this location is 40 Ma.

(a) (ii) The mean spreading rate is given by distance divided by time: $\upsilon = \frac{D}{t}$

$$\text{Therefore } \upsilon = \frac{1600 \times 10^6 \text{ mm}}{40 \times 10^6 \text{ y}} = \frac{1600 \text{ mm}}{40 \text{ y}} = 40 \text{ mm y}^{-1}$$

(b) This is a half spreading rate because it refers to one side of the constructive margin only.

Question 3.5

(a) Ocean depth increases with the square root of lithosphere age so the correlation is positive.

(b) The departure from linearity occurs between 8 and 9 on the horizontal scale, which is equivalent to an age of 64–81 Ma.

(c) For crust older than 64 Ma, ocean depth is less than predicted, i.e. the ocean floor is shallower, implying that it is warmer than predicted by the boundary-layer model.

Question 3.6

Compare your answer with Table 3.5.

Table 3.5 Completed Table 3.1.

	Relative plate motion/mm y^{-1}	Direction of plate motion
American Plate fixed	40	east
Australian Plate fixed	40	west

If the South American Plate is stationary, then the spreading rate across the Mid-Atlantic Ridge means that the African Plate is moving at a rate of 40 mm y^{-1} to the east, because the ridge is migrating east at the half spreading rate while the African Plate is moving at the full spreading rate. Conversely, if the Australian Plate is held stationary, then the Carlsberg Ridge migrates to the west at a rate of 20 mm y^{-1} and the African Plate moves to the west at a rate of 40 mm y^{-1}.

Question 3.7

You should recall that:

$$\text{speed} = \frac{\text{distance}}{\text{time}}$$

but the graph is plotted as time against distance, so the gradient of the graph as it is plotted is $\frac{\text{time}}{\text{distance}}$. To calculate the gradient, and hence the speed, choose two points on the best-fit straight line that are some distance apart and find their coordinates. For example, at a distance of 5500 km the age is 63 Ma, and at a distance of 1000 km the age is 11 Ma. The gradient of the line is:

$$\frac{63 \text{ Ma} - 11 \text{ Ma}}{5500 \text{ km} - 1000 \text{ km}} = \frac{52 \text{ Ma}}{4500 \text{ km}} = 0.0116 \text{ Ma km}^{-1}$$

The speed is therefore $\dfrac{1}{0.0116 \text{ Ma km}^{-1}}$

$= 86.5 \text{ Ma km}^{-1} = 87 \text{ Ma km}^{-1}$ (to 2 sig. figs)

This is equivalent to 87 mm y^{-1}. The current overall direction of migration of volcanism is towards the southeast, therefore the plate is moving to the northwest. However, prior to about 43–45 Ma volcanism migrated more or less due south, therefore the Pacific Plate prior to 43–45 Ma was moving to the north.

Question 3.8

Table 3.6 Completed Table 3.2.

Force	Acts as a driving force	Acts as a resistive force	Might act as *either* a driving force *or* a resistive force
oceanic drag			✓
continental drag		✓	✓
ridge-push	✓		
transform fault		✓	
slab-pull	✓		
slab resistance		✓	
trench suction	✓	✓	

Question 3.9

Table 3.7 Completed Table 3.4.

	Positive association	No clear association	Negative association
Total plate area		✓	
Total continental area			✓
Percentage length of effective ridge boundary	✓		
Percentage length of effective trench boundary	✓		
Percentage length of effective transform fault boundary		✓	

The clearest positive association is between real plate speed and percentage length of effective trench boundary, suggesting that the slab-pull force (F_{SP}) is a major plate driving force.

Question 3.10

The spreading rate is equivalent to an ocean width increase of 30 km Ma^{-1}. For the continents to meet on the opposite side of the globe, the ocean will need to have opened to half the circumference of the Earth less the original width of the continent (i.e. 20 000 km – 5000 km = 15 000 km). The time taken to achieve this will be:

$$\text{time} = \frac{\text{distance}}{\text{speed}} = \frac{15\,000 \text{ km}}{30 \text{ km Ma}^{-1}} = 500 \text{ Ma}$$

Question 4.1

(a) Olivine, pyroxene, garnet and plagioclase feldspar.

(b) Spinel.

(c) Garnet, plagioclase feldspar and spinel.

(d) Olivine, pyroxene, garnet and spinel. (*Note*: pyroxene and garnet also contain calcium.)

Question 4.2

The total charge of one Si^{4+} ion (4+) and four O^{2-} ions (8–) is 4–. The formula for the tetrahedron can be written as $(SiO_4)^{4-}$.

Question 4.3

(a) $$\text{Mg\#} = \frac{\text{Mg}}{\text{Mg} + \text{Fe}^{2+}} = \frac{1.8}{1.8 + 0.2} = \frac{1.8}{2.0} = 0.90 \text{ or } 90\%$$

(b) $$\text{Mg\#} = \frac{\text{Mg}}{\text{Mg} + \text{Fe}^{2+}} = \frac{1.2}{1.2 + 0.4} = \frac{1.2}{1.6} = 0.75 \text{ or } 75\%$$

Question 4.4

The composition represented by X is 80% A and 20% B ($A_{80}B_{20}$). The composition represented by point Y is 20% A, 60% B and 20% C ($A_{20}B_{60}C_{20}$).

Question 4.5

(a) The average oceanic peridotite has the composition of a harzburgite.

(b) Reading from Figure 4.12, the composition AO has roughly 75% olivine, 21% orthopyroxene and 4% clinopyroxene.

Question 4.6

According to the phase diagram for KLB-1 (Figure 4.15), spinel lherzolite is stable at pressures equivalent to between about 25 km and 85 km. (*Note*: given that the sample came from below the Moho and that the crust is typically about 35 km thick, then the minimum depth would actually be about 35 km.)

Question 4.7

Spinel lherzolites are typical of mantle xenoliths found in basalts, whereas garnet lherzolites are typical of kimberlites (see Section 4.2.2). So, according to the phase diagram in Figure 4.15, kimberlites must come from within the garnet stability field, whereas basalts are more likely to come from the more shallow spinel lherzolite field. The presence of diamonds within some kimberlites is also consistent with these magmas coming from great depth.

Question 4.8

Reading from Figure 4.15, KLB-1 is solid at 1.5 GPa below about 1350 °C and liquid above about 1830 °C. KLB-1 is a mixture of crystals and liquid between these temperatures.

Question 4.9

Statements A, C, E, F, G and J are correct.

A is correct because, according to Table 4.3 and Figure 4.16, the per cent liquid changes from 0 to greater than 0 between 1300 °C and 1400 °C. The phase diagram in Figure 4.15 shows the solidus close to 1350 °C.

C and E are correct according to Table 4.3.

F and G are correct according to Figure 4.16.

J is correct because peridotite KLB-1 has an MgO content of about 40% (the composition of totally molten KLB-1) whereas the liquid produced nearer the solidus has a lower MgO content. At the solidus of MM3 the liquid has about 10% MgO, and extrapolating the results from KLB-1 to its solidus temperature implies a similar MgO content in that case too.

The wrong statements should read (modifications in italics):

B The first phase to disappear with increasing temperature is *spinel*.

D *Olivine* is the mineral phase that survives to the highest temperature.

H The Fo content of olivine becomes *higher* with increasing amount of melting.

I The MgO content of the liquid *increases* with increasing temperature.

Question 4.10

This sample plots in the solid forsterite + liquid field. The composition of the liquid coexisting with solid Fo is defined by the liquidus at 1800 °C, i.e. $Fo_{58}Di_{42}$.

Question 4.11

(a) Melting first occurs at the eutectic temperature of 1580 °C.

(b) At temperatures just above the eutectic temperature the sample will plot in the solid diopside + liquid field, meaning that forsterite is the first mineral to be consumed. This is because the bulk sample has a composition that falls on the Di-rich side of the eutectic composition.

(c) The first liquid has the composition of the eutectic, i.e. $Fo_{19}Di_{81}$.

(d) The sample becomes completely molten at the liquidus temperature for $Fo_{10}Di_{90}$. This is read from the phase diagram (with care, because the liquidus curve has a shallow slope) as 1585 °C.

Question 4.12

The behaviours described in statements D, E and F are due to there being more than one mineral present and are therefore similar to the behaviour of a eutectic system.

The behaviour described in statement B is caused by solid solution between Fo and Fa in olivine.

The behaviours described in statements A and C are encountered in both eutectic and solid-solution systems. In the case of statement C, MgO increases because the composition of peridotite is MgO rich, and Mg-rich olivine is the last mineral to melt completely.

Question 4.13

The mantle with the higher T_p crosses the peridotite solidus at a greater depth and reaches a final temperature that is further above the solidus than the mantle with the lower T_p. The hotter mantle is therefore the one that yields the greater amount of liquid.

Question 4.14

From Figure 4.25a, a crustal thickness of 27 km implies a potential temperature of about 1480 °C.

Question 4.15

From Figure 4.29, the liquidus of mid-ocean ridge basalt with Mg# = 65% is about 1200–1230 °C and for Mg# = 50% about 1140–1160 °C, so typical eruption temperatures are in the range 1140–1230 °C.

Question 4.16

The order of crystallisation is: olivine; olivine + plagioclase; olivine + plagioclase + clinopyroxene.

Question 4.17

(a) Using Equation 4.1 (see Box 4.4) the pressure due to the weight of 3 km of seawater $= (1.0 \times 10^3)\ \text{kg m}^{-3} \times 9.8\ \text{m s}^{-2} \times (3 \times 10^3)\ \text{m} = 3 \times 10^7\ \text{Pa}$ (to 1 sig. fig.) = 30 MPa.

(b) The pressure 30 km below the sea floor is the sum of the pressure due to the water and the rocks. The pressure due to the water and rocks is 30 MPa $+ (3 \times 10^3)\ \text{kg m}^{-3} \times 9.8\ \text{m s}^{-2} \times (30 \times 10^3)\ \text{m} = 9 \times 10^8\ \text{Pa}$ (to 1 sig. fig.) = 900 MPa.

Question 4.18

(a) Statements (i) and (ii) are true. Statement (iii) is false – none of these minerals is hydrous.

(b) (i) albite: Al_2O_3 and Na_2O

(ii) chlorite: Al_2O_3, FeO_t and MgO

(iii) epidote: Al_2O_3, FeO_t and CaO

(iv) actinolite: FeO_t, MgO, CaO and moderate amounts of Al_2O_3.

(c) Water ranges from less than 2% to more than 6%.

Question 5.1

$X_{\text{olivine}} = 0.6; X_{\text{orthopyroxene}} = 0.2; X_{\text{clinopyroxene}} = 0.1; X_{\text{garnet}} = 0.1$

$D_{\text{olivine}} = 0.001; D_{\text{orthopyroxene}} = 0.003; D_{\text{clinopyroxene}} = 0.001; D_{\text{garnet}} = 0.01$
(from Table 5.2)

Therefore, using Equation 5.3:

$$D_{\text{bulk}} = (0.6 \times 0.001) + (0.2 \times 0.003) + (0.1 \times 0.001) + (0.1 \times 0.01)$$
$$= 0.0023$$

Question 5.2

$C_0 = 25$, $D = 0.402$ and $F = 0.2$ (remember the amount of crystallisation is $1 - F$). For equilibrium crystallisation (using Equation 5.6):

$$C_{\text{L}} = \frac{25}{0.402 + (1 - 0.402) \times 0.2}$$
$$= 47.9 \text{ ppm}$$
$$= 48 \text{ ppm to two significant figures.}$$

And for fractional crystallisation (using Equation 5.7):

$C_{\text{L}} = 25 \text{ ppm} \times 0.2^{(0.402-1)}$

Start by working out the the value in the brackets:

$0.402 - 1 = -0.598$

You can calculate $0.2^{-0.598}$ using the x^y key on a calculator:

$0.2 \; \boxed{x^y} \; -0.598 = 2.618$

You can then add this to the original equation:

$C_{\text{L}} = 25 \text{ ppm} \times 2.618 = 65.45 \text{ ppm}$

$= 65 \text{ ppm}$ to two significant figures.

This could also be done in a spreadsheet.

Question 5.3

Using simple geometry, the distance that the slab has to travel from the trench to get to 100 km depth is:

$$\text{depth} = \text{distance travelled} \times \sin \theta$$

where θ is the angle of subduction.

Rearranging this gives:

$$\text{distance travelled} = \frac{\text{depth}}{\sin\theta}$$

$$\text{distance travelled} = \frac{100\ \text{km}}{\sin 45}$$
$$= 141.42\ \text{km}$$

$$\text{time} = \frac{141.42\ \text{km}}{60 \times 10^{-6}\ \text{km y}^{-1}}$$
$$= 2.36 \times 10^{6}\ \text{y} = 2.4\ \text{Ma or} \sim 1.6\ \text{half-lives of}\ ^{10}\text{Be}.$$

Therefore, the plate takes 2.4 Ma (to two significant figures) to reach 100 km depth. For a plate travelling at only 10 mm y^{-1} it will take six times as long (~14 Ma).

This simple calculation indicates that subduction zones with relatively rapid convergence rates will have the best chance of preserving some ^{10}Be atoms in the slab.

Question 5.4

The composition of the lavas clearly reflects changes in the composition of the sediment input. Within 2.5 Ma of the change in the sediment U/Th ratio there is a corresponding increase in the U/Th ratio of the arc lavas. This also implies that U is derived from the sediment rather than the basaltic slab in this arc. The data require a total transit time of ~2.5 Ma from sediment subduction to melt eruption. These data are consistent with other pieces of information about this arc. For example the high ^{10}Be ratios in Central America (Figure 5.25e) require rapid (<2 Ma) sediment subduction, consistent with the Cocos Plate being subducted at ~80 mm y^{-1}.

Question 6.1

(i) For Rb: $\frac{C_L}{C_0} = 2$.

From the curve for $D = 0.1$ in Figure 5.21, $F = 0.43$.

(ii) For Sr: $\frac{C_L}{C_0} = 0.15$.

For $D = 3.5$, $\frac{C_L}{C_0}$ does not reach such low values, even at $F = 0$.

These results can be calculated more precisely using Equation 5.6, giving $F = 0.44$ for Rb. For Sr, $\frac{C_L}{C_0}$ is negative.

Question 6.2

For Rb: $\frac{C_L}{C_0} = \frac{360}{200} = 1.8$. From Figure 5.21, $F = 0.1$ (Equation 5.6 gives 0.11).

For Sr: $\frac{C_L}{C_0} = \frac{65}{120} = 0.54$. From Figure 5.21, F lies between 0.1 and 0.2

(Equation 5.6 gives 0.15), so 10–15% melting would be required.

Question 6.3

(a) (i) Geotherm A does not intersect the melting reaction and so no melting occurs.

(ii) Geotherm B crosses the melting reaction (Equation 6.3) at depths of about 23 km, and this is where melting will occur. No melting is possible where the geotherm crosses the wet granite solidus (at about 18 km) because of the absence of both H_2O and alkali feldspar in the sediments.

(b) At 24 km, geotherm B lies within the sillimanite field, which is the polymorph that is observed in the Himalayan metasediments (Figure 6.3c).

Question 7.1

(a) The continental lithosphere beneath North America and Australia is generally seismically fast (blue coloured) in the top 110 km. This is entirely consistent with ancient (Archaean to Proterozoic) lithosphere being present in these continents, as the lithosphere has had a long time to cool. The western edge of North America is clearly seismically slow (red). In continental Africa the region along the East African Rift valley is seismically slow (red), whereas the interior of Africa is seismically fast (blue). This reflects the difference between a region of active volcanism and extension (East African Rift) and cooler Archaean and Proterozoic cratons.

(b) The East Pacific Rise has a small (~0.5%) slow anomaly (yellow) located beneath the position of the ridge to a depth of 600 km. This implies that warmer mantle material is convected upwards from great depth beneath the mid-ocean ridge, even though melting dominantly occurs in the upper 60 km of the mantle.

(c) Down to depths of 200 km there is a strong seismic slow region (red) that defines a narrow strip along the volcanic arc. This seismic slowing relates to arc magmatism. By 300 km depth the slow band is less clear, which may relate to some slab dehydration reactions. Between 400 and 600 km depth a broader zone of seismic slow velocities appears. Because it takes tens of millions of years for the slab to descend through the upper mantle, these images provide evidence for long-term subduction along the Pacific Rim. This is a case where seismic tomography provides more than a simple snapshot. Seismic fast zones located at great depths (>1000 km) have been used to infer plate dynamics back to 60 Ma.

(d) There are clearly zones of seismically slow material (red) that broadly extend down through the mantle to the CMB and which are associated with known hot spots or at centres of volcanism. These include volcanic centres in the southwest Pacific associated with super-swell, Hawaii, East Africa and some Atlantic Ocean islands that have anomalies that extend through the mantle. This is best seen for the southwest Pacific, whereas the other localities mentioned above have weaker signals between 1300 km and 2300 km depth, suggesting a possible lack of continuity of the buoyant zone, or just poor resolution. Hot spots on continental North America generally have little, if any, slow signal in the lower mantle, suggesting they are an upper mantle phenomenon.

Question 7.2

Incorporating Equation 7.5 ($u = \eta/\rho$) into Equation 7.4 gives:

$$Ra = \frac{\alpha g d^3 \Delta T \rho}{\kappa \eta} \tag{7.7}$$

Hence, for $\Delta T = 1$:

$$Ra = \frac{2\times10^{-5}\ \mathrm{K^{-1}} \times 1\ \mathrm{K} \times 10\ \mathrm{m\ s^{-2}} \times (2.9\times10^{6}\ \mathrm{m})^3 \times 3300\ \mathrm{kg\ m^{-3}}}{10^{-6}\ \mathrm{m^2\ s^{-1}} \times 10^{21}\ \mathrm{kg\ m^{-1}\ s^{-1}}}$$

$$= \frac{1.6\times10^{4}\ \mathrm{K\ K^{-1}\ m\ s^{-2}\ m^3\ kg\ m^{-3}}}{\mathrm{m^2\ s^{-1}\ kg\ m^{-1}\ s^{-1}}}$$

The units cancel out, showing that Ra is indeed a dimensionless number, therefore $Ra = 1.6 \times 10^4$ to two significant figures. So if $\Delta T = 10^2$ K, then $Ra = 1.6 \times 10^6$.

Question 7.3

$$Pe_{\mathrm{t}} = \frac{0.01\ \mathrm{m\ s^{-1}} \times 50\times10^{3}\ \mathrm{m}}{10^{-6}\ \mathrm{m^2\ s^{-1}}}$$

so $Pe_{\mathrm{t}} = 5 \times 10^8$.

This calculation again illustrates how the units used to calculate a dimensionless number cancel out.

Pe_{t} is clearly much greater than unity, so the mantle beneath an ocean ridge must advect.

Question 7.4

Using the equation for equilibrium partial melting (Equation 5.6), the concentration of water in the melt phase is given by:

$$C_{\mathrm{L}} = \frac{C_0}{D + F(1 - D)}$$

where C_{L} is the concentration in the melt, C_0 is the concentration in the source, D is the bulk partition coefficient and F is the melt fraction.

But $C_r = C_L D$, where C_r is the concentration in the residue. Therefore:

$$C_r = \frac{C_0 D}{D + F(1 - D)}$$

And so for H_2O with $D = 0.005$, $F = 0.01$ (1% melting) and $C_0 = 1$ wt%

$$C_r = \frac{1 \text{ wt\%} \times 0.005}{0.005 + 0.01\,(1 - 0.005)}$$
$$= 0.33 \text{ wt\%}$$

Therefore 33 wt% of the H_2O remains in the residue.

The values for F = 0.05 (5% melting) and F = 0.1 (10% melting) are calculated in a similar way. So:

for $F = 0.01$, $C_r = 0.33 = 33\%$

for $F = 0.05$, $C_r = 0.09 = 9\%$

for $F = 0.1$, $C_r = 0.047 = 4.7\%$.

Question 7.5

The H_2O content of MORB is $7.5 \times 200 = 1500$ ppm.

You have calculated previously that 10% melting extracts 95% of the H_2O from the source (Question 7.4). By proportion, the amount of H_2O in the source can, therefore, be easily calculated to be ~160 ppm using a mass balance equation.

$0.1 \times 1500 \text{ ppm} = 0.95 \times X$; where X is the concentration of H_2O in the mantle.

$$\text{Therefore } X = \frac{150 \text{ ppm}}{0.95} = 157.9 \text{ ppm} = 160 \text{ ppm (to 2 sig. figs).}$$

Alternatively, the concentration of H_2O can be calculated from $C_0 = C_L \times (D + F(1 - D)) = 157.9$ ppm = 160 ppm (to 2 sig. figs).

Question 7.6

Mass of H_2O in the mantle = H_2O concentration in mantle × mass of mantle

$$= (100 \times 10^{-6}) \times (40 \times 10^{23}) \text{ kg}$$
$$= 0.4 \times 10^{21} \text{ kg}$$

If the H_2O concentration in the mantle is 500 ppm, then the H_2O mass is 2.0×10^{21} kg.

This compares with 1.4×10^{21} kg water in the ocean.

Appendix A The elements

Table A1 The chemical elements, their abundances and condensation temperatures during formation of the Solar System. (Data from Lodders, 2003)

Atomic number, Z	Name	Chemical symbol	Relative atomic mass, A_r	Solar System abundance		CI chondrite abundance by mass/ppm	T_c (50%)[1] (K)
				Solar photosphere by number relative to Si = 1×10^6	CI chondrites by number relative to Si = 1×10^6		
1	hydrogen	H	1.01	2.88×10^{10}	5.5×10^6	21 000	–[d]
2	helium	He	4.00	2.29×10^9	0.604	9×10^{-3}	–[d]
3	lithium	Li	6.94	0.363	55.5	1.46	1142
4	beryllium	Be	9.01	0.407	0.74	0.025	1452
5	boron	B	10.81	14.5	17.3	0.71	908
6	carbon	C	12.01	7.7×10^5	7.1×10^6	35 000	40
7	nitrogen	N	14.01	1.95×10^6	5.54×10^4	2940	123
8	oxygen	O	16.00	1.41×10^7	7.55×10^6	458 000	180
9	fluorine	F	19.00	1047	841	60	734
10	neon	Ne	20.18	2.15×10^6	2.36×10^{-3}	1.8×10^{-4}	9.1
11	sodium	Na	22.99	5.75×10^4	5.75×10^4	5010	958
12	magnesium	Mg	24.31	1.00×10^6	1.04×10^6	95 900	1336
13	aluminium	Al	26.98	8.51×10^4	8.31×10^4	8500	1653
14	silicon	Si	28.09	1.0×10^6	1.0×10^6	107 000	1310
15	phosphorus	P	30.97	8913	7833	920	1229
16	sulfur	S	32.07	4.63×10^5	4.50×10^5	54 100	664
17	chlorine	Cl	35.45	9120	5240	704	948
18	argon	Ar	39.95	1.03×10^5	9.06×10^{-3}	1.33×10^{-3}	47
19	potassium	K	39.10	3800	3580	530	1006
20	calcium	Ca	40.08	6.61×10^4	5.97×10^4	9070	1517
21	scandium	Sc	44.96	42.7	34.2	5.83	1659
22	titanium	Ti	47.88	3020	2422	440	1582
23	vanadium	V	50.94	288.4	288.4	55.7	1429
24	chromium	Cr	52.00	1.26×10^4	1.31×10^4	2590	1296
25	manganese	Mn	54.94	7079	9170	1910	1158
26	iron	Fe	55.85	8.13×10^5	8.63×10^5	183 000	1334
27	cobalt	Co	58.93	2400	2250	502	1352
28	nickel	Ni	58.69	4.78×10^4	4.78×10^4	10 600	1353
29	copper	Cu	63.55	468	527	127	1037
30	zinc	Zn	65.39	1202	1250	310	726
31	gallium	Ga	69.72	21.9	36	9.51	968

Atomic number, Z	Name	Chemical symbol	Relative atomic mass, A_r	Solar System abundance		CI chondrite abundance by mass/ppm	T_c (50%)[1] (K)
				Solar photosphere by number relative to Si = 1×10^6	CI chondrites by number relative to Si = 1×10^6		
32	germanium	Ge	72.61	110	121	33.2	883
33	arsenic	As	74.92	–	6.09	1.73	1065
34	selenium	Se	78.96	–	65.8	19.7	697
35	bromine	Br	79.90	–	11.3	3.43	546
36	krypton	Kr	83.80	55.2	1.64×10^{-4}	5.22×10^{-5}	52
37	rubidium	Rb	85.47	11.5	6.57	2.13	800
38	strontium	Sr	87.62	24.0	23.3	7.74	1464
39	yttrium	Y	88.91	4.68	4.54	1.53	1659
40	zirconium	Zr	91.22	11.2	11.5	3.96	1741
41	niobium	Nb	92.91	0.759	0.752	0.265	1559
42	molybdenum	Mo	95.94	2.40	2.80	1.02	1590
43	technetium	Tc[a]	98.91	–[b]	–[c]	–[c]	–[c]
44	ruthenium	Ru	101.07	2.00	1.81	0.692	1551
45	rhodium	Rh	102.91	0.38	0.36	0.141	1392
46	palladium	Pd	106.42	1.41	1.46	0.588	1324
47	silver	Ag	107.87	(0.25)	0.49	0.201	996
48	cadmium	Cd	112.41	1.70	1.58	0.675	652
49	indium	In	114.82	1.05	0.18	0.079	536
50	tin	Sn	118.71	2.88	3.73	1.68	704
51	antimony	Sb	121.76	0.29	0.33	0.152	979
52	tellurium	Te	127.60	–	4.82	2.33	709
53	iodine	I	126.90	–	1.00	0.48	535
54	xenon	Xe	131.29	5.39	3.5×10^{-4}	1.74×10^{-4}	68
55	caesium	Cs	132.91	–	0.367	0.185	799
56	barium	Ba	137.33	4.27	4.44	2.31	1455
57	lanthanum	La	138.91	0.389	0.441	0.232	1578
58	cerium	Ce	140.12	1.10	1.169	0.621	1478
59	praseodymium	Pr	140.91	0.148	0.174	0.093	1582
60	neodymium	Nd	144.24	0.912	0.836	0.457	1602
61	promethium	Pm[a]	146.92	–[c]	–[c]	–[c]	–[c]
62	samarium	Sm	150.36	0.2818	0.254	0.145	1590
63	europium	Eu	151.96	0.096	0.095	0.0546	1356
64	gadolinium	Gd	157.25	0.380	0.332	0.198	1659
65	terbium	Tb	158.93	0.055	0.059	0.036	1659
66	dysprosium	Dy	162.50	0.398	0.386	0.238	1659
67	holmium	Ho	164.93	0.098	0.090	0.056	1659
68	erbium	Er	167.26	0.246	0.255	0.162	1659

Atomic number, Z	Name	Chemical symbol	Relative atomic mass, A_r	Solar System abundance		CI chondrite abundance by mass/ppm	T_c (50%)[1] (K)
				Solar photosphere by number relative to Si = 1×10^6	CI chondrites by number relative to Si = 1×10^6		
69	thulium	Tm	168.93	(0.029)	0.037	0.024	1659
70	ytterbium	Yb	170.04	0.347	0.248	0.163	1487
71	lutetium	Lu	174.97	0.033	0.036	0.024	1659
72	hafnium	Hf	178.49	0.219	0.170	0.115	1684
73	tantalum	Ta	180.95	–	0.021	0.014	1573
74	tungsten	W	183.85	(0.372)	0.128	0.089	1789
75	rhenium	Re	186.21	–	0.053	0.037	1821
76	osmium	Os	190.2	0.813	0.674	0.486	1812
77	iridium	Ir	192.22	0.692	0.645	0.470	1603
78	platinum	Pt	195.08	1.59	1.36	1.00	1408
79	gold	Au	196.97	(0.30)	0.196	0.146	1060
80	mercury	Hg	200.59	–	0.413	0.314	252
81	thallium	Tl	204.38	(<0.36)	0.185	0.143	532
82	lead	Pb	207.2	2.88	3.26	2.56	727
83	bismuth	Bi	208.98	–	0.14	0.110	746
84	polonium	Po[a]	209.98	–[c]	–[c]	–[c]	–[c]
85	astatine	At[a]	209.99	–[c]	–[c]	–[c]	–[c]
86	radon	Rn[a]	222.02	–[c]	–[c]	–[c]	–[c]
87	francium	Fr[a]	223.02	–[c]	–[c]	–[c]	–[c]
88	radium	Ra[a]	226.03	–[c]	–[c]	–[c]	–[c]
89	actinium	Ac[a]	227.03	–[c]	–[c]	–[c]	–[c]
90	thorium	Th[a]	232.04	–	0.035	0.0309	1659
91	protoactinium	Pa[a]	231.04	–[c]	–[c]	–[c]	–[c]
92	uranium	U[a]	238.03	<0.01	0.0093	0.0084	1610

[1]Temperature at which 50% of the element has condensed at 10^{-4} bars total pressure from a nebula with a CI composition.

[a] No stable isotopes.

[b] Detected in spectra of some rare evolved stars.

[c] No naturally occurring long-lived isotopes.

[d] No condensation temperature has been determined.

SI fundamental and derived units

Appendix B

Table B1 SI fundamental and derived units.

Quantity	Unit	Abbreviation	Equivalent units
mass	kilogram	kg	
length	metre	m	
time	second	s	
temperature	kelvin	K	
angle	radian	rad	
area	square metre	m^2	
volume	cubic metre	m^3	
speed, velocity	metre per second	$m\ s^{-1}$	
acceleration	metre per second squared	$m\ s^{-2}$	
density	kilogram per cubic metre	$kg\ m^{-3}$	
frequency	hertz	Hz	(cycles) s^{-1}
force	newton	N	$kg\ m\ s^{-2}$
pressure	pascal	Pa	$kg\ m^{-1}\ s^{-2}$, $N\ m^{-2}$
energy	joule	J	$kg\ m^2\ s^{-2}$
power	watt	W	$kg\ m^2\ s^{-3}$, $J\ s^{-1}$
specific heat capacity	joule per kilogram kelvin	$J\ kg^{-1}\ K^{-1}$	$m^2\ s^{-2}\ K^{-1}$
thermal conductivity	watt per metre kelvin	$W\ m^{-1}\ K^{-1}$	$m\ kg\ s^{-3}\ K^{-1}$

Appendix C The Greek alphabet

Table C1 The Greek alphabet.

Name	Lower case	Upper case	Name	Lower case	Upper case
alpha	α	Α	nu (new)	ν	Ν
beta (bee-ta)	β	Β	xi (cs-eye)	ξ	Ξ
gamma	γ	Γ	omicron	ο	Ο
delta	δ	Δ	pi (pie)	π	Π
epsilon	ε	Ε	rho (roe)	ρ	Π
zeta (zee-ta)	ζ	Ζ	sigma	σ	Σ
eta (ee-ta)	η	Η	tau (torr)	τ	Τ
theta (thee-ta; 'th' as in theatre)	θ	Θ	upsilon	υ	Υ
iota (eye-owe-ta)	ι	Ι	phi (fie)	Φ	Φ
kappa	κ	Κ	chi (kie)	χ	Χ
lambda (lam-da)	λ	Λ	psi (ps-eye)	ψ	Ψ
mu (mew)	μ	Μ	omega (owe-me-ga)	ω	Ω

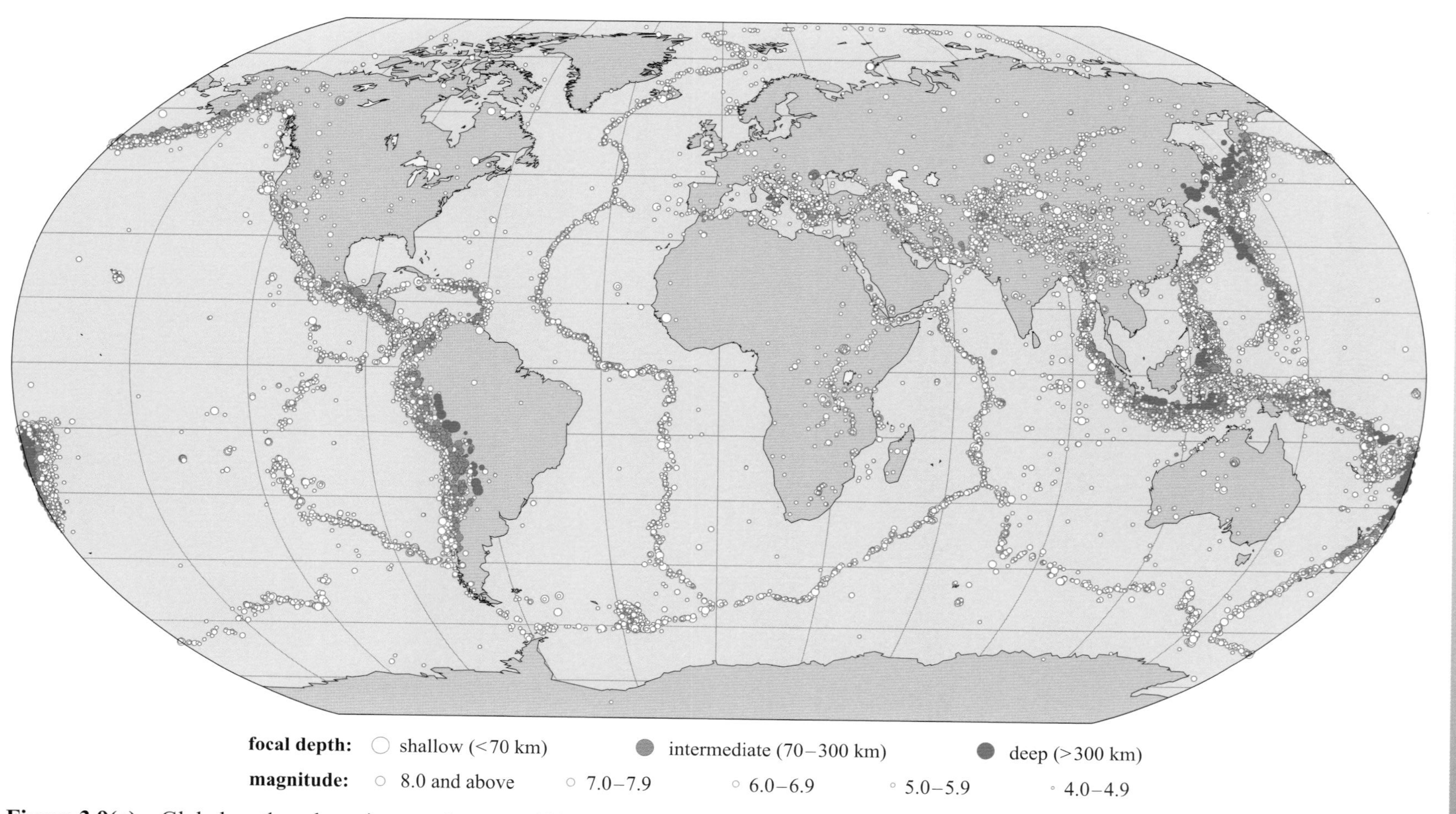

Figure 3.9(a) Global earthquake epicentres between 1980 and 1996. Only earthquakes of magnitude 4 and above are included. (BGS)

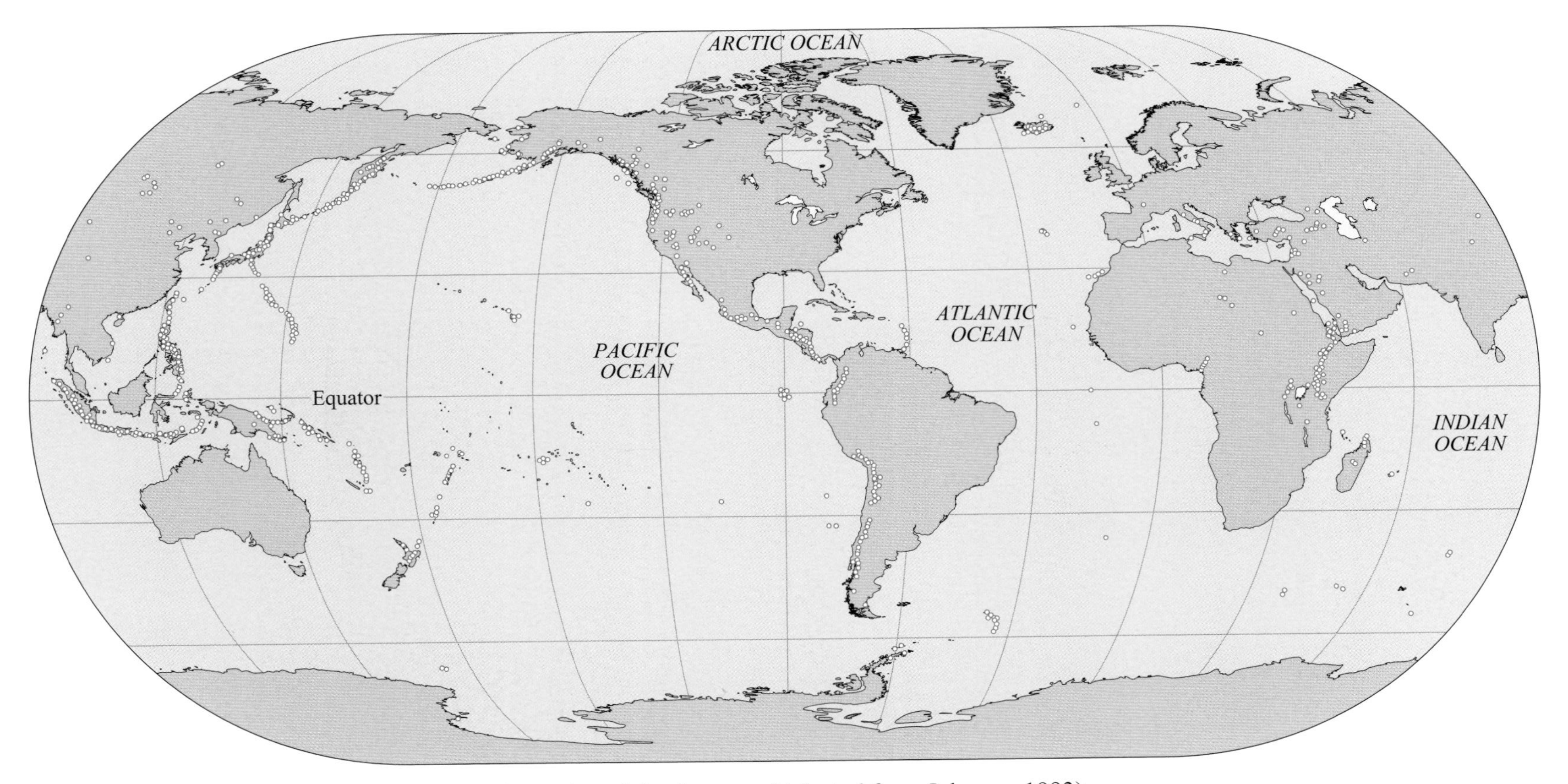

Figure 3.9(b) Map showing locations of active, sub-aerial volcanoes. (Adapted from Johnson, 1993)

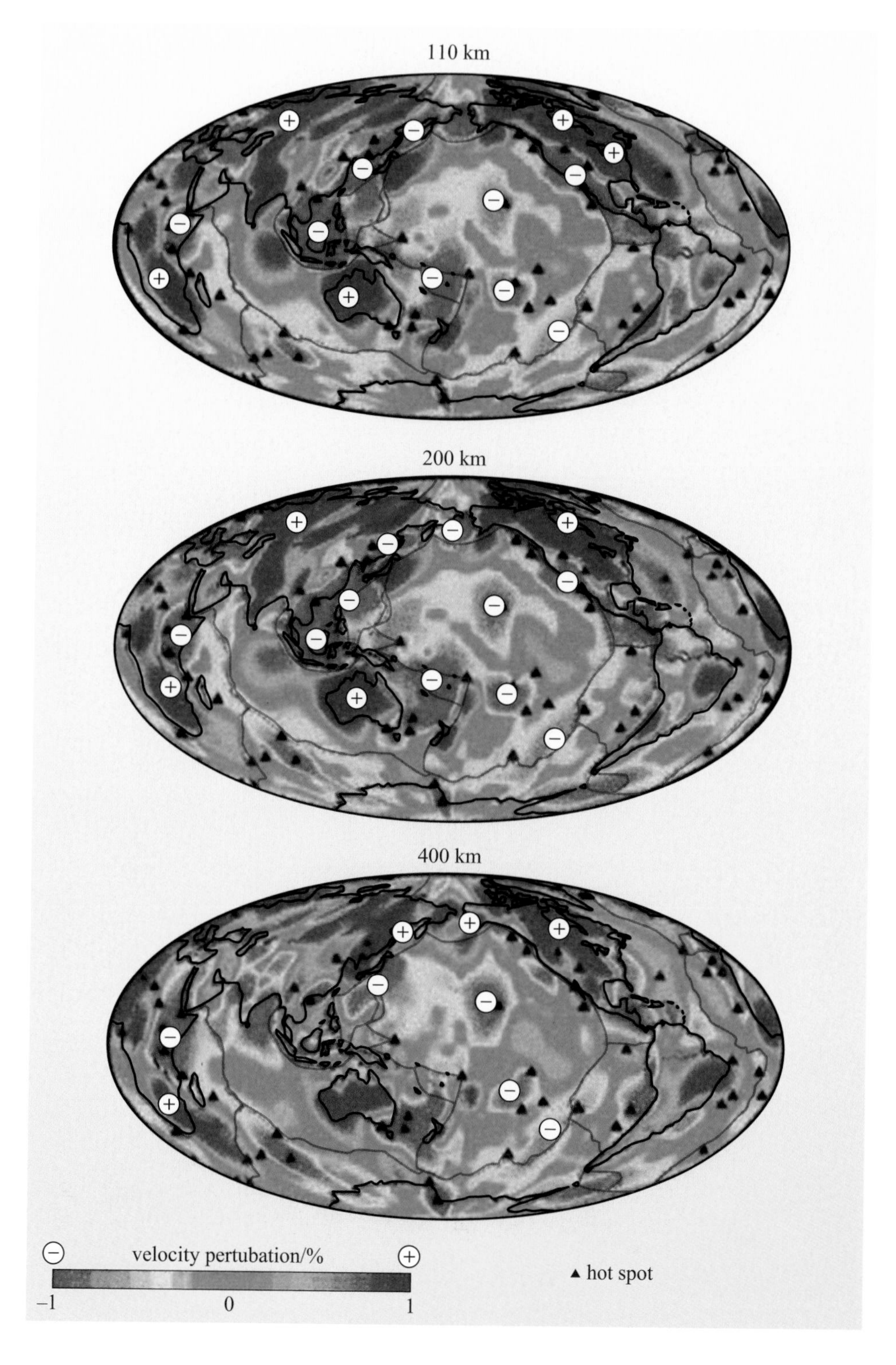

Figure 7.6 Global P-wave tomographic images sliced for six selected depths through the Earth. Red and blue colours denote the slow- and fast-velocity anomalies respectively. Minus and plus signs have been added to key areas to emphasise key regions of slow- and fast-velocity anomalies respectively. Dark lines are continental coastlines and paler lines are tectonic plate boundaries. Note the scalebar is for the upper mantle slices. (Continued overleaf)

Figure 7.6 (continued) Global P-wave tomographic images sliced for six selected depths through the Earth. Red and blue colours denote the slow- and fast-velocity anomalies respectively. Minus and plus signs have been added to key areas to emphasise key regions of slow- and fast-velocity anomalies respectively. Dark lines are continental coastlines and paler lines are tectonic plate boundaries. Note the first scalebar is for the upper mantle slice and the second scalebar is for the two lower mantle slices. (Lei and Zhao, 2006a).

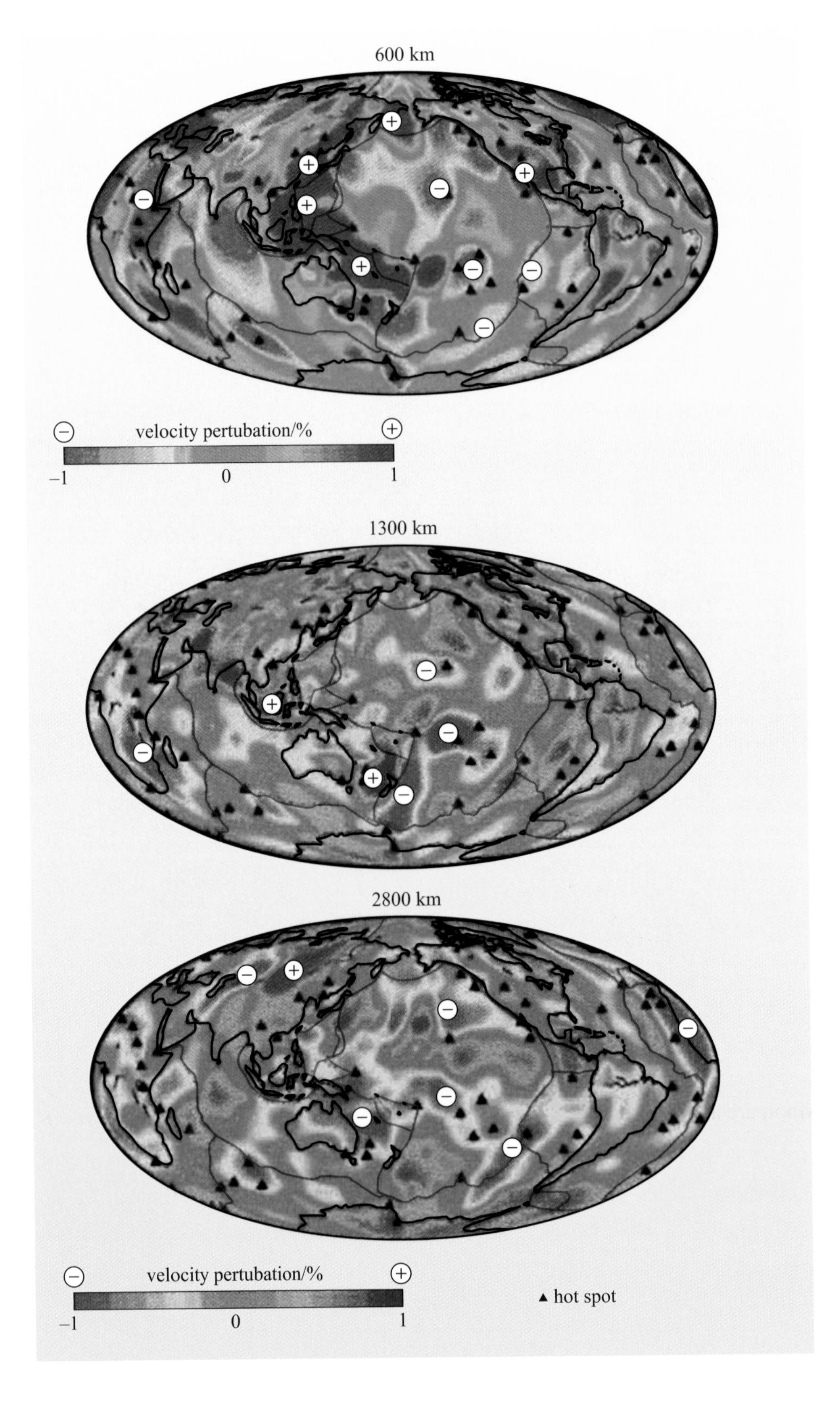

Glossary

abyssal plain Flat region of the deep ocean floor with slopes of less than 1:1000.

accretionary prism A wedge of sediment scraped off the **down-going plate** at a **subduction zone**, or faulted regions containing serpentinised **peridotites**, within the **fore-arc region**.

achondrites Meteorites that do not contain **chondrules**.

adakite Hydrous melt found at **subduction zones**.

adiabatic From the Greek for 'impassable' – a reference to the fact that heat passes neither from nor to a substance undergoing an adiabatic process.

advection Upward transfer of heat by physically moving material, e.g. magma or fluids, passing through fractures in the **lithosphere**.

albite The pure Na endmember for **alkali feldspar** and **plagioclase**.

alkali feldspar Feldspar rich in Na and/or K.

aluminosilicate A silicate in which aluminium substitutes for some of the silicon in the SiO_4 tetrahedra, e.g. the polymorphs of Al_2SiO_5.

andalusite A brown or yellow **aluminosilicate** forming at moderate temperatures and low pressures; one of three polymorphs with the same chemical formula, $A1_2SiO_5$, (the others being **kyanite** and **sillimanite**).

Andean margin A **destructive plate boundary** where oceanic **lithosphere** is subducted beneath a continental margin, typified by the western seaboard of South America.

angular momentum A property of rotating systems that depends upon mass and its distribution, angular velocity, and the radius of rotation. The angular momentum of the Earth–Moon system incorporates the Earth's rotation and the Moon's orbital motion.

anorthite The pure Ca endmember in the **plagioclase** series.

anorthosite A plutonic igneous rock composed of over 90% **plagioclase** feldspar (usually labradorite) plus minor pyroxene. The oldest igneous rocks from the Moon are anorthosite.

apparent polar wander Path generated by plotting the apparent positions of the poles through time. Useful in charting the rifting and suturing of continents.

arc lithosphere The crust and the lithospheric mantle at an island arc.

arc-parallel flow **Asthenosphere** flowing parallel to the length of the **island arc**.

aseismic ridge A submarine ridge that is a fragment of continental crust or anomalously shallow ocean crust, e.g. a **hot spot**.

aspect ratio The ratio of the width to height of a convective cell.

asteroid A small rocky or metallic body orbiting the Sun, usually a few km in diameter.

asteroid belt The region between the orbits of Mars and Jupiter where the majority of **asteroids** are found.

asthenosphere The mobile mantle that deforms like a viscous fluid stressing response to the prevailing stress extending from the base of the **lithosphere** to the 670 km discontinuity.

atmophile A subgroup of elements that tends to be gaseous at the Earth's surface, notably H, C, N, O and the noble gases.

atmosphere The envelope of gases that surrounds the Earth.

β spinel Olivine that has undergone a phase change at ~14 GPa.

back-arc basin A small ocean basin opened up between an **island arc** and the adjacent continent.

banded iron formation (BIF) Rocks containing finely banded dark-brown, iron-rich layers alternating with light-coloured iron-poor cherts that may contain information about the oxidation state of the Earth's environment when they formed.

basalt Dark-coloured, fine-grained mafic igneous rock, usually extrusive but locally intrusive, composed chiefly of calcic **plagioclase** and **clinopyroxene**.

basalt–andesite–dacite–rhyolite association An association of volcanic rock types common in island-arc settings, generated by the progressive fractional crystallisation of a basaltic parent magma.

bending resistance The resistance of the **down-going plate** to flexing at the trench before it begins to slide beneath the opposing plate.

BIF *See* **banded iron formation**.

binary system System with two components.

biosphere All forms of life, particularly on Earth, but possibly on other planets.

blueschist facies metamorphic rocks The metamorphic mineral assemblage containing sodic amphiboles representing low-temperature and high-pressure conditions, e.g. produced in **subduction zones**.

body wave A seismic wave that travels through the interior of the Earth with either longitudinal or transverse movement.

Bouguer gravity anomaly Variations in gravity across the surface of the Earth due to topography, rock type and latitude.

boundary-layer model A model that proposes that the **lithosphere** does not have a constant thickness but thickens as it cools through loss of heat from the underlying **asthenosphere**, progressively cools below the temperature at which it can undergo solid-state creep and is transformed from **asthenosphere** to **lithosphere**.

bulk silicate Earth The hypothetical chemical composition of the Earth's mantle after the formation of the core but before separation of the crust.

buoyancy forces The upward forces on an object produced by the surrounding material in which the object is (partly) immersed due to the pressure difference between the top and bottom of the object.

calcium (Ca)-perovskite A mineral dissociated from **majorite** at 20–25 GPa.

carbonaceous (or C-) chondrites Chondrites containing non-biogenic carbon-rich organic compounds in addition to silicate minerals.

cation A positively charged ion.

chalcophile Elements that frequently occur bound with sulfur.

channel-flow model A model for the mechanical and thermal behaviour of continental **lithosphere** that predicts the formation and movement of a low-viscosity layer within the crust, for example by **melt weakening**.

chemical variation diagram A graphical plot, usually bivariate, of the chemical composition of a series of rocks. Typical examples include the **total alkalis (versus silica; TAS) diagram** used for the classification of igneous rocks, but also used to show igneous differentiation of mafic to felsic compositions, and plots of major and trace element concentrations against magnesium oxide (MgO) to illustrate the differentiation of basaltic and intermediate lavas.

chemosynthetic Description of basic nutrients that have been formed by the breakdown of chemicals such as methane or hydrogen sulfide rather than by photosynthetic processes with light and chlorophyll.

chondrites Meteorites containing **chondrules**.

chondrules Small, roughly spherical globules of silicate minerals between 0.1 mm and 2 mm in diameter.

Clapeyron slope The slope of the **phase boundaries** on pressure–temperature diagrams which has a significant effect on the movement of dense material downwards (slabs) and hot material upwards (plumes).

clinopyroxene Pyroxenes with more than about 10% Ca that have an atomic structure with only two symmetry axes because of the distortions required to fit the large Ca^{2+} ions into the spaces between the chains of silicate tetrahedra.

collisional resistance force A force generated by collision of plates and the associated deformation processes. This acts in the opposite direction within each converging plate, but it is equal in magnitude in both.

comet A minor body, usually with an elongate orbit, composed of mainly water-ice and rocky material.

compatible element Elements that partition into a solid phase ($D \geq 1$).

complete ionic substitution Substitution of two types of atoms in a molecule, e.g. Fe and Mg in olivines.

compositional convection Density-driven convection within a **magma** chamber where density contrasts are generated by local differences in magma composition.

conduction Spread of heat energy through a solid, e.g. the heating of a saucepan handle when the pan is heated.

Conrad discontinuity A marked increase in seismic velocities at ~15 km depth separating the upper crust from the lower crust.

conservative plate boundary Area where lithospheric plates move past each other without creating or destroying crust. Associated with transform faults where motion is largely horizontal rather than vertical.

constructive plate boundary Zone of creation of new crust material where two or more tectonic plates are moving apart.

continental drag force A resistant force that slows plate motion caused by the interaction of the deep lithospheric keel beneath old continental regions, which can be up to 150 km thick, with the underlying asthenosphere.

continental drift The opening and closure of ocean basins, and the associated movement of continents.

continental shelf The area of seabed around a large land mass where the sea is relatively shallow compared with the open ocean. The continental shelf is geologically part of the continental crust.

convection The efficient movement of hot material through fluids from regions that are hotter to those that are cooler. During this transfer cooler regions are heated.

convective thinning Following **lithospheric** thickening, the deep, cold root becomes unstable to convection in the **asthenosphere** and is recycled back into the **convecting** mantle.

corner flow The movement of material that is dragged down by the slab at a subduction zone and is replaced at higher levels by **asthenosphere** from the **back-arc region**.

cosmogenic isotope Isotopes produces by **spallation,** e.g. Be.

coupled substitution Substitution of two atoms by two others to maintain electrical balance, e.g. CaAl by NaSi.

craton Area of old, stable continental crust.

critical angle In seismic refraction, the angle of incidence of a ray path for which the angle of refraction is 90º, i.e. $\frac{v_1}{v_2} = \sin i$.

critical Rayleigh number (Ra_c) The Rayleigh number at which a fluid begins to convect.

cryosphere The part of the hydrosphere locked up in the form of ice.

cumulate A rock containing accumulated crystals lying on the floor of a **magma** chamber, the crystals having settled out by gravity.

Curie point The temperature above which thermal agitation prevents spontaneous magnetic ordering.

decarbonation reaction A chemical reaction that produce CO_2.

decay constant The probability that a given atom of the radionuclide will decay within a stated time.

decompression melting Change from solid rock to partially molten rock caused by a decrease in pressure.

dehydration reaction A chemical reaction by which H_2O is released.

dense hydrated magnesium silicate (DHMS) minerals Minerals that are stable at mantle pressures and temperatures of up to 1000 °C.

depth of compensation A horizontal plane within the earth below which other horizontal planes are isobaric, i.e. at constant pressure.

derivative magma Magma that has been derived from a more primitive (usually more mafic) parent composition by processes such as fractional crystallisation, contamination or magma mixing.

destructive plate boundary Plate boundary where **lithosphere** is destroyed by two or more tectonic plates moving together.

diamond anvil A small anvil formed from two diamonds cut with perfect facets allowing experimentalists to squeeze rock powder to extremely high pressures.

diapir A spherical, elliptical or tear-drop shaped body of **magma** rising due to its lower density.

differentiated magma Material that has undergone an amount of chemical processing, usually by **fractional crystallisation**.

differentiation The process by which different rock types form from a single magma.

dihedral angle, θ The angle formed by the liquid in contact with two solid grains.

dimensionless number A number in which all of the values (time, length, viscosity, etc.) are scaled to a dimensionless form, i.e. all the units cancel out.

diopside The Mg endmember of **clinopyroxene**.

discharge zone Part of the hydrothermal system where reactions occur from the hot fluid rising from the **reaction zone** to the surface.

discontinuity A break in a seismic signal.

distribution coefficient A number representing the partitioning of elements between different phases.

down-going plate *See* **subducted slab**.

dunite A mantle rock that is a variety of **peridotite** composed almost completely of olivine.

dynamic viscosity, η Another name for viscosity, the resistance to flow within a fluid. Units are $N\ s\ m^{-2} = Pa\ s$.

Earth Reference Model The important physical properties within the Earth from the surface to the core.

E-chondrites *See* **enstatite chondrites**.

eclogite A high-pressure metamorphic rock, with the bulk composition of a basic igneous rock, containing sodium-rich **clinopyroxene** and garnet.

elastic modulus (According to Hooke's law and for materials that deform elastically) the ratio of stress to its corresponding strain under given conditions of load.

electronegativity (***E***) The tendency of an element to attract electrons and hence form negative ions (cations) usually expressed on a scale of 0–4. Elements with low negativity include the alkali and alkaline Earth metals, whereas the halogens (F, Cl, I and Br) have high electronegativity.

element partitioning The process of differentiation involving exchange of elements between solid and molten states.

endothermic reaction A chemical reaction that requires heat in order to take place. Units are kJ mol^{-1}.

enstatite The Mg endmember of the **orthopyroxene** group.

enstatite (or E-) chondrites Chondrites rich in the magnesium silicate mineral enstatite ($MgSiO_3$).

enthalpy A thermodynamic variable that is defined as the sum of the internal energy of a body plus the product of it volume multiplied by the pressure, measured in joules.

epicentral angle, Δ The angle subtended by the centre of the Earth between an earthquake focus and a seismic receiving station.

equilibrium crystallisation Crystallisation in which crystals formed on cooling continually react and re-equilibrate with the liquid.

erosion Process by which the materials at the surface of the Earth are dissolved or worn away and transported from place to place.

escape velocity The minimum velocity that enables a small body (including molecules and atoms) to escape from the gravitational field of a more massive body.

eutectic composition The composition of the material at the **eutectic point**.

eutectic point The minimum melting point of any mixture between the components of a chemical system; on a **binary phase diagram** it is the only point where three phases coexist.

evolved magma Also known as **derivative magma** or **fractionated magma**. Magmas furthest removed from **primary magmas**.

exothermic reaction A chemical reaction that releases heat. Units are kJ mol^{-1}.

exsolution The process in which one solid crystal separates into two minerals of distinct composition.

extensional tectonics The response of rocks to extensional forces in terms of their deformation and structure.

extract The bulk composition of the minerals being crystallised (extracted) from a melt. For example, if a melt is crystallising 100% olivine, the extract is simply the composition of olivine. If, however, the melt crystallises 50% olivine, 30% clinopyroxene and 20% plagioclase, the extract is a mixture of the three minerals weighted by their proportions. The composition of the extract can often be derived from variation diagrams.

fayalite Silicate mineral that is the iron endmember of the olivine group.

Fe-magnesiowüstite A mineral dissociated from **γ spinel** at ~24 GPa.

ferromagnesian silicate minerals Dark minerals containing silica, iron and magnesium. Also known as **mafic minerals**.

ferrosilite The Fe endmember of the **orthopyroxene** group.

flexural bulge An elevation of the ocean floor by as much as 0.5 km caused by the flexure of the **lithosphere** in response to its entry into a subduction zone.

fore-arc basin A sediment-filled depression between the volcanic arc and the shelf break.

fore-arc high An elevated zone beyond the **fore-arc basin**.

fore-arc region The region between the subduction-related **trench** and the **volcanic arc**.

forsterite Silicate mineral that is the magnesium endmember of the olivine group.

forward-modelling (In seismic studies) the use of an initial estimate of the depth–velocity profile from earlier inversion results to predict the arrival times of a given seismic event.

fractional crystallisation Continuous physical separation of crystals as they grow from a liquid thus changing the composition of the liquid that remains.

fractionated magma A melts that is the result of **fractional crystallisation**.

fracture zones Oceanic transform faults that link adjacent mid-ocean segments.

free-air gravity anomaly Variations in gravity across the surface of the Earth from measurements taken above the sea.

full spreading rate The combined rate of divergence on both sides of a spreading ridge.

γ spinel Mineral produced from **β spinel** at 17–18 GPa.

garnet A group of silicate minerals including almandine, grossular and pyrope with the formula: $A_3B_2(SiO_4)_3$ where A may be Ca^{2+}, Mg^{2+}, Fe^{2+} or Mn^{2+} and B may be Al^{3+}, Fe^{3+}, Mn^{3+}, Cr^{3+} or V^{3+}. Commonly found in metamorphic rocks.

geobarometer A mineral, or assemblage of minerals, that is stable over a limited range of pressures and therefore provides information on the pressure of formation.

geomagnetic timescale Timescale using information on both the ocean's palaeomagnetic polarity and its absolute age gathered from magnetic and oceanographic surveys of the ocean floor.

geosphere The solid Earth, which makes up the bulk of the planet by mass.

geotherm A curve representing increasing temperature with depth below the Earth's surface.

geothermometer A mineral, or assemblage of minerals, that is stable over a limited range of temperatures and therefore provides information on the temperature of formation.

global subducted sediment (GLOSS) An estimate of the chemical composition of the average sediment currently being subducted.

gneiss Banded high-grade metamorphic rocks in which light and dark minerals are segregated as layers or lenses.

granite An intrusive rock that (i) is made up of at least 75% quartz and feldspars and (ii) contains quartz, **plagioclase** and **alkali feldspar** in roughly equal proportions.

granite minimum The lowest temperature on the entire **liquidus** surface of the Qz–Ab–Or ternary system.

granodiorite A type of granite containing **plagioclase** and **alkali feldspar** in roughly equal proportions, at least 20% quartz and small amounts of biotite and hornblende.

granulite A metamorphic rock that has been subjected to high pressures and temperatures such that it has lost most of its volatile components (largely water through **dehydration reactions**): mineralogy is dominated by **plagioclase** feldspar, pyroxenes and garnets.

greenstones Altered ocean crust from tectonically disrupted sections of sea floor typically pale or dull green in colour.

Hadean The period of time on Earth between its formation at 4.55 Ga and the ages of the earliest rocks at ~3.9 Ga.

half-life The time required for half of the parent atoms to decay.

half spreading rate The rate of movement away from the ridge axis on one side only.

harzburgite A mantle rock that is a variety of **peridotite** composed almost completely of olivine and **orthopyroxene**.

H-chondrites A sub-group of the ordinary chondrites with high contents of iron in the reduced (metallic and sulfide) state.

Hedenbergite **Clinopyroxene** iron endmember.

highlands *See* **lunar highlands**.

hot spot An area of high volcanic activity. Some hot spots, e.g. Iceland, are located on constructive margins; others, e.g. Hawaii, lie within lithospheric plates.

hot-spot volcanism Volcanism that occurs at a **hot spot** and is associated with broad surface up-doming in the mantle.

hydrosphere All of the water on the planet in the oceans, seas, lakes and rivers.

hydrothermal circulation The circulation of seawater along oceanic ridge systems that penetrates the young hot crust to depths of at least a few kilometres. During its passage through the crust the water is heated before being cycled back to the oceans, forming black smokers.

hypsometric plot A histogram of topographic height over the whole planet, with heights determined above or below the median surface.

incompatible elements Elements largely or entirely excluded from the solid phase and concentrated in the silicate liquid ($D<1$).

initial ratio The atomic ratio of two isotopes of an element in a rock or mineral at the time of crystallisation of the mineral(s).

inverse geotherm Any **geotherm** that cools with increasing depth; for example, the geotherm towards the base of the **mantle wedge** where it cools towards the subducted slab.

iron meteorites Meteorites primarily composed of metallic iron.

irreversible process A process with an end result that cannot be returned to its former state, e.g. the generation of continental crust (i.e. the differentiation of the planet).

island arc A belt of active volcanoes which, in the case of intra-oceanic margins, form chains of islands.

island arc basalt (IAB) General term for any basaltic magma generated at a destructive plate boundary.

isochron The line on an **isochron plot** representing a suite of co-genetic rocks or minerals having the same age.

isochron plot Also known as an **isochron diagram** or **isotope evolution diagram**. A plot of isotope data where rocks or minerals of the same age will form a linear array.

isomorphism Literally 'equal form', identical atomic structures.

isomorphous series Two or more crystalline substances that display **isomorphism,** e.g. olivine.

isostasy The condition of mechanical equilibrium, comparable to floating, as shown by rigid blocks of the Earth (usually taken to be either the crust or **lithosphere**) to be buoyantly supported in an underlying fluid medium (usually taken as either the mantle or **asthenosphere**).

isotherms Lines connecting areas of equal temperature.

isotope evolution diagram See **isochron plot**.

isoviscous Having the same viscosity throughout.

kimberlite A rock formed from a rare type of CO_2-rich magma that carries a cargo of xenoliths and crystals from the mantle, including diamond.

kinematic viscosity, ν The dynamic viscosity of a fluid divided by its density. Units are $m^2\ s^{-1}$ = Pa s.

komatiite Ultramafic lava often exhibiting spinifex texture.

kyanite A blue **aluminosilicate** forming at moderate to high pressures, one of three polymorphs with the same chemical formula, $A1_2SiO_5$, (the others being **andalusite** and **sillimanite**).

large ion lithophile elements (LILE) Elements with a low valency and large ionic radii that tend to be soluble in aqueous fluids.

'late veneer' model A model of Earth accretion in which oxidised and volatile-bearing material was added to the Earth after it had differentiated into a mantle and core. The chief evidence for this is to be found in the lack of equilibrium between the mantle concentrations of the highly **siderophile** elements (e.g. the platinum-group elements, PGEs) and those expected for the core.

lawsonite A hydrous calcium aluminium silicate mineral with formula $CaAl_2Si_2O_7(OH)_2{\cdot}H_2O$.

L-chondrites A sub-group of the ordinary chondrites with approximately equal contents of iron in the reduced (metallic and sulfide) and oxidised (silicate) states.

leucogranites **Granites** that are light in colour being composed almost entirely of quartz, **plagioclase** and **alkali feldspar**. Additional minerals present in minor amounts may include muscovite and/or biotite.

lherzolite An **ultramafic** plutonic rock composed mainly of olivine with **orthopyroxene** and **clinopyroxene**.

liquidus The curve, or surface, in a **phase diagram** above which the system is completely liquid where 'above' means 'at higher temperature'.

lithophile Elements that preferentially bond with oxygen, especially in silicate or oxide structures.

lithosphere The rigid, outermost surface layer of the Earth, including the crust and the uppermost mantle, in which heat is transferred by conduction.

lithostatic pressure The force exerted on rock inside the Earth due to the weight of overlying rock.

LL-chondrites A sub-group of the ordinary chondrites with low contents of iron in the reduced (metallic and sulfide) state and relatively high contents of oxidised iron (silicates).

long-lived radiogenic nuclides Isotopes of the elements uranium (U), thorium (Th) and potassium (K) (which all have particularly long **half-lives**) responsible for most of the radiogenic heating that has occurred throughout the history of the planet.

low-velocity zone The region in the mantle beneath the Moho and down to 220 km in which the velocity of seismic waves is lower than expected.

lunar highlands Pale-coloured regions on the Moon with a dense covering of impact craters.

lunar maria Darker, flatter regions than the **lunar highlands** with fewer craters.

mafic minerals *See* **ferromagnesian silicate minerals**.

magma Liquid rock.

magma ocean The molten mantle of material covering the very early Earth.

magnesiowüstite (Mg-wüstite) A mineral dissociated from **γ spinel** at ~24 GPa.

magnesium number (Mg-number or Mg#) The proportion of the Mg endmember (relative to the total amount of the Mg and Fe^{2+} endmembers).

magnesium (Mg)-perovskite A mineral dissociated from **γ spinel** at ~24 GPa.

magnetic reversals The exchange of the positions of the north and south magnetic poles.

major element An element that makes up more than 5% of a rock.

majorite A purple or yellowish-brown mineral of the garnet group that forms at pressures of about 5 GPa.

mantle plume Localised, hot, buoyant material rising through the mantle. Mantle plumes are thought by some geologists to rise beneath hot spots, causing domal uplift.

mantle transition zone A region within the mantle where the crystal structure of the constituent minerals, olivines and pyroxenes are transformed into a dense phase, known as **perovskite**.

mantle wedge A triangle of **asthenosphere** below the **overriding plate**.

marginal basin *See* **back-arc basin**.

maria *See* **lunar maria**.

mechanical boundary layer The upper rigid part of a plate.

melt weakening The variation inviscosity of partially melted granite between its **liquidus** and **solidus** temperatures.

metamorphism Changes in texture and/or mineralogy that affect existing rocks when they are subject to a change in pressure and/or temperature.

metapelites Metamorphosed aluminium-rich sediments.

mid-ocean ridge *See* **oceanic ridge systems**.

mid-ocean ridge basalt (MORB) **Basalt** erupted at a sea-floor spreading axis.

migmatites Metamorphic rock made up of both dark layers of **ferromagnesian** minerals and paler, quartz–feldspar layers or lenses, usually formed by melting at high temperatures.

modal melting Melting in which minerals on the **solidus** enter the melt phase in the same proportion that they are found in the rock.

Mohorovičić discontinuity (Moho) Mohorovičić seismic discontinuity that separates the crust from the mantle.

MORB (mid-ocean ridge basalt) **Basalt** erupted at a sea-floor spreading axis.

multi-anvil Experimental equipment used in order to replicate the extreme pressures and temperatures expected in the deep interior of Earth.

negative buoyancy force The gravity-generated force that pulls the whole oceanic plate down as a result of the negative buoyancy of the slab.

negative feedback In a system, the mechanism by which a process is limited internally (i.e. it slows the system down).

Newtonian fluid A fluid in which the strain rate is proportional to the stress, with the constant of proportionality being the dynamic viscosity.

non-modal melting A melting process during which phases enter the melt in proportions that differ from those in which they are present in the original rock.

Non-Newtonian fluid A fluid in which the viscosity varies as a function of the strain rate.

obduction The tectonic emplacement of segments of oceanic **lithosphere** onto continental and arc crust by convergent plate motions.

ocean drag force The force acting along the bottom surface of an oceanic plate.

ocean driving force The force acting along the bottom surface of the oceanic plate when the lithospheric plate is transported along by a faster moving **asthenosphere**.

ocean island basalt (OIB) General term for any basaltic magma generated at an oceanic intra-plate setting. e.g. Hawaii, Azores and St Helena.

ocean resistant force The force acting along the bottom surface of the oceanic plate when the **asthenosphere** is moving slower than the plate in the direction of plate movement.

oceanic ridge systems Immense, continuous chains of volcanic mountains running down the ocean basins, indicating the position of constructive plate margins.

Oldham-Gutenberg discontinuity The seismic discontinuity that marks the boundary between the core and the lower mantle.

open-system behaviour Isotope systems where losses are due to processes other than radioactive decay (e.g. infiltration or loss of fluids, or diffusional exchange between minerals after their crystallisation).

ophiolite Rocks that make up **obducted** fragments of oceanic lithosphere including deep-sea sediments, basaltic pillow lavas, basaltic dykes, gabbro and peridotite.

ordinary chondrites The most abundant form of **chondrites**.

orogenic belts A linear or arcuate zone, on a regional scale, that has undergone compressional tectonics. For example large mountain chains resulting from collisions between plates.

orography The physical features of mountains, usually related to their relief.

orthoclase The K endmember of **alkali feldspar**s.

orthopyroxenes Pyroxenes in which less than 10% of the **cations** are Ca. The atomic structure has three orthogonal symmetry axes.

overriding plate The plate that lies above the **subduction zone**.

overriding plate resistance Frictional resistance resulting from the pushing of the subducting slab against the overriding plate. This causes both shallow and deep earthquakes at subduction zones.

Pacific Ring of Fire The area around the margin of the Pacific Ocean containing the majority of the world's **volcanic arcs**.

Pangaea A **supercontinent** comprising all the continental crust of the Earth, postulated to have existed in late Paleozoic and Mesozoic times before it fragmented into Gondwana and Laurasia.

partial ionic substitution Instead of **complete ionic substitution** in atoms of unequal size.

partial melting The process by which different minerals within a rock melt at different temperatures, an important process in **magma** formation.

partition or **distribution coefficient** A number representing the way in which elements distribute themselves between different phases.

Péclet number (Pe_t) A number indicating the importance of heat transport by convection relative to that by conduction.

peridotite An **ultramafic** rock rich in the minerals olivine and pyroxene.

perovskite A dense phase found in the mantle transition zone, originating as olivines and pyroxenes.

petrological Moho The top of the mantle often identified as the contact between layered and unlayered **peridotite**.

petrology The study of the composition, texture and structure of rocks.

phase A homogeneous entity (solid, liquid or gas) with a particular chemical composition and molecular structure.

phase boundary The line that forms the boundary between **stability fields**.

phase diagrams Diagrams that illustrate the conditions of temperature, pressure and chemical composition under which different phases exist.

phengite A high-silica variety of muscovite mica.

phenocrysts Large crystals in the matrix of an igneous rock formed during early stages of the magma's cooling.

plagioclase feldspar Feldspars rich in Na and/or Ca.

planetary satellite A body that orbits a planet, also known as a moon.

planetesimals Tiny planets about 1 km in diameter.

plate model A model proposing that the **lithosphere** is produced at a mid-ocean ridge with constant thickness and that the temperature at the base of the plate corresponds to its temperature of formation.

plate tectonics The movement and change in size and shape of lithospheric plates that comprise the outer shell of the Earth.

pole of rotation Every displacement of a plate from one position to another on the Earth's surface can be described by a simple rotation of that plate about a unique axis.

polymerisation The increase in connectedness of atoms in a melt.

polymorphs Different forms of the same chemical substance, e.g. graphite and diamond.

positive feedback In a system, it is the mechanism by which a process intensifies or accelerates, as each cycle of operation establishes conditions that favour a repetition.

post-perovskite phase Mineral transformed from **Mg-perovskite** at extreme pressures (134 GPa), equivalent to 2800 km depth, with a 3% increase in density.

potential energy The energy possessed by a body by virtue of its position or state.

potential temperature The temperature a parcel of mantle would have if it rose to the surface along an adiabatic gradient without melting.

preliminary Earth reference model (PREM) An average velocity structure for the Earth.

primary crust A chilled skin on the surface of the **magma ocean**.

primary magma The liquid produced by **partial melting**.

primitive magma Magma that has undergone only a small amount of fractional crystallisation and is therefore still fairly close in composition to that of a **primary magma**.

proxy A geochemical indicator that varies in direct response to variations in the output flux.

P-waves The transmission of energy through both solid and fluid layers in the Earth by compression and dilation (expansion), in which the particles of the medium vibrate backwards and forwards in the direction of wave propagation.

quartz diorite *See* **tonalite**.

Rayleigh number (*Ra*) A measure of the ratio of heat carried by convection relative to heat carried by conduction.

Rayleigh–Bénard convection The convection that occurs in a tank of Newtonian viscous fluid that is uniformly heated from below and cooled from above.

reaction zone The hottest, deepest part of the hydrothermal system where most of the chemical changes that determine the composition of the hydrothermal effluent occur.

recharge zone The broad area on either side of the plate boundary where seawater penetrates the fractured oceanic crust and is drawn down into deeper, hotter crust.

refracted Change in direction of travel of a wave (e.g. light or seismic wave).

remnant palaeomagnetism Traces of ancient magnetism frozen into very old rocks.

reversible cycle A process involving recycling of material, e.g. the generation of oceanic crust and **lithosphere**, which will ultimately be returned to the mantle.

ridge-push force Force that is a result of gravity acting down the slope of the ridge.

ridge resistance Frictional resistance to **ridge-pull forces**.

ringwoodite Another name for **γ spinel**.

schist Coarsely crystalline metamorphic rocks with an aligned micaceous texture.

sea-floor spreading The creation of new ocean **lithosphere** where lithospheric plates are moving apart along **ocean ridge systems**.

secondary crust Oceanic crust generated by partial melting of the mantle.

segregation (As applied to the formation of the Earth) the physical mechanism by which metal separates from a silicate mantle and accumulates at the centre of the Earth.

seismic Moho The shallowest occurrence of peridotite.

seismic tomography A method of imaging the internal structure of the Earth using seismic energy from earthquakes.

seismology The study of earthquakes and seismic waves.

self-compression The process by which the Earth is held together by its own gravity, which pulls it into its nearly spherical shape and thus compresses the interior; the changes in gravity and density within the Earth brought about by the overlying layers only.

serpentinite A rock consisting almost entirely of serpentinite-group minerals, e.g. antigorite and lizardite.

shadow zones Absence of seismic waves from areas of the Earth's surface.

sheeted dykes A layer within the oceanic lithosphere made up almost entirely of vertical basaltic dykes.

short-lived extinct nuclide Radionuclides not replenished by the decay of other isotopes, that may be entirely depleted.

siderophile Elements that, because of chemical properties, bond more easily with iron than they do with silicates. Elements that are found in the metallic phase of a natural system.

silicate tetrahedron One Si^{4+} and four O^{2-} ions arranged in a pyramid shape where each has the Si^{4+} ion in the centre of the pyramid and an O^{2-} at each corner.

silicic A rock with at least 65% silica, e.g. granite and rhyolite.

sillimanite A brown or white **aluminosilicate** forming at the highest temperatures; one of three polymorphs with the same chemical formula, $A1_2SiO_5$ (the others being **kyanite** and **andalusite**).

slab-pull force The component of the negative buoyancy force that is transmitted to the plate.

slab resistance The combined resistive force of the frictional drag on the slab's upper and lower surfaces and from the viscosity of mantle material that is being displaced.

slate Metamorphic rock resulting from small increase in pressure and temperature acting on a mudstone.

Snell's law The relationship between the angles of refraction and reflection of waves at a boundary between two layers and the velocities of the waves within those layers: $\frac{\sin i}{\sin r} = \frac{v_1}{v_2}$.

solar photosphere The outer atmosphere of the sun.

solid-solution series Minerals that experience complete ionic substitution between endmembers.

solid-state convection Flow of the mantle when subject to temperature differences.

solidus The curve or surface on a **phase diagram** below which the system is entirely solid (where 'below' means 'at lower temperature').

spallation The bombardment of the nuclei of O and N atoms by cosmic rays producing Be in the upper atmosphere.

β spinel Olivine that has undergone a phase change at ~14 GPa.

γ spinel Mineral produced from β spinel at 17–18 GPa.

stability field Area on a **phase diagram** in which a mineral is stable.

stable triple junctions A **triple junction** that maintains its form over time.

stony-iron meteorite A hybrid of stony and iron meteorites containing variable amounts of silicates and metal.

stony meteorite Meteorite consisting dominantly of silicate minerals.

subducted slab (also known as **subducted plate** or **down-going plate**) The subducted plate at depth.

subduction The destruction of cold oceanic lithosphere beneath the **overriding plate** at a **subduction zone**.

subduction zone An area where the process of subduction is taking place, e.g. the Marianas Trench.

subsolidus A chemical system that is below its melting point and in which reactions only occur in solid state.

supercontinent A very large continental plate that subsequently fragments into several smaller tectonic plates.

S-waves The transmission of energy through the solid Earth by shear displacement, the motion of particles in the medium is at right angles (perpendicular) to the direction of wave propagation.

TAS diagram *See* **total alkalis diagram**.

ternary diagram A triangular-shaped diagram to show the proportions of three components in any mixture.

terrigenous input Sediment input made up of clastic or volcaniclastic material of continental origin.

tertiary crust The oldest surviving crust, found in the continents.

thermal boundary layer The lower viscous part of a plate.

thin viscous sheet A description of the large-scale behaviour of the continental **lithosphere** under compression.

tie line Drawn between phases that coexist at any one time.

tonalite (or quartz diorite) A rock similar to granite but with no, or very little, **alkali feldspar** and a higher proportion of **ferromagnesian** minerals such as hornblende and biotite.

total alkalis (TAS) diagram A plot of total alkalis (wt% $Na_2O + K_2O$) versus wt% SiO_2, which is used to classify magma compositions.

trace element An element that makes up less than 1% of a mineral.

transform fault resistance The resistance encountered along transform faults to movement between plate segments sliding past each other, resulting in a series of earthquakes.

transition zone The area marked by increases in seismic velocity between ~410 km and ~670 km depth.

travel-time curve Plot of arrival time for different seismic waves against Δ (delta).

trench The depression of the sea floor associated with the subduction zone.

trench suction force Convection which sucks more mantle into the wedge due to cooling of the **mantle wedge** against the upper surface of the subducting plate.

triple junction A point where three lithospheric plates are in contact.

true plate motion Plate movement in relation to the hot-spot reference frame.

ultramafic rocks Rocks with very high proportions of mafic minerals.

unstable triple junctions A **triple junction** that can only exist briefly before it evolves into another plate configuration.

variation diagram A plot in which geochemical data for two elements are plotted against each other. Commonly used for plotting the composition of a suite of lavas that are related by crystallisation. The composition of the possible minerals that are crystallised can also be plotted on the variation diagram. For example, a plot of wt% CaO versus wt% MgO allows the assessment of whether a suite of lavas are related by the crystallisation of some combination of olivine, clinopyroxene and plagioclase.

viscosity A measure of how freely material is able to flow.

volcanic arc *See* **island arc**.

volcanic-front The position on a map of the **island arc**.

volcaniclastic sediment A sediment derived from the erosion of existing volcanic rocks.

Wadati–Benioff zone An inclined plane of earthquakes associated with a deep ocean trench.

wadsleyite A mineral produced with β spinel, involving a 10% increase in density.

weathering The chemical and physical effects on surface rocks resulting from their interaction with precipitation and changes in temperature.

websterite A pyroxenite containing both **orthopyroxene** and **clinopyroxene**.

wehrlite A mantle rock that is a variety of **peridotite** composed almost completely of olivine and **clinopyroxene**.

Wilson cycle The hypothesis proposed by the geophysicist Tuzo Wilson that an ocean develops through several distinct stages driven by the movement of crustal plates, beginning with initial opening, through a widening phase, then closure and its ultimate destruction.

wollastonite $Ca_2Si_2O_6$ considered a separate mineral to pyroxenes because of its distinct structure despite compositionally being the Ca endmember.

xenocryst A crystal foreign to the igneous rock in which it occurs, and hence derived from a source other than the host rock's magma.

xenoliths Rock fragments foreign to the igneous rock in which they occur. For example, lumps of mantle rock that have been picked up and carried by rapidly rising magma.

zoned crystal A crystal with chemically distinct zones indicative of different stages in its growth, for example concentric zones in plagioclase each with a distinct anorthite content.

Further reading

Chapter 1

Drake, M.J. and Righter, K. (2002) 'Determining the composition of the Earth', *Nature*, vol. 416, pp. 39–44.

Fowler, C.M.R. (2005) *The Solid Earth*, Cambridge University Press, p. 685.

McDonough, W.F. and Sun, S.-S. (1995) 'The composition of the Earth', *Chemical Geology*, vol. 120, pp. 223–253.

Watts, A.B. (2001) *Isostasy and Flexure of the Lithosphere*, Cambridge University Press, p. 458. (Chapter 1 gives a good introduction to the principles behind isostasy.)

Chapter 2

Canup, R.M. and Righter, K. (eds) (2000) *Origin of the Earth and Moon*, Tuscon, University of Arizona Press.

Wood, B.J., Walter, M.J. and Wade, J. (2006) 'Accretion of the Earth and segregation of its core', *Nature*, vol. 441, pp. 825–833.

Chapter 3

Fowler, C.M.R. (2005) *The Solid Earth*, Cambridge University Press, p. 685.

Mussett, A.E. and Khan, M.A. (2000) *Looking into the Earth*, Cambridge, Cambridge University Press.

Chapter 4

Nicolas, A. (1990) *The Mid-Oceanic Ridges*, Berlin, Springer-Verlag.

Chapter 5

Bebout, G.E., Scholl, D.W., Kirby, S.H. and Platt, J.P. (1996) *Subduction: Top to Bottom*, American Geophysical Union Monograph Series, vol. 96, p. 384. (A collection of papers (high-level reading) on various aspects of subduction zones.)

Chapter 6

Beaumont, C., Jamieson, R.A., Nguyen, M.H. and Lee, B. (2001) 'Himalayan tectonics explained by extrusion of a low-viscosity crustal channel coupled to focused surface denudation', *Nature*, vol. 414, pp. 738–742. (A highly technical explanation of the channel flow model.)

Clarke, D.B. (1992) *Granitoid Rocks*, Chapman & Hall, p. 283. (The complete guide to granites: geochemistry, mineralogy, field relations and origin.)

Harris, N.B.W. (2007) 'Channel flow of the Himalayan–Tibetan orogen – a critical review', *Journal of the Geological Society of London*, vol. 164, pp. 511–523.

Hodges, K.V. (2000) 'Tectonics of the Himalaya and southern Tibet', *Geological Society of America Bulletin*, vol. 112, pp. 324–350. (A comprehensive review of the Himalayan orogen, but pre-dating the channel-flow model.)

The Open University (2001) S339 *Understanding the continents*, Block 4 *Mountain building*, Section 7 'The Himalaya and Tibet: a case study of collision', Milton Keynes, The Open University. (The only available comprehensive (but pre-channel flow) account of magma formation in collision zones.)

Yardley, B.W.D. (1989) *An Introduction to Metamorphic Petrology*, Longman Earth Science Series, p. 248. (The best available introduction to metamorphic petrology.)

Chapter 7

Carlson, R.W. (2005) *The Mantle and Core: 2 (Treatise on Geochemistry)*, Oxford, Elsevier, p. 586. (High-level, but a very up-to-date view of the geochemistry and evolution of the Earth's mantle and core.)

Fowler, C.M.R. (2005) *The Solid Earth: An Introduction to Global Geophysics*, Cambridge, Cambridge University Press.

Sparks, R.S.J. and Hawkesworth, C.J. (2004) *The State of the Planet: Frontiers and Challenges in Geophysics*, American Geophysical Union Monograph Series, vol. 150, p. 410. (Some readable reviews of some key geophysical issues.)

van der Hilst, R.D., Bass, J.D., Matas, J. and Trampert, J. (2005) *Earth's Deep Mantle: Structure, Composition, and Evolution*, American Geophysical Union Monograph Series, vol. 160, p. 334. (A collection of the most recent thoughts on the deeper structure of the planet. High-level, but some readable overviews.)

Chapter 8

Bryson, B. (2003) *A Short History of Nearly Everything*, London, Doubleday.

Acknowledgements

The production of this book involved a number of Open University staff to whom we owe considerable thanks for their professional contributions and willingness to accommodate the requests and vagaries of the academic authors. Jennie Neve Bellamy managed the project, including the Open University course associated with this book, with undying optimism and consummate professionalism and ensured that the project kept to deadlines (both original and revised). Ashea Tambe efficiently styled the text for handover to Pamela Wardell, who copy-edited all the chapters with an uncanny attention to detail. Artwork was coordinated and perceptively executed by Sara Hack and design and layout undertaken by Sarah Hofton. The index was prepared by Jessica Bartlett and the production process was managed by James Davies. We are grateful to Christianne Bailey (Open University) and Susan Francis (Cambridge University Press) for steering us successfully through the mysteries of co-publication.

We are grateful to Arlene Hunter for her continued vigilance in reading early drafts of the chapters and keeping us in line with learning outcomes. We also thank Nick Petford (University of Bournemouth), who acted as external assessor and constructive critic on both this book and the associated Open University course, and the anonymous reviewers appointed by CUP for their comments during the publisher's review process.

We also acknowledge and thank the authors (in particular Andrew Bell and Peter Smith) of earlier Open University courses, notably S267 *How the Earth Works*, which laid the foundations of this book.

Grateful acknowledgement is made to the following sources for permission to reproduce material in this book.

Cover image Stephen and Donna O'Meara/Science Photo Library.

Figures 1.1a, 1.1b and 1.1c USGS/Cascades Volcano Observatory; *Figures 1.2a, 1.2b, 1.2d, 1.2e, 1.4a, 1.13* NASA; *Figure 1.2c* United States Geological Survey; *Figure 1.3* NOAA/NASA; *Figure 1.4b* NASA Jet Propulsion Laboratory (NASA-JPL); *Figure 1.5* Watts, A.B. (2001) *Isostasy and Flexure of the Lithosphere*, Cambridge, Cambridge University Press; *Figure 1.11* Earle, P.S. and Shearer, P.M. (1994) 'Characterization of global seismograms using an automated-picking algorithm', *Bulletin of the Seismology Society of America*, vol. 84, p. 366; *Figure 1.15* Gill, R.C.O. (1982) *Chemical Fundamentals of Geology*, Chapman and Hall; *Figure 1.16* Morgan, J.W. and Anders, E. (1980) 'The chemical composition of Earth, Venus and Mercury', *Proceedings of the National Academy of Science*, vol. 77, p. 6973; *Figure 1.17* Copyright © Natural History Museum, London; *Figure 1.19 and 1.20* McSween, H.Y. (1987) *Meteorites and Their Parent Planets*, Cambridge, Cambridge University Press.

Figure 2.5 Rushmer, T., Minarik, W.G. and Taylor, G.J. (2000) 'Physical processes of core formation' in Canup, R.M. and Righter, K. (eds) *Origin of the Earth and Moon*, Tucson, University of Arizona Press; *Figure 2.7* Walter, M.J. et al. (2000) 'Siderophile elements in the Earth and Moon: metal/silicate partitioning and implications for core formation' in Canup, R.M. and Righter, K. (eds) *Origin of the Earth and Moon*, Tucson, University of Arizona Press;

Figure 2.11 Snyder, G.A. et al. (2000) 'Chronology and isotopic constraints on lunar evolution' in Canup, R.M. and Righter, K. (eds) *Origin of the Earth and Moon*, Tucson, University of Arizona Press; *Figure 2.15* Courtesy of Steve Moorpath; *Figure 2.16* Image courtesy of Keiko Kubo; *Figure 2.17* Reprinted with permission from Watson, E.B. and Harrison, T.M. (2005) 'Zircon thermometer reveals minimum melting conditions on earliest Earth', *Science Magazine*, vol. 308, May 2005. Copyright 2005 AAAS; *Figure 2.18* Bill Bachman/Science Photo Library; *Figure 2.19* Adapted from Abe, Y. et al. (2000) 'Water in the early Earth' in Canup, R.M. and Righter, K. (eds) *Origin of the Earth and Moon*, Tucson, University of Arizona Press.

Figure 3.1a Christian Darkin/Science Photo Library; *Figure 3.1b* M-Sat Ltd/ Science Photo Library; *Figure 3.3* Hallam, A. (1975) *Alfred Wegener and the Hypothesis of Continental Drift*, Scientific American Inc.; *Figure 3.4* Creer, K.M. (1965) 'Palaeomagnetic data from gondwanic continents' in Blacket, P.M.S. et al. (eds) *A Symposium on Continental Drift*, The Royal Society; *Figures 3.5a, b and c* Mussett, A.E. and Khan, M.A. (2000) *Looking into the Earth: an introduction to geological geophysics*, Cambridge, Cambridge University Press; *Figure 3.6* Hiertzler, J.R. et al. (1966) 'Magnetic anomalies over the Reykjanes Ridge', *Deep Sea Research*, vol. 13, Elsevier Science Limited; *Figure 3.8* Adapted from Bott, M.H.P. (1982) *The Interior of the Earth: Its Structure, Constitution and Evolution*, Edward Arnold; *Figure 3.9a* Adapted from the British Geological Survey World Seismicity Database, Global Seismology and Geomagnetism Group, Edinburgh; *Figure 3.9b* Adapted from Johnson, R.W. (1993) *AGSO Issues Paper No. 1, Volcanic Eruptions and Atmospheric Change*, Australian Geological Survey Organisation; *Figure 3.11* Parsons, B. and Sclater, G. (1977) 'An analysis of the variation of ocean floor bathymetry and heat flow with age', *Journal of Geophysical Research*, vol. 82, no. 5, February 1977. © 1977 American Geophysical Union; *Figure 3.13* Stein, C. and Stein, S. (1992) 'A model for the global variation in oceanic depth and heat flow with lithospheric age', *Nature*, vol. 359, pp. 123–129; *Figure 3.14* Fowler, C.M.R. (2005) *The Solid Earth*, Cambridge, Cambridge University Press; *Figure 3.16* Sykes, R. (1966) 'The seismicity and deep structure of island arcs', *Journal of Geophysical Research*, vol. 71, no. 12. © American Geophysical Union; *Figure 3.17* Watts, A.B. (2001) *Isostasy and Flexure of the Lithosphere*, Cambridge, Cambridge University Press; *Figure 3.18* Westbrook, G.K. (1982) 'The Barbados ridge complex', Special Publication of *The Geological Society of London*, Blackwell, © The Geological Society; *Figure 3.21* Dalrymple, G.B. et al. (1973) 'The origin of the Hawaiian islands', *American Scientist*, vol. 61, Scientific Research Society (Sigma XI); *Figure 3.22* Adapted from Clague, D. and Dalrymple, G.B. (1987) USGS Professional Paper 1350, United States Geological Survey; *Figure 3.27* Morgan, W. J. (1968) 'Rises, trenches, great faults and crustal blocks', *Journal of Geophysical Research*, vol. 73, pp. 1959–1982; *Figure 3.28* Forsyth, D. and Uyeda, S. (1975) 'On the relative importance of the driving forces of plate motion', *Geophysical Journal of the Royal Astronomical Society*, vol. 43, pp. 163–200; *Figure 3.30* Adapted from Livermore, R. et al. (2005) 'Paleogene opening of the Drake Passage', *Earth and Planetary Science Letters*, vol. 236, pp. 459–470.

Figure 4.1a Professor R.M. Haymon; *Figures 4.1b and 4.35* Dr Ken MacDonald/ Science Photo Library; *Figure 4.2* Andy Tindle (The Open University); *Figure 4.4* Adapted from Auzende, J.M. et al. (1989) 'Direct observation of a section through slow spreading oceanic crust', *Nature*, vol. 337. Copyright © Nature Publishing Group; *Figure 4.5* Nigel Harris (The Open University); *Figure 4.6* Gass, I.G. (1990) *'Ophiolites', The Earth: Tectonics of Continents and Oceans*, W.H. Freeman and Co.; *Figure 4.7a* White, R.S., McKenzie, D. and O'Nions, R.K. (1992) 'Oceanic crustal thickness from seismic measurements and rare earth element inversions', *Journal of Geophysical Research*, vol. 97, no. B13. The American Geophysical Union; *Figure 4.14* Herzberg, C.T. (1987) 'Magma density at the high pressure part 1: the effect of composition on the elastic properties of silicate liquids', *Magmatic Processes: Physiochemical Principles, No. 1*, The Geochemical Society; *Figure 4.15* Takahashi, E. (1986) 'Melting of mantle peridotite to 14 GPa', *Journal of Geophysical Research*, vol. 91. © American Geophysical Union; *Figure 4.26* Based on Roeder, P.L. and Emslie, R.F. (1970) 'Olivine-liquid equilibrium', *Contributions in Mineralogy and Petrology*, vol. 29. Springer-Verlag.

Figure 5.2 Adapted from Scotese, C.R., Gahagan, L.M. and Larson, R.L. (1998) 'Plate tectonic reconstructions of the Cretaceous and Cenozoic ocean basins', *Technophysics*, vol. 155, Elsevier Science, and Fowler, C.M.R. (1990) *The Solid Earth: An Introduction to Global Geophysics*, Cambridge University Press; *Figure 5.3* Hasagawa, A. et al. (1978) 'Double-planed deep seismic zone', *Geophysical Journal of the Royal Astronomical Society*, vol. 54, Royal Astronomical Society; *Figure 5.4* D. van der Hilst, Department of Earth, Atmospheric and Planetary Sciences, MIT; *Figure 5.5* Nakajima, J. et al. (2001) 'Three-dimensional structure of V_p, V_s and V_p/V_s', *Journal of Geophysical Research*, vol. 106, no. B10, American Geophysical Union; *Figure 5.7* Peacock, S.M. (2003) 'Thermal structure and metamorphic evolution of subducting slabs' in Eiler, J.M. (ed.) *Inside the Subduction Factory*, Geophysical Monograph Series, American Geophysical Union, Washington, DC. pp. 7–22. *Figure 5.8* van Keken, P.E., Kiefer, B. and Peacock, S.M. (2002) 'High-resolution models of subduction zones: implications for mineral dehydration reactions and the transport of water into the deep mantle', *G3 – An Electronic Journal of the Earth Sciences*, vol. 3, no. 10. The American Geophysical Union; *Figures 5.9, 5.11 and 5.20* Poli, S. and Schmidt, M.W. (2002) 'Petrology of subducted slabs', *Annual Review of Earth and Planetary Sciences*, vol. 30. Copyright © 2002 by Annual Reviews, www.annualreviews.com; *Figure 5.10* Johnson, M.C. and Plank, T. (1999) 'Dehydration and melting experiments constrain the fate of subducted sediments', *G3 – An Electronic Journal of the Earth Sciences*, vol. 1, 13 December, The American Geophysical Union; *Figure 5.13* Le Bas, M.J. et al. (1986) 'A chemical classification of volcanic rocks', *Journal of Petrology*, vol. 27, no. 3, Oxford University Press; *Figures 5.21 and 5.22* Adapted from Cox, K.G., Bell, J.D. and Pankhurst, R.J. (1979) *The Interpretation of Igneous Rocks*, Unwin Hyman; *Figures 5.27 and 5.28* Plank, T. and Langmuir, C.H. (1993) 'Tracing trace elements from sediment input to volcanic output at subduction zones', *Nature*, vol. 362. Copyright © Nature Publishing Group.

Figures 6.3, 6.9, 6.20 and 6.22 Photographs courtesy of Nigel Harris; *Figure 6.7* Photograph courtesy of Tom Argles; *Figure 6.11* MacKenzie, W.S. and Guilford, C. (1980) *Atlas of Rock-forming Minerals in Thin Section*, Longman.

Figures 7.1 and 7.8 Fowler, C.M.R. (2005) *The Solid Earth: An Introduction to Global Geophysics*, Cambridge University Press. Copyright © Cambridge

University Press 2005; *Figures 7.2a and 7.9* Poli, S. and Schmidt, M.W. (2002) 'Petrology of subducted slabs', *Annual Review of Earth and Planetary Sciences*, vol. 30. Copyright © 2002 by Annual Reviews. www.annualreviews.com; *Figure 7.2b* Helmberger, D. et al. (2005) 'Deep mantle structure and the perovskite phase transition', *Proceedings of the National Academy of Sciences*, vol. 102, no. 48, 29 November 2005. Copyright © 2005 The National Academy of Sciences of the United States of America, all rights reserved; *Figures 7.4 and 7.5* Lei, J. and Zhao, D. (2006) 'Global P-wave tomography: on the effect of various mantle and core phases', *Physics of the Earth and Planetary Interiors*, vol. 156, pp. 44–69, Elsevier Science; *Figure 7.6* Reprinted from *Physics of the Earth and Planetary Interiors*, vol. 156, Lei, J. and Zhao, D. 'Global P-wave tomography: on the effect of various mantle and core phases', pp. 61–62. Copyright 2006, with permission of Elsevier; *Figure 7.7* White, D.B. (1988) 'The planforms and onset of convection with a temperature dependent on viscosity', *Journal of Fluid Mechanics*, vol. 191, Cambridge University Press; *Figures 7.10 and 7.11* Davies, G.F. and Richards, M.A. (1992) 'Mantle convection', *Journal of Geology*, vol. 100, University of Chicago Press. Copyright © 1992 by The University of Chicago. All rights reserved; *Figure 7.12* Reprinted from *Geochimica et Cosmochimica Acta*, vol. 66, no. 17, Davies, G.F. 'Stirring geochemistry in mantle convection models with stiff plates and slabs', p. 3132. Copyright 2002, with permission of Elsevier; *Figure 7.14* Reprinted from *Earth and Planetary Science Letters*, vol. 241, Lei, J. and Zhao, D. 'A new insight into the Hawaiian Plume', p. 449. Copyright 2006 with permission of Elsevier; *Figure 7.15* Hauri, E.H. (1996) 'Major element variability in the Hawaiian mantle plume', *Nature*, vol. 382. Copyright © Nature Publishing Group; *Figure 7.16* Hauri, E.H. and Hart, S.R. (1993) 'Re–Os isotope systematics of HIMU and EMII oceanic island basalts from the South Pacific Ocean', *Earth and Planetary Science Letters*, vol. 114, pp. 353–371, Elsevier Science; *Figure 7.18* Hirth, G. and Kohlstedt, D.L. (1996) 'Water in the oceanic upper mantle: implications for the rheology, melt extraction and the evolution of the lithosphere', *Earth and Planetary Science Letters*, vol. 144, Elsevier Science.

Figure 8.1 Hawkesworth, C.J. and Kemp, A.I.S. (2006) 'Evolution of the continental crust', *Nature*, vol. 443. Copyright © Nature Publishing Group.

Every effort has been made to contact copyright holders. If any have been inadvertently overlooked, the publishers will be pleased to make the necessary arrangements at the first opportunity.

Sources of figures, data and tables

Abe, Y. et al. (2000) 'Water in the early Earth' in Canup, R.M. and Righter, K. (eds) *Origin of the Earth and Moon*, Tucson, University of Arizona Press.

Allègre, C.J. and Rousseau, D. (1984) 'The growth of continents through geological time studied by Nd isotope analysis of shales', *Earth and Planetary Sciences Letters*, vol. 67, pp. 19–34.

Alt, J.C. (1995) 'Subseafloor processes in mid-ocean ridge hydrothermal systems' in Humphris, S.E., Zierenberg, R.A., Mullineaux, L.S. and Thompson, R.E. (eds) *Seafloor Hydrothermal Systems*, American Geophysical Union Geophysical Monograph 91, pp. 85–114.

Anders, E. and Grevasse, N. (1989) 'Abundances of the elements: meteoritic and solar', *Geochimica et Cosmochimica Acta*, vol. 53, pp. 197–214.

Auzende, J.M., Bideau, D., Bonatti, E., Cannat, M., Honnorez, J., Lagabrielle, Y., Malavieille, J., Mamaloukas-Frangoulis, V. and Mevel, C. (1989) 'Direct observation of a section through slow-spreading oceanic crust', *Nature*, vol. 337, pp. 726–29.

Beaumont, C., Jamieson, R.A., Nguyen, M.H. and Medvedev, S. (2004) 'Crustal channel flows: 1, Numerical models with applications to the tectonics of the Himalayan–Tibetan orogen', *Journal of Geophysical Research*, vol. 109, B06406, doi:10.1029/2003JB002809.

Boher, M., Abouchami, W., Michard, A., Albarède, F. and Arndt, N.T. (1992) 'Crustal growth in West Africa at 2.1 Ga', *Journal of Geophysical Research*, vol. 97, pp. 345–369.

Boyet, M. and Carlson, R.W. (2005) '^{142}Nd Evidence for early (>4.53 Ga) global differentiation of the silicate Earth', *Science*, vol. 309, pp. 576–581.

Caro, G., Bourdon, B., Birck, J.-L. and Moorbath, S. (2003) '^{146}Sm–^{142}Nd evidence from Isua metamorphosed sediments for early differentiation of the Earth's mantle', *Nature*, vol. 423, pp. 428–431.

Cox, K.G., Bell, J.D. and Pankhurst, R.J. (1979) *The Interpretation of Igneous Rocks*, London, Unwin Hyman.

Davies, G.F. (2002) 'Stirring geochemistry in mantle convection models with stiff plates and slabs', *Geochimica et Cosmochimica Acta*, vol. 66, pp. 3125–3142.

Davies, G.F. and Richards, M.A. (1992) 'Mantle convection', *Journal of Geology*, vol. 100, pp. 151–206.

Detrick, R.S., Buhl, P., Mutter, J., Vera, E., Harding, A., Kent, G. and Orcutt, J. (1987) 'Multichannel seismic imaging of a crustal magma chamber along the East Pacific Rise', *Nature*, vol. 326, pp. 35–41.

Dietz, R.S. and Holden, J.C. (1970) 'The break-up of Pangea', *Scientific American*, vol. 223, No. 4, pp. 30–41.

Drake, M.J. and Righter, K. (2002) 'Determining the composition of the Earth', *Nature*, vol. 416, pp. 39–44.

Dziewonski, A.M. and Anderson, D.L. (1981) 'Preliminary reference Earth model', *Physics of the Earth and Planetary Interiors*, vol. 25, pp. 297–356.

Earle, P.S. and Shearer, P.M. (1994) 'Characterization of global seismograms using an automated-picking algorithm', *Bulletin of the Seismology Society of America*, vol. 84, pp. 366–76.

Elthon, D., Ross, D.K. and Meen, J.K. (1995) 'Compositional variations of basaltic glasses from the Mid-Cayman Rise spreading center', *Journal of Geophysical Research*, vol. 100, pp. 12497–12512.

Falloon, T.J., Green, D.H., Danyushevsky, L.V. and Faul, U.H. (1999) 'Peridotite melting at 1.0 and 1.5 GPa: an experimental evaluation of techniques using diamond aggregates and mineral mixes for determination of near-solidus melts', *Journal of Petrology*, vol. 40, pp. 1343–1375.

Foley, C.N., Wadha, W., Borg, L.E., Janney, P.E., Hines, R. and Grove, T.L. (2005) 'The early differentiation of Mars from ^{182}W–^{142}Nd isotope systematics in SNC meteorites', *Geochimica et Cosmochimica Acta*, vol. 69, pp. 4557–4571.

Forsyth, D. and Uyeda, S. (1975) 'On the relative importance of the driving forces of plate motion', *Geophysical Journal of the Royal Astronomical Society*, vol. 43, pp. 163–200.

Fowler, C.M.R. (2005) *The Solid Earth: An introduction to global geophysics*, Cambridge, Cambridge University Press.

Gill, R.C.O. (1996) *Chemical Fundamentals of Geology*, London, Chapman & Hall.

Grove, T.L. and Bryan, W.B. (1983) 'Fractionation of pyroxene–phyric MORB at low pressure: an experimental study', *Contributions to Mineralogy and Petrology*, vol. 84, pp. 293–309.

Grove, T.L., Kinzler, R.J. and Bryan, W.B. (1990) 'Natural and experimental phase relations of lavas from Serocki Volcano' in Detrick, R. et al. (eds) *Proceedings of the Ocean Drilling Program Scientific Results*, vol. 106–109, pp. 9–17.

Grove, T.L., Kinzler, R.J. and Bryan, W.B. (1992) 'Fractionation of mid-ocean ridge basalt' in Phipps Morgan, J., Blackman, D.K. and Sinton, J.M. (eds) 'Mantle flow and melt generation at mid-ocean ridges', *American Geophysical Union Geophysical Monograph*, vol. 71, pp. 281–310.

Hasagawa, A. (1978) 'Double-planed deep seismic zone', *Geophysical Journal of the Royal Astronomical Society*, vol. 54, pp. 281–296.

Hauri, E.H. (1996) 'Major element variability in the Hawaiian mantle plume', *Nature*, vol. 382, pp. 415–419.

Hauri, E.H. and Hart, S.R. (1993) 'Re–Os isotope systematics of HIMU and EMII oceanic island basalts from the South Pacific Ocean', *Earth and Planetary Science Letters*, vol. 114, pp. 353–371.

Hawkesworth, C.J. and Kemp, A.I.S. (2006) 'Evolution of the continental crust', *Nature*, vol. 443, pp. 811–817.

Helmberger, D., Lay, T., Ni, S. and Gurnis, M. (2005) 'Deep mantle structure and the post-perovskite phase change', *Proceedings of the American Academy of Science*, vol. 102, pp. 17257–17263.

Herzberg, C.T. (1987) 'Magma density at high pressure, Part 1: the effect of composition on the elastic properties of silicate liquids' in Mysen, B.O. (ed.) *Magmatic Processes: Physicochemical Principles*, Geochemical Society Special Publication 1, pp. 25–46.

Hirth, G. and Kohlstedt, D.L. (1996) 'Water in the oceanic upper mantle: implications for the rheology, melt extraction and the evolution of the lithosphere', *Earth and Planetary Science Letters*, vol. 144, pp. 93–108.

Humphris, S.E. and Thompson, G. (1978) 'Hydrothermal alteration of oceanic basalts by seawater', *Geochimica et Cosmochimica Acta*, vol. 42, pp. 107–25.

Johnson, M.C. and Plank, T. (1999) 'Dehydration and melting experiments constrain the fate of subducted sediments', *Geochemistry, Geophysics, Geosystems*, vol. 1, doi:10.1029/1999GC000014.

Juster, T.C., Grove, T.L. and Perfit, M.R. (1989) 'Experimental constraints on the generation of FeTi basalts, andesites, and rhyodacites at the Galapagos spreading center, 85° W and 95° W', *Journal of Geophysical Research*, vol. 94, pp. 9251–9274.

Kleine, T., Munker, C., Mezger, K. and Palme, H. (2002) 'Rapid accretion and early core formation on asteroids and the terrestrial planets from Hf–W chronometry', *Nature*, vol. 418, pp. 952–955.

Kramers, J.D. and Tolstikhin, I.N. (1997) 'Two major terrestrial Pb isotope paradoxes, forward modelling, core formation and the history of the continental crust', *Chemical Geology*, vol. 139, pp. 75–110.

Langmuir, C.H., Klein, E.M. and Plank, T. (1992) 'Petrological systematics of mid-ocean ridge basalts: constraints on melt generation beneath ocean ridges' in Morgan, J.P., Blackman, D.K. and Sinton, J.M. (eds) *Mantle Flow and Melt Generation at Mid-Ocean Ridges*, American Geophysical Union Monograph, vol. 71, pp. 183–280.

Le Bas, M.J., Le Maitre, R.W., Streckeisen, A., Zanettin, B. and IUGS Subcommission on the Systematics of Igneous Rocks (1986) 'A chemical classification of volcanic rocks', *Journal of Petrology*, vol. 27, pp. 745–750.

Lee, D.-C., Halliday, A.N., Leya, I., Weiler, R. and Wichert, U. (2002) 'Cosmogenic tungsten and the origin and earliest differentiation of the Moon', *Earth and Planetary Science Letters*, vol. 198, pp. 267–274.

Lei, J. and Zhao, D. (2006a) 'Global P-wave tomography: on the effects of various mantle and core phases', *Physics of the Earth and Planetary Interiors*, vol. 156, pp. 44–69.

Lei, J. and Zhao, D. (2006b) 'A new insight into the Hawaiian plume', *Earth and Planetary Science Letters*, vol. 241, pp. 438–453.

Livermore, R. et al. (2005) 'Palaeogene opening of the Drake Passage', *Earth and Planetary Science Letters*, vol. 236, pp. 459–470.

Lodders, K. (2003) 'Solar System abundances and condensation temperatures of the elements', *The Astrophysical Journal*, vol. 591, pp. 1220–1247.

MacKenzie, W.S. and Guilford, C. (1980) '*Atlas of rock-forming minerals in thin section*, London, Longman.

McDonough, W.F. and Sun, S.-S. (1995) 'The composition of the Earth', *Chemical Geology*, vol. 120, pp. 223–253.

McKenzie, D. and Bickle, M.J. (1988) 'The volume and composition of melt generated by extension of the lithosphere', *Journal of Petrology*, vol. 29, pp. 625–79.

McSween, H.Y. (1987) *Meteorites and Their Parent Planets*, Cambridge, Cambridge University Press.

Minster, J.F. and Allègre, C.J. (1980) '^{87}Rb–^{87}Sr dating of LL chondrites', *Earth and Planetary Science Letters*, vol. 56, pp. 89–106.

Morgan, J.W. and Anders, E. (1980) 'The chemical composition of Earth, Venus and Mercury', *Proceedings of the National Academy of Science*, vol. 77, pp. 69–73.

Morgan, W. J. (1968) 'Rises, trenches, great faults and crustal blocks', *Journal of Geophysical Research*, vol. 73, pp. 1959–1982.

Murty, V.R. and Patterson, C.C. (1962) 'Primary isochron of zero age for meteorites and the Earth', *Journal of Geophysical Research*, vol. 67, pp. 1161–1167.

Nakajima, J., Matsuzawa, T., Hasegawa, A. and Zhao, D. (2001) 'Three-dimensional structure of V_p, V_s and V_p/V_s', *Journal of Geophysical Research*, vol. 106, pp. 21843–21858.

Nelson, K.D., Zhao, W., Brown, L.D., Kuo, J., Che, J., Liu, X., Klemperer, S.L., Makovsky, Y., Meissner, R., Mechie, J., Kind, R., Wenzel, F., Ni, J., Nabelek, J., Leshou, C., Tan, H., Wei, W., Jones, A.G., Booker, J., Unsworth, M., Kidd, W.S.F., Hauck, M., Alsdorf, D., Ross, A., Cogan, M., Wu, C., Sandvol, E. and Edwards, M. (1996) 'Partially molten middle crust beneath southern Tibet: synthesis of project INDEPTH results', *Science*, vol. 274, pp. 1684–1695.

Pan, Y. and Batiza, R. (2003) 'Magmatic processes under mid-ocean ridges: a detailed mineralogic study of lavas from East Pacific Rise 9° 30′ N, 10° 30′ N, and 11° 20′ N', *Geochemistry Geophysics Geosystems*, vol. 4, no. 11, doi: 10.1029/2002GC000309.

Peacock, S.M. (2003) 'Thermal structure and metamorphic evolution of subducting slabs' in Eiler, J.M. (ed.) *Inside the Subduction Factory*, Geophysical Monograph Series, American Geophysical Union, Washington, DC. pp. 7–22.

Plank, T. and Langmuir, C.H. (1993) 'Tracing trace elements from sediment input to volcanic output at subduction zones', *Nature*, vol. 362, pp. 739–743.

Poli, S. and Schmidt, M.W. (2002) 'Petrology of subducted slabs', *Annual Review of Earth and Planetary Sciences*, vol. 30, pp. 207–235.

Roeder, P.L. and Emslie, R.F. (1970) 'Olivine–liquid equilibrium', *Contributions to mineralogy and petrology*, vol. 29, p. 281.

Rudnick, R.L. and Gao, S. (2005) 'Composition of the continental crust', *Treatise on Geochemistry*, vol. 3, pp. 1–64.

Rushmer, T., Minarik, W.G. and Taylor, G.J. (2000) 'Physical processes of core formation' in Canup, R.M. and Righter, K. (eds) *Origin of the Earth and Moon*, Tucson, University of Arizona Press.

Scotese, C.R., Gahagan, L.M. and Larson, R.L. (1988) 'Plate tectonic reconstructions of the Cretaceous and Cenozoic ocean basins', *Tectonophysics*, vol. 155. pp. 27–48.

Sherstein, A., Elliott, T., Hawkesworth, C., Russell, S. and Masarik, J. (2006) 'Hf–W evidence for rapid differentiation of iron meteorite parent bodies', *Earth and Planetary Science Letters*, vol. 241, pp. 530–542.

Sinton, J.M. and Detrick, R.S. (1992) 'Mid-ocean ridge magma chambers', *Journal of Geophysical Research*, vol. 97, pp. 197–216.

Stein, C. and Stein, S. (1992) 'A model for the global variation in oceanic depth and heat flow with lithospheric age', *Nature*, vol. 359, pp. 123–129.

Takahashi, E. (1986) 'Melting of a dry peridotite KLB-1 up to 14 GPa: implications on the origin of peridotitic upper mantle', *Journal of Geophysical Research*, vol. 91, pp. 9367–9382.

Takahashi, E., Shimazaki, T., Tsuzaki, Y. and Yoshida, H. (1993) 'Melting study of a peridotite KLB-1 to 6.5 GPa, and the origin of basaltic magmas', *Philosophical Transactions of the Royal Society of London, Series A*, vol. 342, pp. 105–120.

Taylor, S.R. and McClennan, S.M. (1995) 'The geochemical evolution of the continental crust', *Reviews of Geophysics*, vol. 33, pp. 241–265.

Tormey, D.R., Grove, T.L. and Bryan, W.B. (1987) 'Experimental petrology of normal MORB near the Kane Fracture Zone: 22°–25° N, mid-Atlantic ridge', *Contributions to Mineralogy and Petrology*, vol. 96, pp. 121–139.

van Keken, P.E., Kiefer, B. and Peacock, S.M. (2002) 'High resolution models of subduction zones: implications for mineral dehydration reactions and the transport of water into the deep mantle', *Geochemistry, Geophysics, Geosystems*, vol. 3, doi:10.1029/2001GC000256

Watts, A.B. (2001) *Isostasy and Flexure of the Lithosphere*, Cambridge, Cambridge University Press.

White, D.B. (1988) 'The platforms and onset of convection with a temperature dependent on viscosity', *Journal of Fluid Mechanics*, vol. 191, pp. 247–286.

White, R.S., McKenzie, D. and O'Nions, R.K. (1992) 'Oceanic crustal thickness from seismic measurements and rare earth element inversions', *Journal of Geophysical Research*, vol. 97, pp. 19683–19715.

Yang, H.J., Kinzler, R.J. and Grove, T.L. (1996) 'Experiments and models of anhydrous, basaltic olivine–plagioclase–augite saturated melts from 0.001 to 10 kbar', *Contributions to Mineralogy and Petrology*, vol. 124, pp. 1–18.

Yin, Q., Jacobsen, S.B., Jamashita, K., Blichert-Toft, J., Telouk, P. and Albarede, F. (2002) 'A short timescale for terrestrial formation from Hf–W chronometry of meteorites', *Nature*, vol. 418, pp. 949–952.

Index

Entries and page numbers in **bold type** refer to where terms defined in the Glossary are printed in **bold** in the text. Page numbers referring only to figures and tables are printed in *italics*.

D

E

O

P

Q

R

S

T